CHEMISTRY OF THE ENVIRONMENT

SECOND EDITION

Thomas G. Spiro
Princeton University

William M. Stigliani
Center for Energy and Environmental Education
University of Northern Iowa

PRENTICE HALL, Upper Saddle River, New Jersey 07458

Library of Congress Cataloging-in-Publication Data

Spiro, Thomas G.
    Chemistry of the environment / Thomas G. Spiro, William M.
Stigliani. -- 2nd ed.
      p. cm.
    Includes bibliographical references and index.
    ISBN 013-754896-6 (hardcover)
    1. Environmental chemistry.   I. Stigliani, William M.   II. Title.
TD193 .S7   2003
540—dc21   2002022048

Editor in Chief, Physical Sciences: John Challice
Project Manager: Kristen Kaiser
Senior Marketing Manager: Steve Sartori
Editorial/Production Supervision: ICC
Assistant Manufacturing Manager: Michael Bell
Manufacturing Manager: Trudy Pisciotti
Copy Editor: Bob Cooper
Art Director: Jayne Conte
Cover Designer: Bruce Kenselaar
Cover Image: Photodisc
The image used on Part and Chapter starting pages is courtesy of Digital Imagery © Copyright 2001 PhotoDisc, Inc.

Prentice
Hall

© 2003, 1996 by Prentice-Hall, Inc.
Upper Saddle River, NJ 07458

Printed in the United States of America

10   9   8   7   6   5   4   3   2   1

ISBN 0-13-754896-6

Pearson Education LTD.
Pearson Education Australia PTY, Limited
Pearson Education Singapore, Pte. Ltd
Pearson Education North Asia Ltd
Pearson Education Canada, Ltd.
Pearson Educación de Mexico, S.A. de C.V.
Pearson Education -- Japan
Pearson Education Malaysia, Pte. Ltd

CONTENTS

PREFACE

This book is about environmental issues and the chemistry behind them. It is not a methods book, nor is it a catalog of pollutants and remediation options. It aims to deepen knowledge of chemistry and of the environment and to show the power of chemistry as a tool to help us comprehend the changing world around us.

In the six years since the first edition of *Chemistry of the Environment* was published, the frontiers of environmental science have advanced rapidly, and the debates on environmental issues have shifted ground. In this new edition, we have updated the various strands of our environmental story by integrating new facts and figures in the text, tables, and diagrams. [Recognizing that no book on environmental themes can stay current for long, we plan to post further updates on our website **http://www.prenhall.com/spiro2**]. Some of the new material [e.g., ocean chemistry and the inorganic carbon cycle (pp. 284–288), or the evolution of the oxygen atmosphere (pp. 316–319)] might have been included earlier, but some topics had not surfaced six years ago. These include genetically modified crops (pp. 401–406), carbon sequestration as a strategy for reducing greenhouse gas emissions (pp. 42–43, 288), and contamination of drinking water by the gasoline additive MTBE (pp. 260, 271, 344).

Chemistry of the Environment can be used in a one- or two-term environmental chemistry course. The instructor in a one-term course will want to pick a limited set of the book's topics for special emphasis; in a two-term course there would be time to address other topics and to explore the underlying chemical principles in more detail. The new edition is also suitable for basic environmental science courses. Readers will find that the biggest change from the first edition is improved accessibility through reorganization and expansion of the basic chemistry. We have separated background material relevant to the understanding of the topic under discussion into boxes marked Fundamentals. We have also added additional basic material, to help those readers without exposure to college chemistry, and to refresh the memories of those who have had such exposure. In addition, we have included worked problems in other boxes, and have added more end-of-chapter problems. A periodic table is now included, as is an Appendix that gives a brief introduction to organic chemical structures. Some of the Fundamentals boxes contain non-chemical background information [e.g., how to relate reservoirs and flows in environmental chemical cycles, p. 285–286].

In addition, we have separated the more advanced or specialized technical information into other boxes called Strategies, which readers can read or skip at their discretion. In this way, the environmental story line is unimpeded by background or technical information. We hope that these changes will make the book easier to read, and also more useful as a textbook.

We are indebted to a number of colleagues for reviewing parts of the manuscript, and/or providing new material: Drs. Michael Bender, Andrew Bocarsly, Harold Feiveson, Robert Goldston, Peter Jaffe, Hiram Levy, Francois Morel, Steve Pacala, Lynn Russel, Jorge Sarmiento, Daniel Sigman, Robert Socolow, Valerie Thomas (all from Princeton University); Trace Jordan (New York University); Bibudhendra Sarkar (University of Toronto); David Walker (University of British Columbia); and Chris Weber (student assistant, University of Iowa). Helen Spiro provided encouragement throughout the writing, and key editorial advice. Thanks also to Marie Stigliani—her companionship on bicycle trails along the Cedar River provided balance to long days at the office.

Supplements

Instructor's Solutions Manual—(0-13-017843-8) Contains the full solution to all end-of-chapter problems and is available to instructors upon adoption.

Companion Website—*http://www.prenhall.com/spiro2* This on-line site brings you abstracts of relevant environmental chemistry articles in newspapers, magazines, and scientific journals.

Thomas G. Spiro
spiro@princeton.edu

William M. Stigliani
stigliani@uni.edu

*Dedicated to
our wives Helen and Marie,
and to our children and their generation*

INTRODUCTION

"Here's a short chemistry lesson," wrote Bill McKibben in a *New York Times Magazine* feature story.* "Grasp it and you will grasp the reason the environmental era has barely begun. . . ." McKibben's lesson is about the difference between two molecules, carbon monoxide (CO) and carbon dioxide (CO_2). Today's automobiles release half a pound of carbon as CO per gallon of gasoline burned, around half as much as they did a generation ago, and the rate is going down with continuing improvements in technology. As a result, the air is now cleaner than it used to be in Los Angeles and in many other cities. But the same gallon of gasoline releases almost five and a half pounds of carbon as CO_2, and this rate cannot be decreased. The atmospheric concentration of CO_2 is increasing world-wide, and bringing global warming with it, according to a consensus of international scientific opinion. The two molecules capture two sides of the environmental coin, local versus global effects of human activity. Environmental quality has improved in many localities, thanks to environmental controls and new technologies, but the global problems are just beginning to be addressed, and they are much more difficult to solve. CO is a side-product of combustion, and is subject to emission controls, but CO_2 is the end-product of combustion and is the inevitable accompaniment of our reliance on fossil fuels. "CO versus CO_2," says McKibben, "one damn oxygen atom, and all the difference in the world."

To us this is a wonderful illustration of the power of chemistry to illuminate environmental issues. Chemistry is all around us, and it really does make a difference. The chemical cycles of the planet are increasingly disturbed by human activities, and these disturbances can degrade the quality of life, as when auto emissions overwhelm the atmosphere's capacity to clean the air over our cities. We are capable of ameliorating these perturbations, as the experience of Los Angeles demonstrates. But first we must understand the chemistry. In the case of Los Angeles, initial attempts to alleviate smog back in the 1960s actually made things worse. Standards were imposed on CO and hydrocarbon levels in auto emissions, and auto makers met those standards by increasing the air/fuel ratio to burn the fuel more completely. But smog levels *increased* because higher air/fuel ratios

*Bill McKibben, "Not So Fast," *New York Times Magazine,* July 23, 1995, pp. 24–25.

made combustion hotter, thereby increasing the nitrogen oxide emissions. Only then was it discovered that nitrogen oxides and hydrocarbons are *both* key actors in smog formation and that both have to be controlled. This kind of surprise is not unusual in environmental affairs. The world is a marvelously complex place, chemically, as in other ways. We are just beginning to understand how it works.

This book tells the environmental story in chemical language. It is grounded in the flows of chemicals and energy through nature on the one hand, and through our industrial civilization on the other. The units of the book, Energy, Atmosphere, Hydrosphere/Lithosphere, and Biosphere, reflect this holistic perspective. Environmental issues frequently cut across these divisions, and the resulting interconnections add richness to the story. For example, leaded gasoline is linked to the issue of auto emission controls, a subject that arises in the Atmosphere section, but it is also a major health hazard, as discussed in the Biosphere section.

Interconnections are even more numerous at the level of the underlying chemistry. For example, the reactivity of dioxygen, O_2, is a leitmotif for all parts of the book. Thus energy flow through industrial civilization (as well as through the biosphere itself) depends on the oxygen-oxygen bond being relatively weak, so that energy is released when O_2 combines with organic molecules. Yet, because of its unusual electronic structure, O_2 is quite unreactive until it encounters a free radical or a transition metal ion. These O_2 activators determine most aspects of atmospheric chemistry, including how smog is formed. They are also vitally important for the biosphere, since O_2 metabolism gone awry is a threat to the integrity of biological molecules and has been implicated as a factor in cancer and aging.

We hope these interconnections fascinate the reader as much as they fascinate us, and we hope the tapestry we weave will provide a satisfying context for understanding the chemical world we live in and the environmental issues we face.

PART I

ENERGY

CHAPTER 1

ENERGY FLOWS AND SUPPLIES

The question of energy use underlies virtually all environmental issues. The harnessing of energy for the manifold needs of industrial civilization has driven economic development, and access to affordable energy has been the key to a better life for people around the world. At the same time the environmental costs of human energy consumption are becoming ever more apparent: oil spills, the scarring of land by mining, air and water pollution, and the threat of global warming from the accumulation of carbon dioxide and other greenhouse gases. Increasingly, maintaining an expanding supply of cheap energy seems to clash with concern for the environmental costs of such expansion. In this part of the book, we explore the background of energy production and energy consumption, and examine the prospects for meeting the energy needs of society while protecting the environment.

1.1 PROLOGUE ON ENERGY AND SUSTAINABILITY

Environmental discussions often revolve around *sustainability,* an evocative and much-debated idea. It arises from the perception that human activity is using up nature's resources at rates beyond the capacity of nature to restore them. Sustainability implies maintaining these resources for future generations.

The concept has many applications. *Sustainable logging,* for example, refers to extracting wood in ways that permit regeneration of forests. *Sustainable agriculture* would feed people without depleting the nutritive capacity of the soil or the biodiversity of natural habitats. A growing number of companies embrace sustainability by protecting the environment in ways that go beyond legal requirements. Examples include voluntarily curbing emissions, switching to environmentally benign materials (known as *green chemistry,* see p. 128), or refraining from using tropical hardwoods in furniture manufacture. The new field of *Industrial Ecology* (see p. 125) contributes to sustainability by finding ways to minimize the consumption of material in industrial society.

However, there is wide argument about how to achieve sustainability. Does sustainable logging mean leaving the wilderness undisturbed and harvesting only tree farms, or does it mean logging wild forests selectively to maintain their health? Does sustainable agriculture mean organic farming, with no inputs of synthetic chemicals, or does it mean using whatever inputs are optimal to maintain continually high soil and crop productivity?

Mining is a good illustration of the slipperiness of the sustainability idea. No activity seems less sustainable than the extraction of minerals from the earth. Yet despite centuries of mining, metals are as available as ever. The reason is that continual improvements in technology permit extraction of lower-grade deposits and increased recycling of metals. The metals market is sustainable, even though high-grade ores have been used up, never to be replenished.

Of course there are impacts of mining other than resource exhaustion. Mining scars the face of the earth and pollutes the local environment, although these impacts can be minimized through regulation. Extraction and processing spread metals around the globe, raising concern about threats to health from toxic elements, such as mercury and lead (see pp. 437, 445). It is not clear whether mining these metals is ultimately sustainable. On the other hand, no one worries about health threats from the spread of common industrial metals like iron and aluminum, which have low toxicity and are abundant in the Earth's crust. There are incentives to recycle these metals, in order to save energy (see p. 116) and minimize pollution, but none to stop using them.

Energy use raises the most dramatic issues of sustainability. Burning up the Earth's deposits of oil, gas, and coal is clearly an unsustainable practice. The energy cannot be recovered (second law of thermodynamics, see p. 98). Yet, is there something intrinsically bad about using up the fossil fuels? After all, they provide us with cheap, abundant energy, and useful commodities (e.g., plastics). The most important reason for curbing fossil fuel use is the rising level of heat-trapping CO_2, the end product of fossil fuel utilization. However, even this effect might be avoided by sequestering the CO_2 in the Earth or under the oceans (see p. 42).

In any event the use of fossil fuels as an energy source is ultimately unsustainable, but how long is ultimately? The time scale could be decades or centuries, depending on energy policies. To replace fossil fuels, we will have to turn to alternative energy sources: nuclear fission (if it can be developed safely, see p. 54), fusion (if it is technically feasible, see p. 68), and/or renewable energy (the harnessing of which on large scales is difficult and expensive, see p. 74). Some people hope to end reliance on fossil fuels as soon as possible, by betting on renewable energy sources now. But it is far from obvious how best to promote wise energy use while protecting the environment.

In sum, the sustainability concept is more a way of raising important questions than a guide to action. The answers involve making moral and political choices, based on a clear understanding of how the material world works. Things are rarely as simple as they seem.

1.2   NATURAL ENERGY FLOWS

It is instructive to view human energy use against a backdrop of the continual and massive flow of energy that occurs at the surface of the Earth. This flow is diagrammed in
Figure 1.1; the magnitudes of the energy fluxes are given in units of 10^{20} kilojoules (kJ)
per year. (See Fundamentals 1.1 for energy units and conversions.) A tiny part of Earth's
energy budget derives from non-solar sources: tidal energy (0.0013×10^{20} kJ), which
arises from the gravitational attraction between the moon and Earth, and geothermal heat
(0.01×10^{20} kJ), which emanates from Earth's molten core. (Human-generated nuclear
energy is non-solar, but it is also minimal compared to total energy flows.) The rest of the
energy on Earth comes from the sun, either directly or indirectly.

   The sun radiates a nearly unimaginable number of kilojoules every year—$1.17 \times$
10^{31}. A very small fraction of the total, 54.4×10^{20} kJ per year, is intercepted by Earth,
which is 150 million kilometers (93 million miles) from the sun. Of this quantity, about
30 percent is reflected or scattered back into space from its atmosphere (26 percent) or surface (4 percent). This fraction is called the *albedo,* and it contributes significantly to the
overall energy balance of Earth.

   The rest of the light is absorbed by Earth's atmosphere (24 percent), land surface
(14 percent), and oceans (32 percent), and is converted to heat before being re-radiated

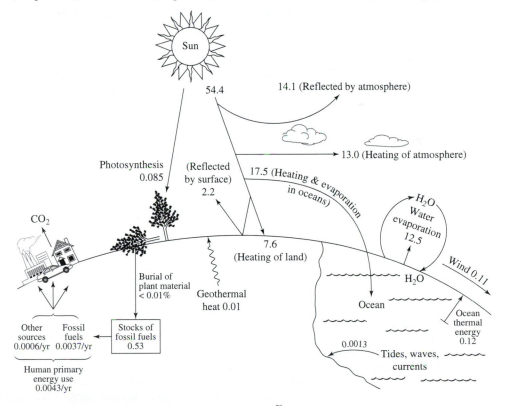

Figure 1.1   Annual energy fluxes on Earth (in 10^{20} kilojoules).

back into space. This heat flow drives Earth's weather system via wind, rain, and snow. About half of the absorbed energy reaching Earth's surface flows through the hydrological cycle, the massive evaporation and precipitation of water upon which we depend for our freshwater supplies. While it takes 4.2 joules (1 calorie) to heat a gram of water by 1°C, much more energy is needed to vaporize the same gram of water; the energy required to vaporize a liquid is called its *latent heat of vaporization*. At 15°C, which is the average annual global temperature, the latent heat of water is 2.46 kJ/g. The latent heat is released again when water vapor condenses into rain. This is the reason that rainfall is associated with storms; even a modest rainfall releases a huge amount of energy.

We extract a very small fraction of the energy in the hydrological cycle through dams and hydroelectric power generation, which constitute an indirect tapping of the solar energy flow.

About 0.34 percent of the sunlight absorbed on Earth's surface (land plus oceans) is used by green plants and algae in photosynthesis. We depend on this fraction of the solar flux for our food supply and for a habitable earth. Some of the energy we use is provided by burning wood and other biomass forms (garbage, cow dung), and most of the rest is obtained by mining the store of photosynthetic products buried in ages past, in the form of fossil fuels.

FUNDAMENTALS 1.1: UNITS AND CONVERSIONS

When considering environmental issues, the overriding question is "how much"? Whether it is exposure to toxic substances, or emission of pollutants, or the amount of energy used, a lot or a little makes the difference between a problem and a non-problem. Often the number need not be precise; knowing something to within a factor of ten (called an order of magnitude) may sometimes be enough, because amounts can vary by many factors of ten.

a. Exponents

In order to avoid having to write out many zeros, we can express numbers in *exponential* notation, $n \times 10^y$, where y is the exponent, i.e., the number of times n is multiplied by ten. If there is a minus sign in front of it, then y is the number of times n is multiplied by 10^{-1}, i.e., by 0.1. For example, 5.1×10^3 means 5,100, while 5.1×10^{-3} means 0.0051.

In addition, we can use prefixes in front of the units in order to modify the amounts. The commonly used prefixes are *kilo (k), mega (M), giga (G), terra (T), peta (P),* and *exa (E)* for $y =$ 3, 6, 9, 12, 15, and 18, respectively; and *centi (c), milli (m), micro (μ), nano (n), pico (p),* and *femto (f)* for $-y =$ 2, 3, 6, 9, 12, and 15, respectively. The letters in parentheses are accepted abbreviations for the prefixes. Thus 5.1×10^3 meters (abbreviated m) is the same as 5.1 kilometers (km), and 5.1×10^{-3} meters is the same as 0.51 centimeters (cm) or 5.1 millimeters (mm).

b. Metric Units

To gauge how much something is, we must specify the units of measurement. For scientific purposes, we use the metric system. Length is specified in meters (m) and kilometers (km), area in hectares (ha) (one hectare is 10,000 square meters), volume in liters (L) (one liter is 1,000 cubic centimeters), and mass in grams (g), kilograms (kg), and metric tons (t) (one metric ton is 1,000 kilograms). This is also common usage around the world, except in the U.S. where English units continue to be used: inches, feet, and miles for length; acres (one acre is 43,560 square feet) for area; pints, quarts, and gallons for volume; ounces, pounds, and tons (one ton, sometimes called a short ton, is 2,000 pounds) for mass. We will use the metric system in this book, but the following conversion factors may sometimes be useful:

$$1 \text{ inch} = 2.540 \text{ centimeters}$$
$$1 \text{ foot} = 0.305 \text{ meters}$$
$$1 \text{ mile} = 1.609 \text{ kilometers}$$
$$1 \text{ acre} = 0.405 \text{ hectares}$$
$$1 \text{ quart} = 0.946 \text{ liters}$$
$$1 \text{ gallon} = 3.785 \text{ liters}$$
$$1 \text{ pound} = 0.454 \text{ kilograms}$$
$$1 \text{ short ton} = 0.9072 \text{ metric tons}$$

Temperature is measured in degrees centigrade (°C) in the metric system, and degrees Fahrenheit (°F) in the English system. At sea level, water boils at 100°C, but 212°F, and freezes at 0°C, but 32°F. Consequently

$$(\text{°F}) = (\text{°C}) \times 9/5 + 32°$$

In scientific calculations we need the absolute temperature (T), which is expressed in Kelvin (K) units. Absolute zero is −273°C, so

$$K = \text{°C} + 273°$$

Time is measured everywhere in seconds (sec), minutes (min), hours (h), days (d), and years (y).

c. Energy Units

Energy comes in many forms, and has historically been measured in many different units. The unit which has been chosen as the standard is the joule (J). The joule was originally defined as a unit of work energy. One joule is the work done by a force that accelerates a 1-g mass at 1 cm/sec^2 for a distance of 1 m. A kilojoule (kJ) is 1,000 joules, and an exajoule (EJ) is 10^{18} joules.

All forms of energy can be given in joules, but many still prefer to express heat energy in the historically used calorie units, because of their intuitive appeal. One calorie (cal) is the heat energy needed to raise the temperature of 1 g of water by 1°C (from 14.5 to 15.5°C). The conversion between the two units is

$$1 \text{ cal} = 4.184 \text{ J}$$

In common usage, calories are most often associated with measuring the energy value of food. Unfortunately, the nutritionists' Calorie (distinguished by the capital letter), is not the same as a calorie, but 1,000 times larger (actually a kilocalorie). So if your daily intake is 2,000 Cal, you are actually consuming 2×10^6 cal.

Another energy unit commonly used in the U.S. is the British Thermal Unit (BTU), defined as the quantity of heat required to raise the temperature of one pound of water one degree Fahrenheit. A quad (Q) is one quadrillion BTUs (10^{15} BTUs). The conversion factors between joules and BTUs are:

$$1 \text{ BTU} = 1,055 \text{ joules} = 1.055 \text{ kilojoules}$$
$$1 \text{ quad} = 1.055 \text{ exajoules}$$

Power is also related to energy units. Power is the rate at which energy is delivered. The standard unit is the watt (W).

$$1 \text{ W} = 1 \text{ J/sec}$$

Conversely, energy is power multiplied by the time over which the energy is delivered. Thus electricity is commonly metered out in kilowatt-hours (kWh).

$$1 \text{ kWh} = (1,000 \text{ J/sec})(3,600 \text{ sec/h})$$
$$= 3.6 \times 10^6 \text{ J}$$

The energy of light waves is proportional to their frequency, which is usually expressed as the wave number ($\nu = 1/\lambda$) in cm^{-1}. For example, blue light with a 500-nm wavelength has a wave number, or "energy," of (500 × 10^{-7} cm)$^{-1}$ = 20,000 cm^{-1}. It is possible to relate the wave number to the equivalent quantity of heat, or chemical energy per mole of photons; thus

$$1 \text{ kJ/mol} = 1463.6 \text{ cm}^{-1}$$

The electron volt (ev) is a unit commonly used by physicists in describing radiation and elementary particles. It is the amount of energy acquired by any charged particle that carries unit electric charge when it falls through a potential difference of 1 volt. It can be related to the equivalent wave number of electromagnetic radiation:

$$1 \text{ ev} = 8,064.9 \text{ cm}^{-1}$$

A commonly used multiple of this unit is the megaelectron volt = Mev = 10^6 ev.

WORKED PROBLEM 1.1: Energy in Rain

Q. *Given that evaporation of 1 mL of water requires 2.46 kJ of energy at ambient temperatures, how much energy is released in a 2-cm rainfall over an area 10 by 10 km? If 1 ton of TNT releases 4.18×10^6 kJ of energy, how many tons of TNT would be equivalent to the energy released in rainfall?*

A. At the end of Fundamentals 1.1 we converted kilowatt-hours to joules. Many problems, this one included, are essentially unit conversion problems. We know that 2.46 kJ of energy are released by condensing 1 mL of water, so we can apply that conversion to the larger volume of water in the rainfall. That volume is found by multiplying the depth times the area, remembering that 1 mL = 1 cm^3, and

that 1 km = 10^3 m = 10^5 cm:

Volume of rain = 2 cm × (10 × 10^5) cm × (10 × 10^5) cm
= 2 × 10^{12} mL
Energy from rain = 2 × 10^{12} mL × 2.46 kJ/mL
= 4.92 × 10^{12} kJ

And since 1 ton of TNT releases 4.18×10^6 kJ of energy, we can divide this factor into the energy from rainfall to obtain the TNT equivalents.

Equivalent TNT = 4.92 × 10^{12} kJ/4.18 kJ per ton TNT = 1.18 × 10^6 tons TNT.

This gives some feeling for the scale of energy in storms. A modest rainfall releases the equivalent of about a million tons of TNT.

1.3   HUMAN ENERGY CONSUMPTION

Compared to the enormous energy flux provided by the sun, human energy utilization is puny (see Table 1.1). In 2000, the total primary energy consumed by all humans amounted to 4.3×10^{17} kJ, equivalent to only 0.017 percent of the solar heat absorbed by Earth's surface. Figure 1.2 shows, however, that world energy consumption more than tripled between 1960 and 2000, and is still growing rapidly.

Between 1960 and 1973, energy consumption grew at 4.2 (U.S.) and 4.5 percent (world) per year. In response to the energy crises in 1973 and again in 1979, the rate of energy growth slowed substantially between 1974 and 1986, particularly in the U.S., where energy consumption was nearly flat. Nevertheless the U.S. economy grew by 40 percent

TABLE 1.1   GLOBAL ENERGY FLUXES

Sources	Rates (10^{20} kJ/year)
Energy radiated by the sun into space	1.17×10^{11}
Solar energy incident on Earth	54.4
Solar energy affecting Earth's climate and biosphere	38.1
Energy taken up in global evaporation of water	12.5
Energy in wind	0.109
Solar energy taken up in photosynthesis	0.0850
Energy taken up in net primary production	0.0430
Energy conducted from Earth's interior to its surface	0.0100
Total primary energy consumed by humans, 2000	0.00430
Energy content of fossil fuels consumed, 2000	0.00369
Energy in tides and waves	0.00130
Total energy consumed in United States, 2000	0.00104
Energy content of food consumed by humans, 2000	0.000260

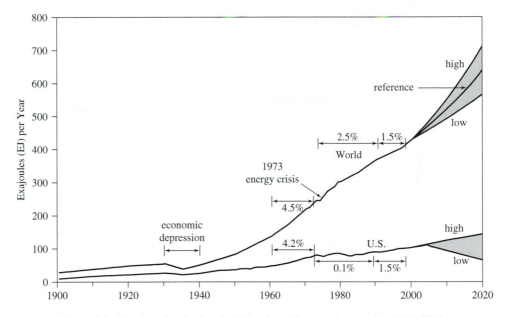

Figure 1.2   Trends and projections in U.S. and world energy consumption, 1900–2020. *Sources:* World data taken from Energy Information Agency, U.S. Department of Energy, *International Energy Outlook 2001—World Energy Consumption,* Washington, DC. U.S. "high" trend taken from same source. U.S. "low" trend assumes U.S. complies with reduction in carbon emissions by 7 percent relative to year 1990 in the period 2008–2012, as prescribed by the Kyoto Protocol on Climate Change.

during this period, and reduced its energy bill by an estimated $150 billion dollars per year. This dramatic saving resulted from major improvements in energy efficiency in buildings, automobiles, and industrial technologies.

However, since 1987, the U.S. growth rate increased to 1.5 percent, the same as the world growth (which declined from 2.5 percent), during this time. A variety of factors contributed to the increased U.S. growth rate relative to the no-growth period between 1973 and 1986. These factors included record low prices of crude oil on the international market. By 2000, the U.S., with 5 percent of the world's population, accounted for 25 percent of the world's energy use.

Three important global trends are not revealed in the overall numbers graphed in Figure 1.2. 1) Most of the wealthy nations succeeded in reducing energy growth by 1.5 percent or more. In Western Europe the average annual growth rate has been only 0.9 percent per year. 2) Energy use was drastically reduced in Eastern Europe and the former Soviet Union after its collapse in 1989, from 79.0 EJ, to 51.9 EJ in 1998. 3) The developing countries are increasing their energy consumption rapidly; growth rates are 3.6, 2.6, and 4.4 percent per year in South and Central America, Africa, and East Asia, respectively. Some of these trends are revealed by the data for individual countries listed in Table 1.2.

Global energy consumption will be greatly influenced by trends in China and India. If their current growth rates, 3.5 and 6.3 percent, were to continue into the future, China would be consuming 100 EJ per year (comparable to energy use in the U.S. today) by 2030, and India by 2033.

TABLE 1.2   ENERGY TRENDS IN TEN MOST POPULOUS COUNTRIES

Country	Population (millions) (2000)	Energy use (EJ) (1998)	Rate of increase (%) (1987–1998)	Doubling time (yrs)
China	1,277.6	36.0	+3.5	19.8
India	1,013.7	13.3	+6.3	11.0
United States	278.4	100.5	+1.5	46.2
Indonesia	212.1	3.8	+7.3	9.5
Brazil	170.1	8.6	+3.8	18.2
Pakistan	156.5	1.8	+5.2	13.3
Russian Federation	146.9	27.5	−4.6	—
Bangladesh	129.2	0.4	+5.9	11.7
Japan	126.7	22.6	+2.6	26.7
Nigeria	111.5	1.0	+3.5	19.8
Totals (1998)	3,622.7	215.5	+1.3	53.3

Country	Population (millions) (2020)	Energy use (EJ) (2020)	Rate of increase (%) (1998–2020)	Doubling time (yrs)
Totals (2020)	4,333.8	380.9*†	+2.6	26.7

*For individual countries, assumes rates of increase during the period 1998–2020 are the same as they were in 1987–1998.
†Assumes Russian Federation consumption is fixed at 27.5 EJ throughout the time period 1998–2020.
Source: Energy Information Agency, U.S. Department of Energy (1999). International tables for population and energy consumption trends, Washington, DC. (http://www.eia.doe.gov/emeu/international/contents.html)

Projections for the next two decades are also graphed in Figure 1.2 (shaded areas). These projections are taken from forecasts prepared by the U.S. Department of Energy's Energy Information Administration (EIA), and by the Intergovernmental Panel on Climate Change (IPCC). They reflect different assumptions about economic growth, the future price of oil, implementation of energy-saving technologies, regional differences in per capita income, and intensity of fossil fuel use. The higher trajectory for the U.S. is based on a continuation of current trends, while the lower trajectory corresponds to the EIA scenario in which U.S. emissions of CO_2 are cut 7 percent by 2020 relative to emissions in 1990, the percentage called for in the Kyoto Protocol on Climate Change (see p. 179).

Exponential growth cannot be maintained indefinitely. Something that increases at a rate of 4 percent a year takes only 18 years to double, and then another 18 years to double again. Nothing in the natural world can grow through too many doubling periods before it runs into some constraint. Eventually the growth must level off (see natural growth curve in Figure 1.3). The present era of rapidly increasing energy use represents a transition to a new level of energy consumption. There remains considerable debate, however, about what that level will be and at what rate it is to be attained. What the future holds is a matter of considerable uncertainty, as the divergent projections in Figure 1.2 indicate.

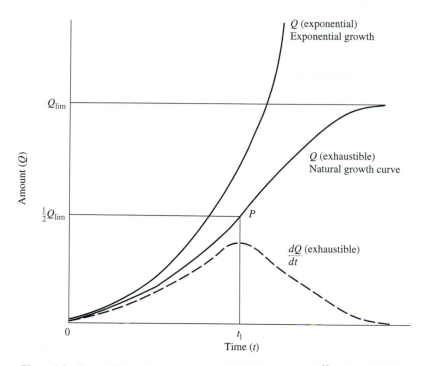

Figure 1.3   Exponential and natural growth curves. For the former, $\frac{dQ}{dt} = kQ$, while for the latter, $\frac{dQ}{dt} = (2\pi)^{-1/2} Q_{\text{lim}}\, e^{-(t-t_1)^2/2}$ (dashed curve). t_1 is the time when half of Q has been used up and the rate of growth shifts from increase to decrease.

FUNDAMENTALS 1.2: EXPONENTIAL GROWTH AND DECAY

When something grows at a constant *percentage* rate, it has a growth history that looks like the graph in Figure 1.3. This graph has the quantity, Q, doubling in constant intervals of time; the doubling time, t_2, is the same no matter how large Q gets. As the number of doublings increases, Q gets very large indeed. This is what happens with compound interest. If you leave your money in the bank, the return is modest at first, but after a number of years it grows impressively. For a given percentage rate of interest, p, you can calculate how long it will take to double from the "rule of 70":

$$t_2 = 70/p \qquad (1.1)$$

Thus a 7 percent interest rate doubles your money in 10 years, while a 1 percent rate requires 70 years for doubling.

What is the connection between constant percentage growth and a constant doubling time? And where does the "rule of 70" come from? We will seldom use calculus in the book, but here it comes in handy. A constant percentage means that Q grows in proportion to how much is present at any instant, a condition that is expressed by the differential equation

$$dQ/dt = kQ \qquad (1.2)$$

The rate of growth is dQ/dt, and k is the proportionality constant (rate constant); it represents the fraction of Q by which Q is growing at any instant. When this equation is integrated, the result is:

$$Q = Q_0 e^{kt} \qquad (1.3)$$

Q_0 is the amount present initially, t is the elapsed time, and e is the natural number 2.71828.... k and t appear in the exponent, and this is why the growth is

called exponential. The doubling time is constant because t has the same value whenever Q/Q_0 is 2, no matter what the starting value of Q_0. To obtain the doubling time, we set $Q/Q_0 = 2$ and take logarithms of both sides of eq. (1.3):

$$\ln 2 = kt_2 \qquad (1.4)$$

(ln stands for *natural* logarithm, i.e., the required exponent of e; thus the logarithm of the right-hand side of equation (1.4) is $\ln(e^{kt}) = kt$.) The value of ln 2 is 0.693, and so

$$t_2 = 0.693/k \qquad (1.5)$$

This is the rule of 70 (approximately; for most purposes 0.7 is as good as 0.693), if we recognize that the percentage growth, p, is $100k$.

The same mathematics also describes a process in which something diminishes at a constant percentage rate; we just put a minus in front of the k:

$$dQ/dt = -kQ \qquad (1.6)$$

and

$$Q = Q_0 e^{-kt} \qquad (1.7)$$

This is exponential decay. Instead of a constant doubling time, there is a constant half-life, $t_{1/2}$, reached when $Q = Q_0/2$. Again the rule of 70 applies, this time to the half-life:

$$t_{1/2} = 0.693/k \qquad (1.8)$$

Instead of a growth curve there is a decay curve (see Figure 1.4), along which Q diminishes by half at constant intervals of time.

We will encounter exponential decay in connection with radioactive materials (see p. 47).

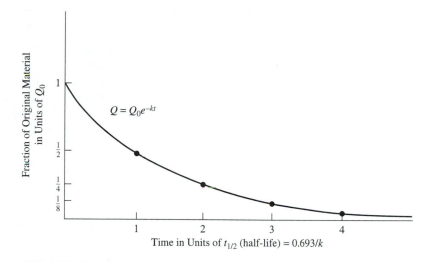

Figure 1.4   Plot of exponential decay.

WORKED PROBLEM 1.2: World Energy Consumption

Q. *If world energy consumption were to increase at an annual rate of 2.8 percent, how long would it take to double?*

A. We can apply the rule of 70 here (see Fundamentals 1.2)

$$t_2 = 70/p = 70/2.8$$

$$= 25 \text{ years}$$

Q. *If this rate of growth continued for a century, how would human energy consumption compare with all the solar energy that is incident on the Earth (see Table 1.1)?*

A. According to Table 1.1, humans used 0.00430×10^{20} kJ/y in 2000, while the incident solar radiation is 54.4×10^{20} kJ/y. The ratio was 7.90×10^{-5}. After a century, consumption would have doubled four times, and therefore be 16-fold larger. The ratio would then be 1.26×10^{-3}, still a small fraction.

Q. *If the same rate of growth continued further, how long would it take to equal the incident solar radiation?*

A. To answer this question, we can no longer just use doubling times, but need the equation for exponential growth, (1.3) from Fundamentals 1.2.

$$Q = Q_0 e^{kt}$$

Taking logarithms, we have

$$\ln(Q/Q_0) = kt$$

where t is the answer we want, and k, the growth rate, is 0.028/y (2.8 percent). When human consumption and solar output are equal, $Q = 54.4 \times 10^{20}$ kJ/y, while Q_0 (in 2000) is 0.00430×10^{20} kJ/y. These numbers give $\ln(Q/Q_0) = 9.95$ (you need a log table or calculator for this), and dividing by k gives $t = 337$ y. So less than four centuries of 2.8 percent annual growth would have us consuming as much energy as the sun delivers to the Earth.

1.4   HUMAN ENERGY SOURCES

What are the sources of our external energy? We are overwhelmingly dependent on the fossil fuels—oil, gas, and coal. Figure 1.5 shows the distribution of energy sources for the United States. A century and a half ago we relied almost exclusively on wood for fuel, but wood was supplanted by coal to power the industrial revolution. Then over the past fifty years, gas and oil took over as the dominant energy sources, which they remain today. (This, too, is a transitory phase of history, because the fossil fuel deposits will one day run out.) After fossil fuels, the largest source of energy is nuclear power, which has made substantial inroads in the past twenty years. Only a small fraction of our energy consumption derives from hydropower, although it is quite important in the regions having large dams. In a few places, geothermal plants and windmills are starting to contribute to the electricity supply, and solar heating and electricity-producing installations are in operation here and there, but the energy totals from these sources are too small to be seen on the graph.

If we compare industrialized and developing countries with respect to energy sources (see Figure 1.6a), we find that 1) the developing countries currently use less energy in aggregate, but their rate of energy growth is high, and they will surpass developed countries in a few years (see Figure 1.6b), and 2) they rely much more on biomass—wood and agricultural wastes—than do the industrialized countries. Nevertheless, the developing world is increasingly dependent on oil and coal, as is the industrialized world. Fossil fuel deposits are not evenly distributed around the globe. The great majority of the

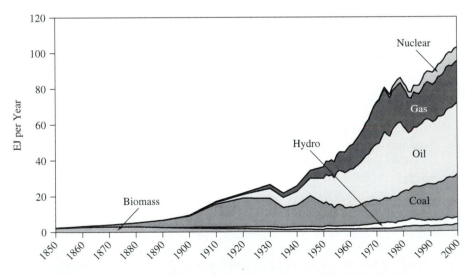

Figure 1.5   Historical trends in U.S. energy consumption, 1850–2000.   *Source:* Energy Information Agency, U.S. Department of Energy, *Annual Energy Outlook 2000*, energy consumption by source, Washington, DC.

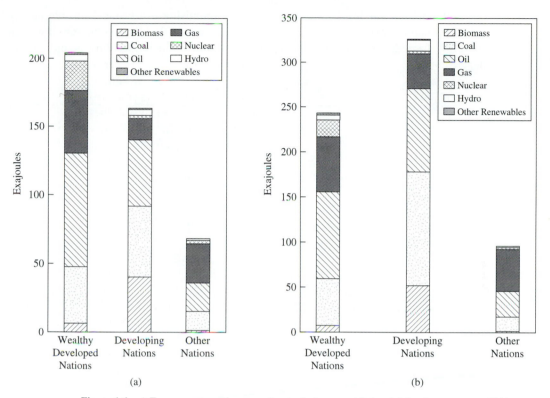

Figure 1.6   a) Energy consumption according to fuel type and industrial development, year 2000. b) Projected energy consumption according to fuel type and industrial development, year 2020. The designation "Other Nations" refers to the former Soviet Union and countries of Eastern Europe. *Source:* Energy Information Agency, U.S. Department of Energy, *Outlook 1998 with Projections through 2020,* Washington, DC.

available oil is in the fields of the Middle East (see Figure 1.7a). Because the industrialized world imports most of its oil from the Middle East, it is an area of great geopolitical importance, as the Gulf War of 1990 reminded us forcefully. The distribution of gas (see Figure 1.7b), and especially of coal (see Figure 1.7c), is somewhat more even around the globe.

What about the future? There are many energy source trade-offs before us. The sun pours enough energy on us to fulfill our needs, if we can learn to extract it efficiently and economically. There are many forms of this "renewable" energy: wind and hydropower, biomass, solar electricity, and direct solar heating. Some of these are already being harnessed on a limited scale, and new developments are occurring rapidly. The geothermal energy flowing from Earth's molten core also has a limited role to play. Nuclear fission can extend our fuel supply substantially by using the energy locked in the uranium nucleus, if we can overcome the serious problems of nuclear safety, proliferation, and waste disposal. And nuclear fusion waits in the wings, promising nearly inexhaustible energy from the fusion of hydrogen atoms, the same energy source that powers the sun, if we can solve the formidable problem of maintaining the operating temperature at millions of degrees.

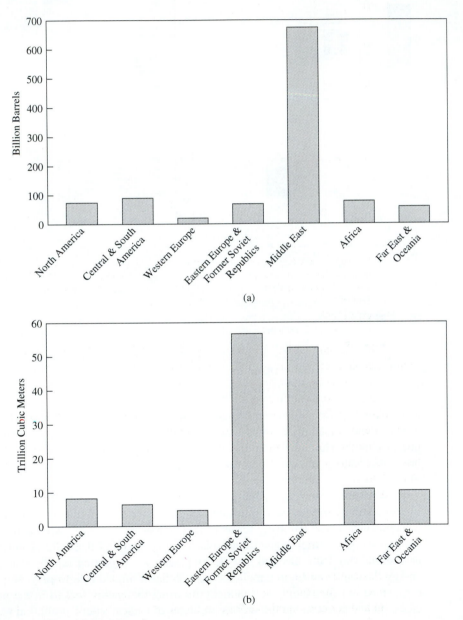

(a)

(b)

Figure 1.7   (*continued*)

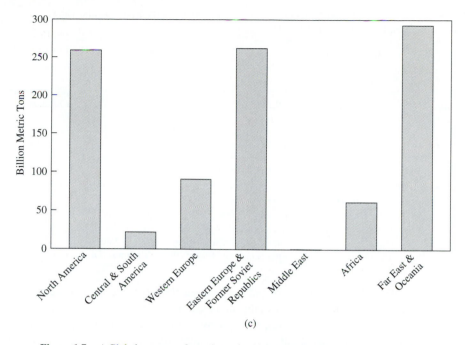

(c)

Figure 1.7   a) Global reserves of petroleum; b) Global reserves of natural gas; c) Global reserves of coal.   *Source:* Energy Information Agency, U.S. Department of Energy (1999). *Energy Reserves,* Table 8.1—World Crude Oil and Natural Gas Reserves, January 1, 2000. Table 8.2—World Estimated Recoverable Coal, January 1, 2000. (http://www.eia.doe.gov/iea/res.html)

Table 1.3 summarizes the current estimates of the amount of energy available from nonrenewable energy sources in the United States. These estimates are necessarily uncertain because what is "available" depends on how much money and energy are required to extract the resource under consideration. For example, as rich pools of oil near the land surface are depleted, it becomes more expensive to obtain the remaining oil in deeper or thinner deposits, or under the oceans. A limit would be reached, of course, when the energy required to extract and refine the oil exceeds the energy content of the oil itself, but long before this point, the process becomes uneconomical compared to other energy options. This limit of practicality itself changes with advancing technology, as exploration methods improve, and extraction techniques become more efficient. As a result, estimates of fossil fuel resources have tended to increase with the passage of time, and may yet increase further. The numbers in Table 1.3 indicate that the amount of oil that can still be recovered economically in the United States is equivalent to around thirteen times the current annual consumption (or about 29 years worth if we continue to import oil at the current rate). Accessible natural gas deposits in the United States are comparable to the petroleum deposits. The rest of the world has much more of these fuels, but it is clear that as world oil and gas consumption continues to rise, the age of petroleum will be measured in decades. Coal is a good deal more abundant, and is likely to last us for centuries. But coal

TABLE 1.3   NONRENEWABLE ENERGY SOURCES IN THE UNITED STATES IN 1999
(BILLION BARRELS OF OIL EQUIVALENTS)

Fuel	Discovered & estimated reserves	Supply years*	Undiscovered/ Uneconomical reserves	Additional supply years	Total supply years
Petroleum	90.2	29.1 (13.1)[†]	89.0	28.7 (13.0)	57.8 (26.1)
Natural gas	75.0	18.6	161	39.9	58.5
Coal	1,261	318	1,067	269	587
Nuclear	34.9	64.7 (21.5)	19.3	35.8 (11.9)	100 (33.4)

*Supply years correspond to number of years under current usage rates of each fuel.

[†]Numbers in parentheses refer to supply if all consumption was derived from domestic production only, rather than current level of domestic production plus foreign imports.

Sources: Oil and natural gas: Energy Information Administration, U.S. Department of Energy (1998). *U.S. Crude Oil, Natural Gas, and Natural Gas Liquids 1998 Annual Report.* Washington, DC. Coal: Energy Information Administration, U.S. Department of Energy (1999). *U.S. Coal Reserves: 1997 Update.* Washington DC. Nuclear: Energy Information Administration, U.S. Department of Energy (2000). *Uranium Industry Annual 1999.* Washington, DC (DOE/EIA-0478(99)).

is a much less versatile fuel and does not easily substitute for petroleum. What do we do when the oil is no longer available at a reasonable cost?

As momentous as this question may seem, it is actually being overshadowed by more immediate concerns about the environmental consequences of using fossil fuels. Burning the reduced carbon of ages past is driving up the CO_2 content of the atmosphere and contributing to greenhouse warming of Earth. And the combustion of fossil fuels produces acid rain, as well as urban and regional air pollution. Increasing determination to clear these pollutants from the air of U.S. urban areas, particularly Los Angeles, is driving the design of cars and the formulation of gasoline. The California Air Resources Board, in its 1990 mandate, required automobile makers to market and sell zero emission vehicles (ZEVs) as a fixed percentage of their overall sales. Currently, the only vehicles qualifying as ZEVs are electric cars powered by batteries, but "partial ZEV" credits will be allowed for low-emission fuel cell vehicles, and for fuel-efficient hybrids that are powered by a combination of electric motors and internal combustion engines. Thus, the need to reduce air pollution is beginning to direct the ways in which fossil fuels are used for transportation.

The issues of energy supply and resource substitution are profoundly influenced by the rate at which energy is used. Reducing energy consumption lowers environmental impacts, stretches out the resource base, and buys time for the development of new, more benign technologies. Different ways of projecting the energy consumption curve produce large differences in the estimate of future energy demand. The high and low projections in Figure 1.2 are both attempts at realistic forecasting, based on quite different assumptions about technology and behavior. The validity of these assumptions is hotly contested, and energy policy will be the focus of intense debate for a long time to come. But the experience of the last twenty years, in which economic activity, at least in the industrialized countries, has increased much faster than energy consumption (see p. 130), leaves little doubt that new technology and conservation measures can substantially increase the efficiency with which we use energy. We will return to this subject at the end of Part I, after looking more closely at the various energy sources.

CHAPTER 2

FOSSIL FUELS

2.1 CARBON CYCLE

Only about 0.3 percent of the energy in the sunlight reaching Earth's surface is converted by photosynthesis to chemical energy in the form of carbohydrates. Carbohydrates are so named because they contain two atoms of hydrogen and one of oxygen for every atom of carbon. Their chemical formula is $(CH_2O)_n$ where n is a definite integer, often a very large one. The overall reaction of photosynthesis is

$$CO_2 + H_2O = CH_2O + O_2 \tag{2.1}$$

The products of this reaction are less stable than are the reactants, by an amount of energy corresponding to about 450 kJ per mole of carbon. This is the energy extracted from the sunlight. It can be released by reversing the reaction, either by combustion, or, in biology, by respiration. Respiration provides aerobic organisms with the energy needed for all vital functions. Green plants use up about half of their carbohydrates for their own energy needs. The remainder is converted to other biological molecules or it accumulates in the growing plant tissue; the energy it represents is the *net primary productivity*.

The processes of photosynthesis and respiration are balanced closely, and the cycling of carbon between the carbon dioxide of the atmosphere and the organic compounds of biological organisms is nearly a closed loop. However, a very small fraction of plant and animal matter, estimated at less than 1 part in 10,000, is buried in the Earth and removed from contact with atmospheric oxygen (see Figure 2.1). Over the millennia this small fraction added up to a large amount of reduced carbon compounds. Some of the buried carbon compounds accumulated in deposits, and were subjected to high temperatures and pressures in

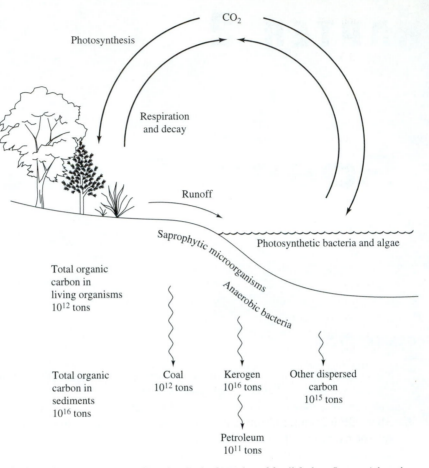

Figure 2.1   Burial of organic carbon in the formation of fossil fuels.   *Source:* Adapted from G. Ourisson et al. (1984). The microbial origin of fossil fuels. *Scientific American* 251(2):44–51. Reprinted with permission from J. Kuhl.

Earth's crust. They became coal, oil, and gas, which we now use to fuel our industrial civilization. We are therefore living off the store of solar energy of past ages. The total energy estimated to be available in easily recoverable fossil fuels is 5.3×10^{19} kJ.

WORKED PROBLEM 2.1: Will We Run Out of Oxygen?

Q. *Some people are concerned that we will use up Earth's oxygen supply if we burn all the fossil fuel. Are they right to worry?*

A. a) To decide this question, first calculate how many moles of O_2 would be used up,

given that the energy available in recoverable fossil fuels is estimated to be 5.3×10^{19} kJ, and that about 407 kJ are released per mole of O_2, on average, when fossil fuel is burned. [This value differs from the 450 kJ absorbed in reaction (2.1), above, because fossil fuels differ in composition from carbohydrates; see Table 2.2, p. 25]:

Dividing 5.3×10^{19} kJ by 407 kJ per mole of O_2 gives 1.3×10^{17} moles of O_2.

b) Next calculate what fraction of the atmosphere's O_2 this represents, using the following facts: the atmosphere weighs 1,000 g for each square centimeter (cm^2) of Earth's surface, and is 22 percent O_2 by weight; the radius of Earth (r) is 6.4×10^6 meters (m).

Since each cm^2 of the Earth's surface accounts for 1,000 g of air, or 220 g of O_2, we need to know the surface area. This can be obtained from the radius using the formula for the area of a sphere,

$$a = (r^2) \times 4\pi$$

r is given as 6.4×10^6 m, which we need to convert to cm by multiplying by 100 cm/m. Then we multiply the area, in cm^2, by the weight of O_2 per cm^2, and finally divide by the molecular weight of O_2, 32, to find the number of moles:

$$\text{moles } O_2 = (6.4 \times 10^6 \times 100)^2 \times 4\pi/3$$
$$\times\ 220/32$$
$$= 118 \times 10^{17}.$$

This is 90 times greater than the answer in part a.

The total buried carbon (see Figure 2.1: 10^{16} t $\times$ 10^6 g/t $\times$ 1/12 mole/g = 0.8×10^{21} mole) greatly exceeds the atmospheric O_2, but most of the buried carbon is widely dispersed in Earth's crust, and only a small fraction is in recoverable deposits. So even if all the fossil fuels were consumed, the loss in O_2 would be hardly noticeable.

However, the situation is very different for the amount of CO_2 produced by burning fossil fuels, because CO_2 constitutes only 0.037 percent of the atmosphere. Although the moles of CO_2 produced by burning fossil fuels are the same as the moles of O_2 consumed, the impact on the atmospheric CO_2 is much greater. (This impact is a major theme of Part II.)

2.2   ORIGINS OF FOSSIL FUELS

Petroleum and natural gas deposits are of marine origin. In the oceans, photosynthesis is estimated to produce 25 to 50 billion tons of reduced carbon annually. Most of this is recycled to the atmosphere as carbon dioxide, but a minute fraction settles to the bottom where there is no access to oxygen. This biological debris is covered by clay and sand particles and forms a compacted organic layer in a matrix of porous clay or sandstone. Anaerobic bacteria digest the biological matter, releasing most of the oxygen and nitrogen. The molecules most resistant to digestion are the hydrocarbon-based *lipids* (see pp. 271–272), and the saturated hydrocarbons found in oil have structural and carbon number distributions similar to those found in the lipids of living organisms. All petroleum deposits contain derivatives of the hydrocarbon hopane ($C_{30}H_{52}$), attesting to the importance of bacterial processing, since bacteria contain hopane derivatives in their membranes (see Figure 2.2). (For a brief review of the basics of organic chemical structures, see the Appendix.)

As the sediment becomes more deeply buried, the temperature and pressure rise. Bacterial action decreases, and organic rearrangement reactions occur. These reactions release large quantities of methane and light hydrocarbons as gases, and the gases accumulate in pockets under impermeable rock. Oil is produced from the remaining heavy organic compounds, which migrate as a water emulsion, from which the water is squeezed

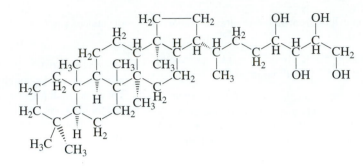

Figure 2.2   The structure of bacteriohopanetetrol.

out as the sediment is compacted. The oil is trapped in porous layers of rock. As shown in Figure 2.3, the process of gas and petroleum formation has spanned hundreds of millions of years. However, the time span of our use and depletion of this resource is likely to be on the order of a century and a half.

A great deal more methane is contained in undersea deposits as *gas hydrates,* or *clathrates.* These are frameworks of hydrogen-bonded water molecules that surround the methane molecules (see pp. 269–270) and form solid materials that are stable at sufficiently low temperature and high pressure. Gas hydrates are found in ocean sediments at depths of 300 meters or more. The methane, derived from microbial decomposition of organic matter in the sediment, is trapped at these depths in the hydrate structures. Estimates of the amount of methane in these deposits are enormous, amounting to perhaps twice as much fuel energy as the Earth's oil, gas, and coal deposits combined. No practical method of extracting this methane is currently available, but a number of development projects are underway. However, attempts at extraction will have to proceed with great caution, because of the possibility of releasing large amounts of methane, a potent greenhouse gas, into the atmosphere (see pp. 164–165).

In contrast to oil and gas, coal is of terrestrial origin. Coal deposits are the remains of plant matter from the huge, thickly wooded swamps that flourished 250 million years ago during a period of mild and moist climate. Woody plants are made up mainly of lignin and cellulose. While aerobic bacteria rapidly oxidize cellulose (a carbohydrate) to carbon dioxide and water when the plant dies, lignin is much more resistant to bacterial action. It is a complex, three-dimensional polymer based on benzene rings (see Figure 2.4). The building units are coniferyl and sinapyl alcohol for lignins from coniferous and deciduous plants, respectively.

In swamps, the lignin accumulates under water, compacting into peat. Over the geological ages, the peat layers of the primeval swamps metamorphosed into coal. Depression and thrusting of Earth's crust buried the deposits and subjected them to high pressures and temperatures for long periods of time. Under those conditions, the lignin gradually lost its oxygen atoms via the expulsion of water and carbon dioxide gas, and additional bonds formed among the aromatic groups, producing a hard, black, carbon-rich material, which we mine as coal. If this metamorphosis goes even further, the eventual product is graphite, a form of pure carbon in which the carbon atoms are arranged in layers of fused benzene rings.

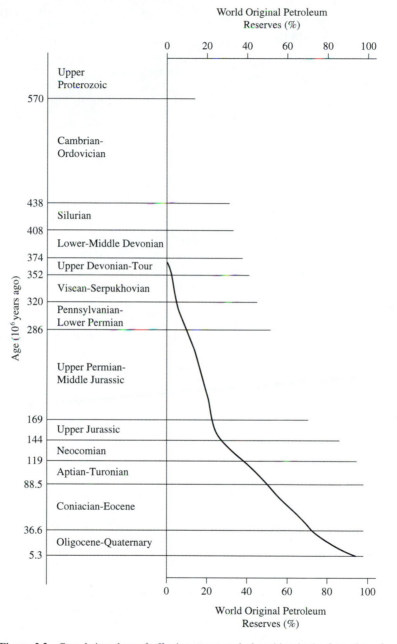

Figure 2.3   Cumulative chart of effective source rock deposition in the formation of petroleum.   *Source:* Adapted from H.D. Klemme and G.F. Ulmishek (1991). Effective petroleum source rocks of the world: Stratigraphic distribution and controlling depositional factors. *The American Association of Petroleum Geologists Bulletin* 75:1809–1851. AAPG Copyright © 1991. Reprinted by permission of the AAPG whose permission is required for further use.

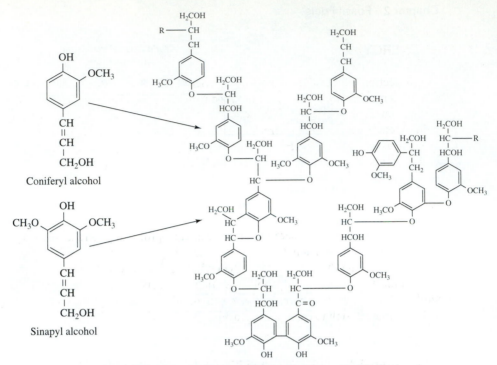

Coniferyl alcohol

Sinapyl alcohol

Representative structure for a part of the lignin polymer

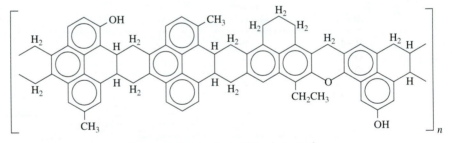

Representative structure of bituminous coal

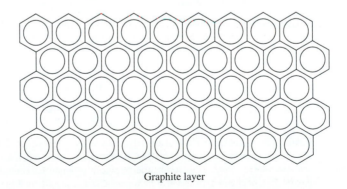

Graphite layer

Figure 2.4   Structural units of coal formation.

2.3   FUEL ENERGY

In this section we consider the chemical energy stored in fuels. Where does this energy reside? It resides in the bonds that hold the atoms together. But there is a seeming paradox in this statement, since breaking bonds does not release energy. Quite the opposite, an input of energy is required to break a bond. A balloon filled with H_2 is quite stable as long as no air is admitted. The H—H bond is strong; it takes 432 kJ to dissociate a mole of H_2 into H atoms. But an explosion occurs if H_2 is mixed with O_2, and a spark is lit. The explosion results from the reaction

$$2H_2 + O_2 = 2H_2O \tag{2.2}$$

Thus, the chemical energy of H_2 resides in its propensity to react with O_2. Two molecules of H_2O are formed in the reaction, each with a pair of O—H bonds, and the energy gained by forming these bonds more than offsets the energy lost by breaking the bonds of two H_2 molecules and one O_2 molecule. Table 2.1 lists several values for bond energies, while Table 2.2 shows the energy account for the reaction. The net energy released is 482 kJ, quite enough to cause an explosion.

TABLE 2.1   SOME AVERAGE BOND DISSOCIATION ENERGIES

Bond	Enthalpy (kJ/mol)	Bond	Enthalpy (kJ/mol)
H—H	432	C≡O	1,071
O=O	494	C—C	347
O—H	460	C=C	611
C—H	410	C⋯C*	519
C—O	360	N=O	623
C=O	799	N≡N	941

*Aromatic, 1.5 bond order

TABLE 2.2   COMBUSTION ENERGETICS ESTIMATED FROM BOND ENERGIES

	Energy content (kJ)				
	Reaction enthalpy	Per mole O_2	Per mole fuel	Per gram fuel	Moles CO_2 per 1,000 kJ
Hydrogen: $2H_2 + O_2 = 2H_2O$	482	482	241	120	0
Gas: $CH_4 + 2O_2 = CO_2 + 2H_2O$	810	405	810	51.6	1.2
Petroleum: $2(—CH_2—) + 3O_2 = 2CO_2 + 2H_2O$	1,220	407	610	43.6	1.6
Coal: $4(—CH—) + 5O_2 = 4CO_2 + 2H_2O$	2,046	409	512	39.3	2.0
Ethanol: $C_2H_5OH + 3O_2 = 2CO_2 + 3H_2O$	1,257	419	1,257	27.3	1.6
Cellulose: $(—CHOH—) + O_2 = CO_2 + H_2O$	447	447	447	14.9	2.2

FUNDAMENTALS 2.1: ENERGY AND BONDS

The reaction of hydrogen and oxygen is represented by the chemical equation (2.2). This equation is balanced chemically: the number of atoms is the same on the right- and left-hand side of the reaction for each element. But we know that the reaction is accompanied by the release of energy. This energy can be included in the equation:

$$2H_2 + O_2 = 2H_2O + 482\,kJ \quad (2.2a)$$

482 kJ is the energy released when two moles of hydrogen are reacted with one mole of oxygen. (Recall that a mole is the number of grams equal to the molecular weight.) If the amounts of the reactants were halved, then half as much energy would be released. Thus the energy released per mole of hydrogen is 241 kJ. (Strictly speaking we are talking here of the amount of heat released, or the enthalpy. The energy of the reaction also includes a contribution from the entropy. However, this contribution is small for combustion reactions. We will consider the entropy more closely when we consider the efficiency of energy conversion processes, see p. 98.)

The reaction energy arises because of the rearrangement of the chemical bonds. Energy is lost when bonds are broken and gained when they are re-formed. The energy released in reaction (2.2a) can be thought of as the resultant of the energy lost when two H_2 molecules dissociate into atoms:

$$2H_2 = 4H - 2E_{H-H} \quad (2.3)$$

and one O_2 dissociates into atoms:

$$O_2 = 2O - E_{O-O} \quad (2.4)$$

and the energy gained when these six atoms assemble into two water molecules:

$$4H + 2O = 2H_2O + 4E_{O-H} \quad (2.5)$$

Here we use E_{X-Y} to represent the energy gained from forming the X—Y bond from the atoms X and Y. If we add up equations (2.3–2.5), the result is:

$$2H_2 + O_2 = 2H_2O + 4E_{O-H}$$
$$- 2E_{H-H} - E_{O-O} \quad (2.6)$$

Comparing equations (2.2a) and (2.6) we see that

$$4E_{O-H} - 2E_{H-H} - E_{O-O} = 482\,kJ$$
$$(2.7)$$

Both E_{H-H} and E_{O-O} can be measured directly, by atomizing H_2 and O_2, and equation (2.7) provides a means for determining E_{O-H}. Tables of bond energies, like Table 2.1, are constructed in this way.

With a table of bond energies, it is possible to estimate the energies of many reactions by adding the energies of the bonds in the product molecules and subtracting the energies of the bonds in the reactant molecules. This is how the numbers in Table 2.2 were obtained. The estimates are not exact values, because the actual bond energies can differ somewhat in different molecules. For example, the O—H bond energy is not exactly the same in water and in ethanol. However, these differences are fairly small.

Examination of Table 2.1 reveals some interesting trends. H—H bonds are strong (because the valence electrons are close to the nuclei), but O—H bonds are even stronger. The reason is that the electrons are drawn closer to the O atom than the H atom because the O atom has a higher nuclear charge. (The O atom is more *electronegative*.) The result is that the bond has ionic character: negative charge accumulates near the O atom and positive charge near the H atom. The attraction of the unlike charges adds to the bond energy; ionic bonds are generally stronger than non-ionic bonds. Another trend is that double bonds are stronger than single bonds. However, the O=O bond is weaker than might be expected, while the C=O bond is very strong (compare the value for the C=C bond). The strong C=O bond is due to O being more electronegative than C, while the weak O=O bond is due to repulsion between the non-bonding electrons on the two O atoms in O_2 (see p. 188). The combination of a relatively weak O=O bond, and strong O—H and C=O bonds, is what makes combustion reactions so energetic.

WORKED PROBLEM 2.2: Methane Combustion

Q. *Check the energy values for methane combustion in Table 2.2, using the bond energies in Table 2.1.*

A. The equation for methane combustion,

$$CH_4 + 2O_2 = CO_2 + 2H_2O \qquad (2.8)$$

has one molecule of methane and two of oxygen as reactants, and one molecule of carbon dioxide and two of water as products. The product bonds are two C=O, worth 2 × 799 kJ, and four O—H, worth 4 × 460 kJ, for a total of 3,438 kJ. The reactant bonds are four C—H, worth 4 × 410 kJ, and two O=O, worth 2 × 494 kJ, for a total of 2,628 kJ. Subtracting reactant energies from product energies leaves 810 kJ for the reaction enthalpy. This is also the energy per mole of methane, but per mole of oxygen the energy is halved, to 405 kJ. To obtain the energy per gram of fuel, we divide 810 kJ by the molecular weight of methane, 16, to obtain 51.6 kJ. One mole of CO_2 is released per mole of methane, but to obtain 1,000 kJ requires 1,000/810 = 1.2 moles CO_2.

The energy release for methane combustion is much larger than for hydrogen combustion, according to reactions (2.2) and (2.8). Does this mean that methane explodes more violently than hydrogen? Not at all. The number of oxygen molecules is different in the two reactions, as written. If we compare the energy released per mole of oxygen reacted, the methane value comes down to 405 kJ and is smaller than the hydrogen value. So the reaction of an oxygen molecule with methane is slightly less violent than the reaction with hydrogen. On the other hand, a mole of methane has a much higher energy content than a mole of hydrogen (see the "per mole of fuel" column in Table 2.2), since one mole of oxygen reacts with two moles of hydrogen, but 0.5 moles of methane. Because a mole of any gas occupies about the same volume (at a given temperature and pressure), a cubic meter of methane has over three times the energy content of a cubic meter of hydrogen. But if it is weight that counts, then hydrogen wins (see the "per gram of fuel" column in Table 2.2). The energy content per gram of fuel is over twice as high for hydrogen as for methane, since its molecular weight is eight times smaller. This is why rockets are fueled with liquid hydrogen. The weight of the fuel is a large fraction of the weight of the rocket, and the lighter the fuel per unit of energy, the better.

The energy content of other fossil fuels can be similarly estimated. Table 2.2 shows schematic reactions for petroleum and coal. Since neither is a pure substance, we choose representative composition and bonding arrangements. Petroleum is largely made up of saturated hydrocarbons (see the Appendix), and we therefore consider the combustion reaction of a representative CH_2 group in a hydrocarbon chain:

$$2(—CH_2—) + 3O_2 = 2CO_2 + 2H_2O \qquad (2.9)$$

In the energy account, we include the C—C bond energy once per CH_2 group, since each of its two bonds joins two neighboring groups. As written, the reaction is estimated to release 1,220 kJ. Per mole of oxygen, however, the energy released is 407 kJ, nearly the same value as for methane. Per gram of fuel, the energy is 43.6 kJ, somewhat less than methane. [The H/C ratio of saturated hydrocarbons is greater than 2/1, especially for shorter molecules, because of the methyl groups at the ends of the chains. On the other hand, petroleum has a significant fraction of aromatic molecules with H/C ratios less than 2/1. For crude oil, the average heating value is 45.2 kJ per gram of oil, very close to the value calculated for reaction (2.9), while for gasoline, the value is slightly higher, 48.1 kJ per gram of gas, reflecting a higher H/C ratio.]

The hydrocarbons in coal are largely aromatic in nature, and the H/C ratio is 1/1 or slightly less. We consider a representative C—H group in an aromatic ring:

$$4(—CH—) + 5O_2 = 4CO_2 + 2H_2O \qquad (2.10)$$

The C atom is connected to neighboring C atoms by bonds with 1.5 bond order (see the Appendix). Again, the C—C bond energy is entered only once in the energy account in order to avoid duplicating bonds between neighbors. The energy released is 2,046 kJ for the equation as written, 409 kJ per mole of O_2, and 39.3 kJ per gram of fuel, slightly less than petroleum. Actual heating values for coal are less than this, mainly because they contain significant amounts of water and minerals. Typical values for a hard coal (bituminous or anthracite) are 29–33 kJ/gram, while soft coals (sub-bituminous, lignite) have heating values near 17–21 kJ/gram.

Bond-energy accounting can be applied equally to biomass-derived fuels (see p. 87), such as ethanol:

$$C_2H_5OH + 3O_2 = 2CO_2 + 3H_2O \qquad (2.11)$$

As shown in Table 2.2, the energy released is 419 kJ per mole of O_2, slightly higher than the fossil fuels, but the energy per gram of fuel, 27.3 kJ, is significantly lower than that for fossil fuels. The reason is that the O atom in ethanol already has strong bonds and does not contribute to the combustion energy, although it does add to the weight. A car will get fewer miles on a tank of ethanol than gasoline, since although the density of ethanol (0.79 g/cc) is about 12 percent higher than gasoline (~0.70 g/cc), the energy density is nearly 40 percent lower.

Finally, we can estimate the combustion or respiration energy for carbohydrates, the reverse of the photosynthesis reaction:

$$(—CHOH—) + O_2 = CO_2 + H_2O \qquad (2.12)$$

In the energy account for this reaction (see Table 2.2), we assume that each fuel unit has one C—H, one C—O, and one O—H bond, and is connected to its neighbors by C—C bonds. This leads to a slight underestimate of the energy, because in carbohydrates up to half of the O atoms are actually connected to two C atoms, and there are fewer C—C but more C—H bonds. For example, the combustion energy for glucose, which has the formula $(CHOH)_6$, is 2,803 kJ per mole of fuel, whereas six times the 447 kJ obtained for reaction (2.12) is 2,682 kJ. The energy per gram for carbohydrates is only about a third of that for hydrocarbons, since there are so many O atoms in the carbohydrate molecules. This is a familiar fact of nutrition and dieting: fats, which are mainly hydrocarbon in composition, have many more Calories per gram than do carbohydrates (see Part IV, pp. 369–370).

2.4   PETROLEUM

a. Composition and refining.   Oil is a complex mixture of hydrocarbons, molecules that contain mostly carbon and hydrogen. There are also small quantities of sulfur (up to 10 percent), oxygen (up to 5 percent), and nitrogen (up to 1 percent), bound in complex organic molecules. Several metallic elements—V, Ni, Fe, Al, Na, Ca, Cu, and U—are present in traces. Most of the hydrocarbon molecules are saturated (no multiple bonds), but an appreciable fraction, around 10 percent, are aromatic (contain benzene rings). The molecules range widely in size and are separated in refineries on the basis of their boiling points. Figure 2.5 is a diagram of the distillation process for dividing petroleum into its various fractions, and indicates the uses to which these fractions are put.

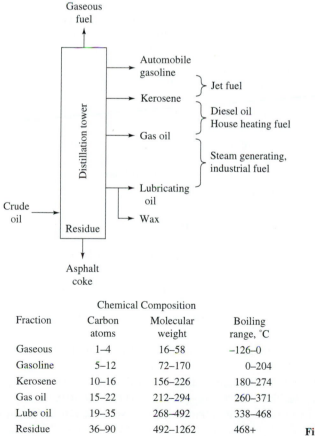

Fraction	Chemical Composition		Boiling range, °C
	Carbon atoms	Molecular weight	
Gaseous	1–4	16–58	–126–0
Gasoline	5–12	72–170	0–204
Kerosene	10–16	156–226	180–274
Gas oil	15–22	212–294	260–371
Lube oil	19–35	268–492	338–468
Residue	36–90	492–1262	468+

Figure 2.5   Crude oil refining.

In addition to distillation towers, oil refineries are equipped with reactors that transform the molecules chemically in order to match the quantities of the various fractions to the needs of the marketplace. Since gasoline is the most valuable fraction, several reactions increase the percentage of the gasoline fraction.

A particularly important chemical transformation is "cracking," whereby a larger hydrocarbon, in the kerosene or gas-oil range, is broken down to two smaller hydrocarbons in the gasoline range. This is accomplished at high temperature (400–600°C) with the aid of a catalyst, an aluminosilicate material impregnated with potassium.

$$C_{(m+n)}H_{2(m+n)+2} = C_mH_{2m} + C_nH_{2n+2} \qquad (2.13)$$

$$\begin{array}{ccc} \text{alkane} & \text{alkene} & \text{alkane} \\ \text{(kerosene or gas-oil size)} & \multicolumn{2}{c}{\text{(gasoline size)}} \end{array}$$

Another way to enhance the gasoline fraction is to build up a mid-size molecule from two smaller ones, in the process of "alkylation."

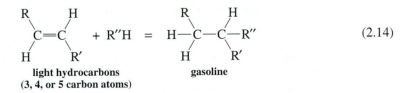

$$\qquad (2.14)$$

light hydrocarbons
(3, 4, or 5 carbon atoms)   **gasoline**

This process is catalyzed by strong acids. The cracking and alkylation reactions increase the gasoline fraction of crude oil, typically 20 percent by volume, to 40–45 percent.

The conditions of the alkylation process are arranged to produce hydrocarbons that have a high degree of branching; these have higher octane ratings (less tendency to pre-ignite, or "knock," during piston compression, see p. 237) than straight-chain hydrocarbons. Even higher octane ratings are obtainable with aromatic hydrocarbons (benzene derivatives), and an additional process carried out in the refinery is catalytic reforming, whereby straight-chain alkanes are converted to aromatics:

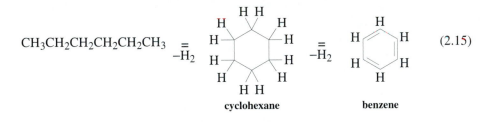

$$\qquad (2.15)$$

cyclohexane   benzene

These reactions are run at high pressure (15–20 atm) and temperature (500–600°C) with a Re-Pt-Al$_2$O$_3$ catalyst.

A final reaction is the controlled oxidation of the hydrocarbons, which can produce oxygen-containing fuel molecules called "oxygenates." These are now required to be added to gasoline in order to reduce carbon monoxide emissions (see p. 239). Until now the most widely used oxygenate has been methyl tertiary-butyl ether (MTBE), which is made by adding methanol to isobutene, using an acidic catalyst.

$$\underset{\underset{\displaystyle CH_3}{|}}{\overset{\overset{\displaystyle CH_3}{|}}{C}}=CH_2 + CH_3OH \ = \ CH_3-\underset{\underset{\displaystyle CH_3}{|}}{\overset{\overset{\displaystyle CH_3}{\diagup}\overset{O}{\diagdown}}{C}}\ CH_3 \qquad (2.16)$$

The reaction is run with excess methanol to suppress isobutene dimerization and at low reaction temperature ($<100°C$) to suppress the formation of dimethyl ether. MTBE is produced in refineries where isobutene is generated in the catalytic cracking units.

Refinery chemistry is shifting rapidly, under the pressure of gasoline regulations. Thus benzene is being eliminated from gasoline, because of its carcinogenisity (see pp. 221, 239), and MTBE is being abandoned because of its pollution of groundwaters (see pp. 259–260).

b. Advantages.   The overwhelming advantage of petroleum is that it is a liquid, and therefore easily transported. The liquid fuels of the petroleum age have made possible the development of efficient means of transport, from the airplane to the automobile to diesel locomotives. Petroleum fuels are relatively clean, since the refinery produces hydrocarbon fractions, leaving most of the sulfur- and metal-containing compounds in the residue. The entire system of oil extraction, transport, refining, and delivery has been developed to a high level of integration and efficiency, creating a formidable challenge for fuel alternatives.

c. Disadvantages

1) Oil spills.   The extraction of petroleum from the ground inevitably produces contamination from spills. Coastal waterways are particularly vulnerable because of the fragility of their ecosystems combined with the importance of shipping in petroleum transport. Figure 2.6 illustrates the geography of oil transport. Tankers ply the shipping lanes continually, loading and unloading their cargo at the ports of the world. Tanker accidents are the most notorious instances of petroleum pollution. Oil washes up on beaches, smothers birds and sea animals, and contaminates fish.

Thanks to improved safety and operating measures, the number of large tanker accidents, as well as the quantities of oil spilled, have been declining since 1970 (see Figure 2.7a,b). However, tanker accidents, although dramatic, are a small part of the problem of spills. According to the U.S. Coast Guard, tanker accidents accounted for 29 percent of oil discharges to American waterways during 1995–1999 (see Figure 2.8). Routine operations of tankers and other oil-burning vessels accounted for nearly as much (26 percent), as did land and waterfront facilities that handle the oil (23 percent), with leaks from on- and off-shore oil pipelines (15 percent) accounting for most of the remainder.

There are even larger sources that are not quantified by the Coast Guard. The greases and oils accumulating for a year on the streets of a city of five million have been estimated to be comparable to the release of a large tanker accident. Some of this grease and oil

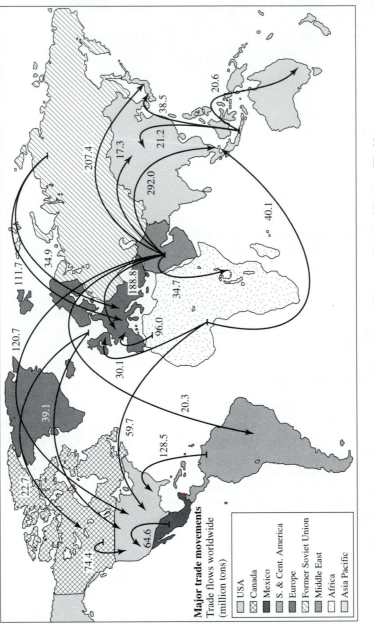

Figure 2.6 Routes of international oil trade. *Source:* BP Amoco, *Statistical Review of World Energy 2000.* (http://www.bp.com/centres/energy/world_stat_rev/oil/trade.asp)

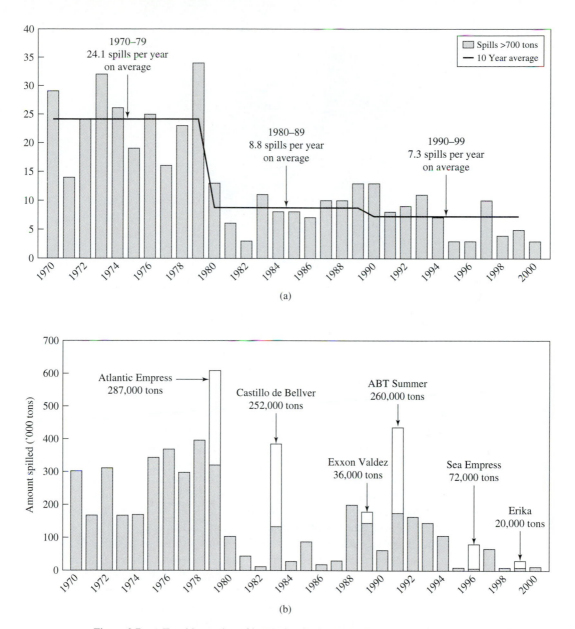

Figure 2.7   a) Trend in number of international oil spills; b) Trend in quantity of oil spills. White sections of bars indicate major tanker oil spills.   *Source:* International Tanker Owners Pollution Federation. *Tanker Oil Spill Statistics* (2000). (http://www.itopf.com/stats/html)

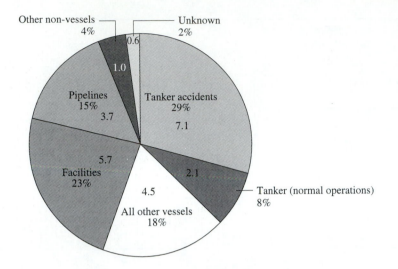

Figure 2.8   Sources of oil inputs to U.S. Waters, 1995–1999 (thousands of metric tons). *Source:* U.S. Coast Guard, *Polluting Incident Compendium, Cumulative Data and Graphics for Oil Spills, 1973–1999, Oil Spill Compendium Data Table, Volume of Spills by Source.* (http://www.uscg.mil/hq/g%2Dm/nmc/response/stats/C8Data.htm)

washes into nearby waterways during large storm events. Changing lubrication oil of the U.S. light-vehicle motor fleet (cars, vans, and sport utility vehicles—200 million vehicles in all) generates about 1.3 billion gallons (4.9 billion liters) of waste oil annually. The National Oil Recyclers Association estimates that over 200 million gallons are not collected, but rather dumped illegally or accidentally into sewers, streams, drains, landfills, and backyards. This quantity is equal to nearly 20 Exxon Valdez spills per year. The volumes of municipal waste and runoff from urban areas entering water bodies are difficult to assess quantitatively, but they are likely to be substantial. In the 1980s, the U.S. National Research Council estimated their annual contribution globally to be about 800,000 tons (equivalent to 22 Exxon Valdez accidents per year), about a third of total inputs.

Interestingly, a significant quantity of petroleum, estimated at 0.2 million tons a year, enters the sea without any help from humans, through natural seeps at continental margins (see Figure 2.9). Thus, oil is a natural constituent of the marine environment. This oil does not accumulate, because it is metabolized by microbes, which have evolved to exploit seepage oil as their food source. In fact, hydrocarbon-metabolizing bacteria are ubiquitous in nature, because hydrocarbons are continuously released by plants and algae. The total hydrocarbon input to the sea from marine biota is estimated to be 180 million tons annually, dwarfing the petroleum inputs from all sources. These same microbes eventually break down the oil molecules spilled by human activity, but the process can take a long time, during which there can be considerable ecosystem damage.

It is possible to speed up the process under favorable conditions. The fate of the oil spilled in the March 1989 Exxon Valdez disaster at Prince William Sound, Alaska, attests

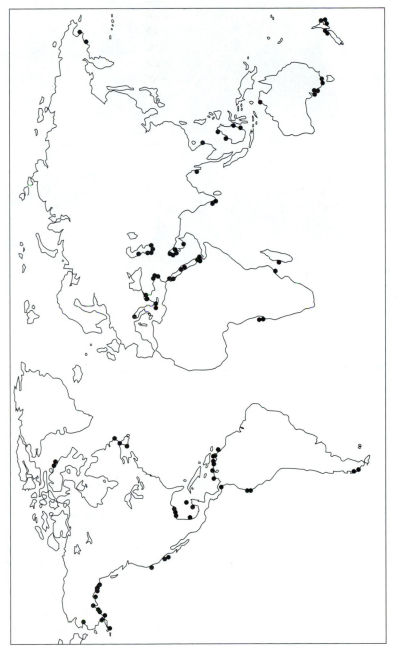

Figure 2.9 Locations of reported marine seeps. *Source:* National Research Council (1985). *Oil in the Sea* (Washington, DC: National Academy Press).

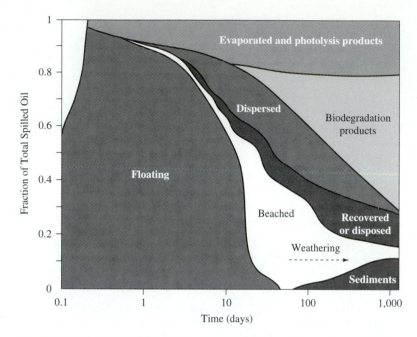

Figure 2.10   Overall fate of oil spilled in the *Exxon Valdez* accident over time period from March 1989 through Autumn 1992.   *Source:* D.A. Wolfe et al. (1994). The fate of the oil spilled from the *Exxon Valdez. Environmental Science and Technology* 28:561A–568A. Reprinted with permission from ES&T. Copyright (1994) American Chemical Society.

to the cleansing power of nature. The spill resulted in the release of 35,500 metric tons of North Slope crude oil into the sound. As shown in Figure 2.10, by the autumn of 1992 all of the original floating oil had disappeared. About 50 percent was biodegraded either on the beaches, or in the water column; 20 percent evaporated and underwent photolysis in the atmosphere; 14 percent was recovered or dispersed; 13 percent remained in subtidal sediments; 2 percent remained beached on intertidal shorelines; and less than 1 percent remained dispersed in the water column. One of the more effective remediation steps taken in the wake of the accident was to fertilize oil-soaked beaches with a preparation designed to stick to the sand and provide the natural bacteria with nitrogen and phosphorus to supplement their hydrocarbon diet. The oil disappeared from these beaches much faster than from untreated beaches.

The lighter hydrocarbon fractions are the most toxic to sea creatures, because they have the highest solubility in water. These lighter molecules are also the most volatile and evaporate rapidly. Between one-third and two-thirds of this fraction can evaporate in a few days. The remaining molecules are heavier and tend to form an emulsion with seawater, sometimes called a "mousse." The emulsion eventually forms tar balls, which can persist for a long time. All of the evaporated molecules, and a substantial fraction of those left behind, are oxidized through the action of sunlight and oxygen (photo-oxidation), processes that we will encounter in discussing air pollution and photochemical smog (see pp. 226–232).

   2) Emissions.   The burning of all fossil fuels produces CO_2 and adds to the greenhouse warming of Earth (see pp. 172–178). Because the C/H ratio declines from about 1/1 for coal to about 1/2 for petroleum and to 1/4 for methane, the amount of CO_2 produced per kJ of energy decreases in the same order. The last column of Table 2.2 lists the moles of CO_2 released per thousand kJ of energy produced by the combustion of each of the fuels; this number is obtained by dividing the number of molecules of CO_2 in each of the equations by the reaction enthalpy and multiplying by 1,000. For the same energy production, coal and petroleum release significantly more CO_2 than methane, by about 60 percent and 33 percent, respectively. For completeness, CO_2 values are also listed in Table 2.2 for ethanol and carbohydrates, but biomass-derived fuels do not alter the atmosphere's CO_2 content, since the released CO_2 exactly balances the CO_2 that was taken from the atmosphere in producing the biomass (although there may be significant time lags).

   All combustion processes produce the pollutant nitric oxide, NO. Although the N_2 and O_2 of the atmosphere do not react at ordinary temperatures, they do react at the high temperatures of a furnace or an automotive engine (see pp. 183–186). Once formed, NO is slowly oxidized further to nitrogen dioxide, NO_2, and then to nitric acid, HNO_3, which is absorbed by raindrops and washed out of the atmosphere. Combustion sources, both stationary and mobile, are major contributors to acid rain, through the production of nitric acid (see pp. 279, 306).

   In addition, nitrogen dioxide is a key ingredient in smog production because of its photochemical activity (see pp. 228–229). The products of incomplete combustion, carbon monoxide (CO) and unburned hydrocarbons, are the other key ingredients in smog production. In both cases the main culprit is the automobile, and enormous effort has gone into curtailing automotive emissions. Because of the ever-increasing volume of traffic, however, these efforts are approaching diminishing returns. As mentioned in the introduction, the requirement that hundreds of thousands of zero- and low-emission vehicles be introduced in Southern California over the next ten years is a harbinger of changes to come.

   In summary, petroleum stands as a major contributor to the greenhouse effect, acid rain, and urban air pollution.

2.5   GAS

Although natural gas represents as large an energy resource as petroleum, it has historically been considered a by-product of oil exploration and production. Indeed, the magnitude of the potential gas supply has only recently become appreciated, thanks to better information about gas-bearing formations, and to improved recovery techniques. Natural gas provides a substantial fraction of the U.S. energy budget (nearly 25 percent). It has mostly been used in heating and cooking, but its use in electricity generation is expanding rapidly (15 percent of total gas consumption in 1999), because of the introduction of the gas turbine electric generator. Stemming from the development of advanced jet engines for aircraft, the gas-fired turbine is a heat engine that can run at very high temperatures, and is therefore capable of high efficiency (see p. 101).

Other new technologies may further expand the role of gas in view of its superior environmental attributes. A number of car and bus fleets now operate on natural gas (3 percent of 1999 consumption). Gas may facilitate the transition to a "hydrogen economy" (see p. 109), since a pipeline distribution system is already in place. Indeed, gas is already the source of 95 percent of the hydrogen currently produced (via steam-reforming, see equation (2.21), p. 41).

a. Advantages. Natural gas is a clean fuel, requiring very little processing. It is readily transported overland through pipelines. Its CO_2 emission rate, per unit of energy, is lower than other fossil fuels, especially coal, as already noted. In addition, gas contributes less to smog formation than does gasoline, because unburned CH_4 molecules are considerably less reactive, with respect to the free radical chemistry responsible for smog, than are hydrocarbon molecules having more than one carbon atom.

b. Disadvantages. Natural gas is much harder to carry around than liquid hydrocarbons. To pack enough of it into a reasonable space for mobile power sources requires high pressures or low temperatures, or both. Compressors and/or refrigerators are required, and the storage tank must be thick-walled and/or insulated. In addition, it requires a distribution system capable of transferring the gas under pressure. These requirements are formidable obstacles to the replacement of petroleum by natural gas in automotive transport. The problems are more manageable for fleets of trucks or buses, which can carry large tanks and are supplied at a central depot. A number of fleets are currently running on natural gas. Countries with abundant gas reserves, notably New Zealand, Canada, and Russia, are actively promoting the use of natural gas in vehicles.

Although natural gas produces less CO_2 than other fossil fuels, methane is itself a potent greenhouse gas. Its infrared absorption bands fall in the window of the CO_2 and H_2O spectra (see pp. 162–164), and, because it is less reactive than other hydrocarbons, it has a long atmospheric lifetime. An additional methane molecule contributes about twenty times as much to the greenhouse effect as an additional CO_2 molecule. Consequently, leaks of methane are a serious environmental concern. These can occur at the gas wells, during transfers, and from power sources. Escape of methane during idling of gas-powered vehicles, for example, could nullify their CO_2-related greenhouse gas advantage.

2.6 COAL

Coal deposits vary significantly in the extent to which the original woody plant tissue has been metamorphosed. Hard coals have undergone greater transformation than soft coals. Table 2.3 gives percentages of various coal constituents for deposits in different regions of the United States. Lignite is the softest coal; its name recognizes the close similarity

TABLE 2.3 COMPOSITION AND HEAT CONTENT OF COMMON COALS FOUND
IN THE UNITED STATES

Rank	Location by state	Chemical analysis				Heating value (kJ/g)
		Moisture	Volatile matter	Fixed carbon	Ash	
Anthracite	Pennsylvania	4.4%	4.8%	81.8%	9.0%	30.5
Bituminous						
Low volatile	Maryland	2.3	19.6	65.8	12.3	30.7
High volatile	Kentucky	3.2	36.8	56.4	3.6	32.7
Sub-bituminous	Wyoming	22.2	32.2	40.3	4.3	22.3
Lignite	North Dakota	36.8	27.8	30.2	5.2	16.2

Source: U.S. Bureau of Mines (1954). *Information Circular No. 769* (Washington, DC: U.S. Department of Interior).

to the parent wood component, lignin. Over a third of the lignite mass is moisture, while the remaining carbonaceous material is almost evenly divided between "volatile matter," hydrocarbons that are released upon heating, and "fixed carbon," the nonvolatile carbon fraction. Sub-bituminous coal is harder than lignite, containing about 20 percent moisture and 40 percent fixed carbon, but softer than bituminous coal, which contains very little moisture. The hardest coal is anthracite, which is about 80 percent fixed carbon. The heating value of the coal varies with the fraction of reduced carbon and hydrogen, and is much lower for soft than hard coals because of their high moisture content.

The different coals have variable amounts of ash, the mineral residue left after complete combustion, reflecting different amounts of minerals incorporated during the metamorphic processes. Some of this mineral is pyrite, FeS_2. In addition, some sulfur is bound in the complex organic molecules of coal. When coal is burned, both inorganic and organically bound sulfur is oxidized to SO_2, a significant air pollutant. Figure 2.11 shows the distribution of the coals in the U.S., and their sulfur content. The high-sulfur bituminous coals are found mostly in Appalachia and the interior of the country, while the low-sulfur sub-bituminous and lignite coals are mainly in the West.

a. Advantages. The coal resource base is very large, and coal is relatively cheap to mine and transport by rail. This is coal's great advantage.

b. Disadvantages. Coal is, of course, much less convenient to carry around and handle than petroleum. Its use in transportation disappeared when diesel replaced the steam locomotive. In technologically advanced countries, coal is no longer used directly for space heating, but in the rest of the world, the burning of coal in stoves and furnaces fouls the air with soot and SO_2, contributing significantly to respiratory distress. The main use of coal is in large electricity-generating plants, where it is burned efficiently and relatively completely. Tall smokestacks disperse emissions widely, lessening the local air pollution. However, the SO_2 and NO emitted from the burning of coal in power plants are the main sources of acid

Figure 2.11 Coal distribution in the United States and average sulfur content of coal.

Average Sulfur Content of Coal by Rank		
Rank	% of total coal reserves	% with sulfur content > 1%
Anthracite	0.9	2.9
Bituminous	46.0	70.2
Sub-bituminous	24.7	0.4
Lignite	28.4	9.3
Total, all ranks	100.0	35.0

Figure 2.11   Coal distribution in the United States and average sulfur content of coal. *Source:* P. Averitt (1960). U.S. Geological Survey, *Bulletin 1136* (Washington, DC: U.S. Department of Interior).

rain (see pp. 157, 279, 306), and of aerosol particles, which play a major role in regulating global climate (see pp. 153–155), and also have significant impacts on human health (see p. 222). As with the other fossil fuels, the emitted CO_2 enters the global greenhouse budget; because of its lower C/H ratio, coal emits more CO_2 per unit of energy produced than either gas or oil (see Table 2.2).

Extraction of coal adds significant costs to the environment and to human health. Traditional coal mines, which follow coal seams deep into the earth, are hazardous places to work, and the coal dust produces black-lung disease in miners. These problems have been alleviated to some extent through improved safety measures, better ventilation, and by spraying water to reduce dust during drilling operations. The drainage from mines, which is highly acidic, once contaminated local streams, but federal regulations in the United States now require collection of drainage in settling and treatment ponds. Strip-mining has permitted surface extraction of shallow seams of coal, at the expense of great gashes in the earth, often on steep and erodible hillsides. United States regulations now require strip-mine operators to restore the original contour of the land, replace topsoil, and replant grasses, legumes, and trees.

STRATEGIES 2.1   **Coal-Derived Fuels**

Technologies are available for converting coal to clean fuels via chemical reactions, to produce gaseous or liquid hydrocarbons. The basic requirement is to increase the H/C ratio of the coal. For example, direct reaction of coal with H_2 can yield methane

$$C + 2H_2 \overset{800°}{=} CH_4 + 74.9 \text{ kJ} \qquad (2.17)$$

But this "hydrogasification" reaction requires a high operating temperature, 800°C, for adequate reaction rates. Moreover, the reaction is inefficient because, being exothermic, it is thermodynamically unfavored at the required high temperature. (Adding heat drives the reaction back to the left.)

A more efficient route is provided by "methanation" of CO:

$$CO + 3H_2 \overset{400°}{=} CH_4 + H_2O + 206.3 \text{ kJ} \qquad (2.18)$$

This reaction is even more exothermic, but it can operate at a lower temperature, 400°C, in the presence of a nickel catalyst. Reaction conditions can be altered to produce liquid hydrocarbons via what is called Fischer-Tropsch chemistry:

$$nCO + (2n + 1)H_2 = C_nH_{2n+2} + nH_2O \qquad (2.19)$$

again using metal catalysts. Still another possibility is to make methanol, via

$$CO + 2H_2 = CH_3OH \qquad (2.20)$$

using still other catalysts and reaction conditions. Methanol is an alternative liquid fuel.

But these conversion reactions require CO and H_2. What is the source of these ingredients? They can be produced by treating coal with water at a very high temperature, 900°C:

$$C + H_2O \overset{900°}{=} CO + H_2 - 131.4 \text{ kJ} \qquad (2.21)$$

This reaction, called the "steam-reforming" reaction, produces only as much H_2 as CO, but extra H_2 can be produced by the "water-gas shift" reaction

$$CO + H_2O = CO_2 + H_2 + 41.4 \text{ kJ} \qquad (2.22)$$

If we multiply equation (2.21) by two and add equations (2.18) and (2.22), the result is:

$$2C + 2H_2O = CH_4 + CO_2 - 15.1 \text{ kJ} \qquad (2.23)$$

In theory, all the heating value of coal can be transferred to methane via reaction (2.23), with an energy expenditure of only 15.1 kJ. In reality, the energy costs are much higher, because the individual stages of the process are not well matched thermodynamically. The heat released in reaction (2.18) cannot be recovered to drive reaction (2.21) because the required temperature is much higher for the latter than for the former. Instead, the energy input to reaction (2.21) must be provided by burning extra coal. The required energy for one mole of methane is twice the enthalpy of reaction (2.21), 262.8 kJ, which is about 32 percent of the energy content of the methane (see Table 2.2). Thus, the energy-conversion efficiency of the process can be no better than 68 percent, and is actually lower than this because of other losses. The cost of coal-derived fuels is therefore high. Moreover, they contribute disproportionately to the greenhouse effect, because excess CO_2 is released by burning the extra coal required for the energy input to the conversion process. More CO_2 is released in the conversion and burning of coal-derived fuels than would be released in the production of equivalent energy from the coal itself.

2.7   DECARBONIZATION

The overarching problem with continued fossil fuel consumption is the build-up of CO_2 in the atmosphere. Why not address this problem by sequestering the CO_2 and storing it out of harm's way? If this could be accomplished economically, we could continue to rely on fossil fuels until they are used up, without worsening the greenhouse effect. This idea seemed fanciful until recently, but it is now receiving serious attention as a strategy for reducing CO_2 emissions.

a. Separation.   The CO_2 could be separated from the exhaust gases after fossil fuel combustion. This could be done with reasonable efficiency using organic amines to absorb the CO_2 from the exhaust gas. CO_2 reacts with amines to form carbamates:

$$R_3N + CO_2 = R_3NCO_2 \tag{2.24}$$

The CO_2 is subsequently trapped by heating the carbamates to release the reactants, thereby regenerating the amines.

Alternatively, the fossil fuel could be converted to CO_2 and H_2, using the chemistry discussed above under "Coal-Derived Fuels" (see Strategies 2.1, p. 41). The advantage of this approach is that the CO_2 could be separated at the source, where it is much more concentrated than in combustion exhaust gases. In addition, the fuel energy can be extracted more efficiently from H_2, using fuel-cell technology (see p. 103), than from fossil fuel combustion. This added efficiency can help to offset the costs of the fossil fuel conversion. The disadvantage of fuel decarbonization is that it requires a new energy infrastructure for H_2 transport, storage, and utilization (see discussion of the hydrogen economy, p. 109).

b. Storage.   Once the CO_2 is sequestered, it must be prevented from escaping into the atmosphere. As shown in Figure 2.12, two options are being considered for long-term storage. One is the deep ocean. At the low temperature and high pressure of the deep ocean, CO_2 becomes a liquid, and at a depth of about 3,500 m, the density of CO_2 becomes greater than that of water. Consequently, if CO_2 were piped to this depth, it would sink to the bottom of the ocean. However, CO_2 also dissolves in water, and at the alkaline pH of the ocean, it reacts to form bicarbonate ion (see p. 286). This reaction is quite vigorous, raising concern about the effectiveness of ocean injection in keeping CO_2 away from the atmosphere. There are also concerns about unknown effects of injected CO_2 on marine life.

The other option is storage underground in geological formations. One possibility is storage in the reservoirs left behind when petroleum and gas are extracted. Indeed, CO_2 is already being injected into some oilfields in order to enhance the recovery of the oil. When liquefied under pressure, CO_2 becomes an effective agent for forcing oil to flow through the reservoir and into oil wells.

In addition to petroleum and gas reservoirs, CO_2 could be stored in deep aquifers. As in the oceans, the CO_2 would react, in this case with carbonate and silicate minerals (see p. 287) in the aquifer, to form bicarbonate. However, the physical stability of the injected CO_2 is not an issue, since the aquifers are trapped by the overlying rock. It is unknown, however, how soon the CO_2 would escape from the aquifer, unless its migration is blocked by

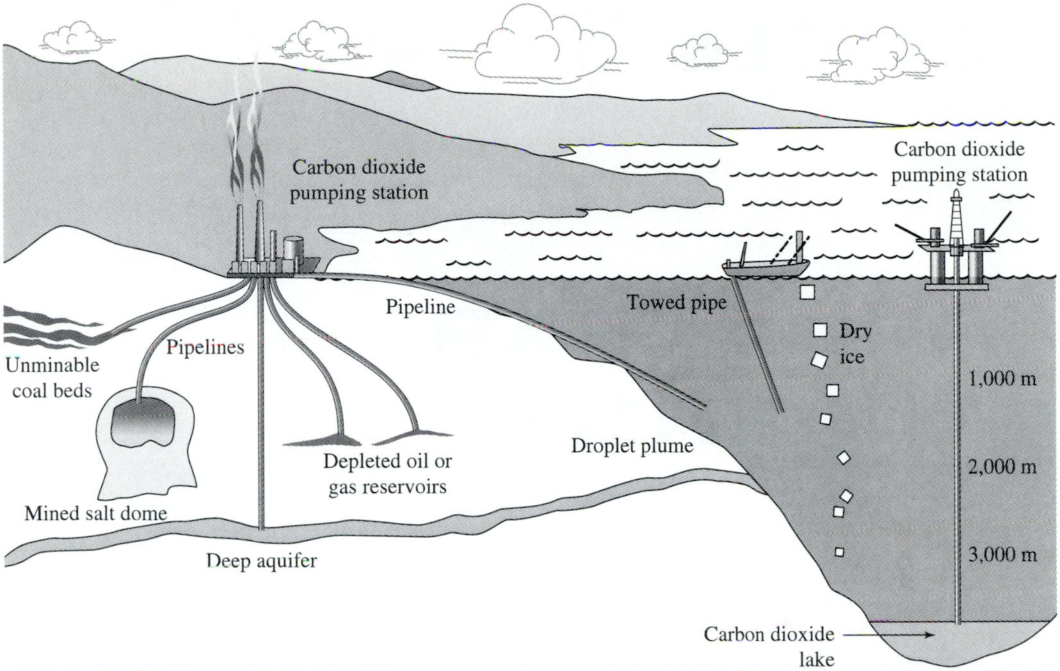

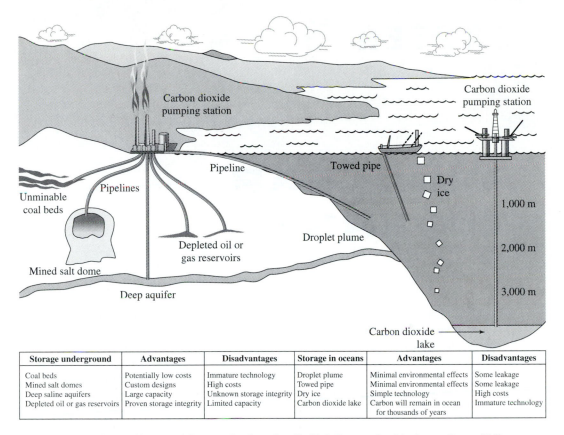

Storage underground	Advantages	Disadvantages	Storage in oceans	Advantages	Disadvantages
Coal beds	Potentially low costs	Immature technology	Droplet plume	Minimal environmental effects	Some leakage
Mined salt domes	Custom designs	High costs	Towed pipe	Minimal environmental effects	Some leakage
Deep saline aquifers	Large capacity	Unknown storage integrity	Dry ice	Simple technology	High costs
Depleted oil or gas reservoirs	Proven storage integrity	Limited capacity	Carbon dioxide lake	Carbon will remain in ocean for thousands of years	Immature technology

Figure 2.12 Potential storage sites for carbon dioxide in the ground and deep sea. *Source:* H. Herzog et al. (2000). Capturing greenhouse gases. *Scientific American* 282(February):72–79. Reprinted with permission from D. Fierstein.

lateral traps. Deep coal beds are another possible storage medium; the injected CO_2 could displace methane adsorbed on the coal, which could be recovered for use as fuel.

A project is currently underway to use an aquifer under the North Sea for CO_2 storage. Many gas deposits naturally contain a substantial amount of CO_2, which is normally vented when the gas is extracted. A Norwegian gas company is engaged in separating this gasfield CO_2 and reinjecting it into a nearby aquifer. The economics of the project rests on the fact that Norway imposes a tax on CO_2 emissions, as part of its contribution to greenhouse gas curtailment.

A final method of storage is to react the CO_2 with basic minerals and convert it to solid carbonates for burial. This scheme amounts fundamentally to acceleration of the natural weathering process (see p. 288), by which the atmospheric CO_2 is controlled on geological time scales. The feasibility of CO_2 sequestration as carbonate depends on the extent to which the normally slow reactions of minerals can be accelerated in reactors.

CHAPTER 3

NUCLEAR ENERGY

Nuclear power is presently the most highly developed alternative to energy supplied by coal. Apart from geothermal energy, the one significant form of energy on Earth that is not related to the sun either directly or indirectly is energy that resides in the nuclei of atoms.

The fundamental basis of nuclear energy is the curve of binding energy, shown in Figure 3.1, which plots the binding energy per nucleon as the mass of the nucleus increases. At first, the binding energy increases strongly, because of the strong nuclear force, which holds the nucleons together. The huge jump from hydrogen to helium is the basis of fusion energy, the energy that powers stars and the goal of the program to create fusion reactors on Earth (see section 3.9). The binding energy continues to climb with increasing nuclear mass, but reaches a maximum in the vicinity of the element iron, and then slowly declines. The reason is that the electrostatic repulsion among the positively charged protons gradually overwhelms the strong nuclear force. Eventually the nuclei become unstable with respect to α emission. (See the next section. The special stability of the helium nucleus, see Figure 3.1, explains why the favored decay route is ejection of α particles.) All elements heavier than bismuth, with 83 protons, are unstable.

An alternative form of nuclear decay is *fission,* in which the nucleus splits into two *daughter* nuclei, with a very large release of energy. This mode of decay is very rare, but it is the basis of the atomic age, ushered in by the atomic bomb, and later by the advent of the fission reactor. There is only one fissionable isotope that is found naturally in the crust of the Earth, the uranium isotope with mass 235, ^{235}U.

3.1 NUCLEI, ISOTOPES, AND RADIOACTIVITY

Nuclei are composed of protons and neutrons (the *nucleons*). They possess essentially the same mass (see Table 3.1), but the proton is positively charged, while the neutron is neutral. The number of protons determines the number of negatively charged electrons that

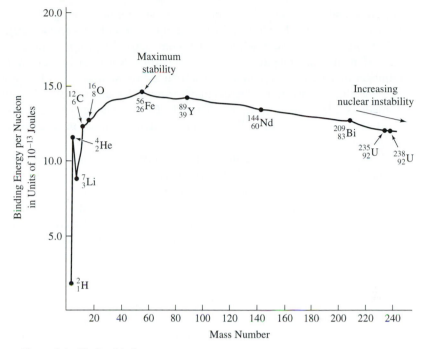

Figure 3.1   Nuclear binding energy curve.

TABLE 3.1   SIMPLE ATOMIC PARTICLES

Type	Schematic representation	Charge	Mass*	Chemical symbol[†]
Neutron		0	1.0087	^1_0n
Proton		+1	1.0078	^1_1p
Electron		−1	0.0009	e^-
Helium-4 (alpha particle)		+2	4.0026	^4_2He

*In atomic mass units (amu), where 1 amu $= 1.6606 \times 10^{-24}$ g.

[†]The superscript for neutron, proton, and helium-4 is the *mass number*, equivalent to the number of protons and neutrons in the nucleus; the subscript is the *atomic number*, equivalent to the number of protons; the superscript of the electron indicates its negative charge.

surround the nucleus; the number of electrons in turn determines the chemical properties of the element. Thus, each element has a specific number of protons; hydrogen has one, helium has two, and so on. This is the atomic number, often indicated by a left subscript next to the symbol for an element (see Figure 3.1).

The neutrons, being neutral, do not alter the number of electrons, and have no effect on chemical properties. Indeed, the nucleus of an element can have variable numbers of neutrons, and this is why the atomic masses are not necessarily close to being integral numbers. For example, chlorine has an atomic mass of 35.453 (see Periodic Table, inside front cover) because while all Cl atoms have 17 protons, three quarters of them have 18 neutrons, while the rest have 20 neutrons. Nuclei of an element having different numbers of neutrons are *isotopes*. The two major isotopes of chlorine are chlorine-35 and chlorine-37. We write the chemical symbol for an isotope by including the mass number as a left superscript, i.e., ^{35}Cl and ^{37}Cl. All elements have several isotopes, but most of them are present in low abundance. For example, most carbon atoms are ^{12}C, but a few (1.03 percent) are ^{13}C.

Not all combinations of protons and neutrons are stable. For a given element the stable isotopes have a somewhat greater number of neutrons than protons. Figure 3.2 shows the stability curve. There is some latitude in the number of neutrons, but nuclei become unstable when there are too many or too few neutrons. Unstable nuclei are transformed into stable ones by undergoing *nuclear decay*. Such a transformation releases a great deal of energy, because the energy stored in nuclei is very large. This release of energy during nuclear decay is called *radioactivity*. *Radioisotopes* are unstable with respect to nuclear decay.

There are several kinds of nuclear decay, resulting in several forms of radioactivity. If there are too many neutrons, one of them may convert to a proton by ejecting an electron:

$$n = p^+ + e^-$$

The electron carries off the energy, and emerges with a very high speed; for historical reasons it is called a beta (β) ray. A new element is created, because the number of protons in the

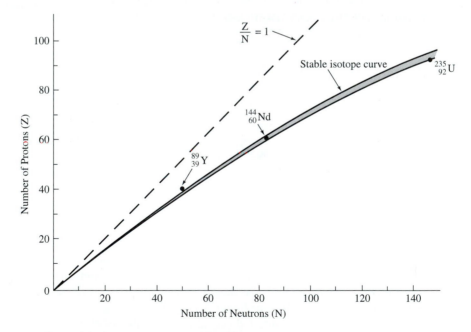

Figure 3.2   The proton-neutron stability curve.

nucleus increases by one. An example is ^{40}K. Potassium has 19 protons, and the mass 40 isotope has 21 neutrons, one too many (^{39}K is stable). This isotope converts to a calcium isotope of the same mass number by ejecting an electron:

$$^{40}K = {}^{40}Ca^+ + e^-$$

If an isotope has too few neutrons, then a proton can convert to a neutron by ejecting a *positron,* e^+. A positron is an anti-electron, and carries a positive charge:

$$p^+ = n + e^+$$

(An ejected positron is also a beta ray; it is given the symbol β^+ to distinguish it from an ejected electron, β^-. Once ejected from the nucleus, a positron is quickly annihilated upon collision with an ordinary electron.) Positron emission likewise creates a new element, with an atomic number one unit lower.

Nuclei can also be unstable by having too many protons, regardless of the number of neutrons. Because protons are positively charged, their mutual repulsion can overcome the glue holding the nucleus together (called the strong nuclear force, one of the fundamental forces of nature) if there are enough of them. When this happens, the nucleus ejects a particle containing two protons and two neutrons, i.e., a helium nucleus. Ejected helium nuclei are called alpha (α) rays.

Finally, a nucleus can simply have too much energy, even if it has a stable number of protons and neutrons. This situation is often encountered in the products of β or α decay. Even though energy is carried off by the β or α particle, the transformed nucleus may be in a nuclear excited state. It can decay to its ground state by emitting a gamma (γ) ray. A γ ray is a form of electromagnetic radiation, like an x-ray, but with even higher energy (a high energy *photon*).

3.2 NATURALLY OCCURRING RADIOISOTOPES

Most unstable isotopes have long since disappeared from the Earth since its formation, some 4.5 billion years ago, but a few decay so slowly that they are still present in significant abundance. The most important of these long-lived radioisotopes are listed in Table 3.2. ^{232}Th, ^{235}U and ^{238}U are α emitters, while ^{40}K and ^{87}Rb are β emitters. ^{232}Th and ^{238}U are both quite abundant in Earth's crust. ^{235}U has become scarce, since 6.4 half-lives have elapsed since the formation of the Earth. ^{40}K is a form of potassium that occurs in low

TABLE 3.2 LONG-LIVED RADIOISOTOPES

Isotope	$t_{1/2}$ (years)
^{238}U	4.5×10^9
^{235}U	7.0×10^8
^{232}Th	1.4×10^{10}
^{87}Rb	4.9×10^{10}
^{40}K	1.3×10^9

abundance (0.001 percent), but because potassium is an important constituent of biological tissues, ^{40}K provides a significant fraction of the background radiation to which we are normally subject.

FUNDAMENTALS 3.1: HALF-LIVES AND ISOTOPE DATING

The rate of nuclear decay varies widely from one radioisotope to another, but the decay process is always exponential, because the number of nuclei diminish in proportion to how many are present, i.e., they diminish at a constant percentage rate (see Fundamentals 1.2). This means that each radioisotope is characterized by a constant half-life. However much is present, the amount will decay to half that value after one half-life (see Figure 1.4, p. 13).

Depending on which radioisotope we are considering, the half-life can range from microseconds to billions of years, but the value never changes. There is nothing one can do to speed up nuclear decay (except to subject the radioisotopes to nuclear reactions, inside a reactor). This is the characteristic that makes disposing of nuclear wastes such a thorny issue. Some of the radioisotopes will remain radioactive for eons. The constant half-life also makes radioisotopes useful for dating various natural processes. The best known example is radiocarbon dating.

The carbon isotope with mass number 14, ^{14}C, has six protons and eight neutrons. It decays by β emission, pro-

ducing ^{14}N, which is a stable isotope. ^{14}C is produced in the atmosphere by cosmic rays, ultra-high energy particles that rain down continuously from outer space. When cosmic rays hit atoms in the atmosphere, they produce an array of nuclear fragments. One of these is ^{14}C, produced by collisions with ^{14}N atoms (N is the most abundant element in the atmosphere). The ^{14}C produced in this way immediately reacts with O_2 molecules, producing $^{14}CO_2$. Thus ^{14}C enters the carbon cycle and is incorporated into all living things. The fraction of living carbon that is ^{14}C is extremely small, but it can be measured quite accurately because of the high energy of the emitted β rays.

When life stops, so does the exchange of carbon with the atmosphere, and the ^{14}C content of preserved organic matter gradually decreases. The half-life of ^{14}C is 5,730 years, so a 6,000-year-old sample of preserved tissue has about half the ^{14}C content of living tissue. This half-life makes radiocarbon dating useful for samples that are several hundred to several thousand years old, an important time span for archeological purposes.

WORKED PROBLEM 3.1: Radiocarbon Dating

Q. *How old is a wooden bowl whose ^{14}C activity is one-fourth that of a contemporary piece of wood?*

A. The activity falls to one-fourth in two half-lives, making the bowl 11,500 years old.

(It is unlikely that the measurement is accurate enough to justify more than three significant figures.)

Q. *At Stonehenge, a charcoal sample was dug up, presumably the remains of a fire; its ^{14}C*

activity was measured at 9.6 disintegrations per minute per gram of carbon. Living tissue has a ^{14}C activity of 15.3, in the same units. When did the Stonehenge fire burn?

A. The ratio of activities is $9.6/15.3 = 0.615$. This is not an integral number of half lives, so we need the equation for exponential decay, (1.7) from Fundamentals 1.2, p. 12:

$$Q = Q_0 e^{-kt}$$

or, in logarithmic form,

$$\ln(Q/Q_0) = -kt$$

Q/Q_0 is the ratio of activities, 0.615, and its natural logarithm is -0.486. k, the decay constant, is related to the half-life by

$$t_{1/2} = 0.693/k \text{ [see Fundamentals 1.2, eq. (1.8)]}$$

The time before the present is then $t = 0.486 \times 5{,}730 \text{ yr}/0.693 = 4{,}020 \text{ yr}.$

3.3   DECAY CHAINS: THE RADON PROBLEM

Alpha emission often leaves a product isotope that is itself unstable with respect to β or further α emission. Consequently, the decay of the heavy elements generally proceeds in a sequential cascade, as shown for ^{238}U in Figure 3.3. The long-lived ^{238}U produces a much

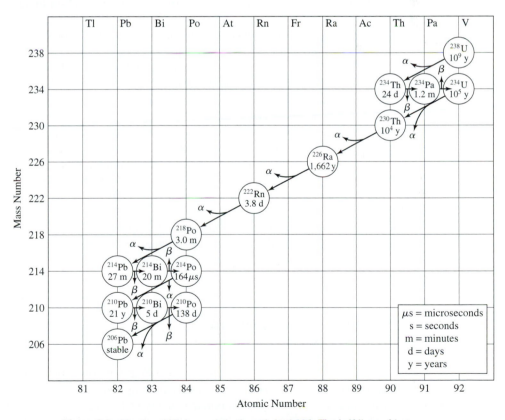

Figure 3.3   Uranium-238 decay chain through lead-206. The half-lives of isotopes are given in the circles.

shorter-lived (24.1 day half-life) isotope of thorium, ^{234}Th. Two successive β decays then produce another uranium isotope, ^{234}U, which is less stable than ^{238}U, but still long-lived ($t_{1/2}$ = 245,000 years). It decays to a succession of more-or-less short-lived isotopes, ending with the stable lead isotope, ^{206}Pb. All of these daughter isotopes accumulate in a sample of matter that contains ^{238}U. Their concentrations build up to a steady state, in which their production rate equals their decay rate. For the rapidly decaying isotopes, this concentration is very low, but for relatively long-lived isotopes, the concentrations can be significant. For example, radium, ^{226}Ra, with a 1,600 year half-life, accumulates suffi- ciently in uranium deposits to support extraction for commercial applications. The other two very long-lived α emitters, ^{235}U and ^{232}Th, have their own sequence of daughter iso- topes, all of which can be found in Earth's crust.

One isotope in the ^{238}U decay chain is of special environmental significance, namely ^{222}Rn. Radon is a noble gas, falling just below xenon in the periodic table. Consequently, radon isotopes do not form chemical bonds, and are free to escape from the site where they are formed. Born in uranium-containing rock, ^{222}Rn can travel an appreciable distance be- fore it decays, with a 3.82 day half-life. It can enter nearby buildings by seeping through foundations, and accumulate to significant levels in the air. It can also infiltrate wells and enter the water supply.

In 1984, a nuclear plant worker in Pennsylvania set off a radiation-monitoring alarm on his way *into* work. His clothes were found to be radioactive and the source of the contamination was traced to his home, in which a high radon level was found. This incident led to the recognition that naturally accumulating radon can be a significant radi- ation hazard. Since uranium is a relatively common element, and is distributed widely in Earth's crust, the problem is widespread. Radon can build up to high levels in houses built on rock with high uranium content, and where the ground is sandy and therefore permeable to gases. The radon level increases when the building foundation is porous, and when the air pressure in the building is less than the pressure of the gases in the soil. This pressure difference, which depends on indoor and outdoor temperatures, wind veloc- ity, and a host of other factors, is a key variable, and probably accounts for large varia- tions in radon levels even among nearby houses. Fortunately, inexpensive tests for radon are readily available, and remediation is generally straightforward. The usual method is to lay vent pipes next to or under the building and to pump soil radon directly out of doors with an exhaust fan. There is continuing debate, however, about just how serious the prob- lem is and how worried people should be.

3.4 RADIOACTIVITY: BIOLOGICAL EFFECTS OF IONIZING RADIATION

When unstable nuclei decay, α, β, or γ rays are released with very high energy, in the range of millions of electron volts (Mev). On encountering molecules in their path, they knock electrons out of the atomic shells. Enough energy is deposited in the ionized mole- cules to break chemical bonds and induce reactions. In biological tissues the result is generalized damage and the production of reactive chemical fragments (for example, free

TABLE 3.3   PATHS OF ENERGETIC PARTICLES IN BIOLOGICAL TISSUE

Type of radiation	Range in biological tissue*	Relative biological effectiveness[†]
alpha	0.005 cm	10–20
beta	3 cm	1
gamma	~20 cm	1

Some hazardous radioactive isotopes			
Element	Type of radiation	Half-life	Site of concentration
$^{239}_{94}$Pu	alpha	24,360 years	Bone, lung
$^{90}_{38}$Sr	beta	28.8 years	Bone, teeth
$^{131}_{53}$I	beta, gamma	8 days	Thyroid
$^{137}_{55}$Cs	beta, gamma	30 years	Whole body

*For a 6 Mev particle.

[†]Accounts for the fact that cell damage increases as the density of the damage sites increases.

radicals, see p. 375). Table 3.3 summarizes the penetration depth for the different rays and their relative effectiveness in producing damage.

Some biological molecules are more susceptible than others to the effects of ionizing radiation. The nucleic acids, which make up each cell's genetic apparatus, are particularly vulnerable. A single ionization in the cell's nucleus can produce an error in the genetic instructions for assembling the protein constituents of the cell. Certain kinds or combinations of such errors are believed to transform normal cells into cancerous ones, and there are well-established correlations between radiation exposure and the incidence of cancer. If the nuclei of reproductive cells are damaged, the result may be genetic mutations and, thus, the transmission of hereditary disorders to succeeding generations. At high levels, radiation damages all cells, particularly those which divide rapidly—white blood cells, platelets, and the cells of the intestinal linings—leading to a variety of symptoms, collectively called radiation sickness.

The potential for damage is crucially dependent on the location of the isotope emitting the radiation, relative to target molecules. If the isotope is external to the body, then the issue is what kind of shielding is needed to absorb the rays. If the isotope is ingested, however, then the issue becomes one of transport and elimination, as well as the decay rate. The most worrisome isotopes are those that lodge in particular tissues and are long-lived enough to do significant damage. Table 3.3 lists four isotopes of particular concern. ^{239}Pu is part of the nuclear fuel cycle, while ^{90}Sr, ^{131}I, and ^{137}Cs are fission products. They all live long enough to do significant damage, and they all concentrate in particular tissues. Plutonium and strontium lodge in bone, primarily, because their chemistry mimics that of calcium. Cesium spreads through all tissues, along with potassium, which it mimics. Iodine is a natural constituent of the thyroid hormone thyroxine, and ^{131}I concentrates in the thyroid gland, which it can damage; indeed, controlled doses of ^{131}I are used therapeutically to counter hyperthyroidism.

a. Alpha rays.   An α ray, being a doubly charged helium nucleus, produces intense damage over a short distance. It ionizes half the atoms in its path, losing about 30 ev (electron volts) per collision. A 6-Mev α ray would therefore produce about 200,000 ionizations among the first 400,000 atoms it encounters, before its energy is dissipated. In a substance with the density of water or biological tissue, the distance traveled, or range, is about 0.05 mm. This is less than the thickness of the skin's protective outer layer of dead cells. If, however, an α emitter is ingested into the body, it produces a high density of localized damage, with an appreciable potential for cancer induction. The likeliest route of ingestion is inhalation of dust particles that carry radioisotopes.

Uranium miners are at high risk of developing lung cancer. The uranium itself is relatively harmless because of its very slow decay. Some of the daughter isotopes, which accumulate in the mines, are much more radioactive and therefore more hazardous. Most of the cancer effect is attributed to radon and its daughters. Radon itself, being an unreactive gas, is expelled from the lungs as fast as it is inhaled (and with a 3.8 day half-life, little of it decays in the lungs). The daughters are isotopes of reactive elements, however (see Figure 3.3), and these are incorporated into dust particles that the miners breathe. The dust particles stick to the lining of the lung, and the radioactive daughter isotopes therefore have time to do their damage. The cancer rate has been correlated with the levels of radon (and its daughters) to which the miners are exposed.

The same mechanism of biological damage applies to radon and its daughters in houses. At the levels found in some houses, the radon exposure is calculated to be close to the low end found in uranium mines (taking into account the longer time people spend in houses than in mines). This is the basis of concern over the cancer-causing potential of radon in houses. Indeed, the extrapolation from concentrations known to cause cancer (uranium mines) to those commonly encountered by the population at large (houses), is less than that for any other known or suspected carcinogen. Nevertheless, there is continuing skepticism as to the extent of the danger, because an association between radon levels in houses and cancer incidence has not yet been established in the epidemiological data. It is often suggested that the uranium miner cancer incidence is not a reliable guide, because mines are much dustier than homes. This debate is representative of those encountered for all environmental carcinogens (see pp. 415–418). The evidence is often equivocal, and extrapolations to usual exposures are uncertain, allowing individuals to evaluate the risks very differently.

b. Beta and gamma rays, and neutrons.   A β ray, being an energetic electron, is much lighter than an α ray and is only singly charged. It also loses about 30 ev per collision, but it ionizes only about 1 in 1,200 atoms in its path. The damage density is therefore lower, but the range of a β ray is larger; a 6-Mev β ray travels 3 cm in water or biological tissue. External β radiation is therefore hazardous, and β-emitting isotopes must be shielded.

A γ ray is a high-energy photon and has a different mode of interaction with matter than do charged particles. The probability of a γ ray hitting an atom in its path is quite low, but when it collides, it transfers a large amount of energy, and the ionized electron carries away enough energy to ionize many other electrons (secondary ionizations). Because of this, γ rays do not have a well-defined range, but rather a distribution of path lengths. For 6-Mev γ rays, the median path length at which half the γ rays have stopped is 20 cm in water or biological tissue. Energetic γ emitters require heavy shielding.

Finally, neutrons interact with matter by penetrating the electron shells and reacting directly with the nuclei of matter, displacing them and causing ionization or producing radioisotopes, which in turn release ionizing radiation. Neutrons decay spontaneously ($t_{1/2} = 12$ minutes) into protons and electrons, and are therefore of concern only in the immediate vicinity of nuclear reactions.

3.5   RADIATION EXPOSURE

Aside from the location of the radiation source and the type of radiation, exposure depends on the concentration of a given radioisotope and its half-life. The shorter the half-life, the greater the disintegration rate, and the more intense the exposure. On the other hand, if the half-life is very short, then the exposure is also brief.

Radioactive disintegrations are measured in *curies* (Ci); one Ci is 3.7×10^{10} disintegrations per second. In the case of radon in houses, the U.S. Environmental Protection Agency has set an action level at 4 pCi/L (4 picocuries per liter), corresponding to 0.15 disintegrations per second for each liter of house air. Exposure to radiation, on the other hand, is measured in several different units. A *roentgen* (R) is the amount of radiation that, on passing through 1 cm^3 of air (at 0°C and 1 atm pressure), would create one electrostatic unit (2.08×10^9 times the charge on an electron) each of positive and negative charges. A *rad* is the amount of radiation that deposits 100 ergs of energy in a gram of material; for biological tissues, 1 R is about equivalent to 1 rad. Finally, a *rem* (roentgen-equivalent-man) is the amount of radiation that produces the same biological effect in a person as 1 R of x-rays. For β and γ rays, 1 rad is equivalent to 1 rem, but for α rays, because of their greater ionizing ability, 1 rad is equivalent to 10–20 rem, depending on their energy. (There are new international units, the *gray,* equivalent to the rad, and the *sievert,* equivalent to 100 rem.)

To how much radiation are we exposed? Table 3.4 lists the average annual doses for people in the United States. The total is about 360 mrem (millirem), of which the great majority is from natural sources. Over half the total comes from radon, with significant additional increments from cosmic rays, radiation from rocks and soil (other than radon), and radioisotopes that occur naturally in the body (principally ^{40}K, see p. 47). The remaining 18 percent is from artificial sources, mainly x-rays and radioisotopes used in medicine. Consumer products, principally building materials, account for 10 mrem, or 3 percent of the total. Fallout from nuclear weapons testing (a serious concern during the period of heavy testing) is now estimated to contribute less than 1 mrem, as does the entire nuclear fuel cycle. A similar small dose is produced by occupational exposures. But this entry in the table is a little misleading, since it is an average dose for the entire U.S. population. In the one million or so workers engaged in the affected occupations, the average annual dose is 230 mrem, about the same as the average radon dose, while for those in particularly hazardous occupations, such as uranium miners, the dose can be much higher.

TABLE 3.4   AVERAGE ANNUAL EXPOSURE (1990) TO RADIATION FOR PEOPLE
IN THE UNITED STATES

Source of radiation	Dose (mrem)	Percent of total dose
Natural		
Radon gas	200	55
Cosmic rays	27	8
Terrestrial (radiation from rocks and soil other than radon)	28	8
Inside the body (naturally occurring radioisotopes in food and water)	39	11
Total natural	**294**	**82**
Artificial		
Medical		
X-rays	39	11
Nuclear medicine	14	4
Consumer products (building materials, water)	10	3
Other		
Occupational (underground miners, x-ray technicians, nuclear plant workers)	<1	<0.03
Nuclear fuel cycle	<1	<0.03
Fallout from nuclear weapons testing	<1	<0.03
Miscellaneous	<1	<0.03
Total artificial	**64**	**18**
Total natural plus artificial	**358**	**100**

Source: National Council on Radiation and Measurement (1990). (Washington, DC: National Academy Press).

There are other variables that can increase an individual's exposure from the average. The large variability in radon levels in houses has already been mentioned. Each dental or chest x-ray adds 10 mrem to an individual's total, while a gastrointestinal tract x-ray adds 200 mrem. The cosmic ray entry in Table 3.4 is for people living at sea level, but the cosmic ray intensity increases at high altitudes. (It is twice as high at 2,000 m and four times as high at 3,000 m.) For this reason individuals living at high altitudes, or who are frequent fliers (3 mrem exposure for a five-hour flight at 9,000 m), have a higher exposure to cosmic rays. None of these factors are terribly worrisome, however, when viewed in the context of the 360 mrem/yr background exposure. This background also puts radiation hazard from the nuclear industry in context. For the population at large, this hazard is seen to be negligible, provided that all the radioisotopes stay where they are supposed to stay. This is, of course, a major proviso.

3.6   FISSION

If a heavy nucleus splits into two lighter ones, the release of energy is very large, because of the greater stability of the lighter nuclei (see Figure 3.1). But fission is an extremely rare event, even among uranium atoms. In the case of ^{235}U, however, fission is induced

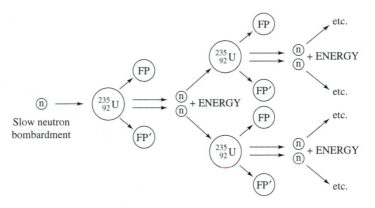

Figure 3.4   Chain reaction induced with thermal (slow) neutrons (FP and FP′ represent various fission products).

when the nucleus absorbs a neutron. Not only does the nucleus then split into two lighter nuclei, but two or three neutrons are also released. An example of a fission pathway is

$$\ce{^1_0 n} + \ce{^{235}_{92}U} = \ce{^{144}_{56}Ba} + \ce{^{89}_{36}Kr} + 3\ce{^1_0 n} \tag{3.1}$$

The lighter nuclei, called *fission products,* can have a range of masses; this is why in some cases two and in some cases three neutrons are released. These newly produced neutrons can then collide with other ^{235}U nuclei, inducing them to fission. Because more than one neutron is released per fission event, there can be a nuclear *chain reaction,* in which the number of fission events rapidly increase (see Figure 3.4). It was the realization among physicists, on the eve of World War II, of the enormous energy that could be released in a fission chain reaction that led to the race to develop an atom bomb, culminating in the explosions over Hiroshima and Nagasaki in 1945.

Whether a chain reaction will be sustained depends mainly on the amount of uranium present, since some neutrons escape from the system before they are absorbed by the uranium nuclei. This tendency decreases as the surface-to-volume ratio of the total sample decreases, that is, the mass of the uranium increases. A *critical mass* is that amount of uranium for which the probability, k, that a neutron produced in a fission reaction induces another fission reaction, is equal to unity. This is the break-even point for a self-sustaining chain reaction. A fission explosion is set off by assembling a *supercritical mass* ($k > 1$) of uranium-235 (or plutonium-239, a man-made fissionable isotope, see p. 63) very rapidly, using a chemical explosive for the trigger. For pure uranium-235, the critical mass is 15 kg, while for pure plutonium-239, it is 4.4 kg. (Smaller masses can be critical if surrounded by material that reflects escaping neutrons, making them available for additional fission events. Reflectors are incorporated into the design of nuclear weapons.)

The fission products are highly radioactive. Because the n/p ratio stability curve has a downward curvature, fission of a heavy element creates daughter nuclei that have too many neutrons, as illustrated in Figure 3.5. They move toward the stability line by β emission. Also, the nuclear chain reaction leads to the production of additional *actinides*

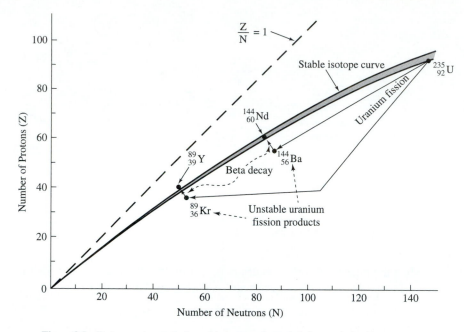

Figure 3.5   Proton-neutron ratio for stable isotopes formed via beta emission [beta particles (β^-) are high-energy electrons emitted by the reaction: $(^1_0n) \rightarrow$ proton $(^1_1p^+) + \beta^-$].

(elements in the uranium row of the periodic table). Not all neutron collisions with uranium nuclei lead to fission. Often the neutrons are simply absorbed, leading to heavier nuclei. These are all unstable nuclei, which decay by α emission, but many are long-lived, with half-lives extending to thousands of years. The combination of intensely radioactive fission products and long-lived actinides produces the uniquely complicated potential for environmental impact that characterizes the nuclear age.

   a. Pressurized light-water reactor.   Since the grim dawn of the nuclear age, scientists have dreamed of harnessing the power of the nucleus for peaceful purposes, and from this impetus the nuclear power reactor program arose. Figure 3.6 shows the operating principles of the most common nuclear reactor design—the pressurized light-water reactor. The fuel rods of this reactor contain pellets of uranium or uranium oxide. Naturally occurring uranium is made up mostly of ^{238}U, which does not fission. ^{235}U constitutes only 1 out of 140 uranium atoms. In order to build up a chain reaction in the light-water reactor, the uranium is concentrated in the ^{235}U isotope to a level of 3–4.5 percent. Control rods containing cadmium or boron, which absorb neutrons effectively, are lowered automatically among the fuel rods to a level that adjusts the neutron flux so that the chain reaction is maintained but does not run out of control.

   Surrounding the fuel and control rods is a bath of water, which acts both as a coolant to carry away the energy generated in the fission reaction, and as a *moderator*, that is, a substance that slows down the neutrons to increase the fission probability. The probability

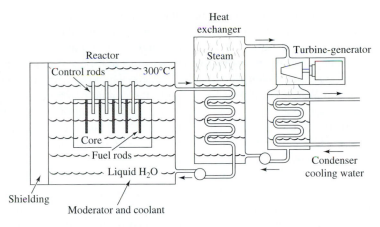

Figure 3.6   Pressurized light-water reactor.

of fission occurring when a neutron collides with a ^{235}U atom is maximal when the neutron energy is close to the *thermal energy,* the average energy of the surrounding molecules. The fission neutrons are released with high energy (velocity), and they must be slowed down to propagate the chain reaction effectively. The lighter the atoms of the moderator, the greater the energy removed per collision. With its two hydrogen atoms per molecule, water is an effective moderator.

An alternative is to use heavy water, D_2O, instead of H_2O, as the coolant and moderator. D_2O is a less effective moderator because of the larger deuterium mass, but deuterium absorbs neutrons much less than hydrogen. As a result, it is possible to maintain a chain reaction even with unenriched uranium (0.7 percent ^{235}U) if D_2O is used as the moderator. This is the operating mode of the CANDU reactor designed in Canada. The relative merits of light and heavy water reactors depend in large measure on the relative costs of separating D_2O from H_2O versus separating ^{235}U from ^{238}U. The costs of the latter have been somewhat hidden by the pre-existence of large ^{235}U separation plants constructed for nuclear weapons production, which have subsequently been used to supply the fuel for commercial reactor programs.

The water in the reactor circulates through a heat exchanger that generates steam from a secondary water coolant, and this steam is then used to drive a turbine to generate electricity. Basically, the reactor is a conventional steam generator in which the heat source is fissioning uranium instead of burning coal. The amount of energy concentrated in uranium is vastly greater than that in coal. Whereas the highest heating value found for coal is about 33 kJ/g, a gram of ^{235}U can release 7.2×10^7 kJ. One gram of ^{235}U is equivalent to about 2.5 metric tons of high-grade coal.

b. Isotope separation.   Because isotopes of a given element have the same chemistry, their separation must be based on the slight differences in their properties resulting from their differences in mass. The greater the mass ratio of the isotopes, the easier it is to exploit these differences. The rates of chemical reactions involving hydrogen, for example, are appreciably faster than those involving deuterium, and electrolysis of water to H_2 and O_2 leaves a liquid that is increasingly rich in D_2O.

The mass ratio of ^{235}U and ^{238}U is too close to unity to utilize reaction rate differences, but physical separations that are sensitive to this ratio can be operated in many successive stages to produce gradual enrichment. The first method to be developed (during the World War II crash program to build the atomic bomb) was gaseous diffusion. The gaseous compound UF_6 is passed through a succession of porous diffusion barriers. At each of these, the lighter $^{235}UF_6$ is enriched by a factor equal to the square root of the $^{238}UF_6/^{235}UF_6$ mass ratio, that is, a factor of 1.0064. The enrichment after n diffusion barriers is $(1.0064)^n$. The factor-of-four enrichment from the natural abundance of ^{235}U, 0.7 percent, to the 2.8 percent minimally needed for light-water reactors, takes 348 diffusion stages (that is, $4 = (1.004)^{348}$). The plants required for this process are very large and expensive, and high inputs of energy are required to force UF_6 through so many barriers.

Gas diffusion has largely been replaced by the gas centrifuge. The application of centrifugal force offers a more efficient route to isotope separation because, for a given centrifugal velocity and radius, the force is directly proportional to the mass, rather than to its square root. Gas centrifuge and gas nozzle techniques are both based on this principle.

The most promising alternative approach, however, is laser isotope enrichment. Lasers are devices that produce light of very well-defined (*monochromatic*) energy. When a photon of appropriate energy is absorbed by a molecule, it undergoes a transition to an excited state, in which it may be considerably more reactive than it is in its ground state. The excited-state energy levels depend slightly on the isotopic composition, and with a sufficiently well-tuned laser it is sometimes possible to excite molecules that contain one isotope, while exciting only a small fraction of the molecules that contain other isotopes. If the excited-state reactivity can be properly exploited, then large enrichments are possible with a single pass. Laser enrichment has the potential of substantially lowering the complexity and cost of ^{235}U separation.

Low-cost isotope separation would improve nuclear power's economics, although the effect is limited to the fuel cost, which is a minor component of the cost of delivering power. It would also place nuclear technology within the means of many currently nonnuclear nations and complicate the problems of nuclear weapons proliferation.

c. Breeder reactor.
Although ^{235}U represents an extremely concentrated form of energy, there is not a great deal of it present in the world. However, it is possible to extend the nuclear fuel supply by converting the dominant uranium isotope, ^{238}U, to another fissionable isotope, ^{239}Pu. This is accomplished simply by irradiating uranium with neutrons via the nuclear reactions diagrammed in Figure 3.7. Absorption of a neutron by ^{238}U produces a very unstable isotope, ^{239}U, which produces ^{239}Np by beta emission. ^{239}Np undergoes a similar transformation to ^{239}Pu. Being fissionable, this isotope can be used in a power reactor just as ^{235}U can. Indeed, conventional reactors actually gain some of their energy from ^{239}Pu, which is inevitably produced when the reactor neutrons encounter ^{238}U, the main uranium isotope present even in enriched reactor fuel.

Not much ^{239}Pu is produced in this way, because the probability of the ^{238}U absorbing a neutron is maximal for fast neutrons, not the thermal neutrons for which a pressurized

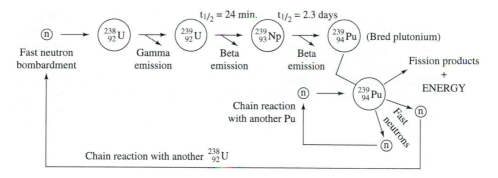

Figure 3.7   Production of Pu-239 from U-238 bombarded with fast neutrons.

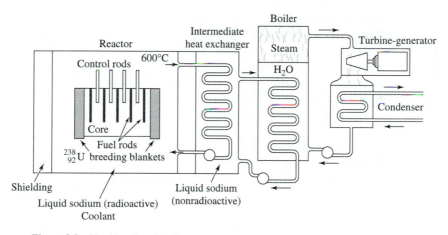

Figure 3.8   Liquid sodium breeder reactor.

light-water fission reactor is designed. Efficient production of ^{239}Pu requires a "breeder" reactor, operating with fast neutrons. Figure 3.8 shows a schematic diagram for a fast breeder reactor. The main difference with respect to an ordinary fission reactor is that the water coolant is replaced by liquid sodium. Being much heavier than hydrogen, the sodium atoms slow the neutrons to a much smaller extent, while liquid sodium efficiently carries away heat from the reactor. The primary sodium coolant transfers its heat to a secondary liquid sodium coolant, which in turn transfers heat to a steam generator that runs the turbine for producing electricity. The purpose of the secondary sodium coolant is to prevent any accidental contact of the primary sodium coolant with water, since some of the sodium atoms do absorb neutrons and become radioactive.

Surrounding the reactor is a blanket of ordinary uranium, in which the ^{239}Pu is bred by the fast neutrons. At the same time, the fast neutrons make the induced fission reaction less efficient, so the reactor fuel must be enriched to the extent of 15–20 percent with a fissionable isotope, either ^{235}U or ^{239}Pu.

There are inevitable losses in the system, but use of the breeder reactor would still stretch the supply of uranium fuel by at least a factor of 50 and would transform nuclear fission into an energy resource much larger than coal. The technology of the breeder reactor is far more complex than that of the ordinary fission reactor, however, and it is still being developed. The most formidable problem seems to be associated with the plumbing of the heat exchanger, since leaks of liquid sodium can be disastrous. Just such a leak resulted in a 1994 explosion at a French breeder reactor, which killed a worker. The French breeder program is the most advanced in the world, but technical difficulties and poor economics have stalled its development.

d. Reprocessing.   The ^{235}U in a fuel rod of a pressurized light-water reactor cannot be completely used up, because of the buildup of fission products, which themselves absorb the neutrons and eventually slow the chain reaction down. After about a year, the fuel rods must be replaced with new ones.

The spent fuel can be reprocessed by chemical extraction of the fission products, separation of the accumulated plutonium, and reconcentration of the uranium. The uranium can be refabricated into new fuel rods. What to do with the plutonium is a difficult issue (see following section on weapons proliferation), in the absence of a functioning breeder program. The current choice is to blend plutonium and uranium into MOX (mixed uranium-plutonium oxide) fuel rods. These can be burned in conventional reactors, although the differing nuclear properties of the two elements restrict the amount of MOX fuel to one-third of the reactor core. However, MOX fuel is currently much more costly to fabricate than conventional ^{235}U fuel.

At the fuel reprocessing plant, the fuel rods are chopped up and dissolved in acid, and the resulting solution is subjected to successive solvent extraction and ion exchange steps in order to separate the elements. The chemistry is straightforward, but the technology is complicated by the need for remote handling of intensely radioactive material. Because of technical and safety problems, as well as policy considerations, no reprocessing plant has been in operation in the United States for over two decades. Currently, the largest reprocessing plants operate in France and England, which reprocess fuel from other countries as well. Russia and Japan reprocess smaller amounts of nuclear fuel.

3.7   HAZARDS OF NUCLEAR POWER

a. Reactor safety: Three Mile Island and Chernobyl.   Under normal operating conditions, the radiation released by nuclear reactors is very low, but the potential for accidental releases is a serious concern because the average reactor contains as much radioactive material as that released by the Hiroshima atomic bomb. The danger is not that a nuclear explosion could be set off; the fissionable material in reactor fuel is too dilute to become explosive itself. But a great deal of heat is generated by an operating fuel rod, even after control rods have been lowered to stop the chain reaction (which happens automatically in case of an accident). This heat is carried away by the water circulating through the reactor. However, if the water leaks out or boils away and is not replenished, then the temperature can rise to disastrous levels, and allow highly radioactive materials to be released.

There are emergency backup systems to replenish the water, but these systems can fail through equipment malfunction or human error.

Both malfunction and human error were responsible for the two major accidents that have marked the era of nuclear power generation. In March 1979, the fuel rods in the nuclear plant at Three Mile Island, near Harrisburg, Pennsylvania, melted down when cooling water was lost. A pump in the primary cooling system failed, the auxiliary pumps were not operational, and the emergency cooling system was momentarily turned off, due to operator error. By the time it was turned back on, a few minutes after the pump failure, a large bubble of hydrogen gas had formed, as a result of high-temperature water attacking the zirconium cladding on the fuel rods:

$$Zr + 2H_2O = ZrO_2 + 2H_2 \qquad (3.2)$$

The bubble prevented the cooling water from reaching the fuel rods, which partially melted. There were fears that the hydrogen would mix with air and explode, breaching the containment wall. But fortunately, this did not happen, and the core eventually cooled after having vented a small amount of radioactive gas.

Much more serious was the accident on April 26, 1986 in the Ukrainian town of Chernobyl, a name now synonymous with nuclear disaster. This time a reactor literally caught fire and blew apart, releasing a great cloud of radioactive debris that rained out radioisotopes over much of Europe and parts of Asia. The reactor continued to burn for about ten days, eventually releasing 10 million curies of radioactivity into the environment. Reactors of the Chernobyl class, which are still in use throughout the former Soviet Union, are of a different design than the light-water reactor described above. They employ graphite, instead of water, as the moderator. Graphite contains only carbon atoms, and it slows neutrons quite effectively; it can sustain a chain reaction when only 1.8 percent of the fuel is the fissionable ^{235}U. The heat is carried away by flowing water around the individual fuel rods. Because the rods are cooled individually, they can be replaced one at a time, without shutting down the reactor. Consequently, this type of reactor has one of the highest rates of productive time online in the nuclear industry.

Unfortunately, a graphite reactor is also inherently hazardous. The plumbing is very complicated, and the graphite, being carbon, is flammable. That is why the Chernobyl reactor burned so vigorously after the accident. Most important, graphite reactors have a dangerous property known as a "positive reactivity coefficient." When power levels dip very low, the reactor becomes unstable, and can race out of control. This is what happened on the day of the accident during a test of the reactor's response to a simulated power failure. The loss of power, in combination with control rod malfunctions and operator misjudgments, allowed the reactor to surge out of control, vaporizing the water, blowing off the roof of the reactor, and igniting the graphite. Two people died in the explosion and a thousand others were injured. Over the next several months, 29 individuals died of radiation effects. Long-term effects of the radiation can be expected but are hard to gauge because of large variations in exposure. However, a substantial rise in the number of childhood thyroid cancers in the region is well documented. Children are particularly susceptible to the ^{131}I fission product, since their thyroids are growing rapidly and concentrating iodine. Fortunately, thyroid cancers are rarely fatal.

Some areas of Europe downwind from the reactor experienced heavy rain as the radioactive cloud was passing overhead, and registered fallout 100 to 1,000 times greater than during the peak years of nuclear weapons testing. The milk from cows eating the fallout-laden grass was contaminated and had to be condemned. Even several years after the accident, the radioactivity in reindeer in Scandinavia and sheep in some areas of northern England exceeded levels permissible for meat.

Taken as a whole, the safety record of the nuclear industry is actually quite impressive. Only one terrible disaster has occurred among the hundreds of nuclear reactors in operation over the last four decades, and the human toll was not worse than in a number of non-nuclear large-scale industrial accidents. Nevertheless, the possibility of another Chernobyl casts a pall over the industry, and the burden of taking heavy precautions to avoid conceivable accidents adds substantially to the costs of nuclear plant construction and maintenance.

There are new reactor designs currently under development that promise to improve safety margins by a factor of ten or more. These reactors would be "passively stable" because they are designed to shut themselves down automatically in the event of an accident. For example, in a "passively stable light-water cooled reactor" (see Figure 3.9), emergency cooling water would be driven by gravity and by nitrogen pressure rather than by electric pumps, and the containment shell would be cooled in an emergency by evaporating water that is fed by gravity from large tanks located above the containment vessel. Moreover, this new reactor has a simplified design that requires far fewer valves, pumps, ducts, and cables than do current reactors. A more radical departure from current practice is the "advanced modular high-temperature gas-cooled reactor," in which the problem of overheated fuel is obviated by coating fuel pellets with a protective layer of very hard material, silicon carbide; these pellets can withstand temperatures exceeding 1,600°C, a temperature higher than the fuel could generate. The reactor would run at 600°C, making

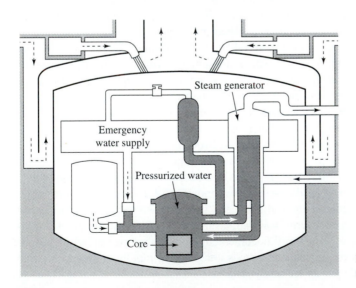

Figure 3.9 Passively cooled light-water reactor.

it thermodynamically efficient (see p. 100), and would utilize helium gas as the heat transfer agent. It would be buried in the ground and would be small enough to ensure that any excess heat would be conducted away by the surrounding soil. A modified version of this reactor may soon be built in South Africa.

b. Weapons proliferation.   Since nuclear energy made its first appearance as the atom bomb, an overriding issue for civilian nuclear power is how to avoid diversion of nuclear fuel into weapons manufacture. ^{235}U is not the major problem in this regard, since a bomb requires highly enriched ^{235}U, greater than 93 percent, whereas conventional fuel rods are only slightly enriched. Weapons-grade ^{235}U fuel requires isotope enrichment, a technically demanding process that involves a major commitment of resources. Since highly enriched uranium is used only in weapons, safeguards against ^{235}U diversion are relatively straightforward. Suspicions that Iraq was planning the clandestine production of a ^{235}U weapon (a program for production of highly enriched uranium was subsequently confirmed) was one of the factors contributing to the 1990 Gulf War.

^{239}Pu is another matter entirely. Plutonium and uranium, being different elements, are readily separated by chemical means. Since ^{239}Pu is the major plutonium isotope created in a uranium reactor, no isotope enrichment is needed to produce weapons-grade fuel. The plutonium extracted in a reprocessing plant can be used in nuclear explosives. India produced and tested a nuclear bomb using plutonium recovered from reprocessed reactor fuel. It has been claimed that as little as 5 kg can be fashioned into a crude but usable weapon by groups or even individuals working from published government documents. (Some of the ^{239}Pu produced in reactor fuel is converted to ^{240}Pu and ^{241}Pu, as the fuel rod continues to be irradiated with neutrons. ^{240}Pu emits neutrons spontaneously and can set off a chain reaction prematurely, reducing the yield of a Pu bomb. But a simple bomb made of reactor-grade Pu could still have a yield of a few kilotons, about a third as powerful as the Hiroshima bomb.)

As nuclear power grows, and fuel reprocessing expands, the amount of plutonium in circulation will increase. Accurately accounting for the plutonium will become more difficult, and opportunities for diversion will multiply. Figure 3.10 compares the amounts of separated Pu (shaded bars), and of Pu remaining in spent fuel rods (unshaded bars) for 16 nations having nuclear power reactors. (Another 16 nations have an additional 10 percent of the world's Pu stock.) About 25 percent of the civilian stores of Pu has been separated (205 tons), almost all of it by France, Germany, Japan, Russia, and the U.K. This Pu is intended for use in nuclear fuel; the preferred technology is to mix it with uranium fuel, in the form of the mixed oxide (MOX). In this form, the Pu is not available for weapons; it would have to be separated chemically once again. About half of the separated Pu has been fabricated into MOX, but little MOX is currently in use, because it is more expensive than ordinary uranium fuel. Even when future plans for the utilization of MOX are taken into account (see Table 3.5), it seems apparent that the supply of separated Pu will outpace the demand.

Ironically, the issue of Pu diversion has been magnified by the fruits of disarmament. During the decades of the Cold War, the United States and the Soviet Union stocked their arsenals with tens of thousands of nuclear weapons. The end of the Cold War has meant that many of these weapons can be dismantled, to the world's vast relief. Yet the dismantled

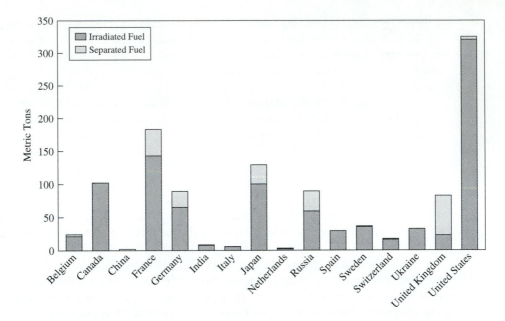

Figure 3.10 Stocks of plutonium in the civilian nuclear fuel cycle. *Source:* Data from Albright and Gorwitz (2000). Tracking civil plutonium inventories: End of 1999. *Plutonium Watch (ISIS),* October 2000.

TABLE 3.5 SEPARATED CIVIL PLUTONIUM INVENTORIES AND PROJECTED INVENTORIES (METRIC TONS)

Status of plutonium management	1998	2010	2015
Countries with plans to use Pu in MOX*	98.2	45–110	45–115
Countries without plans to use Pu in MOX[†]	92.5	130	140
Countries with plans to dispose of civil Pu with excess military Pu[‡]	4–5	5	5
Total	**195**	**180–245**	**190–260**

*Belgium, France, Germany, Japan, Sweden, and Switzerland.

[†]China, India, Italy, Netherlands, Spain, Russia, United Kingdom.

[‡]United States.

Source: Albright and Gorwitz (2000). Tracking civil plutonium inventories: End of 1999. *Plutonium Watch (ISIS)* October 2000.

weapons add large stockpiles of bomb-grade Pu to the amounts generated in the civilian cycle. The amounts are compared in Figure 3.11a. Military stockpiles currently hold 107 tons of Pu in excess of military needs. Disarmament has also resulted in a much larger surplus (770 tons) of highly enriched uranium (see Figure 3.11b). However, the uranium can be readily diluted with unenriched uranium, for use in conventional nuclear fuel.

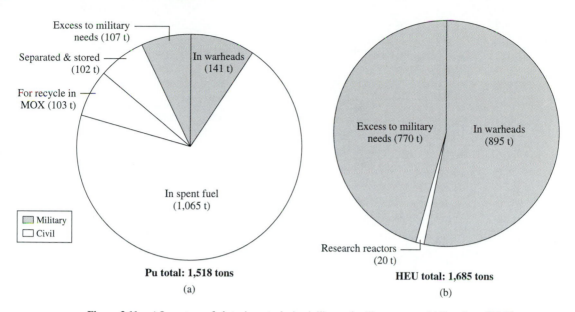

Figure 3.11 a) Inventory of plutonium stocks in civilian and military sectors; b) Supplies of highly enriched uranium in warheads and in excess due to disarmament after end of Cold War. *Source:* Data from Albright and Gorwitz (2000). Tracking civil plutonium inventories: End of 1999. *Plutonium Watch (ISIS),* October 2000.

There is no current consensus on what to do with the accumulating Pu. Nuclear power advocates want to burn it in reactors, thereby extending the nuclear fuel supply. The governments of France, Japan, and Russia, in particular, are committed to this course as a method of assuring energy independence. At present, however, Pu is costlier than uranium as a nuclear fuel, and this situation is unlikely to change in the near future. For this reason, it is argued (in many quarters) that the Pu should be disposed of as waste, along with the high-level nuclear wastes from reprocessing plants (see next section). Indeed, it is further argued that uranium fuel should not be reprocessed at all, but used only on a once-through basis, in order to avoid the Pu safeguard problem and because reprocessing is currently uneconomical. This course of action would, of course, forego a significant source of energy. Whether this energy will be needed, in light of future alternative sources, is a matter of conjecture and debate.

c. Nuclear waste disposal. Even with safer reactor designs and an effective means of combating the spread of nuclear weapons, nuclear power remains potentially hazardous because it produces radioactive wastes.

There are several points in the entire cycle of nuclear fuel where the dispersal of radioactive materials is an actual or potential problem (see Figure 3.12). At the beginning of the cycle, uranium mining is, itself, a hazardous occupation. Miners have a high risk of developing lung cancer because they inhale dust and radon. Even more pressing in terms of public health are uranium mine tailings, fine particles that remain after uranium ore

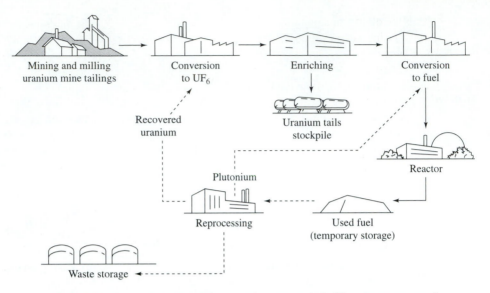

Figure 3.12   Fuel cycle for the light-water nuclear reactor; dashed lines are not yet part of the U.S. cycle.

is extracted from the rocks. Although the levels of radioactivity are low, the amount of material is huge: 75 million tons in the United States. Moreover, the material will pose a hazard for millennia since one of the important nuclides, thorium-230, has a half-life of 80,000 years. Because the radioactivity is dilute, its danger went unrecognized for decades. The tailings were left in huge piles to be dispersed by wind and leached by rain. Some material was even used as fill under buildings, exposing the occupants of some buildings to levels of radioactivity exceeding the legal limits for uranium miners. Safe disposal of tailings remains a major task.

Further into the cycle, dispersal of radioactive materials is a hazard at the fuel reprocessing plant, where the spent fuel rods are dissolved in acid to separate the fission products and the heavy element contaminants before the uranium or plutonium is sent back to fuel fabrication plants. Reprocessing produces high-level waste (HLW) that is highly radioactive and of moderately high temperature. Since no method has been approved of disposing of HLW permanently, it has been accumulating at reprocessing sites for almost half a century. Ultimately, it is planned to collect the HLW and mix it with glass for burial at a designated geological site. The intense radioactivity and high temperature place severe technical demands on this process.

Particularly large amounts of HLW have accumulated at military reprocessing sites where Pu for nuclear weapons has been manufactured. In the U.S., a major site is the 560-square-mile Hanford Nuclear Reservation in Washington state, where millions of gallons of HLW sit in 177 tanks. Most tanks have been in use for decades longer than planned, and every year several tanks begin to leak. Cleanup of the Hanford site is complicated by some lax recording practices over many years. Many of the tanks are filled with unknown mixtures of chemicals, some of which are turning out to be highly explosive. Military reprocessing sites in the former Soviet Union are reported to have severe contamination problems.

Most spent fuel rods have not been reprocessed (see Figure 3.10) and in the U.S. reprocessing is not contemplated any time soon. Instead, the intention is to place the rods in caskets and bury them in a geologically stable repository. But the problem is extraordinarily contentious because the waste must be isolated from the human environment for exceedingly long periods of time. The length of time for safety can be assessed by the rule of thumb that radiation sinks to negligible levels after ten half-lives: the total amount of radioactive material, x, is reduced to $x/2^{10}$, or about $1/1,000$. For fission products, this interval is a few hundred years, but plutonium and other heavy elements have half-lives of tens of thousands of years; ten half-lives is on the order of a quarter million years. It is difficult to find a burial site where disturbance by earthquakes and ground-water infiltration can be excluded for this period of time. The U.S. has chosen Yucca Mountain in Nevada as its burial site, but development of the site has been slow. Technical issues of long-term stability and safety linger, and the project is strenuously opposed by Nevada residents. Other countries, too, have proceeded slowly toward geological disposal.

In the meantime, spent fuel rods are being stored temporarily at the reactor site, in pools of water to dissipate the heat. As reactor operators fill their pools to capacity, they are making room by transferring the spent fuel to dry-storage in casks, made of concrete or steel. These casks are designed for up to a century of storage. However, the accumulation of spent fuel rods, and the difficulty of implementing permanent geological storage, casts a shadow over the future of fission energy.

3.8   IS NUCLEAR POWER PART OF THE FUTURE?

Nuclear power currently supplies about 23 percent of the electrical power in the United States, but the plants are aging, and no new nuclear reactors have been ordered since 1979. The most recent plants have estimated lifetimes of 30 years, on average, and building a new nuclear power plant requires an average of ten years. Nuclear plant construction has slowed in many other countries as well. It is hard to avoid the conclusion that the nuclear industry is in decline, defeated by high prices and public mistrust (the two are connected through the regulatory process).

Still, it is too soon to count nuclear power out. It is a vital source of energy for Japan, France, and Russia, and significant expansion is planned in East Asian countries. In the U.S., too, surging electrical demand and deregulation of the power market have brightened the prospect for fission energy. Old nuclear plants have become marketable, and are being acquired by large power companies, which hope to make them more competitive. In a world where CO_2 emissions may become regulated, nuclear power has a big advantage. As discussed above, new plant designs could effectively meet safety concerns. The big remaining problems are weapons proliferation and waste disposal. It has been estimated that if nuclear power were to make a significant dent in the CO_2 emissions of the U.S., it would have to grow at a rate that would require the equivalent of one Yucca Mountain repository per year over the next several decades for the spent fuel rods. In view of the enormous difficulties in establishing the first repository, this is a daunting prospect.

3.9 FUSION

Nuclear energy can potentially be obtained not only from the fission of nuclei, but from their fusion. As we saw in Figure 3.1, the most stable nuclei have intermediate masses; very small nuclei are less stable. In particular, helium nuclei are considerably more stable than hydrogen nuclei. When two hydrogen nuclei fuse to form a helium nucleus, an enormous amount of energy is released. However, fusing the nuclei requires extreme reaction conditions in order to overcome the huge energy barrier caused by the repulsion between the two protons. Such extreme conditions are found in the centers of stars, including our sun, whose energy outputs are due to fusion reactions. Similar conditions are also obtainable in hydrogen bombs, which use the power of a fission explosion as a trigger for hydrogen fusion.

The *ignition temperatures* required for fusion reactions are on the order of 100 to 1,000 million °C. At such high temperatures, all earthly materials vaporize; hence, one of the challenges in developing fusion technology has been to devise a way to contain the fusing nuclei. Two approaches appear promising: magnetic confinement, in which the fusing nuclei are suspended in a magnetic field, and inertial confinement, in which the fusing nuclei are forced together by the impact of high-power lasers, or ion beams.

a. Fusion reactions.   Fusion cannot be carried out on ordinary hydrogen atoms, because two protons cannot fuse without neutrons being available to stabilize the product nucleus. This requirement for neutrons means that heavy isotopes of hydrogen must be used, deuterium (^{2}H) or tritium (^{3}H). An additional constraint is that the reaction must produce two (or more) particles, in order to carry off the fusion energy. If a single particle were produced, it would immediately fission again.

The available fusion reactions are shown on Figure 3.13. Reaction 1, the fusion of deuterium with tritium, has by far the lowest ignition temperature, 100 to 200 million °C. Also, at achievable pressures of a high temperature ionized gas (called a *plasma*), the power density from the deuterium-tritium (D-T) reaction is much higher than from other fusion reactions. This makes it the most practical approach to fusion, and the focus of current development efforts.

However, the D-T reaction has the disadvantage that it requires tritium as a fuel. Tritium is a radioactive gas that releases low-energy β particles. These are not dangerous when emitted externally to the body, but are dangerous enough when tritium is inhaled, or ingested in water. The half-life of tritium is 12.3 years, so there is no natural source of the isotope. It must be synthesized from lithium by reactions 6 and 7 in Figure 3.13.

Another disadvantage of the D-T reaction is that it produces one high-energy neutron per nuclear fusion. Because neutrons are electrically neutral, they do not respond to the magnetic fields guiding the plasma in a tokamak reactor (see p. 71). Instead, these neutrons fly out of the plasma at high speed. They bombard the structural materials of

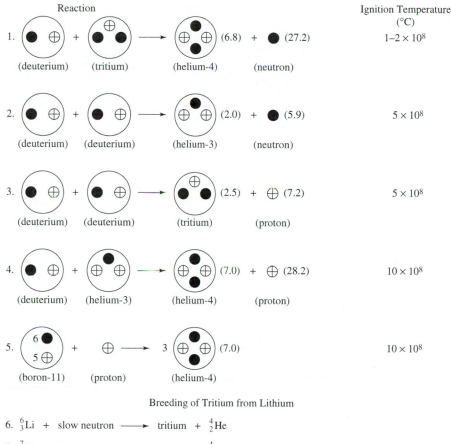

Reaction   Ignition Temperature
(°C)

1. (deuterium) + (tritium) ⟶ (helium-4) (6.8) + (neutron) (27.2)   1–2 × 10^8

2. (deuterium) + (deuterium) ⟶ (helium-3) (2.0) + (neutron) (5.9)   5 × 10^8

3. (deuterium) + (deuterium) ⟶ (tritium) (2.5) + (proton) (7.2)   5 × 10^8

4. (deuterium) + (helium-3) ⟶ (helium-4) (7.0) + (proton) (28.2)   10 × 10^8

5. (boron-11) + (proton) ⟶ 3 (helium-4) (7.0)   10 × 10^8

Breeding of Tritium from Lithium

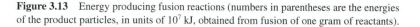

6. $^6_3$Li + slow neutron ⟶ tritium + $^4_2$He

7. $^7_3$Li + fast neutron ⟶ tritium + $^4_2$He + neutron

Figure 3.13   Energy producing fusion reactions (numbers in parentheses are the energies of the product particles, in units of 10^7 kJ, obtained from fusion of one gram of reactants).

the reactor, causing *neutron activation* and rendering the materials radioactive. These radioactive materials are a substantial problem for disposal; the volume of material to be discarded from the core of a fusion reactor is likely to be comparable to the volume of radioactive wastes generated by a fission reactor. Most fusion wastes will be relatively short-lived compared to fission wastes (see Figure 3.14). However, the most benign materials, vanadium alloys and silicon carbide, are expensive; less expensive ferritic steels require about 100 years to decrease their levels of radioactivity significantly. Nonetheless, the contrast with the very long-lived radioactivity from a fission reactor is considerable.

Deuterium is the most abundantly available fuel, constituting 0.8 percent of the hydrogen on Earth (mostly in the oceans). Thus, the deuterium-deuterium (D-D) reactions (2 and 3) provide a much greater long-term energy potential. In addition, neutron damage is less of a problem, since neutrons are produced only in reaction 2, and are far less energetic

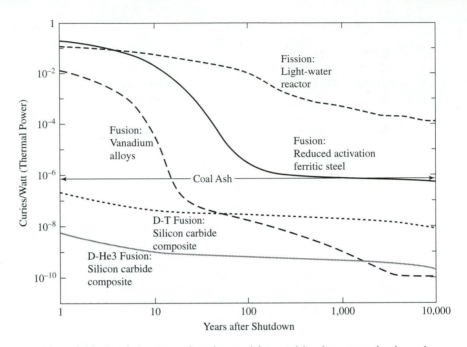

Figure 3.14   In a fusion power plant, the materials comprising the vacuum chamber and other components facing the plasma will become radioactive as a result of the energetic neutrons produced in fusion reactions within the plasma. The decrease in radioactivity with time after shutdown is shown for a number of candidate materials for plasma-facing components. The residual radioactivity in these materials is compared with that found in a light-water reactor fission power plant, and with the level of radioactivity released by a typical coal-fired power plant.   *Source:* Princeton Plasma Physics Laboratory, Princeton, New Jersey.

than those produced in reaction 1. However, reaction 3 produces tritium, contributing to radioactivity, and some of this tritium can also react with the deuterium, via reaction 1, producing high-energy neutrons. However, the main obstacle to D-D reactors is the much higher ignition temperature, 500 million °C.

Still better from an environmental perspective would be the D/^{3}He and H/^{11}B reactions (4 and 5 in Figure 3.13), which produce no neutrons (although, if deuterium is present in the heated fuel, some of it will react with itself via the D-D reaction, producing some neutrons). However, both of these reactions require ignition temperatures that are five to ten times greater than for the D-T reaction. It is questionable whether these reactions will ever be technologically feasible.

b. Fusion power reactors

1) Magnetic confinement.   In the magnetic confinement method, the fusion fuels are suspended in free space by using powerful magnets. At the high temperatures under consideration, atoms separate into their charged species, producing *plasmas,* or clouds of

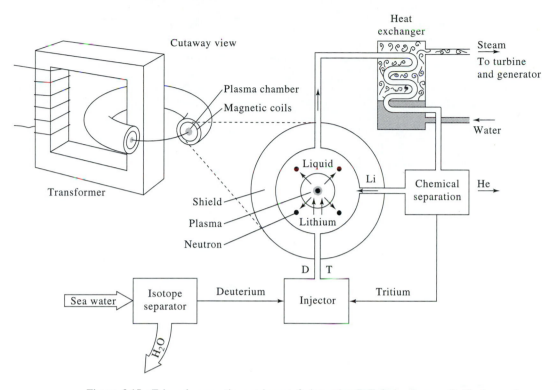

Figure 3.15   Tokamak magnetic-containment fusion using D-T fuel.   *Source:* G. Gordon and W. Zoller (1975). *Chemistry in Modern Perspective* (Reading, Massachusetts: Addison-Wesley).

nuclei and electrons. Because the plasma is charged, it can be controlled with magnetic fields. The magnets keep the heated, charged plasma away from the walls of the reactor, preventing the particles from colliding with the walls and cooling down.

The most successful magnetic design is the Tokamak reactor, diagrammed in Figure 3.15. In this design, an electrical current is generated within the plasma; the current causes the magnetic field to bend around in a donut shape or *torus,* which confines the fusion reaction. The reactor walls are lined with light materials such as carbon, boron, or beryllium to minimize reactions of energetic particles that escape the plasma. Surrounding the walls is a blanket of lithium, in either liquid or salt form, which captures neutrons given off in the reaction and produces tritium for additional fuel (see reactions 5 and 6, Figure 3.13).

For fusion to generate power, the energy input needed to sustain the plasma conditions must be less than the energy output from the reactor. The break-even point where input and output are balanced has yet to be achieved in any experimental reactor. The break-even point depends upon three parameters: the density of the plasma (n), its temperature (T), and the confinement time (τ), or the length of time it would take the plasma to cool substantially. The product of these three, called the Lawson Triple Product, must be 10

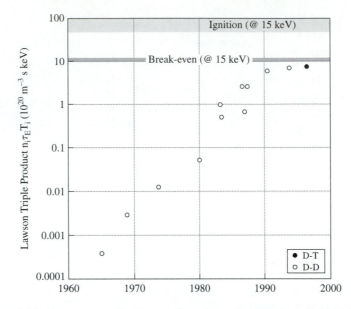

Figure 3.16   Progress toward ignition (a self-sustaining plasma) is shown in this plot as a function of the "Lawson Triple Product," which is a measure of plasma confinement quality in a fusion plasma. It is a multiple of the ion density (n_i), energy confinement time (τ_E), and ion temperature, T_i.   *Source:* Princeton Plasma Physics Laboratory, Princeton, New Jersey.

(in units of 10^{20} m^{-3} s keV) for break-even at a target energy of 15 keV (see Figure 3.16). Since the magnetic fusion program began, the Lawson Triple Product has been increased by four orders of magnitude in experimental reactors, and is approaching break-even. There is still a considerable distance to go, however, to achieve ignition conditions for practical power generation.

2) Inertial confinement.   In the inertial confinement method, a target pellet containing deuterium and tritium is subjected to a very short and intense pulse of energy from lasers or ion beams. The intense energy vaporizes the outer surface of the particle, ejecting matter outward. The laws of physics require that the force of the outwardly moving material be matched by inwardly moving material; hence, the remaining part of the pellet is imploded, reaching very high densities and temperatures of up to 100 million °C, conditions sufficient for fusion. The D-T pellet is confined by inertia for a fraction of a nanosecond, long enough to allow about 30 percent of the pellet to fuse before it comes apart.

The feasibility of inertial confinement depends on advances in laser and ion beam technology. If inertial fusion is to produce enough energy for power generation, it must deliver energy to the pellets in tens of nanoseconds, and by beams with precisely determined shapes. The energy-transfer efficiency of currently available lasers is very far from what will be required, as is the intensity of current ion beams.

The pulsed nature of inertial fusion also creates problems in designing the reaction chamber. The systems needed to recover energy and breed fuel in the chamber would have to withstand explosions equivalent to 100 or more kilograms of TNT several times a second. At present, inertial confinement fusion is much less well-developed than magnetic confinement fusion.

c. Is fusion the energy source of the future?   Fusion's greater appeal compared to nuclear fission rests mainly in fusion's greater safety. Unlike fission reactors, fusion reactors cannot sustain runaway reactions. Fuel must be continuously injected, and the amount contained within the reactor vessel at any given time can operate the reactor only for a matter of seconds. As a result, fusion reactors cannot blow themselves apart, and should require simpler post-shutdown or emergency cooling systems than fission reactors. Moreover, the radioactivity created in the reactor, although comparable to that of a fission reactor, is much shorter-lived. Furthermore, the only radioactive material that could be released in an accident is tritium, because the other radioactive materials would be largely bound up in the structural elements.

However, substantial technical challenges remain before an operating fusion power plant can be built. Until that happens, the economic competitiveness of fusion will remain unknown. There are optimistic projections about the eventual price of fusion power, but development costs are very large, and will be spread over time. It will be some decades before we know what part fusion power can play in the world's energy supply.

CHAPTER 4

RENEWABLE ENERGY

The major alternative to fossil and nuclear fuels is to harness renewable energy sources that, with the exception of geothermal energy, are derived directly or indirectly from sunlight. The renewable energy resources of the U.S. are shown in Table 4.1. The annual energy deposited by sunlight on the continental United States is nearly 600 times the total annual U.S. energy consumption in 1999. Enough sunlight falls yearly on each square meter to equal the energy content of 190 kilograms of high-grade bituminous coal. The sun already provides us, in fact, with our most basic energy needs, including heat, fresh water, and plant life. There is plenty left over for our other energy needs if we could learn how to use it effectively. The difficulty is that sunlight is diffuse and intermittent. The technologies required to harness and store solar energy are currently expensive; it is cheaper to extract and consume fossil and nuclear fuels from the ground. However, the increasing environmental burdens associated with these fuels are sparking heightened interest in a variety of renewable energy technologies. A number of these are commercially viable in some circumstances, and there is reason to think that with further technological improvements, renewable forms can play a substantial role in human energy utilization.

Renewable energy currently accounts for 14 percent of global energy consumption, mostly from biomass, and mostly in developing countries (see Figure 4.1a,b). Traditional societies have always relied on firewood, charcoal, crop residues, and animal wastes for cooking, heating, and village-based industries. However, the traditional burning of biomass is associated with severe environmental and health problems, due to deforestation and air pollution. The gathering of biomass is labor intensive, and the stoves and fireplaces in which it is burned are typically inefficient and smoky. Small wonder that, as development proceeds, biomass fuels are abandoned in favor of cleaner and more convenient forms of energy.

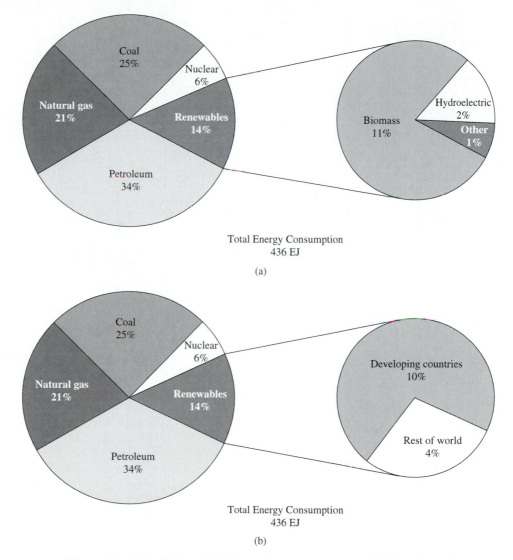

Total Energy Consumption
436 EJ

(a)

Total Energy Consumption
436 EJ

(b)

Figure 4.1   a) Distribution of fuels in world primary energy use including share of renewable energy contribution (2000); b) Distribution of renewable fuels in developing countries and rest of world.   *Source:* Energy Information Agency, U.S. Department of Energy, *Outlook 1998 with Projections through 2020,* Washington, DC.

However, the introduction of new technologies can make renewable energy sources attractive for developing as well as developed countries. The dispersed nature of renewable sources can be an advantage for countries lacking a highly developed electrical grid or fuel transportation system. The economics of small generators powered by sunlight or wind, or biomass-derived fuels, becomes more attractive if the alternative is to extend power lines

TABLE 4.1   RENEWABLE SOURCES OF ENERGY IN THE UNITED STATES

Currently attainable

Resource	Electricity		Gaseous fuel (hydrogen)*		Liquid fuel (methanol)†	
	Trillion kWh/yr	Fraction of annual electricity demand	Trillion cubic feet natural gas equivalents/yr	Fraction of annual natural gas demand	Billion barrels oil equivalents/yr	Fraction of annual petroleum demand
Wind	3.08	0.850	7.35	0.314	—	—
Solar	0.21	0.055	0.50	0.021	—	—
Biomass	1.77	0.489	4.22	0.180	1.64	0.239
Geothermal	0.16	0.044	0.38	0.016	—	—
Hydropower	0.31	0.086	0.74	0.032	—	—
Total	**5.52**	**1.524**	**12.18**	**0.562**	**1.64**	**0.239**

Potentially attainable

Resource	Electricity		Gaseous fuel (hydrogen)		Liquid fuel (methanol)	
	Trillion kWh/yr	Fraction of annual electricity demand	Trillion cubic feet natural gas equivalents/yr	Fraction of annual natural gas demand	Billion barrels oil equivalents/yr	Fraction of annual petroleum demand
Wind	9.80	2.707	23.39	0.999	—	—
Solar	1.51	0.417	3.60	0.154	—	—
Biomass	3.90	1.077	9.31	0.397	3.62	0.528
Geothermal	0.81	0.224	1.93	0.083	—	—
Hydropower	0.32	0.088	0.76	0.033	—	—
Total	**16.23**	**4.483**	**38.74**	**1.65**	**3.62**	**0.528**

*Except for biomass, refers to production of hydrogen via electrolysis of water, assuming 70 percent of electric energy is retained in hydrogen fuel. In case of biomass, hydrogen is produced directly via biomass gasification at an efficiency of 60 percent.

†Methanol is produced directly from biomass thermal processes at an energy conversion efficiency of 70 percent.

Sources: Wind: Adapted from American Wind Energy Association (2000). *Wind Energy Projects in the U.S.* Washington DC. (http://www.awea.org) Low estimate (3.08 trillion kWh/yr) is based on wind potential in 24 states, where each state has installed the same percentage of the total potential that California currently has in place (31.4 percent). High estimate (9.80 trillion kWh) is based on the 24 states using full extent of their annual wind potential.

Geothermal: Energy Efficiency and Renewable Energy Network, U.S. Department of Energy (2000). *Geothermal Energy Program: Geothermal Electricity Production.* Washington, DC. (http://www.eren.doe.gov/geothermal)

Biomass: Adapted from T. B. Johansson et al. (eds.) (1993). *Renewable Energy, Sources for Fuels and Electricity.* Island Press: Washington, DC. Appendix A (p. 1094). For lower estimate, it was assumed that only excess agricultural land would be used for energy plantations. Higher estimate assumed 10 percent of U.S. forests/woodlands + cropland + permanent pasture would be used for energy plantation (approx. 100 million hectares).

Solar: Lower estimate from climate stabilization scenario (p. 87). *In:* Alliance to Save Energy, American Council for an Energy-Efficient Economy, Natural Resources Defense Council, Union of Concerned Sciences (1991). *America's Energy Choices, Investing in a Strong Economy and a Clean Environment.* Union of Concerned Scientists: Cambridge, MA. Higher estimate is from: Idaho National Engineering Laboratory, Los Alamos National Laboratory, Oak Ridge National Laboratory, Scandia National Laboratories, and the Solar Energy Research Institute (1990). *The Potential for Renewable Energy,* an interlaboratory white paper, prepared for the Office of Policy, Planning, and Analysis of the U.S. Department of Energy, SERI/TP-260-3674, Washington, DC (as cited in *Renewable Energy, Sources for Fuels and Electricity* (p. 1128).

Hydropower: Adapted from: Energy Information Administration, U.S. Department of Energy (1998). *The Kyoto Report—Electricity Supply,* Washington, DC. Report # SR/OIAF/98-03. Energy Information Administration, U.S. Department of Energy (1999). *U.S. Coal Reserves: 1997 Update.* Washington, DC.

over long distances. In developed countries as well, distributed power generation is gaining adherents, as technology for integrating small generators into a power grid becomes better developed, and as the consequences of power failures and blackouts become painfully clear (as in the California power crisis in 2000–2001), in the age of computers and electronic controls.

4.1   SOLAR HEATING

The most convenient and straightforward application of solar energy is for heating of residential and commercial buildings. Currently, space heating and hot water constitute 20–25 percent of U.S. energy needs.

Solar heating can be accomplished by either passive or active designs. A passive design maximizes the capture of direct solar energy to provide most of a building's heating and daylighting needs. To achieve this, the building and its windows are oriented so that solar radiation is admitted during the winter and avoided during the summer. Dense materials such as concrete and stone can provide thermal storage by absorbing solar radiation and storing it as a buffer against temperature fluctuations. An interesting example, coming into increasing use in solar homes, is the "Trombe Wall," a 20–40-cm-thick masonry wall coated with a dark, heat-absorbing material faced with single- or double-layer glass, separated by a 2–4 cm airspace. Heat from sunlight passing through the glass is absorbed by the dark surface, stored in the wall, and conducted slowly inward through the masonry. Because of the slow heat transfer, rooms remain comfortable through the day and receive slow, even heating for many hours after the sun sets, thereby reducing the need for additional night heating.

The U.S. Department of Energy has conducted extensive studies of heating efficiencies of buildings with passive solar design in different climate regions. For an average indoor temperature of 19.5°C (67°F), passive solar buildings require, on average, auxiliary heating equivalent to about half the heating requirement of new conventional buildings, and about one-fourth that of the existing housing stock.

Active solar energy systems are another alternative. They are termed *active* because, in contrast to *passive* systems, a source of energy other than solar energy is required to drive the system. One such design is shown in Figure 4.2a,b. Water circulates through the flat-plate collectors on the roof, where it is heated by the sun and pumped into a storage tank, which provides heat and hot water. Such buildings also function best when the demand for heat and hot water is minimized by proper insulation and water conservation.

Many improvements in efficiency can be accomplished quite inexpensively, with short pay-back periods. For example, wrapping an insulating blanket around the tank of the water heater can reduce standby heat loss by 25 percent to 45 percent. These easily installed blankets cost $10–25, an investment that pays for itself within a year. Water-saving showerheads (under $10) use two to three gallons of water per minute, while conventional showerheads use five to eight gallons. Washing machines and dishwashers, the two major water-consuming appliances, are available in new designs that use only a fraction of the water of conventional models. Depending on the climate, a well-designed and properly sized solar water heater can provide up to two-thirds of a household's hot-water needs. It can save 50 percent to 85 percent of the hot-water portion of monthly electricity utility bills if the backup element is kept at 50°C (122°F).

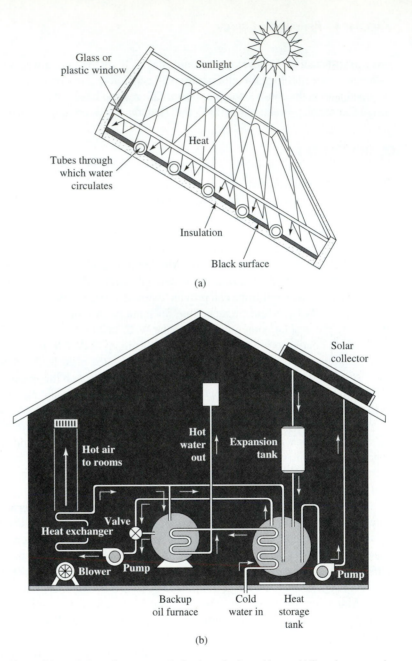

(a)

(b)

Figure 4.2   a) Solar collector on roof of active solar-heated house; b) Pumping system in active solar-heated house for provision of space heating and hot water.   *Sources:* a) B. J. Nebel (1993). *Environmental Science: The Way the World Works* (Upper Saddle River, New Jersey: Prentice Hall). Copyright © 1993 by Pearson Education. Reprinted with permission; b) B. Anderson and M. Riordan (1976). *The Solar Home Book: Heating, Cooling and Designing with the Sun* (Harrisville, New Hampshire: Brick House Publishing). Copyright © 1976 by Brick House Publishers. Reprinted with permission.

4.2   SOLAR THERMAL ELECTRICITY

A more demanding application is the generation of electricity from sunlight. It would take an area of 10,000 square miles (less than 1/4 of the area covered by U.S. roads and streets) in the Nevada desert to supply the total electricity needs of the United States by flat-plate solar collectors converting sunlight to electricity at 10 percent efficiency. In the initial stages of development, solar electricity could complement existing power plants, especially in areas of plentiful sunshine. In this regard, it is an advantage that daylight hours coincide with high electricity demand, for heating, air-conditioning, and many other purposes, so that solar generators are well-matched to the peak loads of electrical utilities.

One approach to solar electricity is to collect the sun's rays directly to heat water and run a steam generator. Three solar thermal designs are currently under development: parabolic troughs, power towers, and dish/engine systems. These technologies generate high temperatures by using mirrors to concentrate the sun's rays up to 5,000 times. Solar troughs (see Figure 4.3a) are the most mature technology today, and the one most likely to be used in the near-term. The trough concentrates the sun's energy onto a receiver pipe located along the focal line of a parabolically curved reflector. Oil flowing through the pipes is heated to about 400°C, and the heat is used to generate electricity in a conventional steam generator. Utility-scale (30 to 80 MW) power plants are already operating in California. The solar heat is supplemented with natural gas (up to 25 percent) to provide power when solar energy is not available.

The power-tower (see Figure 4.3b) uses a field of heliostats to direct the sun's rays to a receiver located on top of a tall tower. Since 1992, a consortium of utilities has operated a pilot plant in Southern California that uses a molten-salt working fluid and thermal storage system. With the addition of thermal storage, the power tower can operate for 65 percent of the year without need of a backup fuel. The molten salt is heated to 565°C in the

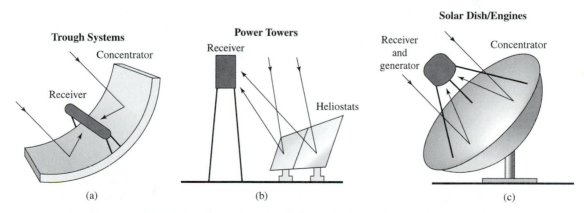

Figure 4.3   Electricity from solar thermal energy. a) Trough system; b) Power tower; c) Solar dish/engines.   *Source:* Sun Lab Snapshot (1998). *Concentrating Solar Power Program, Overview.* U.S. Department of Energy. (http://www.nrel.gov/docs/fy99osti/24649.pdf)

tower and stored in a hot tank for use in the steam generator as needed. Power plants in the 30 to 200 MW range are envisaged.

Dish/engines concentrate the sun's energy at the focal point of a parabolically shaped dish, where it directly powers an engine generator, producing 5 to 50 kW of electricity (see Figure 4.3c). Temperatures of up to 800°C are achieved, making dish/engine systems the most efficient (see pp. 99–101) of all solar technologies (29.4 percent solar-to-electricity conversion at peak efficiency). The modular design of dish/engine systems makes them a good match for remote power needs in the kilowatt range, and also in grid-connected utility applications in the megawatt range.

Solar thermal electricity is costlier than electricity from fossil fuel plants, but is becoming increasingly attractive for peak power additions to the electrical supply. The hot days when electricity demand for air conditioning is highest are also best suited for solar thermal production. As costs diminish with further experience, and the price of fossil fuels rises, solar thermal electricity will become increasingly competitive.

4.3 PHOTOVOLTAIC ELECTRICITY

Sunlight can be transformed directly into electricity via the photovoltaic (PV) effect. When light is absorbed in a PV material, positive and negative charges are created, which can be collected by electrodes at either side, and passed to an external electric circuit. The most highly developed PV material, and the one most likely to succeed in large-scale applications, is silicon, the same material on which the electronics industry is based. The silicon solar cell was developed for the space program and has undergone extensive development over the past four decades.

The great advantage of PV power is its simplicity and versatility. Solar cells are portable and can be set up in numbers required for local use. In locations remote from power grids, PV electricity is often cheaper than building grid extensions. This is particularly true for developing countries, in which many areas are not served by a grid. In these areas, solar cells could provide the small amounts of power required for basic human needs at relatively low cost. Essential services that could be powered by solar cells include:

- Pumping water to villages, livestock and irrigation
- Refrigeration, particularly to preserve vaccines, blood, and other health care perishables
- Lighting and appliances for homes and community buildings
- Recharging batteries for flashlights and portable appliances

The problem, of course, is to find financing for such projects, even when they make sense economically.

In developed countries, too, new PV applications are emerging as low-cost alternatives. The rapidly expanding wireless phone industry is using fields of solar panels to power remote towers and stations. In Arizona, PV is cost competitive with the electric grid for low power uses (e.g., irrigation control equipment) requiring line extensions of over 0.5 miles. Protection from supply interruptions can be a motivating factor for installing

PV power. The recent electricity crisis in California was followed by a tripling of U.S. sales of inverters, devices that convert the direct current electricity produced by solar cells and batteries to the alternating current required by appliances.

At current electricity prices on the grid, PV power is uncompetitive by a factor of about three. This represents a dramatic improvement since 1972, when PV electricity costs were over 100-fold higher. Worldwide, PV capacity exceeded one billion watts (1 GW) in 1999, and it is increasing at a 17 percent annual rate (4-year doubling time). Costs will decrease as sales increase. A U.S. Department of Energy study forecasts an 18 percent decline in cost per doubling period. One likely improvement is in the production of silicon, which represents 40–60 percent of the cost of a solar panel. Currently, Si is purchased from the excess capacity of computer chip manufacturers. However, the ultra-high purity required for the chips, whose attainment requires high-temperature vacuum processing, is not needed for solar cells. A new, cooler process for solar panel-grade Si has been developed, which reduces energy consumption by 80 percent.

An important issue for the future of PV power is "net metering," whereby the output of PV installations can be fed back into the grid. In this way, solar panels, from individual rooftops to commercial installations, become extensions of the grid. The power can be used locally when needed, or sold to the utility when not needed. Utilities have generally resisted net metering because of perceived threats to their economics, but some of them are now supporting the idea, hoping to exploit PV power themselves.

The electricity from solar cells could alternatively be used to electrolyze water to produce hydrogen, which can be shipped for use as a fuel (see discussion of the "hydrogen economy," p. 109). Or hydrogen can be generated directly from sunlight, in *photoelectro-chemical* cells (see p. 87). Such cells are not yet practical, but encouraging research results are being reported.

a. Principles of the PV cell.   PV materials are semiconductors, solids with electrical properties between metals and insulators. In solids, the electronic levels spread into energy *bands,* owing to the mutual interactions of the levels on all the atoms in the solid lattice. The filled levels together produce the *valence* band, while the empty levels produce the *conduction* band, as illustrated in Figure 4.4. If an electron is injected into the conduction band of a solid, it can move freely throughout the lattice, since all the orbitals in the band are empty. Likewise, if an electron is removed from the valence band, the resulting hole, a center of positive charge, can move freely throughout the lattice via motions of the valence band electrons to fill the successive vacancies. An electron in the conduction band and a hole in the valence band are called *mobile carriers* of electricity.

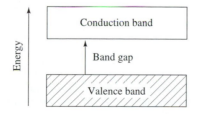

Figure 4.4   Valence and conduction bands in a semiconductor.

The energy difference between the top of the valence band and the bottom of the conduction band is called the *band gap*. Insulators have very large band gaps; the energy cost for promoting an electron from the valence to the conduction band is prohibitive. Consequently, an insulator will not carry a current. In a metal, the band gap is zero. There is no barrier to inhibit an electron from leaving the top of the valence band and entering the conduction band, leaving a hole behind. Therefore, metals carry electric current whenever an electrical *potential* is applied.

Semiconductors have band gaps of intermediate energies, energies that are well matched to solar photons. When light is absorbed by a semiconductor, an electron is promoted from the valence to the conduction band. This is the basis of the PV effect. To be absorbed, however, the light must have energy equal to or greater than the band gap. Light from the sun has a distribution of wavelengths, which peaks near 500 nanometers (nm) in the green-blue region of the spectrum (see Figure 4.5). The distribution has a broad tail, extending far into the infrared region, but two-thirds of the photons have wavelengths shorter than 1,140 nm, which corresponds to the energy of the silicon band gap, 1.09 electron volts (ev). Consequently, silicon can capture most of the solar photons.

It is not enough, however, simply to shine sunlight on a piece of silicon. The electrons, which are promoted to the conduction band, have no reason to travel in any particular direction. After a short time, they fall back to the valence band, refilling the holes that had been created by the light. This wasteful process is called *electron-hole recombination*.

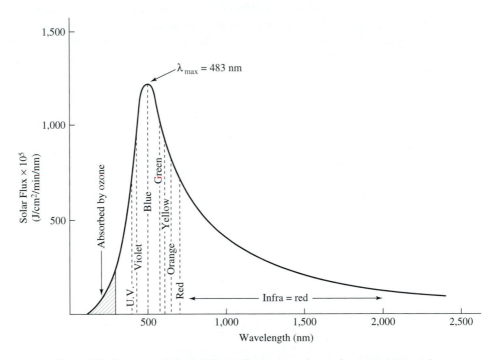

Figure 4.5 Spectrum of solar radiation on Earth's atmosphere; solar constant (area under curve) = 8.16 kJ/cm^2/min (energy of radiation increases with decreasing wavelength).

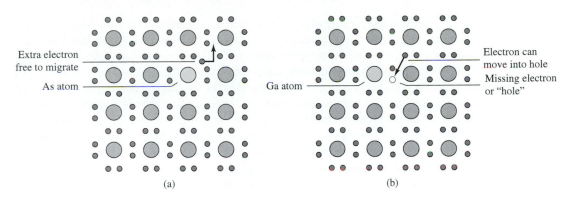

Figure 4.6 a) An arsenic-doped *n*-type silicon semiconductor; b) A gallium-doped *p*-type silicon semiconductor. *Source:* C. L. Stanitski et al. (2000) *Chemistry in Context, Applying Chemistry to Society* (third edition) (New York: McGraw-Hill). Copyright © 2000 by The McGraw-Hill Companies. Reprinted with permission.

In order to produce a flow of electrons in an external circuit, a PV material must have a built-in electrical potential, which is created by an intrinsic charge asymmetry. This charge asymmetry is produced by doping the semiconductor with foreign atoms that have either more valence electrons than the native atoms, or fewer valence electrons. As illustrated in Figure 4.6, silicon can be doped with arsenic, which has five valence electrons instead of four, or with gallium, which has only three valence electrons. The foreign atoms enter the Si lattice and are surrounded by four Si atoms. The four bonds utilize four valence electrons, however, leaving the As atom with an extra electron, or the Ga atom with a hole. Because of repulsion from the bonding electrons, the extra electron on the As atom has a high energy, close to that of the Si conduction band. It takes only a little energy for this electron to enter the conduction band and move freely through the lattice. A fixed positive charge is left on the As atom. Doping silicon with As therefore produces an *n-type* semiconductor, because negative mobile carriers are created. Doping with Ga, on the other hand, produces a *p-type* semiconductor, having positive mobile carriers. The missing electron on the Ga atom can be supplied by the surrounding Si lattice, creating a hole in the valence band, which becomes the mobile carrier. A fixed negative charge is now left behind on the Ga atom.

The charge asymmetry required by the PV cell is created by joining a *p*-type with an *n*-type semiconductor, forming a *p-n junction* (see Figure 4.7). The mobile carriers migrate

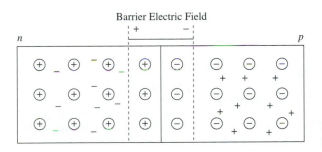

Figure 4.7 The *p-n* junction in a photovoltaic cell.

away from the junction because of repulsion from fixed charges of the same sign. Thus, the mobile electrons in the *n*-type semiconductor are repelled by the adjacent fixed negative charges on the Ga atoms of the *p*-type semiconductor, while the mobile holes in the *p*-type semiconductor are repelled by the adjacent positive charges on the As atoms of the *n*-type semiconductor. The result is an electric potential at the junction, created by the fixed negative charges on one side and the fixed positive charges on the other. If a photon is now absorbed in the *p*-type semiconductor, the electron that is injected into the conduction band will have a chance of being accelerated across the junction before it recombines with a hole. Likewise, a photo-generated hole in the valence band of the *n*-type semiconductor can be accelerated in the opposite direction before it recombines with an electron. This directed movement of the photo-generated mobile charges produces an electric current in an external circuit.

The actual geometry used in a solar cell is shown in Figure 4.8. Which semiconductor is on top is immaterial, since the current is produced by both electrons and holes. But it is important to keep the top layer very thin, in order to avoid absorbing the photons before they reach the immediate region of the junction, where the charge separation occurs. The charges produced by photons absorbed outside this region will recombine before they can be accelerated across the junction.

Even in the absence of recombination, solar energy cannot be converted at 100 percent efficiency because of its spectral distribution. About a third of the photons have energies below the band gap and cannot be utilized. For the remainder, 1.09 ev is the maximum energy that can be extracted; energy increments above the band gap are wasted as heat. Consequently, only half the solar energy can be converted to photoelectrons. From this must be subtracted the efficiency losses due to recombination.

The recombination probability increases if there are imperfections in the silicon lattice. Consequently, the best solar cells are made of crystalline silicon. Efficiencies as high as 28 percent have been achieved with such cells. Unfortunately, crystalline silicon is expensive to manufacture in quantity; it must be cast in ingots and then cut into wafers. Interest has therefore turned to amorphous silicon, which can be manufactured continuously in thin films at a much lower cost. Because of lattice imperfections, the conversion

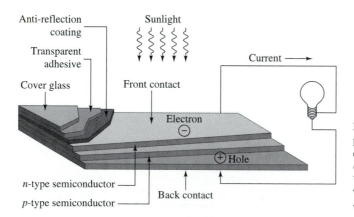

Figure 4.8   Schematic diagram of a photovoltaic cell.   *Source:* C. L. Stanitski et al. (2000) *Chemistry in Context, Applying Chemistry to Society* (third edition) (New York: McGraw-Hill). Copyright © 2000 by The McGraw-Hill Companies. Reprinted with permission.

efficiency is reduced for amorphous silicon, but intensive development has raised the achievable efficiency to 16 percent. The lowered efficiency is more than made up by the lower costs and ease of manufacture.

b. Photosynthesis and photoelectrochemistry. Hundreds of millions of years before the invention of the *p-n* junction, nature solved the problem of harnessing photo-induced charge separation to produce useful energy. In photosynthesis, as in the silicon solar cell, electrons and holes are created by the absorption of solar photons, and are induced to travel in opposite directions. Instead of a *p-n* junction, nature has evolved a photo-reaction center, at which charge is separated across a biological membrane, as illustrated in Figure 4.9. The key components of the photo-reaction center are a collection of chlorophyll molecules (see Figure 4.10a) capable of absorbing sunlight and generating electrons and holes in their photo-excited states. The electrons hop from one molecule to another along an energy gradient of empty energy levels. This gradient serves the same function as the charge asymmetry in a *p-n* junction, namely, to separate electrons and holes physically. Instead of an electric current, the charge separation leads to the production of energy-storing chemicals. Because the photo-reaction center is held in a membrane, the electron and hole are able to react separately with molecules that are located on opposite sides of the membrane. The electron is taken up in a series of biochemical steps that lead to the reduction of CO_2 to carbohydrates, while the hole is involved in another series of steps that oxidize water to O_2.

Because of its extended system of conjugated double bonds, chlorophyll absorbs light strongly in the 400–700 nm region of the spectrum (see the absorption spectrum in Figure 4.10b). This region contains about 50 percent of the total energy in solar photons. A green plant absorbs about 80 percent of the incident photons in this range (the rest are lost to reflection, transmission, and absorption by other molecules), leaving about 40 percent of

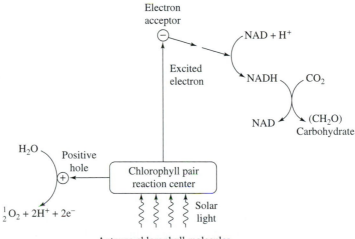

Figure 4.9 The photo-reaction center and energy storage in photosynthesis.

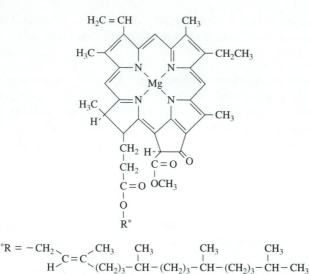

*R =
Structural Formula of Chlorophyll *a*

(a)

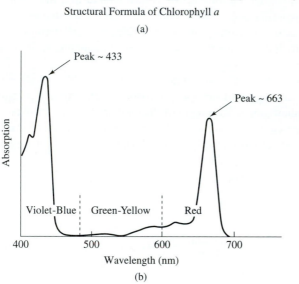

Peak ~ 433

Peak ~ 663

Absorption

Violet-Blue | Green-Yellow | Red

400   500   600   700
Wavelength (nm)

(b)

Figure 4.10   a) The chemical structure of *chlorophyll a*; b) absorption spectrum of *chlorophyll a*.

the total solar energy available for photosynthesis. Of this, 28 percent actually ends up in carbohydrates; the rest is lost at the various electron transfer and chemical steps. Thus, the conversion efficiency from sunlight to carbohydrates is $0.5 \times 0.8 \times 0.28 = 0.11$. However, the plant uses about 40 percent of the energy for its own metabolic needs, leaving only $0.11 \times 0.6 = 0.067$ as the fraction of sunlight stored as photosynthetic energy. This maximum conversion efficiency applies to the so-called C_4 plants, in which the first product of photosynthesis is a 4-carbon sugar; these include corn, sorghum, and sugar cane, and some other plants that grow best in hot climates. C_3 plants, in which the first photosynthetic product is a 3-carbon sugar, include wheat, rice, soybeans, trees, and other plants that dominate temperate climates and account for 95 percent of global plant biomass. These plants are about half as efficient in photosynthesis as the C_4 plants. To these inefficiencies must be added the limitations on plant growth imposed by temperatures that are too low or too high, insufficient water, and insufficient nutrients. These limitations and inefficiencies explain why only about 0.3 percent of the global insolation reaching Earth's surface is used by green plants and algae in photosynthesis.

It is possible to carry out chemical processes analogous to photosynthesis with a *photoelectrochemical* cell. In such a cell, a *photosensitizer,* usually an organic dye, analogous to chlorophyll, captures sunlight and transfers the photoexcited electrons and holes to a pair of electrodes, where they react separately with molecules. If the electrons and holes can be induced to react with water, then hydrogen and oxygen are generated, in a reversal of the fuel cell reaction (see p. 103). The same result can be achieved by using the electricity of a solar cell to electrolyze water, but the photoelectrochemical cell eliminates the need for an electrolyzer, and could in principle be more efficient. The basic difficulty is that the electrodes carrying out the reactions tend to degrade in time; lifetimes (and efficiencies) are improved when catalysts are used to speed up the electrode reactions. Research is leading to improved electrode materials and catalysts.

4.4   BIOMASS

Although the process of industrial development leads societies to switch from traditional biomass fuels to fossil fuels, the available biomass still represents a large energy resource. Because biomass carbon is continuously converted to CO_2 by respiration, burning it to extract energy does not add to the net CO_2 content of the atmosphere (although there may be issues of time scale in the case of slow-growing trees). The energy density of biomass is variable, and is substantially lower than that of fossil fuels, because of relatively high abundances of elements other than carbon and hydrogen (see Table 2.2, cellulose, p. 25). For this reason, and because of the difficulty of collecting and processing the biomass, it has not been economically competitive with fossil fuels. If, however, the emission of CO_2 becomes regulated or taxed, as it already is in several countries, then the economic prospects of biomass fuels will brighten.

It is likely that closer attention will be paid to the waste streams of forests, farms, and municipalities, which represent a substantial energy resource. The wood products industry already uses most of its mill residues to generate steam and electricity for the mills. On farms, large quantities of wheat straw, corn stalks, and other plant residues are left to decompose in the fields. Much of this is needed to replenish the soil organic matter and to limit erosion, but an estimated 40 percent could be harvested for energy production if it made economic sense. Municipal waste also contains a substantial fraction of combustible matter. Much of this is land-filled, but there are also many municipal incinerators that burn the trash instead. An increasing number of these (about 100 in the U.S., delivering 23 billion kWh per year) are using the heat of combustion to generate electricity. The resulting income offsets some of the cost of trash collection, but incinerators are contentious because of concerns about toxic emissions (e.g., dioxins and mercury). Emissions are minimal in a properly run modern incinerator, but the operating condition may not always be optimal.

Beyond the utilization of wastes, there is the possibility of planting fast-growing trees and grasses expressly as energy crops. Switchgrass is a prime candidate, since it can grow vigorously for ten years, resist drought, and thrive with low fertilizer application. Fast-growing trees, such as poplar, willow, cottonwood, and green ash, already serve other environmental purposes as well. They are planted next to crops in buffer strips along stream banks to mitigate erosion and water pollution. They can also be planted around live-stock feedlots to control odors and intercept runoff. When these trees are harvested, the stumps grow new sprouts, allowing the trees to regrow for several rotations before weak-ening rootstocks must be replaced. The harvesting of switchgrass or woody crops for en-ergy production would be an opportunity for farmers with marginal lands that are unsuit-able for row-crops. Many such U.S. farmers are currently enrolled in the Conservation Reserve Program, which pays them not to farm marginal land.

a. Ethanol from biomass.   There is much interest in converting biomass to liq-uid fuels, which could be used in automotive transport. Although there are several schemes for chemical conversion of biomass to a variety of liquids, the only conversion currently being utilized in quantity is the well-known biological fermentation of sugars to ethanol. Brazil has been the leader in ethanol fuel production, exploiting its abundant sunshine and fast-growing sugarcane to produce ethanol for transport. At one point, half its automotive fleet was running on ethanol. In the U.S., corn is used for ethanol production, which is blended in with gasoline to a level of 10 percent (E10), or 85 percent (E85). Several policies support the ethanol industry, including mandates for government vehicles to use ethanol-blended fuel, and state and federal tax exemptions. The rationale for these policies has been energy independence, encouragement of renewable resources, and pollution prevention.

The U.S. EPA has mandated oxygen-containing additives (oxygenates) for gasoline, in order to reduce emissions of carbon monoxide and other smog-forming pollutants (see p. 230). The two additives vying for the gasoline market have been ethanol and methyl *t*-butyl ether (MTBE). Gasoline producers opted for MTBE (produced in refineries, see pp. 30–31), but MTBE has now fallen from favor because of groundwater contamination (see pp. 259–260). When California recently banned MTBE and applied to the EPA for an oxygenate exemption, it was turned down. This ruling was a major boost for ethanol, the only alternative to MTBE. Ethanol currently accounts for 1.2 percent of the U.S. gasoline supply, and will grow rapidly.

Corn for fuel production has been criticized as wasteful, since its economics rests on farm subsidies and tax breaks, and since its production requires external energy. Indeed, the energy available from a gallon of ethanol is not much greater than the energy required to produce it, when the inputs for fertilizer, farm machinery, and the energy for fermentation and distillation are taken into account. The ratio of energy output to input has improved with recent technological advances and farm efficiencies, but still stands at only 1.24.

The situation would be improved if the entire corn plant could be converted to ethanol, instead of just the kernels. The reason it cannot is that the kernel is made mostly of starch, while the rest of the plant (called "stover," consisting of stalk, leaves, and cob), is

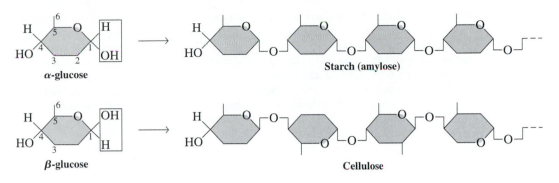

Figure 4.11   Structures of starch (amylose) and cellulose. [Starch has several different structures, the most common of which are amylose and amylopectin. The former (shown here) has a linear structure, while the latter has a branched structure.]   *Source:* R.A. Wallace, G. P. Sanders, and R. J. Ferl (1991). *Biology: The Science of Life.* Harper Collins: New York. Copyright © 1991 by Addison Wesley Longman. Reprinted with permission.

mostly cellulose. Both starch and cellulose are polymers of sugars, but the way the sugar molecules are strung together is different (see Strategies 4.1 and Figure 4.11). This difference makes cellulose much harder to break down into sugars than starch. Only ruminants are able to digest cellulose, and only because their rumens contain bacteria that harbor *cellulase* enzymes, which are capable of breaking down the cellulose polymer. If there were a practical method for converting cellulose to ethanol, many plants besides corn could be harvested for the purpose. Cellulose is the primary structural component of plants, and is the most abundant organic compound in the biosphere, typically comprising 30–50 percent of the plant mass. To give an idea of the scale of cellulose production, if all the cellulose in the annual plant growth, estimated to be 90–130 billion dry tons, were converted to ethanol, the energy equivalent would be 4–6 times larger than global petroleum consumption in 2000.

STRATEGIES 4.1   ETHANOL FROM STARCH OR CELLULOSE?

Starch and cellulose (see Figure 4.11) are both composed of glucose monomer units ($C_6H_{12}O_6$), but the monomers differ in the orientation of two of their bonds.

Glucose has a six-membered ring of five carbons and one oxygen atom. Four of the ring carbons have hydrogen and hydroxyl substituents, while the fifth ring carbon has a hydrogen and a —CH_2OH group attached. In the

α-isomer, the hydroxyl on the carbon numbered 1 is on the opposite side of the ring from the CH_2OH group on carbon-5, while in the β-isomer, the OH is on the same side. Starch contains α-glucose, while cellulose contains β-glucose. In both cases the polymers are made by connecting the hydroxyl on carbon-1 with carbon-4 of another monomer, eliminating water in the process:

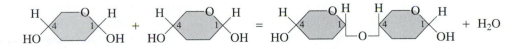

Organisms break down the polymers by catalyzing the reverse of this reaction, using enzymes. Because of the differing orientation of the connecting bonds in starch and cellulose, the enzymes must have different placements of the catalytic groups in their active sites. The enzyme for starch is *amylase,* while the enzyme for cellulose is *cellulase.* Many microorganisms have amylase, but only a few—certain fungi and bacteria—have cellulase. Termites and ruminants can digest cellulose, because their digestive systems harbor cellulase-producing bacteria.

In the industrial production of ethanol from corn, the starch is first degraded to glucose with readily available amylase, and the resulting glucose is then fed to yeasts, which metabolize the glucose and release ethanol (fermentation) (see Figure 4.12). Cellulose conversion, however, requires cellulase, which is difficult to obtain in industrial quantities. The fungi or bacteria that produce cellulase are hard to grow.

An additional obstacle to cellulose conversion is that the cellulose in plants is protected by

a protective sheath of *hemicellulose* (a polymer of five-carbon sugars) and lignin (see Figure 2.4, p. 24). This sheath must first be disrupted, requiring treatment with dilute sulfuric acid at elevated temperature and pressure. This treatment breaks down hemicellulose to its five-carbon sugar monomers, the most abundant of which is xylose:

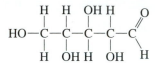

Thus, the glucose produced by the cellulase is mixed with xylose and other five-carbon sugars. However, the common yeasts that ferment glucose are unable to ferment five-carbon sugars. Other organisms can be used to ferment five-carbon sugars, but a separate fermentation unit is required. Researchers are currently trying to develop strains of bacteria capable of utilizing all the sugars in a single fermentation process.

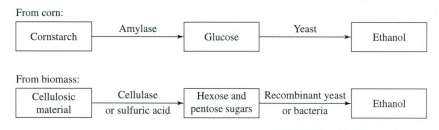

Figure 4.12   Two biocatalytic routes lead to ethanol.   *Source:* M. Mcoy (1998). Biomass ethanol inches forward. *Chemical & Engineering News* 76(49):29–32. Reprinted with permission from C&EN. Copyright (1998) American Chemical Society.

---

b. Methane from biomass.   When oxygen is removed from dead biomass, anaerobic bacteria convert the carbon to methane. This happens in landfills and manure piles, both of which produce copious quantities of methane. It also happens in the rumens of ruminants, where the bacteria that break down cellulose also release methane. Figure 4.13 shows that the methane released into the atmosphere from these three sources exceeds the methane released by coal mining and by drilling for oil and gas.

The released methane is a wasted energy resource, as well as a significant contributor to global warming. Each additional methane molecule has 23 times the global warming potential of a CO_2 molecule (see p. 165). This factor is recognized in schemes to regulate greenhouse gases, including the Kyoto Protocols (see p. 179). If these take effect, the in-

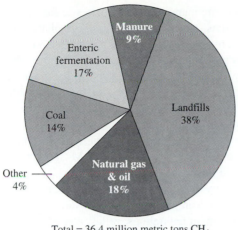

Total = 36.4 million metric tons CH_4
[includes emissions (30.3 tons) and recovery (6.1 tons)]

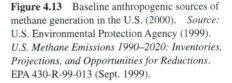

Figure 4.13   Baseline anthropogenic sources of methane generation in the U.S. (2000).   *Source: U.S. Environmental Protection Agency (1999). U.S. Methane Emissions 1990–2020: Inventories, Projections, and Opportunities for Reductions.* EPA 430-R-99-013 (Sept. 1999).

centive to collect the methane from landfills and manure piles will increase substantially. (Collecting the methane from ruminants is not practical, but research is being directed to altering the composition of their diet to reduce methane production.)

Landfills have been collecting methane for some time, originally because of the risk of explosions. Before 1990 there were 40 recorded landfill fires in the U.S., and ten deaths. Subsequently, gas was collected and vented to the atmosphere, but the so-called Landfill Rule, promulgated under the 1996 Clean Air Act, required large landfills to collect non-methane organic compounds and prevent their emission to the atmosphere, because of their contribution to smog formation (see p. 228). The collected gas, containing both methane and non-methane organics, was generally flared, but recently the gas streams have increasingly been used for energy production. The collected gas contains considerable quantities of CO_2, as well as methane and other organics, so extraction of pure methane for delivery to gas suppliers is impractical. However, the gas may have sufficient energy content to make electricity production worthwhile. By 2000, captured gas from landfills came to around 12 percent (5 million tons) of total methane production in the U.S.

Methane from livestock manure is estimated at 3.2 million tons annually, and is expected to increase substantially because of implementation of liquid or slurry manure management systems, which produce more methane. None of this methane is currently recovered, but it could be, via installation of anaerobic digesters. There is potential of significant energy recovery from such systems, especially if greenhouse gas regulations are implemented. It is worth noting that anaerobic digesters are being used in some rural areas of developing countries, where they produce gas for village heat and power.

4.5 HYDROELECTRICITY

Hydroelectric plants utilize part of the energy of the solar-driven hydrological cycle. The continents of the world receive more rain than the water they lose by evapotranspiration, and the excess runs off in rivers to the ocean. The running water can be used to turn a turbine and generate electricity. Although there are many small hydroelectric facilities that utilize riverflow directly, the larger installations, which account for most of the available hydropower, rely on dams to increase the hydraulic head (water pressure), and to even out the flow, thereby allowing the continuous production of electricity. Dams serve other purposes as well, including the provision of water for residential, industrial, and agricultural purposes, facilitating flood control and/or navigation, and providing recreational facilities. In fact, most dams do not generate electricity, although many could be retrofitted to do so.

The share of world electricity provided by hydropower is about 20 percent. This is a small fraction of the available hydrological potential, but probably represents a third to a fourth of the potential sites that could be developed economically. Most of the undeveloped sites are in the former Soviet Union and in developing countries.

Like other forms of solar energy, hydroelectricity adds no CO_2 nor other emissions to the atmosphere, but it is not without environmental costs. The water backing up behind dams floods the shoreline, inundating human habitations, relics, and ecosystems, and the character of the river is permanently altered. The effects are particularly devastating on migratory fish, and are only partially remedied by fish ladders. The large dams on the Columbia River have contributed to the collapse of the salmon population in the northwest of the United States. Dams also hold back silt, which can have deleterious effects downstream. The most celebrated example is Egypt's Aswan dam, which stopped the annual flooding of the Nile valley, resulting in diminished nutrient input to crops, and salinization and subsidence of the Nile delta.

The silt behind the dam must eventually be flushed out before it engulfs the turbine intakes. The still waters behind the dam are subject to eutrophication (overfertilization, see p. 319), and, especially in the tropics, can promote the spread of disease-carrying organisms. Most of these problems can be minimized by appropriate site selection and management, but public opposition now limits the number of available dam sites significantly. Small hydroelectric plants that work on river flow avoid most of these problems, but they are less efficient and are subject to the seasonal fluctuation of river levels.

4.6   WIND POWER

The wind provides another form of solar power, since wind results from air-temperature differences associated with different rates of solar heating. A global circulation of air, the Hadley circulation, is created by moist hot air rising at the equator and being replaced by drier air flowing in from the region of 30 degrees north and south latitude. At higher latitudes, the air flows toward the poles and is deflected westward by Earth's rotation, creating a wavelike pattern known as the Rossby circulation. Regional variations in atmospheric temperature superimpose smaller circulation systems on the global pattern. Locally strong winds are created by sharp temperature differences between, for example, the land and the sea, and they can be channeled by mountains and valleys. Many regions have steady prevailing winds as a result of these conditions.

Wind technology is nearly as old as recorded history. Well over 2,000 years ago, windmills were used to pump water in China and to grind grain in Persia. Europe was introduced to windmills in the eleventh century by veterans of the Crusades; the windmills were improved in the succeeding centuries, especially in Holland and England. By the eighteenth century there were 10,000 windmills in Holland alone. In the United States, windmills were a vital asset of early settlers of the Great Plains, and there were 50,000 backyard windmills pumping water and generating electricity until 1950, after which they were phased out by the Rural Electricity Administration.

The 1973 oil crisis rekindled interest in wind energy and in other alternative energy sources. A number of development programs were set in motion, and costs have fallen steadily thanks to the development of new technology. Wind electricity is now the lowest-cost alternative to electricity from fossil fuel and nuclear plants, and is growing rapidly. Between 1990 and 1999, the annual growth rate averaged 24 percent (three-year doubling time). Installed capacity by the end of 2000 was 17,000 MW worldwide.

Current advanced turbines are of the horizontal axis design (see Figure 4.14). They capture the wind's energy with two or three propeller-like blades, which are mounted on a

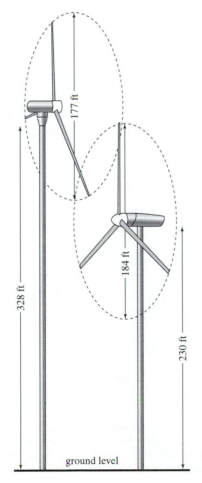

Figure 4.14   State-of-the-art wind turbines with high hub heights and large rotor diameters.   *Source:* U.S. Department of Energy, Wind Energy Program. (http://www.eren.doe.gov/wind/large.html)

rotor to generate electricity. The turbines sit atop tall towers, taking advantage of increased wind speed and decreased turbulence with altitude. Strong winds are critical since power generation increases with the cube of wind speed. It increases linearly with the area swept by the propellers, which is proportional to the square of their length. Consequently there is a premium on using long propellers mounted on tall towers. The limiting factor on size is the possibility of structural failure in high winds. Innovative tower designs using stronger light-weight materials now allow taller towers to be built at reduced cost. Hub heights currently average 30 meters and are expected to reach 70 meters or higher by 2005.

Average wind speeds vary greatly from place to place, and determine the practicality of wind power. Figure 4.15 is a map of the U.S. showing regions having class 3 or higher winds. (Class 3 winds have average power densities of 150–250 W/m^2 at a 10 m height, and 300–400 W/m^2 at 50 m.) Although wind power was first developed in California, the potential is much higher in the Great Plains, where a swath of class 3 or higher winds extends from the Dakotas to Texas. The total wind resource of this region is nearly three times as large as U.S. electricity generation in 1999.

The growth of wind power is being assisted by government policies. In California, customers of the state's three major utilities receive rebates of up to 50 percent of the cost of a small-scale home turbine. Sales of these units soared during the state's electricity crisis of 2000–2001. A 1994 Minnesota law requires the state's largest utility to install 425 MW of wind power by 2002, as a condition for the right to store nuclear waste from its power plants. Iowa passed a law in 1983 requiring utilities to obtain 2 percent of their electricity from renewable sources. The program has been popular with Iowa farmers, who receive $2,000 per year as rent for each turbine, which typically occupies 1/4 acre (1/10 hectare). Since a farmer growing corn has a profit margin of about $300 per acre, the turbines offer a

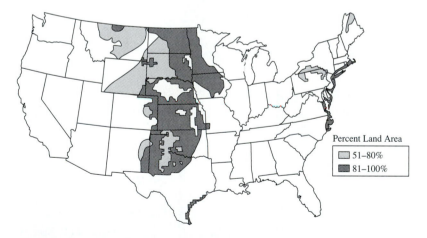

Figure 4.15   Percentage of land area estimated to have class 3 or higher wind power in the contiguous United States. *Source:* National Renewable Energy Laboratory, U.S. Department of Energy (1986). *Wind Energy Resource Atlas of the United States.* (http://rredc.nrel.gov/wind/pubs/atlas/maps/chap2/2-10m.html)

TABLE 4.2   TEN COUNTRIES WITH LARGEST WIND ENERGY PRODUCTION (2000)

Country	Installed wind capacity (MW)	Wind energy production (million MWh)	Total electricity production (million MWh)	Total provided by wind (percent)
Germany	6,113	11.50	495.2	2.3
United States	2,554	5.10	3,235.9	0.2
Spain	2,250	4.50	189.6	2.4
Denmark	2,140	4.28	32.9	13.0
India	1,167	2.33	424.0	0.6
Netherlands	449	0.90	97.8	0.9
Italy	420	0.84	272.4	0.3
United Kingdom	400	0.80	333.0	0.2
China	265	0.53	1,084.1	<0.1
Sweden	226	0.45	128.8	0.4
Total	**15,984**	**31.2**	**6,294**	**0.5**

Source: American Wind Energy Association, news release (February 9, 2001).

handsome return on the land. In 1999, Iowa inaugurated the world's largest wind power farm, with 257 turbines with a combined capacity of 193 MW.

However, most of the wind power is installed in the countries of Western Europe (see Table 4.2), which have committed themselves to the Kyoto Protocol on greenhouse gases, and therefore pursue renewable energy more assiduously than the U.S. For example, the Danish government imposed a carbon tax of about $14 per ton of CO_2 back in 1992, and today supplies 13 percent of its electricity from wind towers. Although in 1990 the U.S. had 75 percent of the world's 1,900 MW of installed wind capacity, the U.S. share of the 17,000 MW installed by 2000 had fallen to 15 percent.

4.7   OCEAN ENERGY

A great deal of energy is stored in the world's oceans in the form of tides and waves, and in gradients of temperature and salt concentration. But the oceans are vast and these energy forms are highly dispersed. Tidal energy is the only form currently being exploited for commercial power generation at an installation on the Brittany coast in France. The tides result from the gravitational attraction of the moon and sun. In the open ocean, the waters rise and fall by about one meter, but at shorelines the tides are higher, and can reach several meters in estuaries due to the funneling effect of the shore. Tidal power can be extracted much like power in falling rivers. A dam is built with sluices to admit the incoming tide to a reservoir, which is then emptied through turbines when the tide recedes. Although several tidal sites, especially in the United Kingdom and France, have potential for power generation (when the economic conditions become favorable), the number of such sites is insufficient for tidal energy to become an important part of the global energy supply.

Waves result from the action of wind on the waters and can carry significant amounts of energy. Where winds blow for long distances over open oceans, waves can be tens of meters high and displace tons of water. The west coasts of Europe and the United States and the coasts of Japan and New Zealand are particularly suitable for extracting the energy of waves. To accomplish this, a variety of mechanical devices have been proposed and are being researched. None are currently close to being practical, however.

The difference in salt concentration between the oceans and fresh water represents a large osmotic pressure, equivalent to a 240-meter head of water. But no practical method of harnessing the mixing energy has been devised.

More interesting are the possibilities for exploiting the energy stored in ocean thermal gradients. The sun heats the surface layers of the oceans, but the deeper layers remain cold. In tropical regions, the surface temperature is as high as 26°C, whereas at a depth of 1,000 meters, the temperature is 5–6°C. Because of the expanse of the ocean, the total energy stored in the surface gradients is enormous, about two orders of magnitude greater than the energy of all the tides and waves. But capturing this energy is not easy. Not only is it widely spread, but the 20°C gradient limits the theoretical heat engine efficiency (see p. 100) to under 7 percent. Consequently, a significant supply of electricity from this source requires a very large plant, and capital costs would be high. Moreover, transmission losses would limit such plants to operation near shores, although plants operating in the open ocean could, in principle, generate hydrogen electrolytically, for transport by ship (see discussion of the "hydrogen economy" p. 109).

In addition to electricity, an ocean thermal energy conversion (OTEC) plant delivers cold seawater to the surface, which can be used to air-condition nearby buildings. Also, the water from the deep ocean is rich in nutrients, and can support agricultural and maricultural enterprises. Finally, if seawater is also used as the working fluid of the plant, freshwater is produced as a byproduct. Opportunities for integrated power, water, air-conditioning, and farming activities may exist at many places on tropical shores. An experimental station in Hawaii is currently exploring this potential.

4.8   GEOTHERMAL ENERGY

Aside from tidal energy, geothermal energy is the one form of renewable energy that is unconnected with the sun. Earth itself generates heat from its molten core, and from the decay of naturally occurring radioisotopes. This heat source is thousands of times weaker than the sun's rays falling on Earth, but in some parts of Earth's crust, geological anomalies permit the concentrated upwelling of heat from below. Hot springs and volcanoes are familiar examples of these anomalies. Hot springs have been valued for bathing and for their curative powers for thousands of years, and the Romans used the hot water to heat their bath houses. A modern geothermal industry began in the nineteenth century, with the extraction of boric acid from hot springs near Laradello, Italy. By 1827, geothermal steam replaced firewood as a fuel for concentrating the boric acid, and the first geothermal power plant began operation in 1913, also in Italy. Since then, power plants have been installed at a number of sites around the world where geothermal steam is available, and currently supply 0.3 percent of the world's electricity.

Geothermal energy can be divided into three temperature regimes: high ($>150°C$), moderate $90–150°C$, and low ($<90°C$). The high-temperature resource is used almost exclusively for electricity generation. Geothermal steam plants operate like any other steam plant, except that special measures are needed to deal with the salts and gases, especially H_2S, entrained in the steam, in order to minimize corrosion and environmental contamination. U.S. geothermal plants currently have a capacity of ~2,200 MW, about the same as four large nuclear power plants.

Geothermal resources in the moderate temperature range can be used in diverse ways, including space heating, industrial process steam, greenhouses, and aquaculture. Current U.S. installed capacity is about 470 MW, or enough to heat 40,000 homes.

The low-temperature resource has found increasing use coupled with heat pumps (see p. 102) for home heating and cooling. The geothermal heat pumps transfer heat from the soil (or groundwater) to the house in winter, and from the house to the soil in summer. The geothermal heat pool (generally tapped at depths of 30–50 m) maintains a relatively steady temperature, allowing the heat pump to work well even in areas with harsh winters. The geothermal heat pump is an emerging technology experiencing rapid market growth; about 40,000 units were sold in 2000.

CHAPTER 5

ENERGY UTILIZATION

Extraction of energy from the various available sources is one side of the human energy equation; utilizing the extracted energy is the other. We need energy for myriad human activities, but the amount of energy required depends strongly on the efficiency with which it is used. The efficiency of energy utilization can vary enormously. It depends on technology, on the integration of energy systems, and on our patterns of living.

When energy is converted from one form to another there is always some loss of useful energy as waste heat. If you push a box across the floor, some of your muscle energy is converted to the mechanical energy needed to move the box, but some of it becomes heat because of the friction with the floor. This friction can be reduced by putting the box on wheels, but it can't be eliminated entirely. The wheels increase the efficiency of the conversion from muscle energy to mechanical energy, but the efficiency can never be 100 percent. A major challenge in energy conversion is minimizing waste heat.

FUNDAMENTALS 5.1: HEAT, TEMPERATURE, AND ENTROPY

Entropy: The highest efficiency attainable in energy conversion processes is set by the second law of thermodynamics. The first law is simply a statement of energy conservation: energy is neither destroyed nor created, provided we remember that heat is a form of energy. (Strictly speaking it is the sum of matter and energy that is conserved. Energy can be converted to matter and vice versa, according to Einstein's famous formula, $E = mc^2$, where E is energy, m is mass, and c is the speed of light. Only in nuclear reactions, however, does this conversion become significant.) The second law of thermodynamics states that the *entropy* always increases in a spontaneous process, i.e., one in which there is no input of external energy.

Entropy is a measure of disorder, and the second law is a statement of the fact that disorder naturally increases. If we allow a handful of white marbles and a handful of blue marbles to roll around in a box, the two colors will mix; they become disordered. The entropy was lower before the mixing, and higher afterward. Moreover, the blue and white marbles will not unmix by themselves, no matter how long they roll around. It takes external energy—e.g., in the form of fingers picking out all the blue marbles—to unmix them, and thereby lower the entropy. This is a simple illustration of the second law.

Heat and Temperature: What is the connection between entropy and heat? Heat arises from the motion of atoms and molecules. When water feels hot on our hand, we sense the rapid motion of the water molecules, imparting their *kinetic* energy to our skin. The temperature of the water is the average kinetic energy of the molecules, while the heat itself is the total kinetic energy of all the molecules in the quantity of water we are dealing with.

One measure of entropy, (S), is the heat energy, (Q), divided by the temperature, (T):

$$S = Q/T \qquad (5.1)$$

In this expression, T is the absolute temperature. Zero on the absolute temperature scale corresponds to zero kinetic energy, i.e., no motion of the molecules. Recall from Fundamentals 1.1 that the absolute temperature is measured in degrees Kelvin, which are the same as degrees Centigrade, but are offset by 273, i.e., 0°C = 273 K. (The degree superscript is omitted from the absolute temperature.)

Equation (5.1) expresses the fact that a given quantity of heat energy produces greater disorder for a cold sample (small value of T) than for a hot sample (high value of T) of matter. The second law says that Q/T increases in a spontaneous process. A familiar example is the cooling of a cup of hot tea. We know that a quantity of heat is transferred from the hot liquid to the cool surroundings. If this quantity is Q_{transf}, then clearly

$$Q_{transf}/T_{hot} < Q_{transf}/T_{cool}$$

Thus entropy has increased in the cooling process.

Heat Engines: From equation (5.1), it is a simple matter to find the maximum efficiency of a power plant, or any other heat engine. Consider the diagram of a steam-to-electricity plant in Figure 5.1.

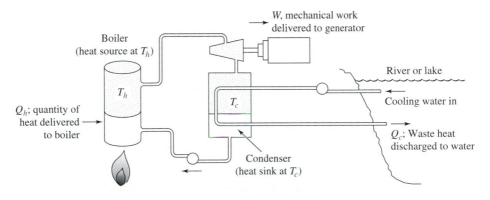

Figure 5.1   Maximum work and waste heat from a steam engine power plant.

Steam is generated by burning fuel in a boiler, and turns a turbine that generates electricity. The spent steam is then condensed to liquid water and returned to the boiler. Thus the energy obtained from combustion is converted to work in the turbine. The work energy must be the difference between the quantity of heat used to boil the water, Q_h, and the quantity of heat transferred to the cooling water in the condenser, Q_c:

$$W = Q_h - Q_c \qquad (5.2)$$

Equation (5.2) is just an expression of energy conservation, the first law of thermodynamics. Dividing both sides by Q_h, we obtain

$$W/Q_h = 1 - Q_c/Q_h \qquad (5.3)$$

W/Q_h is the efficiency by which heat is converted to work.

But the second law tells us that entropy must increase, and therefore

$$Q_c/T_c > Q_h/T_h \qquad (5.4)$$

where T_h is the temperature of the heat source, and T_c is the temperature of the condenser. Multiplying both sides by T_c/Q_h gives

$$Q_c/Q_h > T_c/T_h \qquad (5.5)$$

If we now substitute (5.5) in (5.3), we see that

$$W/Q_h < 1 - T_c/T_h \qquad (5.6)$$

Equation (5.6) has far-reaching implications. It means that no matter how the power plant is designed, its efficiency is constrained by T_c/T_h, the ratio of the condenser temperature to the temperature of the heat source, in this case the boiler. The lower this ratio, the closer the efficiency approaches unity. Since the condenser temperature is constrained by the available cooling water, i.e., the temperature of the environment, the ratio mainly depends on how hot the boiler can be heated.

The same equation holds for all heat engines, whether a steam plant for electricity or an internal combustion engine. It stresses the critical importance of the temperature at which the heat energy is delivered. Although energy is always conserved, the efficiency with which heat can be converted to work is fundamentally constrained by its temperature. The maximum conversion efficiency goes up with the temperature of the heat source. High-temperature heat is high-quality energy, because its entropy (per unit of energy) is low.

5.1 HEAT ENGINE EFFICIENCIES

An industrialized society relies on converting heat to work. Transport, manufacturing, mining, and construction systems all utilize the energy of fossil fuels in internal combustion engines, while electricity is produced in power plants, where heat (from fossil fuels or nuclear reactors) produces steam to drive turbines (see Figure 5.1). The electricity lights, and sometimes heats, buildings, and powers innumerable appliances and motors. This vast array of heat-driven processes offers many opportunities for improved energy efficiency.

The design of power plants has improved substantially, but the heat-to-electricity efficiency is still only about 40 percent at best. In a modern coal-fired plant, the boiler temperature reaches about 550°C (823 K). The condenser temperature depends somewhat on the available cooling water, but cannot be far from 300 K (27°C). The maximum theoretical efficiency, according to equation (5.6), is $1 - 300/823 = 0.64$, or 64 percent.

However, additional heat losses from the boiler, the turbine, and the electrical generator all conspire to reduce the actual efficiency well below the theoretical efficiency. Thus about 60 percent of the heating value of the coal is lost to waste heat. For a nuclear power plant, the theoretical efficiency is lower, because the temperature of the boiler is restricted for safety reasons; the overall efficiency of a nuclear power plant is about 30 percent.

Higher efficiencies could be achieved by boosting the temperature of the heat source. A boiler cannot be heated much higher than 550°C, but a gas turbine can be run at temperatures up to 1,260°C. Gas turbines were developed as jet engines for aircraft, but have recently been adapted to run electricity-generating turbines. Despite the high operating temperature, the efficiency of a gas turbine is modest, because the temperature at which the heat is transferred to the environment, i.e., the exhaust gas, is still quite high, typically 500°C. These values for T_h and T_c give a maximum efficiency of 50 percent; the actual efficiency is only about 33 percent, about the same as a nuclear plant. But the hot gases can themselves be used to run a steam turbine (see Figure 5.2). This second stage operates at

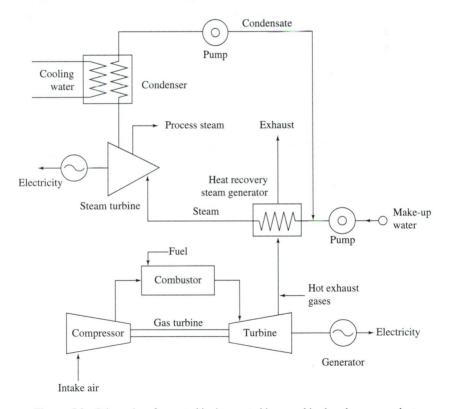

Figure 5.2 Schematic of gas turbine/steam turbine combined-cycle power plant. *Source:* R. H. Williams and E. D. Larson (1993). Advanced gasification-based biomass power generation. In *Renewable Energy, Sources for Fuels and Electricity.* T. B. Johansson et al., eds. (Washington, DC: Island Press). Copyright © 1993 by Island Press. Reprinted with permission.

about the same conversion efficiency as a regular steam plant. By combining these two stages, one can extract close to 80 percent of the energy in the fuel. Many coal-fired power plants are now being fitted with a natural gas turbine "topping cycle," in order to boost efficiency.

The same thermodynamic considerations apply when energy is "consumed." The energy does not disappear, of course. Rather it is converted to "waste" heat, raising the entropy of the universe. En route to this destination, the energy can be utilized more or less efficiently. Heating a house provides an illustrative example. A gas furnace is more efficient for this purpose than is electrical heating, because most of the energy in the fossil fuel can be transferred directly to the house (assuming a highly efficient furnace), whereas with electrical heating, about two-thirds of the energy has been thrown away at the power plant to produce the electricity that is then converted back to house heat.

Electricity can be converted to heat in a much more efficient way, however, by using it to run a heat pump. This device (see Figure 5.3) is a small version of a power plant run in reverse. Mechanical work is used to condense a working fluid at the temperature of the heat sink. The fluid is then allowed to evaporate, thereby absorbing heat from the heat source. In this way, heat can be pumped from lower to higher temperatures by the expenditure of work. This is how refrigerators and air-conditioners operate. A house can be heated in cold weather by reversing the direction of the air-conditioner.

The conversion of work to heat is governed by the same entropy considerations as the reverse process. The maximum degree of conversion is simply given by the inverse of equation (5.6)

$$Q/W < T_h/(T_h - T_c) \qquad\qquad (5.7)$$

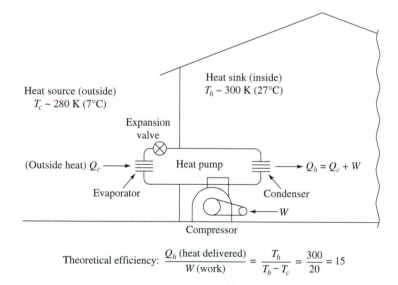

Figure 5.3   The heat pump: An efficient means of residential space heating and cooling.

but T_h and T_c are reversed, since the heat source (T_c) is now colder than the heat sink (T_h). If the inside temperature is about 300 K, and the outside temperature is 20°C colder, then the amount of pumped heat is potentially $300/20 = 15$ times greater than the amount of work. One kJ of electricity could transfer as much as 15 kJ of heat. Does this mean that we have violated the first law of thermodynamics? Not at all. It just means that we need to pay attention to the quality as well as the quantity of the energy. The higher the temperature, the higher the quality; high-grade heat can be converted to a larger quantity of low-grade heat, increasing the entropy in the process. Electricity is very high-quality energy, and this quality is largely wasted by converting it directly to heat at low temperature; the heat pump allows this quality to be more fully utilized. The high theoretical efficiency cannot be achieved in practice because of resistances to heat transfer, but a heat/work ratio of 2 is easily attainable, thereby recapturing a substantial fraction of the fossil fuel calories that are lost at the power plant.

In practice, heat pumps work best when the weather is moderate, and may require backup heating on cold winter days. As noted above, the use of geothermal reservoirs as the heat source can substantially improve heat pump performance.

5.2   FUEL CELLS

Combustion is not the only way to extract useful energy from chemical fuels. Electricity can be obtained directly with the aid of a fuel cell. Instead of burning hydrogen to produce heat, the same reaction

$$2H_2 + O_2 = 2H_2O \tag{5.8}$$

can be carried out at two electrodes, with an electric current flowing between them, as illustrated in Figure 5.4. The electrode reactions are

$$2H_2 = 4H^+ + 4e^- \tag{5.9}$$

$$4e^- + O_2 + 4H^+ = 2H_2O \tag{5.10}$$

Hydrogen is oxidized at the *anode,* where electrons are removed and passed through the external circuit to the *cathode,* where oxygen is reduced. The two electrode reactions together add up to reaction (5.8). But the energy is released mostly as electricity instead of as heat. Electricity production via the fuel cell is just the reverse of the familiar electrolysis process, in which hydrogen and oxygen are produced at a pair of electrodes immersed in water when an electric current is passed between them.

Because the reaction energy is transformed directly into electricity, the efficiency is not limited by the heat engine formula. As discussed in Fundamentals 5.2, p. 108, the theoretical efficiency is given by the ratio of the free energy change to the enthalpy change, $\Delta G/\Delta H$, which is about 80 percent, for reaction (5.8).

However, the theoretical efficiency can only be reached if no current is actually drawn from the fuel cell. When a current does flow through the external circuit, several

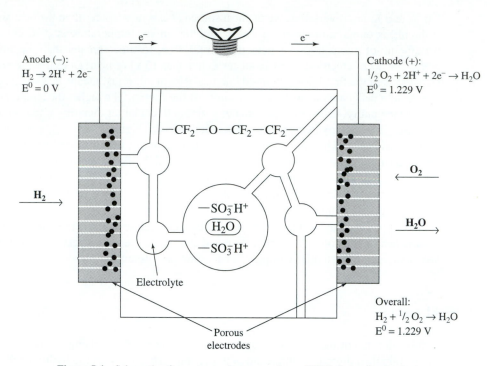

Figure 5.4   Schematic of a proton exchange membrane (PEM) fuel cell using hydrogen as fuel.

resistances build up that reduce the efficiency, and increase the fraction of the energy that is converted to heat. These include: 1) the electrical resistance of moving positive ions from the region of the anode to the region of the cathode to counterbalance the electron flow (electrolyte resistance); 2) the resistance to the movement of reactant molecules to the electrodes and product molecules away from the electrodes (mass transfer resistance); and 3) the resistance to the chemical reactions themselves due to slow reaction kinetics (activation barriers). These resistances can be minimized, but not eliminated, by optimizing the fuel-cell design. Thus, the reaction kinetics can be speeded up with catalysts (platinum metal is most commonly used); mass transfer can be enhanced by increasing the surface area of the electrodes, for example, by making them porous; and electrical resistance can be minimized by connecting the electrode compartments with an electrolyte having a high conductance. All three resistances are reduced at elevated temperature, which increases rates for chemical reactions and for molecular and ion transport. Consequently, successful fuel-cell designs have tended to have elevated operating temperatures. However, the proton exchange membrane (PEM) fuel cell, operating in a relatively low temperature range (80–100°C), has emerged as the most suitable design for transport. Characteristics of the five most prominent fuel-cell technologies are listed in Table 5.1.

TABLE 5.1 COMPARISON OF FIVE FUEL-CELL TECHNOLOGIES

Fuel-cell type	Electrolyte	Operating temperature (°C)	Electrochemical reactions	Applications	Advantages	Disadvantages
Alkaline (AFC)	Aqueous solution of potassium hydroxide soaked in a matrix	90–100	Anode: $H_2 + 2OH^- \rightarrow 2H_2O + 2e^-$ Cathode: $\frac{1}{2}O_2 + 2H_2O + 2e^- \rightarrow 2OH^-$ Cell: $H_2 + \frac{1}{2}O_2 \rightarrow H_2O$	• Military • Space	• Cathode reaction faster in alkaline electrolyte—so high performance	• Expensive removal of CO_2 from fuel and air streams required
Polymer electrolyte membrane (PEM)	Solid organic polymer poly-perfluorosulfonic acid	60–100	Anode: $H_2 \rightarrow 2H^+ + 2e^-$ Cathode: $\frac{1}{2}O_2 + 2H^+ + 2e^- \rightarrow H_2O$ Cell: $H_2 + \frac{1}{2}O_2 \rightarrow H_2O$	• Transportation • Electric utility • Portable power	• Solid electrolyte reduces corrosion & management problems • Low temperature • Quick start-up	• Low temperature requires expensive catalysts • High sensitivity to fuel impurities
Phosphoric acid (PAFC)	Liquid phosphoric acid soaked in a matrix	175–200	Anode: $H_2 \rightarrow 2H^+ + 2e^-$ Cathode: $\frac{1}{2}O_2 + 2H^+ + 2e^- \rightarrow H_2O$ Cell: $H_2 + \frac{1}{2}O_2 \rightarrow H_2O$	• Transportation • Electric utility	• Up to 85% efficient in co-generation of electricity and heat • Impure H_2 as fuel	• Pt catalyst • Low current and power • Large size/weight
Molten carbonate (MSFC)	Liquid solution of lithium, sodium, and/or potassium carbonates, soaked in a matrix	600–1000	Anode: $H_2 + CO_3^{2-} \rightarrow H_2O + CO_2 + 2e^-$ Cathode: $\frac{1}{2}O_2 + CO_2 + 2e^- \rightarrow CO_3^{2-}$ Cell: $H_2 + \frac{1}{2}O_2 + CO_2 \rightarrow H_2O + CO_2$	• Electric utility	• High temperature advantages*	• High temperature enhances corrosion and breakdown of cell components
Solid oxide (SOFC)	Solid zirconium oxide to which a small amount of yttria is added	600–1000	Anode: $H_2 \rightarrow 2H^+ + 2e^-$ Cathode: $\frac{1}{2}O_2 + 2H^+ + 2e^- \rightarrow H_2O$ Cell: $H_2 + \frac{1}{2}O_2 \rightarrow H_2O$	• Electric utility	• High temperature advantages* • Solid electrolyte advantages (see PEM)	• High temperature enhances breakdown of cell components

*High temperature advantages include higher efficiency, and the flexibility to use more types of fuels and inexpensive catalysts as the reactions involving breaking of carbon-to-carbon bonds in larger hydrocarbon fuels occur much faster as the temperature is increased.

Source: S. Thomas and M. Zalbowitz (1999) *Fuel Cells: Green Power.* Los Alamos National Laboratory: Los Alamos, NM.

PEM fuel cells operate on hydrogen fuel at relatively low temperatures because proton conduction from anode to cathode is facilitated by a thin polymer electrolyte membrane. The polymer is a fluorocarbon with covalently attached sulfonate ($-SO_3^-$) groups. These sulfonate groups form a network of water-filled channels, through which protons pass easily (while anions are excluded by the negatively charged sulfonates). Because of the high proton conductivity, the PEM fuel cell starts up quickly, has a high power density, and the output can vary quickly to meet shifts in power demand. These characteristics make it suitable for automotive transport. The efficiency of PEM fuel-cell engines can approach 60 percent, compared with 25 percent for an internal combustion engine (see p. 124). And, of course, there are no emissions (except water) from a hydrogen fuel cell. Thus, PEM fuel cell technology is attractive for zero-emission vehicles.

However, there are difficulties with PEM fuel cell vehicles. The lack of a hydrogen distribution system and the difficulty of on-board hydrogen storage (see p. 111) are obvious issues. A less obvious one is that, because of the low temperatures, highly active platinum (Pt) catalysts must be used. These are sensitive to poisoning by a variety of chemicals, and particularly by carbon monoxide, which binds tightly to the Pt. This would not be a problem for hydrogen generated by electrolyzing water, but this method of preparation is expensive (because of the electricity cost). For the foreseeable future, hydrogen will continue to be produced by steam-reforming fossil fuels (see p. 41), particularly methane. CO is an inevitable byproduct of this process, and must be extracted to very low levels if the hydrogen is to be used in PEM fuel cells. This separation step adds to the fuel cost, and the possibility of catalyst poisoning puts vehicle owners at risk of having to replace the fuel cells prematurely. Current development efforts are aimed at raising the PEM operating temperature somewhat (a challenging task, since the polymer electrolyte dehydrates at high temperatures, losing its proton conductivity) in order to make the catalyst less sensitive to poisoning.

An option for fuel-cell vehicles is to generate the hydrogen fuel from methanol, via steam-reforming (see p. 41). Methanol is more easily reformed than are hydrocarbons, and the reaction can be run at sufficiently low temperature that it is possible to design cars and trucks with a compact, on-board reforming unit in order to run a fuel cell with methanol. This option would have the considerable near-term advantage that it would eliminate the need to transport and store hydrogen. A number of car manufacturers are touting methanol as the fuel of choice for fuel-cell vehicles. However, on-board reformers add complexity and cost to the vehicle, and it is unclear whether they can deal with the catalyst poisoning problem. Also, methanol has only half the energy density of gasoline, and the fuel tank would have to be doubled for the same driving range. Finally, methanol is highly corrosive; each methanol fueling station would require a new underground storage tank, costing about $50,000.

The alkaline fuel cell (AFC) is another relatively low-temperature alternative to the PEM fuel cell, and is being tested as the basis for taxicab engines. Instead of a polymer electrolyte, a potassium hydroxide solution provides high conductivity via the hydroxide ion. The reduction of O_2 at the cathode is faster in alkali, boosting performance. However, CO_2 must be removed from the air supply, because CO_2 reacts with hydroxide, forming carbonate. This separation step diminishes efficiency and adds to the cost.

Phosphoric acid fuel cells (PAFCs) use phosphoric acid for proton conduction. They operate at higher temperatures (175–200°C), and the catalyst can withstand relatively impure hydrogen streams. They are large and heavy, and are used as stationary sources for electricity generation. Fuel is provided by reforming natural gas (or even gas from landfills, see pp. 90–91) on site. Over 200 PAFCs, each generating 200 kW, are in operation worldwide, powering hospitals, nursing homes, hotels, office buildings, schools, airport terminals, and even a municipal waste dump. PAFCs generate electricity at more than 40 percent efficiency; also, the steam produced in the fuel-cell reaction can be used for cogeneration (see p. 109).

Fuel cells operating at still higher temperatures, 600–1000°C—the molten carbonate and solid oxide cells—are promising for electric utilities because of the conversion efficiencies available from the high-temperature heat, as well as the primary electrode reactions. The heat can be used to run the on-site reformer for gas-to-hydrogen conversion, and the hot exhaust gas can run a gas turbine generator, for a second stage of electricity production. As the names imply, molten carbonates or solid oxide ceramics are the fuel-cell elements separating the anode from the cathode, and providing efficient ion conduction at the high operating temperatures.

All of these fuel cells operate on hydrogen, because electrode reactions of carbon fuels have been too slow to provide acceptable currents. However, researchers have recently reported direct oxidation of hydrocarbon fuels in a modified version of the solid oxide fuel cell (see Figure 5.5). They replaced the currently favored nickel and zirconia anode with copper and ceria. Nickel is poisoned by hydrocarbons, which it turns into graphite, but copper avoids this side reaction because, unlike nickel, it is inert to C—H bond breaking. Ceria was selected because of its high activity for hydrocarbon oxidation, as well as its high ionic conductivity. It is unclear whether this fuel cell will have improved overall performance relative to the conventional solid oxide cell with reformer.

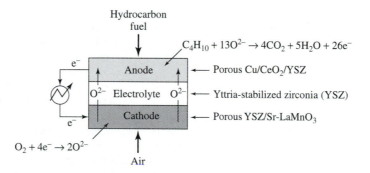

Figure 5.5   Direct oxidation of hydrocarbons in a fuel cell.   *Source:* M. Jacoby (2000). New electrode oxidizes hydrocarbons directly in fuel cell. *Chemical & Engineering News* 78(12):11–12.

FUNDAMENTALS 5.2: ENTROPY AND CHEMICAL ENERGY

Chemical reactions are associated with changes of entropy, as well as energy, and the entropy limits the amount of work that can be extracted, just as it does in the case of heat engines. We have already seen (see Fundamentals 2.1) that the reaction of H_2 with O_2

$$2H_2 + O_2 = 2H_2O \qquad (5.11)$$

releases 482 kJ of energy, as heat, per mole of O_2. This reaction also results in a decrease in entropy, because there are only two product molecules, but three reactant molecules. There are fewer ways to arrange two molecules (fewer *degrees of freedom*) than three molecules. Also, if the reaction is run at a temperature below the boiling point of water, the product is a liquid while the reactants are gases. Liquids take up less space (for a given number of molecules) than do gases, and therefore have lower entropy.

The decrease in entropy limits the amount of energy that can in principle be extracted from the reaction, consistent with the second law of thermodynamics. The amount of energy available for work is called the *free energy* change of the reaction, symbolized as ΔG. If the heat release is symbolized as ΔH (also called the *enthalpy* change of the reaction), then the expression for the free energy change is

$$\Delta G = \Delta H - T\Delta S \qquad (5.12)$$

ΔS, the entropy change of the reaction, is multiplied by the absolute temperature in order to express the energy deficit that it produces. (Recall from Fundamentals 5.1 that $S = Q/T$, so that $T \times S = Q$, the heat value of the entropy.) Since ΔG is the maximum chemical energy that can be converted to work, the maximum efficiency for energy conversion is

$$\Delta G/\Delta H = 1 - (T\Delta S)/\Delta H \qquad (5.13)$$

$T\Delta S$ increases with increasing temperature, and therefore the efficiency goes down as the temperature goes up, in contrast to the efficiency of a heat engine. However, ΔS can also change with temperature, especially when there is a phase change, as when water vaporizes. In addition, ΔH is not entirely independent of temperature, so the relationship of the efficiency to the temperature depends on the particulars of the reaction.

An additional, point of interest is that for some reactions, the entropy increases. In that case, the theoretical efficiency exceeds 100 percent.

The convention for writing energy changes is to subtract the energy of the reactants from the energy of the products. Thus if heat is released (an *exothermic* reaction) the reactants have higher energy than the products, and ΔH is *negative*. Likewise ΔG is negative, if the reaction proceeds spontaneously as written. And ΔS is negative if the entropy of the products is lower than the entropy of the reactants. For an exothermic reaction, a negative ΔS makes ΔG less negative (smaller) than ΔH, as it should.

For the H_2/O_2 reaction above, $\Delta H = -476$ kJ at 1,000 K, while $T\Delta S = -84$ kJ, giving $\Delta G = -392$ kJ, for a theoretical efficiency of 82 percent. At 300 K, the entropy change is greater, because of the condensation of liquid water, and $T\Delta S = -116$ kJ, but ΔH also increases, to -590 kJ/mol. ΔG is now -474 kJ, and the theoretical efficiency remains at 80 percent.

5.3   SPACE HEATING, COGENERATION

Much of our energy supply goes into direct heating of houses, offices, and factories. This heat is eventually dissipated to the environment, but the longer it is held in place, the less fuel is required. There are many opportunities for energy savings from improved insulation and windows, as well as improved delivery of the heat to needed areas. "Smart" buildings are being designed to tailor heat, light, and air-conditioning to the needs of the occupants through sensors and control systems.

In addition, energy savings are available through *cogeneration,* combining power and heat generation. The waste heat of a power plant can be put to use by piping it into the heating systems of buildings. Because heat cannot be transported over long distances, these plants must be in close proximity to the buildings being heated. Factories and large institutions are good candidates for cogeneration plants. Cogeneration is potentially attractive for high-density housing.

5.4   ELECTRICITY STORAGE: THE HYDROGEN ECONOMY

Electricity storage and conversion are critical issues for many power systems at both large and small scales. In many cases, either the source of electricity or the need for it is intermittent, producing a mismatch between supply and demand. Efficient electricity storage is then needed to optimize the system. For example, the main drawback to the production of direct solar or wind electricity is the intermittent nature of the energy source. Electricity is produced only when the sun shines or the wind blows. On the other side of the ledger, power companies must cope with large fluctuations in the daily demand for electricity. Extra generating capacity is required to meet peak demands during daylight hours, particularly during the summer when air conditioners draw a heavy load; the extra capacity is left idle much of the time. To some extent these fluctuations can be coordinated, since the solar flux also peaks during daylight hours and during the summer, but there is still an important need for efficient energy storage.

A few electricity companies have developed water pump storage, in which excess electricity is used to pump water uphill to reservoirs, and the water running back downhill can then be used to run turbines to meet peak demands. Because of the continually fluctuating water levels, these reservoirs create ecological problems, and are especially contentious if they displace multiple-use ponds. Other energy storage schemes under consideration include storage of compressed air in caverns, mechanical energy storage in flywheels, and direct electrical storage in large superconducting magnets.

On the small-scale end of electricity utilization, electricity storage is also a critical issue for the development of the electric car. At present, electric cars run on current from the lead-acid storage battery. In this battery (see Figure 5.6), electricity is stored in the chemical conversion of Pb^{2+} ions to metallic Pb at one electrode, and to PbO_2 at the other. The overall reaction,

$$2Pb^{2+} + 2H_2O = Pb + PbO_2 + 4H^+ \qquad (5.14)$$

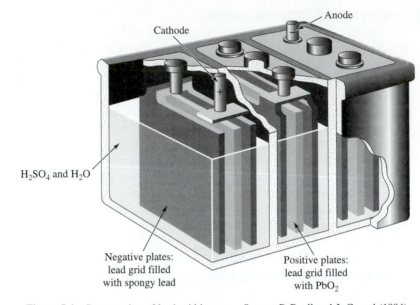

Figure 5.6   Cross-section of lead-acid battery.   *Source:* P. Buell and J. Gerard (1994). *Chemistry in Environmental Perspective* (Upper Saddle River, New Jersey: Prentice Hall). Copyright © 1994 by Pearson Education. Reprinted with permission.

is energetically uphill. When current is drawn, both electrode reactions are reversed, and the reaction is allowed to run downhill. The lead-acid battery performs these energy conversion steps very efficiently, and can be charged and discharged many times before it is worn out through competing chemical processes. It is used in all automotive vehicles as a portable store of electricity for auxiliary needs.

In electric vehicles, however, the need is not auxiliary; electricity is the source of power for locomotion itself. The lead-acid battery has two serious drawbacks as the main power source: 1) it is heavy and adds significantly to the weight of the vehicle, thereby lowering the efficiency and the driving range; and 2) it takes several hours to charge, making "refueling" an inconvenient operation. Consequently, much effort has been devoted to developing alternative batteries that give more power with less weight and easier charging. Nickel-cadmium batteries, already in use for many appliances, offer greater power, energy density, and longer life than lead-acid batteries, but are substantially more expensive. Promising alternatives that are currently under development include nickel-metal hydride, sodium-sulfur, lithium-iron disulfide, and lithium-polymer batteries. But the reliable lead-acid battery, itself undergoing continuing improvements, will be hard to beat, despite its disadvantages.

The hydrogen fuel cell offers an attractive alternative to any of the storage batteries. Indeed, it can itself be thought of as a battery, in which the storage medium is hydrogen and oxygen, instead of Pb and PbO_2. The same drive train can be utilized with a battery or a fuel cell. But the fuel cell is not required to run itself backward with external electricity in order to store its energy. A canister of hydrogen, or alternatively, a tank of methanol with an onboard reformer (see p. 106) is all that is required. Consequently, refueling is as fast as it is for gasoline-powered cars.

However, on-board storage of hydrogen is problematic because it must be contained at high pressures or low temperatures. The size and weight of the required tank constrains the design and efficiency of the vehicle (as does the size and weight of storage batteries required for an electric car). Hydrogen can also be adsorbed on metals, such as palladium, thereby reducing the volume, but not the weight of the storage container. Research is being directed at finding effective lightweight adsorbents for H_2; carbon fibers show promise in this regard. On-board storage of methanol is much easier, but problems of corrosiveness and of on-board reformer performance (see p. 106) must be dealt with.

Power companies could also use hydrogen for electricity storage, once appropriate fuel cells are commercialized. Electricity is convertible to hydrogen through the electrolysis of water (the reverse of the fuel-cell reaction), with efficiencies as high as 85 percent. Thus a combination of electrolysis and fuel cells, with tank storage of the hydrogen, could provide relatively efficient load-leveling capacity.

Moreover, hydrogen can be transported more efficiently than electricity. The cost of electrical transmission over long distances is high. Hydrogen transport by pipeline would be much more efficient and less expensive. The areas with the greatest amount of sunlight, where solar plants would be most efficient, are often far from centers of population. Transmission problems for ocean-based generating plants are even more severe. Instead of electricity, remote plants could generate hydrogen, which could be shipped or piped to urban centers.

These considerations have led to the concept of the hydrogen economy (see Figure 5.7) in which hydrogen gas would become the main energy currency. It would be consumed directly for electrical generation and heating, either by combustion or by fuel cells. For transportation, it could also be used directly, via fuel-cell electric vehicles, or it could

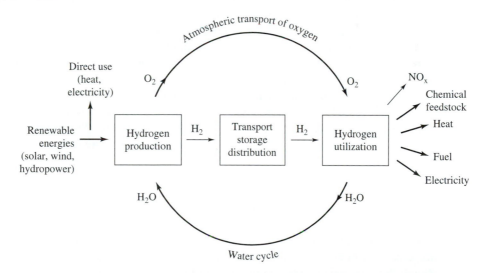

Figure 5.7   A solar-based hydrogen economy.   *Source:* J. M. Ogden and J. Nitsch (1993). Solar hydrogen. In *Renewable Energy, Sources for Fuels and Electricity*. T. B. Johansson et al., eds. (Washington, DC: Island Press). Copyright © 1993 by Island Press. Reprinted with permission.

be used to synthesize liquid fuels by chemistry similar to that applied for coal conversion to methanol and liquid hydrocarbons (see Strategies 2.1, p. 41).

The first country to test the concept is Iceland, whose current energy portfolio is 39 percent geothermal, 19 percent hydropower, 38 percent oil, and 4 percent coal. It has pledged to wean itself from fossil fuels by converting to a full hydrogen economy by 2030. The first step is to replace the Reykjavik city bus fleet (100 buses) with hydrogen fuel-cell buses, serviced by a hydrogen filling station. This would be followed by converting other city bus fleets, and introducing hydrogen-based fuel-cell cars for private transport. In parallel, the government is supporting the development of hydrogen fuel cells for powering fishing vessels, with the goal of fitting the entire fishing fleet with fuel-cell engines. Large international companies have invested in the Icelandic program in order to test the feasibility of fuel-cell transport.

Hydrogen is often thought to be a particularly hazardous substance. Ever since the Hindenburg disaster of 1937, when a dirigible filled with hydrogen caught fire and crashed, killing half the people on board, hydrogen has haunted the popular imagination. But hydrogen is not particularly more dangerous than natural gas or gasoline. The percentage of fuel in air that can sustain a fire (*lower flammability limit*) is only a little lower for hydrogen, 4 percent, than it is for methane, 5 percent, and is substantially higher than for gasoline, 1 percent. Since hydrogen is much lighter than air, it disperses rapidly if there is a leak. Methane does also, but gasoline fumes, being heavier than air, tend to accumulate in the vicinity of a leak and are more likely to catch fire. There is actually a history of hydrogen use in home heating, because, before the widespread availability of natural gas, many utilities supplied "town-gas," manufactured from coal or wastes, which was rich in hydrogen. Process industries have long used hydrogen, and hydrogen pipelines several hundred kilometers long have operated safely in Germany, England, and the United States.

Despite this experience in industrial settings, the widespread availability of hydrogen will no doubt be some time in coming, and the introduction of transfer facilities at vehicle filling stations will take much longer. One can imagine intermediate stages in the development of the hydrogen economy that would smooth the transition. There might initially be increased utilization of natural gas, as petroleum stocks dwindle and the demand for clean fuels increases. The methane could be burned directly, in homes, power plants, and new gas-powered vehicles, or it could be converted to methanol, which could serve as an intermediate fuel in fuel-cell vehicles with onboard reformers. Methanol from coal and biomass could add to the liquid fuel supply. Eventually, the pipelines used for natural gas could be converted to hydrogen transport, as solar hydrogen becomes more feasible and demand for hydrogen increases.

5.5   THE MATERIALS CONNECTION

The development of stronger, lighter, more durable materials has an enormous impact on the efficiency of energy utilization. In the case of cars, the substitution of strong but lightweight plastics and composites (mixtures of different structural elements such as glass or carbon fibers with resins, to increase strength) for metal body parts has decreased the weight required for the same carrying capacity. New materials have decreased the weight of many consumer and industrial products, thereby decreasing the energy costs for their transportation, and frequently for their manufacture as well. This trend is augmented by miniaturization of many products, also made possible by advanced materials. More durable materials also mean longer life for the products and a lower production rate. These trends are sometimes called the "dematerialization" of industrial societies, as advanced materials and better information make it possible to make do with less stuff.

In addition, materials able to withstand high temperatures have a direct impact on energy efficiency by improving the performance of heat engines. The most notable example is the steady improvement of jet engines for aircraft, through the development of alloys and ceramics that permit the engines to operate at higher temperatures. These engines have improved aircraft fuel economy substantially. The same technology, in the form of gas turbines, is now being introduced to improve power plant efficiencies (see pp. 101–102).

a. Materials properties: Paper versus plastics.   The choice of materials can also affect energy efficiency in more mundane ways. For example, some communities have banned disposable styrofoam containers because of concerns that, being nonbiodegradable, discarded styrofoam leaves unsightly trash and fills up overburdened landfills. Whether styrofoam is actually inferior to other disposable materials, principally paper, is a matter of debate, since biodegradation of paper can be quite slow, especially in sanitary landfills, and since styrofoam is probably more amenable to recycling, although not much of it is actually being recycled at the moment. But an important consideration was left out of the early debates on paper versus styrofoam, namely the environmental costs of producing the two materials. Table 5.2 compares energy and water use, as well as emissions, for manufacturing paper and styrofoam cups. When the values are compared on a *per cup* basis, styrofoam is found to have considerably lower environmental impacts (see problem 32, Part I). The main reason for the difference is that a styrofoam cup of a given capacity weighs less than a sixth as much, on average, as a paper cup. Styrofoam is stronger than paper, especially because, being a hydrocarbon material, it is not wet by aqueous liquids. In contrast, paper—which is made of cellulose, a molecule covered with hydroxyl groups (see Figure 4.11)—interacts with water via hydrogen bonds, and is gradually dissolved (see discussion of water and hydrogen bonding, pp. 266–269). Consequently, the paper cup requires more material to maintain its integrity while in use, and its production has a much larger impact on energy use and the environment.

b. Recycling.   The recyclability of paper versus polystyrene is just one of the many complex issues around recycling. For most people, recycling arises in the context of solid waste disposal. The more the trash is recycled, the less there is to dump in landfills, which are rapidly being filled, and the less pressure there is to build incinerators as an alternative disposal method. In fact, there have been significant gains in recycling in the U.S. over the last 40 years, but this positive trend has been more than offset by the increase in trash generated. In 1960, per capita trash generation was 444 kg, of which 416 kg was landfilled or incinerated, corresponding to a recycling rate of 6.3 percent. By 1998, per capita trash generation had increased to 739 kg, and 530 kg ended up as municipal waste, corresponding to a 28 percent recycling rate. It is unclear whether increased rates of recycling in

TABLE 5.2   RAW MATERIALS, UTILITY, AND ENVIRONMENTAL SUMMARY FOR HOT
DRINK CONTAINERS

Item	Paper cup*	Polyfoam cup[†]
Per cup		
Raw materials		
Wood and bark	25 to 27 g	0 g
Petroleum fractions	1.5 to 2.9 g	3.4 g
Other chemicals	1.1 to 1.7 g	0.07 to 0.12 g
Finished weight	10.1 g	1.5 g
Per metric ton of material		
Utilities		
Steam	9,000 to 12,000 kg	5,500 to 7,000 kg
Power	980 kWh	260 to 300 kWh
Cooling water	50 m^3	130 to 140 m^3
Water effluent		
Volume	50 to 190 m^3	1 to 4 m^3
Suspended solids	4 to 16 kg	0.4 to 0.6 kg
BOD	2 to 20 kg	0.2 kg
Organochlorines	2 to 4 kg	0 kg
Metal salts	40 to 80 kg	10 to 20 kg
Air emissions		
Chlorine	0.2 kg	0 kg
Chlorine dioxide	0.2 kg	0 kg
Reduced sulfides	1 to 2 kg	0 kg
Particulates	2 to 3 kg	0.3 to 0.5 kg
Chlorofluorocarbons	0	0[‡]
Pentane	0 kg	35 to 50 kg
Sulfur dioxide	~10 kg	3 to 4 kg
Recycle potential		
To primary user	Possible. Washing can destroy.	Easy. Negligible water uptake.
After use	Possible. Hot melt adhesive or coating difficulties.	Good. Resin reuse in other applications.
Ultimate disposal		
Proper incineration	Clean	Clean
Heat recovery	20 MJ/kg	40 MJ/kg
Mass to landfill	10.1 g/cup	1.5 g/cup
Biodegradable	Yes, BOD to leachate, methane to air.	No. Essentially inert.

*Uncoated fully bleached kraft paper cup.

[†]Molded polystyrene foam bead (seamless) cup.

[‡]Many producers of foamable beads have never used CFCs.

Source: Updated and adapted by M. B. Hocking, from original article in M. B. Hocking (1991). Paper versus polystyrene: A complex choice. *Science* 251:504–505.

the future can catch up with the ever-increasing amounts of trash generated, for there are many barriers to recycling—political, economic, and technical.

The technical barriers can be thought of in terms of the second law of thermodynamics. When materials are mixed together in products, and then mixed again when discarded products are mingled in trash, entropy is increased. Unmixing the materials requires a decrease in the entropy of the materials, and this requires the input of energy. The trash has to be collected and sorted (the degree of cooperation of the populace in presorting the trash is a key variable), and the materials in the sorted products may have to be separated mechanically or chemically. The difficulty depends on the product and on the material. For example, lead-acid batteries are recycled at a high rate. They are easy to collect, and the lead is readily extracted. Aluminum cans are also recycled at a high rate. They are easily separated from trash and contain little except aluminum; they can be said to have relatively low entropy.

Industries generate large quantities of processing wastes containing metals in dilute concentrations. Often these wastes are classified as "hazardous," and disposed of at high costs under stringent prescribed protocols. In a detailed analysis of industrial hazardous waste streams, researchers compared concentrations of the metals in the waste to those found in the ores from which they are derived (see Figure 5.8), and concluded that large amounts of metal resources that are currently discarded could be recycled at a profit.

Plastic consumer products are another matter. Although pure plastics can be readily recycled, many different plastics are mixed in trash; often a given product contains more than one plastic. Separating them completely may be prohibitively costly, and often they

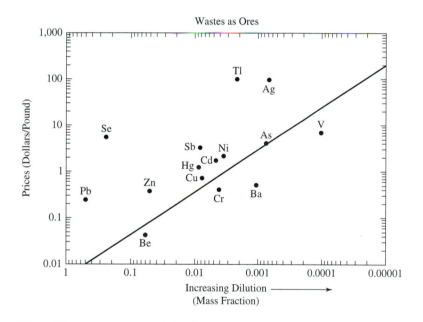

Figure 5.8   T. K. Sherwood empirically identified a relationship between the selling prices of materials and their dilution (or degree of distribution in the initial matrix from which they are separated). The diagonal line denotes this empirically observed linear relationship. The data points indicate the minimum concentrations of metal wastes typically recycled as a function of metal price. Points lying above the line indicate the existence of metals in wastes typically not recycled, even though their concentration exceeds those found in virgin ores.   *Source:* D. T. Allen and N. Behamanesh (1994). Wastes as raw materials. In *The Greening of Industrial Ecosystems.* B. R. Allenby and D. J. Richards, eds. (Washington, DC: National Academy Press).

cannot be processed together. For example, a little polyvinylchloride, the ingredient of plastic films, can ruin the recyclability of polyesters, the ingredient of soft-drink bottles. Another recycling issue is the quality of the recycled product. Paper is recycled in substantial quantities, but the cellulose fibers are degraded in the process and lose strength; they can be recycled no more than four times, before dissolving completely.

Despite these problems, recycling has a significant impact on energy efficiency. Although energy is required to restore materials that have been dispersed, this energy is likely to be substantially less than the energy required to produce the materials in the first place. The bar graph in Figure 5.9 shows the amount of energy required to produce a metric ton of steel, paper, and aluminum from primary materials, compared with the energy required for production from recycled materials. We see that aluminum production from ore is particularly energy-intensive. Recycled aluminum requires only 5 percent as much energy to process as primary aluminum ore. Even so, enough aluminum is thrown away in the United States to rebuild its commercial aircraft fleet every three months. The savings from using steel scrap and paper waste are not as high—52 percent and 70 percent, respectively. Nevertheless, the potential for energy conservation is substantial when we consider the volume of these materials produced annually. The production of steel, aluminum, and paper consumes more than 20 percent of the total industrial energy.

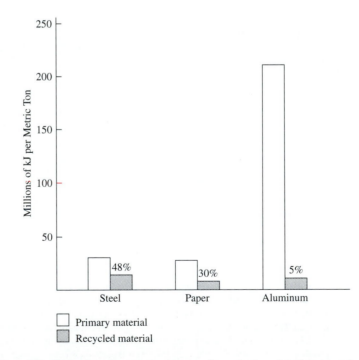

Figure 5.9 Comparison of energy requirements for production of steel, paper, and aluminum from primary and recycled materials.

FUNDAMENTALS 5.3: ENERGY COST OF EXTRACTION: Al VERSUS Fe

Why are the energy savings from recycling so much greater for aluminum than for steel (see Figure 5.9)?

Many factors go into the energy accounting for recycled versus primary materials, but in this case the outstanding difference is the much greater energy requirement for refining aluminum versus iron ore. Both metals occur in the Earth's crust as the oxides, Al_2O_3 and Fe_2O_3, which must be reduced in order to recover the elemental metal. The energy required for this step is much higher for Al_2O_3 than for Fe_2O_3. This can be seen from a comparison of the *enthalpies of formation,* which are $-1,670$ and -822 kJ/mol. The enthalpy of formation is minus the heat released when the elements combine to form a compound, in this case:

$$2M + \tfrac{3}{2}O_2 = M_2O_3 \quad (M = Al \text{ or } Fe)$$
$$(5.15)$$

Almost twice as much heat is released by Al than by Fe in forming the oxide. A correspondingly larger amount of energy is required to reduce the oxide to the metal. This prodigious energy requirement is why aluminum refineries are generally located close to a plentiful source of energy, such as a hydroelectric power plant. It is also why the energy savings are so large in recycling aluminum.

Why is Al_2O_3 so much more stable than Fe_2O_3, relative to the metals? The principal reason is simply that the Al^{3+} ion is much smaller than the Fe^{3+} ion (0.45 versus 0.64 Å), and has a larger electrostatic attraction to the oxide ions. Al is in the second row of the periodic table, whereas Fe is in the third row. Fe has an extra shell of electrons to shield its valence electrons from the nucleus.

c. Dematerialization.   The development of new materials, better suited to their task, means that less material needs to be produced in the first place. The example of paper versus plastics, above, is an illustration, but the trend is pervasive throughout the world of industrial products. Stronger lightweight materials are evident everywhere. At the same time, many items are getting smaller. Computers are the most dramatic example. Currently available laptop computers have much more computing power than a roomful of computing machinery did not many years ago.

These trends mean that less material needs to be extracted from the earth, and less of it goes into manufacturing, with concomitant savings in energy.

Countering these trends, however, is the desire of consumers to have more things, many of them bigger than before. Particularly in the United States, new houses have steadily become bigger, and have more furnishings and appliances. Cars are more numerous, and are also getting bigger, especially with the popularity of sport utility vehicles.

It appears that these countervailing trends have left overall materials use at a high level. Keeping an inventory of materials is difficult because of the complexities of the world economy and the inadequacy of data, but an analysis of the materials flows in four industrial countries (see Figure 5.10) suggests that the totals have not changed much over the 1975–1994 period. However, the totals tend to be dominated by the large quantities of

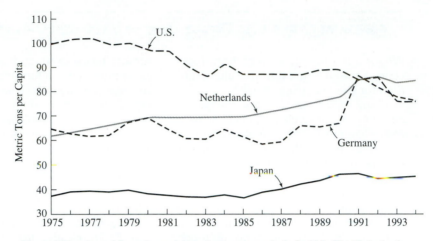

Figure 5.10   Annual flow of material requirement per capita in the U.S., The Netherlands, Germany, and Japan over the period from 1975 to 1994.   *Source:* A. Adriaanse et al. (1997). *Resource Flows: The Material Basis of Industrial Economies.* (Washington, DC: World Resources Institute).

earth materials that are moved about. For example, the decline in U.S. flows are attributed largely to reductions in soil erosion and to the completion of the federal interstate highway system. Nevertheless, it is encouraging that total materials use does not increase in proportion to economic growth in these countries. When the flows are divided by gross domestic product, the ratio is seen to decline steadily over time (see Figure 5.11). (The upturn for Germany and the Netherlands in the 1990s results from the impact on Western Europe of German reunification.)

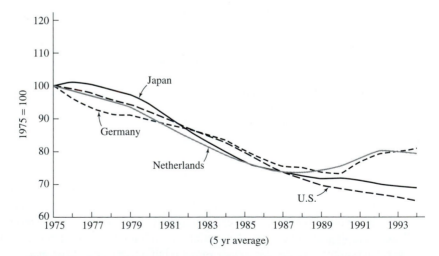

Figure 5.11   Materials intensity as measured by total materials use divided by gross domestic product for the U.S., The Netherlands, Germany, and Japan over the period from 1975 to 1994.   *Source:* A. Adriaanse et al. (1997). *Resource Flows: The Material Basis of Industrial Economies.* (Washington, DC: World Resources Institute).

5.6   SYSTEMS EFFICIENCY

To better understand the possibilities for increased energy efficiency, we need a more comprehensive view of the way energy is utilized in society. Figure 5.12 is a diagram of how energy actually flowed through the U.S. economy in 2000. On the left side are the inputs from coal, petroleum, natural gas, nuclear, and renewables, while on the right side, the end uses are broken down into residential, commercial and industrial, and transportation categories. The units are EJ (10^{15} kJ). In 2000, the United States consumed 103.9 EJ but produced only 75.9 EJ from its own resources; the balance was made up mainly of petroleum imports, which accounted for 63 percent of the petroleum used (up from 46 percent in 1990). Electric utilities used 33.2 EJ (see Figure 5.13), nearly one-third of total consumption, supplying 12.2 EJ of electricity (including purchase of 2.1 EJ from nonutility power producers), which was split nearly equally among industrial, residential, and commercial uses.

About 66 percent of the energy consumed by the electric utilities is "lost" as waste heat, due in part to the inherent inefficiency of heat engines, as discussed above. But this is far from the only loss in the system. Much of the end-use energy is also lost, in the sense that it does not accomplish its intended use before ending up as waste heat. A liter of heating oil, for example, produces exactly its heating value, wherever it is burned, but the effect on the house being heated depends on the efficiency of the furnace and heating system, on the size

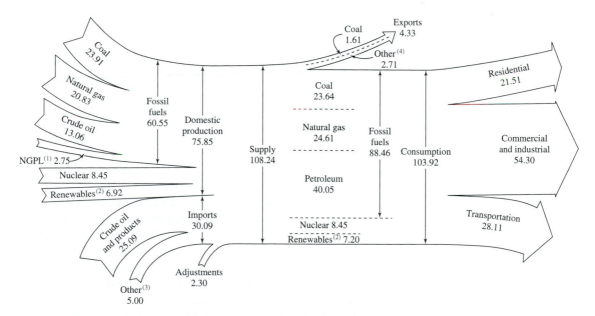

(1) Natural gas plant liquids.
(2) Conventional hydroelectric power, wood, waste, ethanol blended into motor gasoline.
(3) Natural gas, coal, coal coke, and electricity.
(4) Crude oil, petroleum products, natural gas, electricity, and coal coke.

Figure 5.12   Flow of energy through the U.S. economy in 2000 (in units of EJ = 10^{15} kJ). Energy Information Administration (2000). *Annual Energy Review 1999*. (Washington, DC: U.S. Department of Energy).

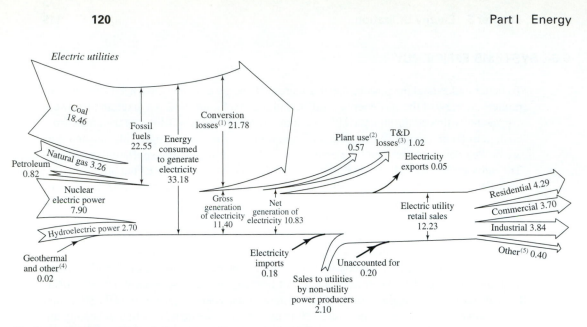

(1)   Approximately two-thirds of all energy used to generate electricity.
(2)   The electric energy used in the operation of power plants, estimated as 5 percent of gross generation.
(3)   Transmission and distribution losses are estimated as 9 percent of gross generation of electricity.
(4)   Wood, waste, wind, and solar power used to generate electricity.
(5)   Public street and highway lighting, other sales to public authorities, sales to railroads and railways,
        and interdepartmental sales.

Figure 5.13   Flow of energy from electricity through the U.S. economy in 2000 (in units of $EJ = 10^{15}$ kJ). Energy Information Administration (2000). *Annual Energy Review 1999.* (Washington, DC: U.S. Department of Energy).

of the house, and on how well it is insulated. Two houses of similar size, side by side, can require quite different amounts of oil or gas to achieve the same indoor temperature. Likewise, the amount of electricity used for heat, light, and appliances depends on the efficiency of electricity conversion to the intended use. For example, compact fluorescent light bulbs are available, which provide light equivalent to a traditional tungsten bulb with 25 percent of the electric current requirement, and sensors are available that turn lights off when no one is in a room. Clearly, the magnitude of the end-use energy requirement for residential purposes could be less than it now appears to be, if the residents adopted energy-saving measures. Much has, in fact, been accomplished in this direction, with improved building codes and enhanced efficiencies in appliance designs, for example, but there is still considerable room for improvement. Likewise, there are many opportunities in the industrial and transportation sectors to improve energy efficiency and reduce the energy utilization rate.

All of the energy flowing through the economy eventually ends up as waste heat. The question is what fraction of it actually accomplishes human ends on its way to its entropic destination. This fraction determines just how much energy we actually need.

a. Transportation.   Transportation is an especially important energy sector, not only because of its high rate of energy consumption (as seen in Figure 5.12, it accounts for more than one-fourth of total energy consumption), but also because of the international

TABLE 5.3   EFFICIENCIES OF FREIGHT AND PASSENGER TRANSPORT IN THE U.S.

Mode	Passenger (1998)	
	PmiT (PkmT)* (millions)	Energy intensity kJ/PmiT (PkmT)
City bus	20,602 (33,154)	2,800 (1,740)
School bus	82,900 (133,408)	920 (572)
Intercity bus	31,700 (51,014)	752 (467)
Commuter train	8,247 (13,269)	2,993 (1,860)
Intercity rail[†]	5,325 (8,569)	2,575 (1,600)
Auto (city traffic)	1,360,330 (2,188,771)	4,380 (2,722)
Auto (highway)	1,112,998 (1,790,814)	3,291 (2,045)
Light truck (city traffic)	551,182 (886,852)	6,646 (4,131)
Light truck (highway)	450,967 (725,606)	4,474 (2,781)
Airplane	464,395 (747,337)	4,219 (2,622)

Mode	Freight (1998)	
	T-km* (millions)	Energy intensity kJ/T-km
Truck	1,475,713	2,195
Rail	1,978,464	268
Ship	966,806	320
Airplane	19,460	13,580[‡]
Pipeline[§]	890,940	188

*PmiT signifies "passenger miles traveled." PkmT signifies "passenger kilometers traveled." T-km signifies "metric ton-kilometers."

[†]Amtrak only

[‡]Intensity is for 1990

[§]For crude oil only

Source: S. C. Davis (2000) *Transportation Energy Data Book.* ORNL-6959, Edition 20 of ORNL-5198. Oak Ridge National Laboratory: Oak Ridge, Tennessee.

economics and politics of oil. About half of the world's oil production goes into transportation, while about 40 percent is used in space heating and industrial processing (including production of petrochemicals), and about 10 percent in electricity production.

Table 5.3 lists the common forms of transport in the U.S., and their energy intensities (the amount of energy required to transport a passenger or a ton of freight, for a given distance). For freight, there are large disparities in the energy intensity. Transport by ship, rail, or pipeline takes far less energy per ton of material per kilometer of travel than does transport by truck, which in turn takes far less energy than transport by airplane. Of course, these different transport modes are appropriate to different kinds of goods of different value and perishability; they are not freely interchangeable. Nevertheless, the growth of trucking, and especially of air freight, is a significant factor in the transportation energy demand.

The data for passenger transport reveal important variations. School and intercity buses are the most efficient people movers, because they have high load factors (passengers per vehicle), 23 and 19, on average. City buses suffer in comparison because of low load factors (about 9). Commuter and intercity trains also have high load factors, 35 and 19, respectively, but efficiency is lowered because of higher energy requirements, due mainly to electric propulsion, especially for commuter trains. Airplanes require even more energy, of course, but because of high load factors, their average energy intensity is only a third higher than commuter trains.

Automobiles in city driving have nearly the energy intensity of airplanes, because most of them carry a single passenger. The average gasoline efficiency of new U.S. cars is 28.1 miles per gallon (mpg) (11.9 km/L), about the same as it was in 1990 (27.8 mpg). There were major efficiency improvements in the 1970s (see Figure 5.14), but these have since leveled off. The same is true of light trucks, a category that includes vans and sport utility vehicles (SUVs). Their mileage remains about 40 percent that of cars, and their energy intensity is about 50 percent higher (see Table 5.3). Because of the popularity of SUVs, the average mileage of the U.S. automotive passenger fleet has actually declined since 1985. This popularity is reflected in the dramatic increase in light truck passenger miles revealed by the trends plotted in Figure 5.15. The total miles traveled per capita increased from 11,200 (18,000 km) in 1970 to 16,200 miles (26,000 km) in 1998, and the share belonging to light trucks increased by over a factor of five, from 5.5 to 29.8 percent. This increase dwarfed even the rising share of air travel, which went from 5.5 percent to 13.7 percent, and was at the expense of cars, whose share fell from 85.7 percent to 53.2 percent. Mass transit, meanwhile, hovered around 3 percent for the entire period.

On a global basis the automobile dominates all other sources of transport. The global car fleet has increased nearly ten-fold over the period from 1950 (when about 55 million cars were in use) to 2000, and it has outstripped the growth in global population by a factor of four (see Figure 5.16). The relatively immature transportation sectors in much of the

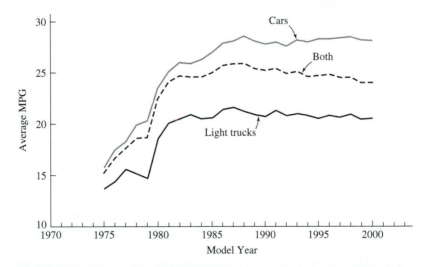

Figure 5.14 Trends in fuel economy of light-duty vehicle transportation in the U.S., 1975–1999. *Source:* U.S. Environmental Protection Agency (1999). *Light-Duty Automotive Technology and Fuel Economy Trends Through 1999.* EPA420-R-99-018. (http://www.epa.gov/OMS/cert/mpg/fetrends/fetrnd99.pdf)

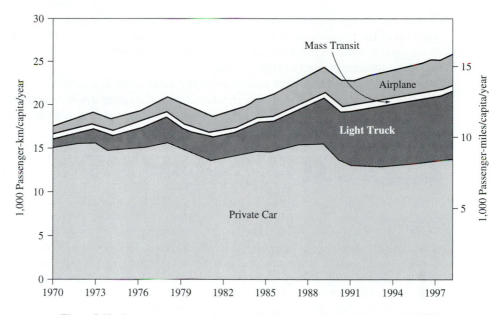

Figure 5.15   Passenger transportation per capita by mode.   *Source:* Data compiled from S. C. Davis (2000). *Transportation Energy Data Book.* Oak Ridge National Laboratory, ORNL-6959, Edition 20 of ORNL-5198.

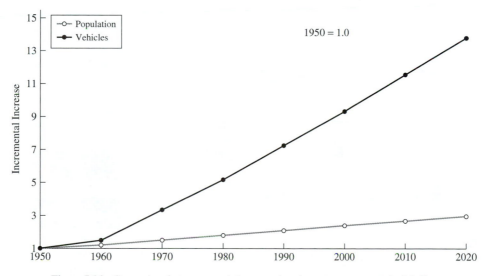

Figure 5.16   Comparison between population growth and passenger car growth globally. *Sources:* Population data are from *World Resources* volumes 1988–89, 1992–93, and 2000–01. (Washington, DC: World Resources Institute). Vehicle data are from U.S. Department of Energy (2000) *World Vehicle Population,* 1960–2020, Fact of the Week, Fact #146, October, 2000. (http://www.ott.doe.gov/facts/archives/fotw146.html)

developing world are expected to expand rapidly as development proceeds. The U.S. Department of Energy projects that energy use for transportation will more than double in the developing world between 1997 and 2020, resulting in a global transport energy increase of about 14-fold compared to 1950.

Consequently, the energy efficiency of the automobile is a critical issue of global dimensions. In addition, system-wide effects on energy consumption are important. For example, an effective mass-transport system might significantly reduce traffic congestion, thereby improving the energy efficiency of all forms of transport in the region.

STRATEGIES 5.1 | ENERGY EFFICIENCY IN AUTOMOBILES

In the conventional automobile, only a small percentage of the energy present in the gasoline actually serves to move the car (see Figure 5.17). The fraction of fuel energy delivered to the driveline is only 25 percent on highways, and 18 percent in urban driving, while the energy delivered to the wheels is even lower, 20 percent and 13 percent respectively. The rest of the energy either becomes waste heat (from the thermodynamic cycle of the engine), is used to overcome frictional losses (such as aerodynamic drag or rolling resistance), or powers auxiliary equipment.

The need for improved automobile fuel efficiency has induced new development efforts. In the U.S., an industry-government partnership, the Partnership for a New Generation of Vehicles (PNGV), was formed in 1993 with a goal of fuel economy up to 80 mpg (34 km/L). To meet this challenge, several auto manufacturers, including Honda, Toyota, and Ford, have developed the hybrid electric vehicle (HEV). It combines the internal combustion engine with the electric motor and battery storage of an electric vehicle. This combination offers extended range and rapid refueling with a significant fraction of the energy and environmental benefits of an electric vehicle. HEVs use regenerative braking, strong light-weight materials, and aerodynamic body shapes to minimize energy

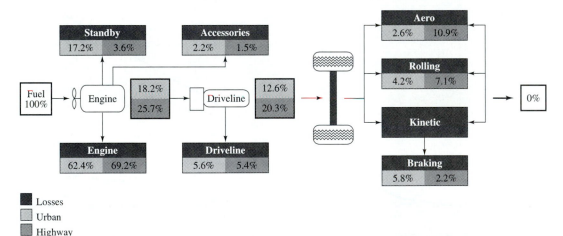

Figure 5.17 Energy losses in automobile transport point to strategies for improving fuel efficiencies in automobiles. *Source:* D. L. Illman (1994). Auto-makers move toward new generation of "greener" vehicles. *Chemical & Engineering News* 72(31):8–16. Reprinted with permission from C&EN. Copyright (1994) American Chemical Society.

losses. The HEV is about twice as efficient as a conventional vehicle. The two cars currently on the U.S. market, the Honda Insight and the Toyota Prius, can go 700 and 500 miles, respectively, on a tank of gas.

There are several candidates for the internal combustion engine of a HEV, including the compression-ignition direct-injection (CIDI) diesel, the turbocharged direct-injection (TDI) diesel, and the spark-ignition direct-injection (SIDI) engine. The CIDI engine has the highest thermal efficiency (40 percent) of any internal combustion engine. The TDI is the turbocharged version of the CIDI engine, popular in Europe. The SIDI engine is the standard spark-ignition gasoline engine also equipped with fuel injection. It is less efficient than the CIDI engine, but has the advantage of burning many alternative fuels.

The electric motor draws energy from a battery pack (currently nickel-metal hydride is the battery of choice) to power the vehicle below a certain minimum speed, and to assist the engine with additional power if extra torque is required. The battery pack is continually recharged by the engine and by the regenerative braking system, which reclaims a portion of the energy otherwise lost to braking. When the driver brakes, the electric motor becomes a generator, using the kinetic energy of the vehicle to generate electricity. Traditional friction brakes are also necessary, requiring computerized electronic controls for blending the two braking mechanisms. To extend the life of the battery pack, high temperatures are avoided with a sophisticated system of heat distribution.

Minimizing the vehicle weight is important for fuel economy. Since 1975, the weight of a typical family sedan has dropped from 4,000 pounds (1,816 kg) to 3,300 pounds (1,498 kg). To achieve the PNGV goal of 80 mpg, re-searchers are working to reduce overall vehicle weight to 2,000 pounds (908 kg), by reducing the mass of both the outer body and chassis by half, and the power train by 10 percent. In this aspect, the CIDI engine is at a disadvantage compared to the SIDI engine, as it is inherently heavier in order to accommodate significantly higher firing pressures. Weight reduction will likely lead to further adoption of strong lightweight advanced materials, e.g., titanium and carbon fiber composites. Because these materials are far more expensive than steel, measures for lowering their costs through advances in production efficiency and recyclability are important considerations.

The HEV is not a "zero emission vehicle." Diesel-based designs emit nitrogen oxides (NO_x) and particulate matter (PM) (see p. 222–225). Advances have been made in lowering emissions, and the CIDI engine certainly burns cleaner than older diesel designs. But as federal and California emission standards become increasingly stringent, the ability of HEVs to meet those standards comes into question. As a consequence, alternative fuels for CIDI engines may be required. Dimethyl ether and methanol have received some attention in this regard.

The ultimate fuel-efficient, zero-emission vehicle is one powered by a fuel cell. As discussed on pp. 103–104, a PEM fuel-cell engine can approach a thermal efficiency of 60 percent, corresponding to a fuel efficiency of 90 mpg (38.3 km/L). The emission product is H_2O when operating on pure hydrogen. Where hydrogen is provided by a reformer that converts methane or methanol to hydrogen, the reformer releases CO_2. But because there are no high-temperature steps in the reforming process, there are no emissions of NO_x, PM, or smog-forming organic gases.

---

b. Industrial ecology.   A systematic way of tracking and controlling energy and materials flows is emerging through the new discipline of *Industrial Ecology (IE)*. IE strives for integrated assessment of the connections among these flows and impacts on the environment. Figure 5.18 provides a diagram of these connections. The assessment

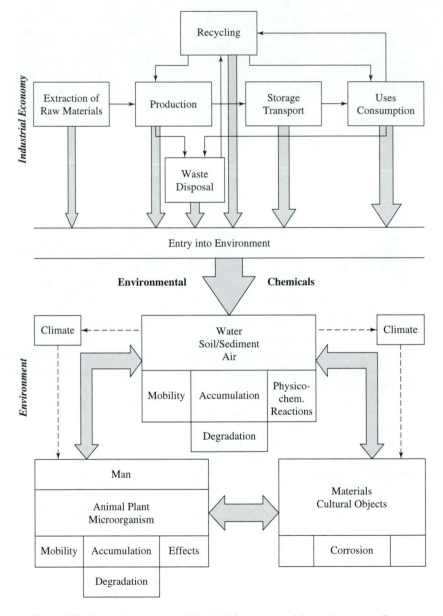

Figure 5.18   Interactions between the industrial economy and the environment.   *Source:*
W. M. Stigliani (1993). The integral river basin approach to assess the impact of multiple
contamination sources exemplified by the River Rhine. In *Integrated Soil and Sediment
Research: A Basis for Proper Protection*. H. J. P. Eijsackers and T. Hamers, eds.
(Dordrecht, The Netherlands: Kluwer Academic Publishers). Reprinted with permission.
Copyright (1993) Kluwer Academic Publishers.

includes "life cycle" analysis of the full impact of various products, from extraction of raw materials, through transport and manufacture, to use and disposal. It also looks for possibilities to recycle materials, turning wastes into raw materials and thereby closing ecological loops.

One application of this approach is the concept of ecoparks, in which industrial facilities are clustered to minimize energy and material wastes. An example is the site in Kalundborg, Denmark, 75 miles west of Copenhagen, housing Denmark's largest power station (1,500 MW, coal-fired), its largest oil refinery (3.2 million tons/yr capacity), the Gyproc plasterboard factory (14 million m^2/yr of gypsum wallboard), and a major pharmaceuticals plant (Novo Nordisk, $2 billion annual sales) (see Figure 5.19).

Waste heat from the power station generates steam for district heating in Kalundborg (replacing 3,500 home furnaces), and to supply the oil refinery and pharmaceutical plant with process steam. The power plant is cooled with salt water from a nearby fiord (reducing withdrawals of fresh water from Lake Tisso). The warmed salt water is sent to a fish farm where it provides heat and water to 57 ponds. The oil refinery, which had previously flared most of its gaseous by-product, now desulfurizes the gas and sends it as fuel to Gyproc and to the power station (reducing its coal consumption).

In addition to energy savings, the ecopark has minimized generation of material wastes. Sludge from the pharmaceutical company and from the fish farm's water treatment plant provides fertilizer to nearby farms (over 1 million tons per year). Fly ash from the power plant's coal is used by a cement factory. SO_2 is removed from the power plant's exhaust by extracting it into calcium carbonate (see p. 233). The calcium sulfate (gypsum) waste product is sent to the wallboard plant, providing it with two-thirds of its raw materials. The oil refinery's desulfurization operation produces elemental sulfur, which is sent to

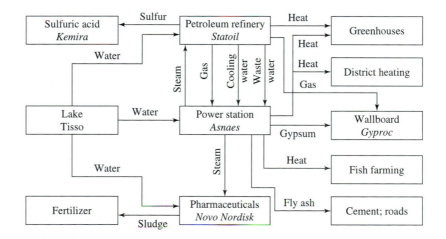

Figure 5.19   A schematic diagram of the industrial ecopark located in Kalundborg, Denmark. The figure shows the industrial firms that occupy the park, the material, and energy flows between them, and the nature and fate of outgoing material and energy streams. Adapted from B. R. Allenby and T. E. Graedel (1994). *Defining the Environmentally Responsible Facility* (Murray Hill, New Jersey: AT&T).

a sulfuric acid producer. Surplus yeast from insulin production at the pharmaceutical company goes to farmers to feed hogs.

This innovative symbiosis among the complementary industries in the park has generated cost savings that far exceed the original $60 million investment in infrastructure to transport energy and materials. In the process, there have been large reductions in pollution to the air, water, and soils, and significant conservation of energy and water. In this sense the park mimics the natural biosphere.

c. Green chemistry.   Awareness of the importance of materials to energy conservation and environmental protection has motivated the development of a "green chemistry" perspective among industrial and academic chemists.*

Industrial production is being examined on a wide front, with a view to reducing environmental impacts by appropriate choice of materials, and their chemical transformations. Some of the goals are to:

- Minimize byproducts in chemical transformations (improved "atom economy") through redesign of reaction sequences
- Improve energy efficiency by developing low pressure and temperature processes, often via improved catalysts
- Develop products that are less toxic and degrade more rapidly in the environment than current products
- Reduce requirements for hazardous or environmentally persistent solvents and extractants in chemical processes
- Develop processes based on renewable (plant-derived) rather than non-renewable (fossil carbon-derived) raw materials
- Develop processes that are less prone to chemical releases, explosions, and fires
- Develop methods to monitor processes continuously for improved control

Some idea of the range of contributions can be seen at the web site http://www. epa.gov/greenchemistry, which lists recipients of the U.S. Presidential Green Chemistry Challenge Awards since 1996. For example, the Argonne National Laboratory won an award for a new selective membrane process that lowers the energy and cost of producing lactate esters from carbohydrate feedstocks; lactate esters are nontoxic and biodegradable liquids, with good solvent properties, which are promising alternatives to many toxic solvents currently used by industry and consumers. Liquid CO_2 is another promising alternative solvent, but most materials have low solubility in CO_2; chemist Joseph DeSimone won an award for developing a series of surfactant polymers that greatly increase the CO_2 solubility of many molecules. In a number of applications, organic solvents have been replaced by water, requiring redesign of materials to overcome problems associated with hydrophobicity. This has been especially important for paints and coatings, since the solvents are

---

*P. T. Anastas and J. C. Warner (1998). *Green Chemistry: Theory and Practice* (New York: Oxford University Press).

vented directly to the atmosphere. The Bayer Corporation won an award for developing a waterborne technology for two-component polyurethane coatings.

Another award recognized chemist Terry Collins for developing a class of tetra-amido metal complexes that activate hydrogen peroxide to break down lignin (see p. 24) and permit bleaching of paper without chlorine (see p. 426). (The chemistry is that of transition metal-catalyzed hydroxyl radical production, see p. 192.) And Biofine Inc., won an award for developing a high-temperature dilute-acid process to convert cellulosic biomass (see p. 89), first to soluble sugars, and then to levulinic acid, which can serve as a building block for a number of chemical products. One of these, methyltetrahydrofuran, could serve as an oxygenate fuel additive, in place of MTBE (see p. 239). Unlike the other alternative, ethanol, methyltetrahydrofuran is miscible only in gasoline, and can be blended at the refinery rather than later in the distribution process.* Thus, methyltetrahydrofuran from sources such as wood chips, agricultural residues, municipal waste, and paper mill sludge, might someday compete with ethanol from corn as a fuel additive.

5.7   ENERGY AND SOCIETY

At this point we pause to ask how much energy humans actually need. The general assumption has been that increasing energy consumption means an increased standard of living. To evaluate this assumption, we can examine Figure 5.20, a plot of per capita energy consumption against per capita gross domestic product in 1999 for many countries. There is a rough correlation between the two measures, with poor countries clustered near the bottom on both scales. At higher GDP, however, the graph fans out. The United States uses twice as much energy per capita as does Japan, Switzerland, or Denmark, although all four have the same high GDP per capita, ~$35,000. (Canada's and Norway's even higher per capita energy use is attributable to the large size of their oil and gas extraction industry, relative to their small population.)

For highly developed countries, it is apparent that more GDP does not necessarily require more energy. This point is illustrated also by the history of Japanese energy use, shown as the solid curve in the figure. From 1960 to 1970, Japanese energy utilization increased steadily in proportion to the GDP. But after 1973, energy use leveled off while GDP kept on growing. Thus the Japanese energy *intensity,* defined as energy consumption per unit of GNP, has been decreasing since 1973. This has been the experience of other economically advanced countries as well. Even the U.S. energy intensity was lower in 1999 than it was in 1973.

The discussion in the preceding sections suggests that there are many opportunities for decreasing energy intensity further. Just how far can one expect to go in this direction? No one knows the answer to that question, but it is possible to sketch some energy scenarios for a variety of economic and technological assumptions. A new element in scenario analysis is the concern about global climate change caused by the burning of carbon-based fuels, and

---

*Ethanol is miscible in both gasoline and water, and will preferentially dissolve in small amounts of water in the gasoline distribution system. It is thus usually blended with gasoline at local storage terminals.

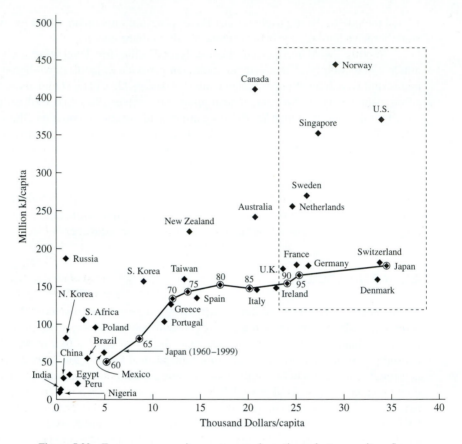

Figure 5.20   Energy use per capita versus gross domestic product per capita.   *Source:*
Energy Information Agency, U.S. Department of Energy (1999). *International tables for
GDP, Population, Energy Consumption*. (Washington, DC: Department of Energy).
(http://www.eia.doe.gov/emeu/international/contents.html)

international actions attempting to limit their emissions. This factor, together with ongoing
concerns about the security of petroleum and gas supplies and the impacts on air quality
from coal and gasoline burning, is reshaping our thinking about energy supply and demand.

Recently, a working group composed of members from the U.S. Department of
Energy's five national laboratories published the report *Scenarios for a Clean Energy
Future*. The scope of their analysis included:

- Measures that reduce the energy intensity of the economy (e.g., more efficient
  lighting, cars, and industrial processes)
- Technology measures that reduce the carbon intensity of the energy used (e.g.,
  renewable energy resources, nuclear power, natural gas, and more efficient fossil
  fueled electricity plants)

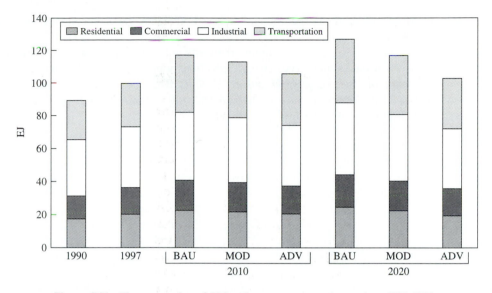

Figure 5.21 Three scenarios of U.S. primary energy use by section, 1990–2020. BAU = business as usual; MOD = moderate; ADV = advanced. *Source:* Interlaboratory Working Group on Energy-Efficient and Clean Technologies (2000). *Scenarios for a Clean Energy Future.* Prepared for Office of Energy, Efficiency, and Renewable Energy (Washington, DC: U.S. Department of Energy).

- Policy measures to reduce carbon intensity of energy consumption (e.g., carbon permit revenues and rebates)

Figure 5.21 shows the working group's energy forecasts for three scenarios. The "business as usual" (BAU) scenario assumes improvements in energy efficiency will continue at the current pace: a 25 percent increase in new residential building shell efficiency in 2020 relative to 1993; a decrease in industrial energy intensity by 1.1 percent annually; and continued innovation in automobile design that improves gasoline efficiency. Despite these improvement factors, U.S. energy consumption in 2020 is projected to be 42 percent higher than it was in 1990. The increase is reduced modestly, to 31 percent, in the moderate (MOD) scenario, which envisages moderately aggressive policies to improve efficiencies in the building, industrial, transportation, and electricity sectors, and increased spending ($1.4 billion per year) for energy-technology R&D.

Even in the advanced (ADV) scenario there is a rather modest increase of 15 percent in energy consumption by 2020. This scenario assumes still more aggressive energy-saving policies, and higher R&D spending ($2.8 billion per year). But the most significant element is a domestic carbon trading system, in which carbon emission permits are auctioned at a price of $50/ton. The permit revenues are assumed to be returned to the public to offset the increased energy cost; incomes would be left intact while changing the relative price of carbon. The projected effect is a substantial reduction in carbon emissions (back to 1990 levels of about 1.35 billion metric tons by 2020), achieved mainly in the building and industrial sectors, via increased efficiency and reduced carbon in fuels used for electricity.

The working group examined the costs and benefits of these scenarios, and concluded that the savings from energy reductions would substantially exceed costs of implementation, including R&D and investments. Thus by 2020, projected costs (relative to BAU) are $38 billion for MOD and $82 billion for ADV, but projected savings are $100 billion and $122 billion, respectively.

Summary

Our survey of sources has shown that energy is supplied to us in abundance by the sun, and this steady energy flow is sufficient to maintain the human population at a steady state, which we must, in any event, achieve over a period of time. Capturing the sun's energy and putting it to useful work are challenges to science and technology. At the moment it is much more expensive to do this than to tap the solar energy that has been stored over the millennia in fossil fuels, which we have used to develop our industrial civilization. Today we are completely dependent on fossil fuels, and particularly on petroleum and natural gas. These fuel stocks are being drawn down, and the easily accessible deposits will be depleted in a matter of decades for oil and gas, and perhaps in a few centuries for coal. In the meantime, the development of atomic science has provided the key for unlocking the very large amounts of energy stored in the nuclei of uranium atoms and potentially in those of hydrogen atoms. The development of nuclear power on a large scale, however, poses unprecedented problems for humanity in terms of the dangers of nuclear weapons proliferation and the long-term hazards associated with massive amounts of radioisotopes.

On the supply side of the energy equation, the key questions are to what extent coal and nuclear fuels can be used safely with acceptable environmental costs, and how long it will take to introduce alternative technologies involving solar power and possibly the fusion reactor. Equally important questions exist on the demand side of the energy equation. A close examination of how energy is actually used shows that significant savings are possible in situations where energy consumption has been predicated historically on the availability of a cheap and plentiful supply. Energy savings have a greater impact on the equation than energy supplies because of the low efficiency with which fuels are used. Each joule of electricity that is not used represents a saving of 3 joules of oil, coal, or nuclear fuels. Savings on the demand side of the energy equation are desirable because they increase the range of choices with respect to the energy supply. The lower the projected energy requirements, the greater the flexibility for providing energy from a variety of sources, and the more time there is to develop the safest, most efficient, and least environmentally harmful technologies.

PROBLEM SET

1. How efficient is photosynthesis? Annual photosynthesis is estimated to produce an average 320 g (dry weight) of plant matter per m^2, 50 percent of which is carbon.
   (a) Calculate the total grams of carbon "fixed" each year as plant matter by an area 1,000 m^2.
   (b) The photosynthetic reaction can be represented by the production of glucose:

$$6CO_2 + 6H_2O = C_6H_{12}O_6 + 6O_2 \qquad (1)$$

   Of the glucose produced, 25 percent is used by the plant to fuel respiration; the rest is converted to plant matter. Given the answer to part (a), calculate the total moles of glucose produced annually in our reference 1,000 m^2 area through reaction (1).
   (c) Each mole of glucose produced represents the absorption of 2,803 kJ of solar energy. If the average energy available from sunlight is 1.527 kJ/cm^2 per day, what percentage of the incident solar energy is converted to chemical energy in our 1,000 m^2 area?
   (d) Would this area produce more energy via biomass, or via solar collectors, assuming a 15 percent collector efficiency?

2. Compare the estimated organic matter burial rate in Figure 1.1 (<0.01 percent) with that implied by Figure 2.1, if oxygenic life dates back 400 million years (and if we assume that the total biomass has stayed the same over that time). For this order-of-magnitude calculation, assume that the energy per mole of buried carbon is the same as that stored in glucose (see p. 28).

3. According to Figure 1.2, the U.S. accounted for one fourth of total world energy consumption in 2000, and both the U.S. and world rates were growing at a 1.5 percent annual rate. Suppose the U.S. cut its growth rate to 1 percent, while the rest of world growth, propelled by the needs of developing countries, continued to increase consumption by 1.5 percent per year. What would the U.S. share of total consumption be in 2050?

4. Follow the logic of Worked Problem 2.1 (see p. 20), and calculate the percentage increase in atmospheric CO_2 if all the fossil carbon ended up in the air.

5. The ratio of H atoms to C atoms is higher in oil than in coal, and even higher in gas. Explain why this is so, and relate your answer to the likely origins of these fuels. What are the consequences for the quality of the energy produced by each?

6. Check the energy values for ethanol combustion in Table 2.2, using the bond energies in Table 2.1. A car driven on ethanol gets lower mileage (miles per gallon) than an equivalent car driven on gasoline. Estimate how much less (assume for the purpose of this question that the densities of ethanol and gasoline are the same).

7. The major component of natural gas used by utility companies is methane, CH_4; the gas used in gas barbecues is propane, C_3H_8. Assuming that each of these gases burns completely, compare the amount of energy released by each (a) in terms of kJ/mole of CO_2 produced, and (b) in terms of kJ/g of fuel. Do the calculations based on chemical bonds broken and formed in the combustion reactions. Use Table 2.1 to obtain bond energies.

8. (a) Compare the costs per unit of energy for electricity and gasoline. Assume current prices are $0.05 per kWh for the former and $1.50 per gallon for the latter. Gasoline weighs 5.51 lb/gal, and its energy release on combustion is 19,000 BTU (British Thermal Units) per pound. (Energy units: 1 calorie = 4.18 joules = 1.16×10^{-6} kWh = 3.97×10^{-3} BTU.)

   (b) Taking into account that gasoline engines are 20–25 percent efficient and that electric cars are 50–80 percent efficient, which is more economical for transport with respect to fuel costs?

9. (a) Consider that in the southwestern United States the average insolation is 270 W per m^2. Calculate the solar energy per m^2 per year in kWh. Given that the energy in a kilogram of coal is 8.14 kWh, the solar energy per m^2 per year is equivalent to how many kg of coal? Calculate the electrical energy that can be produced from a 1-meter flat-plate photovoltaic cell, assuming an efficiency of 15 percent. Calculate the electrical energy produced in a coal-fired power plant from the coal equivalent to the solar energy per m^2 per year, assuming an efficiency of 33 percent.

   (b) Assume that the coal contains 72 percent carbon and 2 percent sulfur. After conversion to electrical energy via insolation and coal combustion, calculate and compare the residual flows of mass and energy from the two sources. (Base calculations on m^2 per year of solar energy, and its coal equivalent.)

   (c) Comment on the relative merits of the two sources of electricity.

10. The well-preserved remains of now extinct wooly mammoths are entombed in the permafrost lands of Siberia. Scientists have conducted carbon-dating experiments on plant debris found in the stomachs of the mammoths. The ^{14}C activity of the youngest mammoths was measured at 4.5 disintegrations per minute per gram of carbon. When did Siberian mammoths become extinct? (Living tissue has a ^{14}C activity of 15.3 disintegrations per minute per gram of carbon.)

11. Describe three major environmental and security problems associated with nuclear power.

12. Plutonium is very damaging when inhaled as small particles because of the ionization of tissue by its emitted alpha particles. One microgram of plutonium is known to produce cancer in experimental animals. From its atomic weight (239) and half-life (24,360 years), calculate how many alpha particles are emitted by a microgram of plutonium over the course of a year. About how many ionizations do the particles produce?

13. A neutron generated by fission typically possesses a kinetic energy of 2 Mev. When such a neutron collides with a hydrogen atom 18 times, its energy is reduced to its thermal energy of 0.025 ev, that is, the kinetic energy it would possess by virtue of the temperature of its surroundings. The same neutron would have to collide with a sodium atom more than 200 times to reduce its energy by the same amount. Explain how these characteristics make water a suitable coolant in the pressurized light-water reactor that utilizes U-235 as the fuel, whereas sodium is the suitable coolant in the breeder reactor.

14. How does a breeder reactor extend the supply of nuclear fuel? What are the problems associated with the breeder design?

15. From the standpoint of weapons proliferation, why is it more dangerous to fuel reactors with plutonium than with uranium in which U-235 is enriched to 2–3 percent?

16. Considering the ^{238}U decay scheme (see p. 49), which of the daughters of uranium are likely to be most abundant in uranium-bearing soil, and why?

17. Any large-scale nuclear process produces radioactive waste. Compare the problems of waste from uranium fission reactors and from tritium fusion reactors.

18. Consider the nuclear fusion reactions listed in Figure 3.13.
    (a) Why is attention focused initially on the D+T reaction?
    (b) Where would the fuel for this reaction come from?
    (c) What advantages and disadvantages would the D+D reactions have?
    (d) What advantages and disadvantages would the D+^{3}He reaction have?

19. Should you install a solar water heater? The average home has a 200-liter (50–60 gallon) hot-water tank, which is effectively drained and replenished three times a day. Assume that the entering tap water is 15°C and is heated to 55°C.
    (a) Given an average energy from sunlight of 1.53 kJ/cm^2 per day, how large would the collection area of a solar water heater need to be if its efficiency is 30 percent?
    (b) Assume that the price of a solar collector is $375/m^2. How much would it cost to install the hot-water system in (a)?
    (c) If the price of oil remains at $0.75/liter for 20 years, how much would the solar collector save? Assume that the heating content of oil is 2.51 × 10^4 kJ/liter and it can be burned with 90 percent efficiency.

20. (a) Explain why a Si solar cell would not work without a *p-n* junction?
    (b) How is a *p-n* junction achieved by doping Si?
    (c) What is the analog of the *p-n* junction in the photosynthetic apparatus of plants?

21. (a) Calculate the energy, in electron volts (ev), of a photon of infrared (IR) light with a wavelength of 1,140 nm. Explain why photons with larger wavelengths cannot excite electrons in a PV cell from the valence band to the conduction band (see Fundamentals 1.1).
    (b) Calculate the energy of a photon of blue light with a wavelength of 483 nm, corresponding to the region of maximum solar flux (see Figure 4.5). Calculate the percentage of energy possessed by the photon that can promote an electron from the valence band to the conduction band of a PV cell. What happens to the residual energy?
    (c) Do the same calculations as in (b) for an ultraviolet (UV) photon of light at 300 nm, the highest energy photon to reach Earth's surface. Since the energy flux of these UV photons is approximately a factor of six less than that of blue photons, how much more energy can be converted into electricity in a PV cell from blue light compared to UV light. The following equation may be helpful:

$$E = hc/\lambda,$$

where E is energy in units of ev, h (Planck's constant) = 4.14 × 10^{-15} ev sec, c (speed of light) = 3 × 10^{10} cm/sec, and λ (wavelength) is in units of cm, where 1 cm = 10^7 nm (nanometers).

22. Why are only the kernels of the corn plant a ready source of ethanol, and not the rest of the plant?

23. (a) An acre (0.40 hectares) of corn can produce about 340 gallons (1,285 L) of ethanol, using only the kernel of the corn for conversion to ethanol. Assume the energy input to produce an acre of corn is 21.5 × 10^6 kJ. If a gallon of ethanol has an embodied energy of about 80,000 kJ, calculate the ratio of embodied energy in the ethanol to the energy required to produce the corn.
    (b) Assume the yield of corn stover (cobs, stalks, and leaves) is 3.6 dry metric tons per acre. The extra energy required to convert the stover to ethanol is about 17,000 kJ/gallon; and one dry

ton of stover can produce about 80 gallons of ethanol. Assuming only 40 percent of the stover is removed (more than that could diminish soil quality) for conversion to ethanol, how would the energy ratio calculated in part (a) change?

(c) If it becomes economically feasible to harvest stover and convert it to ethanol, the stover could be collected from all corn-growing farmlands, rather than just on lands dedicated to corn for conversion to ethanol. The amount of stover available in the U.S. from all corn-growing areas (taking only 40 percent of the total) is estimated to be about 110 million metric dry tons. Using the information provided in part (b), calculate the gallons of ethanol that could be produced, and compare this to the 1.4 billion gallons of ethanol production in 1998.

24. (a) Assume a wind turbine with a hub 50 meters above the ground, a rotor diameter of 50 meters, and a wind-conversion efficiency of 25 percent. The turbine operates in an area with an annual average wind-power density of 500 watts/m^2 at 50 meters altitude. How much electrical energy (in kWh) can the turbine generate per year?

(b) Wind densities equal to or greater than 500 watts/m^2 at an altitude of 50 meters are exploitable with today's technologies; about 1.2 percent of the land area of the contiguous United States possesses such wind densities. If, on average, windfarms contain eight turbines per km^2, what is the U.S. potential for electrical-energy production from wind power? (Assume a uniform power density of 500 watts/m^2, and the same specifications as for the turbine in part (a).) In 2000, 3,236 TWh of electricity were produced in the United States. What percentage of U.S. electricity consumption could be met by wind power?

(c) Continuing technical advances will allow wind-power generation on lands where the wind power density is 300 watts/m^2 at 50 meters. In this case, wind power could be harvested on 21 percent of U.S. land. Assuming that one-third of the land were covered by windfarms (again with a density of eight turbines per km^2), how much electricity would be generated? What percentage of the U.S. demand could be met? (Assume a uniform power density of 300 watts/m^2, and wind turbines with the same design as in part (a).) (The area of the contiguous U.S. is 7,827,989 km^2; 1 TWh $= 10^9$ kWh; 1 kWh $= 3.6 \times 10^3$ kJ.)

25. Assume that due to dwindling supplies of crude oil and natural gas, the United States embarks on a plan to implement the "hydrogen economy," for which hydrogen will be generated by electrolysis of water. The source of electricity is from photovoltaic conversion of sunlight in the southwestern United States.

(a) Assume that a fixed, flat-plate PV system is used, with a solar conversion efficiency of 15 percent, and a hydrogen-production efficiency of 80 percent. Assume that the average annual insolation in the southwest is 270 watts per m^2. Calculate the annual electrical energy produced per m^2 in kWh and in kJ. Calculate the energy content of the H$_2$ produced per m^2, as well as the number of moles, and the weight of H$_2$. (See Table 2.2, p. 25, for H$_2$ data.)

(b) The United States consumed 40.1 EJ of petroleum in 2000. How many square meters of PV collectors would be needed to supply the equivalent amount of energy in H$_2$? The area of the southwest (the states of New Mexico, Arizona, Colorado, Utah, and Nevada) is 1,386,370 km^2. What percentage of the land area would be covered by PV collectors?

(c) How much water per year would be required to produce the hydrogen? Assume a conversion efficiency of 80 percent. What percentage of total national water use would this correspond to, considering that water use in the United States is about 4.7×10^{14} liters?

26. An electric hot-water heater has a standard efficiency rating of 90 percent (that is, for every 10 kJ of electricity consumed, 9 go into heating the water). If the heater normally heats the water from the ambient temperature, 20°C (68°F), to 80°C (176°F), what is the *second law* efficiency (that is, what is the ratio of the amount of energy that an ideal heat pump would use to do the same job, to the amount of energy actually consumed)?

27. Calculate the theoretical efficiency of a heat pump when the room temperature is 20°C (68°F) and the outside temperature is −20°C (−4°F). If the heat pump were drawing heat from a water tank at 5°C (41°F), what would be the theoretical efficiency? Describe how solar energy in conjunction with a heat pump can provide an efficient means of heating a house.

28. **(a)** Consider a PEM fuel cell producing electrical energy at 80°C from the reaction

$$H_2 + \tfrac{1}{2}O_2 = H_2O$$

Calculate the theoretical efficiency of conversion of heat to electrical energy, assuming $\Delta H = -286$ kJ/mole H_2 and $\Delta S = -0.163$ kJ/K.

(b) Calculate the ideal PEM fuel-cell voltage (ΔE) at 80°C from the relationship

$$\Delta E = -\Delta G/nF$$

where n is the number of moles of electrons involved in the reaction per mole of H_2, and F is the Faraday constant = 96.5 kJ/volt.

29. Although $T\Delta S$ increases with increasing temperature, there is little difference in theoretical efficiency for a H_2/O_2 fuel cell operating at 80°C or one operating at 1,000°C. Explain.

30. Petroleum consumption in the United States in 1999 was 39.8 EJ. Of this quantity, 23.8 EJ (60 percent) was supplied by foreign sources. The transportation sector accounted for 26.2 EJ (66 percent) of total petroleum use. The passenger car fleet, numbering 132 million, consumed 9.6 EJ, and light trucks (including vans, pickup trucks, and sport utility vehicles), numbering 72 million, consumed 6.7 EJ. The average fuel efficiency of the passenger car fleet in 1999 was 21.4 miles per gallon (mpg), and for light trucks it was 17.1 mpg. Calculate how much less petroleum would have to be imported if the national stock of cars was replaced by hybrid vehicles, in which passenger car efficiency would be 60 mpg and light truck efficiency would be 40 mpg.

31. A large power plant is being constructed that will produce 6.7×10^{10} kJ/day of electrical energy. For every kJ of energy produced, 2 kJ of waste heat is discharged. If a plant draws 2×10^9 liters/day of river water at 20°C into its cooling condensers, calculate the rise in temperature of the cooling water. If the river has a flow rate of 10^{10} liters/day, estimate the rise in temperature of the water downstream from the plant.

32. **(a)** As presented in Table 5.2, making a metric ton of paper requires 980 kWh of energy, whereas making a metric ton of polystyrene requires about 300 kWh. Given that an average 8 oz paper cup weighs 10.1 g and an average polystyrene cup weighs 1.5 g, what is the ratio of the power requirements on a per-cup basis?

(b) The table also lists the amount of heat recovered from incinerating each type of cup; 20 MJ/kg (megajoule per kilogram) for paper; 40 MJ/kg for polystyrene. The heat could be converted to electricity at a power plant with an efficiency of about 30 percent. Compare the amount of electrical power available from incinerating discarded paper and polystyrene cups with the amount of energy needed to produce them. (1 kWh = 3.6×10^6 joules; 1 MJ = 10^6 joules.)

(c) A fast-food restaurant on a busy street uses polystyrene cups to serve coffee at a rate of 2.5 gross/day (1 gross = 12 dozen = 144). Reacting to pressure from an environmental advocacy group, the restaurant switches over to using paper cups. What is the effect of this decision in terms of kWh of power per week, if the town has no incinerator? What is the effect if the town sorts and incinerates paper and plastic for electricity production?

(d) Write a paragraph explaining to the proprietors of the restaurant in question (c) what they might do if they want to be as environmentally responsible as possible.

SUGGESTED READINGS

Chapter 1: Energy Flows and Supplies

Energy Information Agency (2000). Energy in the United States: A brief history and current trends. In *Annual Review of Energy 1999* (Washington, DC; U.S. Department of Energy). (http://tonto.eia.doe.gov/FTPROOT/multifuel/038499.pdf)

V. Smil (2000). Energy in the twentieth century: Resources, conversions, costs, uses, and consequences. *Annual Reviews of Energy & Environment* 25:21–51.

Energy Information Agency (2001). *Annual Energy Review* (Washington, DC: U.S. Department of Energy). (http://www.eia.doe.gov/aer)

British Petroleum Amoco (2001). *The Statistical Review of World Energy 2001; The Statistical Review of U.S. Energy 2001.* (http://www.bp.com/centres/energy/index.asp)

Chapter 2: Fossil Fuels

Office of Investigations and Analysis, U.S. Coast Guard (2000). Cumulative data and graphics for oil spills (1973–1999). In *Pollution Incidents In and Around U.S. Waters, a Spill/Release Compendium: 1969–1999.* (Washington, DC: Department of Transportation). (http://www.uscg.mil/hq/g%2Dm/nmc/response/stats/aa.htm)

S. Holloway (2001). Storage of fossil fuel-derived carbon dioxide beneath the surface of the Earth. *Annual Reviews of Energy & Environment* 26:9010.

Chapter 3: Nuclear Energy

J. Johnson (2001). Up from the dead[*]. *Chemical & Engineering News* 79(36):29–32.

P. W. Beck (1999). Nuclear energy in the twenty-first century: Examination of a contentious subject: *Annual Reviews of Energy & Environment* 24:113–137.

Committee on the Remediation of Buried and Tank Wastes, Board on Radioactive Waste Management, National Research Council (2000). *Long-Term Institutional Management of U.S. Department of Energy Legacy Waste Sites* (Washington, DC: National Academy Press).

[*]Refers to nuclear energy.

Chapter 4: Renewable Energy

H. S. Shapouri, J. A. Duffield, and M. S. Graboski (1995). *Estimating the Net Energy Balance of Corn Ethanol.* Agricultural Economic Report Number 721 (Washington, DC: U.S. Department of Agriculture).

J. DiPardo (2000). *Outlook for Biomass Ethanol Production and Demand.* Energy Information Agency (Washington, DC: U.S. Department of Energy). (http://www.eia.doe.gov/oiaf/analysispaper/biomass.html)

C. Wyman (1999). Biomass ethanol: Technical progress, opportunities, and commercial challenges. *Annual Reviews of Energy & Environment* 24:189–226.

H. S. Kheshgi, R. C. Prince, and G. Marland (2000). The potential of biomass fuels in the context of global climate change: Focus on transportation fuels. *Annual Reviews of Energy & Environment* 25:199–244.

M. E. Walsh, R. L. Perlack, A. Turhollow, D. de la Torre Ugarte, D. A. Becker, R. L. Graham, S. E. Slinsky, and D. E. Ray (2000). *Biomass Feedstock Availability in the United States: 1999 State Level Analysis.* (Oak Ridge, Tennessee: Oak Ridge National Laboratory). (http://bioenergy.ornl.gov/resourcedata/index.html)

Energy Information Agency (2001). *International Energy Outlook: Hydroelectricity and Other Renewable Resources.* (Washington, DC: U.S. Department of Energy). (http://www.eia.doe.gov/oiaf/ieo/hydro.html)

J. G. McGowan and S. R. Connors (2000). Windpower: A turn of the century review. *Annual Reviews of Energy & Environment* 25:147–197.

Chapter 5: Energy Utilization

S. Thomas and M. Zalbowitz (1999). *Fuel Cells–Green Power.* Los Alamos National Laboratory (Los Alamos, New Mexico: U.S. Department of Energy). (http://www.education.lanl.gov/resources/fuelcells/fuelcells.pdf)

S. Srinivasan, R. Mosdale, P. Stevens, and C. Yang (1999). Fuel cells: Reaching the era of clean efficient power generation in the twenty-first century. *Annual Reviews of Energy & Environment* 24:281–328.

J. M. Ogden (1999). Prospects for building a hydrogen energy infrastructure. *Annual Reviews of Energy & Environment* 24:227–279.

S. Dunn (2001). *Hydrogen Futures: Toward a Sustainable Energy System.* Worldwatch Paper 157 (Washington, DC: Worldwatch Institute).

S. Dunn (2000). *The Hydrogen Experiment* (Hydrogen Economy in Iceland). *World Watch Magazine:* November/December: 14–25.

I. K. Wernick and N. J. Themelis (1998). Recycling metals for the environment. *Annual Reviews of Energy & Environment* 23:465–497.

D. T. Allen and N. Behmanesh (1994). Wastes as raw materials. In B. R. Allenby and D. J. Richards (eds.). *The Greening of Industrial Ecosystems* (Washington, DC: National Academy Press).

G. Matos and L. Wagner (1998). Consumption of materials in the United States, 1900–1995. *Annual Reviews of Energy & Environment* 23:107–122.

J. C. Ryan and A. T. Durning (1997). *Stuff: The Secret Lives of Everyday Things.* NEW Report No. 4 (Seattle, Washington: Northwest Environment Watch).

A. Adriaanse, S. Bringezu, A. Hammond, Y. Moriguchi, E. Rodenburg, D. Rogich, and H. Shütz (1997). *Resource Flows: The Material Basis of Industrial Economies* (Washington, DC: World Resources Institute; Wuppertal, Germany: Wuppertal Institute; The Hague, Netherlands: Netherlands Ministry of Housing, Spatial Planning, and Environment; Tsukuba, Japan: National Institute for Environmental Studies).

A. H. Rosenfeld (1999). The art of energy efficiency: Protecting the environment with better technology. *Annual Reviews of Energy & Environment* 24:33–82.

W. J. Fisk (2000). Health and productivity gains from better indoor environments and their relationship with building energy efficiency. *Annual Reviews of Energy & Environment* 25:537–566.

L. R. Glicksman, L. K. Norford, and L. V. Greden (2001). Energy conservation in Chinese residential buildings: Progress and opportunities in design and policy. *Annual Reviews of Energy & Environment* 26:9007.

S. Dunn (2000). *Micropower: The Next Electrical Era.* Worldwatch Paper 151 (Washington, DC: Worldwatch Institute).

Office of Air and Radiation (2000). *Light-Duty Automotive Technology and Fuel Economy Trends 1975 through 2000.* Report EPA-420-R00-008 (Washington, DC: U.S. Environmental Protection Agency).

S. C. Davis (2000). *Transportation Energy Data Book: Edition 20—2000.* Report ORNL-6959 (Oak Ridge, Tennessee: Oak Ridge National Laboratory).

E. H. Decker, S. Elliot, F. A. Smith, D. R. Blake, and F. S. Rowland (2000). Energy and material flow through the urban ecosystem. *Annual Review of Energy and Environment* 25:685–740.

S. K. Ritter (2001). Green chemistry. *Chemical & Engineering News* 79(29):27–34.

L. Schipper, F. Unander, S. Murtishaw, and M. Ting (2001). Indicators of energy use and carbon emissions: Explaining the energy economy link. *Annual Reviews of Energy & Environment* 26:9004.

Interlaboratory Working Group (2000). *Scenarios for a Clean Energy Future* (Oak Ridge, Tennessee: Oak Ridge National Laboratory; Berkeley, California: Lawrence Berkeley National Laboratory), ORNL/CON-476 and LBNL-44029, November.

PART II

ATMOSPHERE

143

CHAPTER 6

CLIMATE

We now turn our attention to environmental issues associated with Earth's atmosphere, a topic that flows naturally from our previous contemplation of energy sources and uses, because the burning of fuels has a large impact on the atmosphere. Air pollution is a major problem in most of the world's cities, and it often takes on regional dimensions. The atmosphere is a repository for emissions from combustion and from many other human activities; the air can be cleansed by natural mechanisms, but these can be overwhelmed by the amounts of pollutants being produced. On a global scale, we are carrying out vast inadvertent experiments on the atmosphere. Human activities are increasing the atmospheric concentration of carbon dioxide and other "greenhouse" gases, thereby altering the way the sun's heat is distributed on the Earth's surface and in the atmosphere. In addition, the stratospheric ozone shield, which protects us from the sun's ultraviolet rays, is threatened by the emission of ozone-destroying chemicals. Although some of the world's leading scientists and the most powerful computers are trying to predict how these experiments will turn out, all scenarios for the future are riddled with uncertainties.

6.1 RADIATION BALANCE

As we emphasized in Part I, the sun provides Earth with an enormous input of energy every day. Earth rids itself of energy at the same rate and thereby maintains a steady state, with a constant average temperature. It loses energy by radiating light. Of course, Earth does not glow the way the sun does. A hot body gives off radiation with a range of wavelengths (see Fundamentals 1.1, p. 6). The distribution of the radiation shifts toward shorter wavelengths with increasing temperature. That is why a piece of iron heated in a furnace glows red and

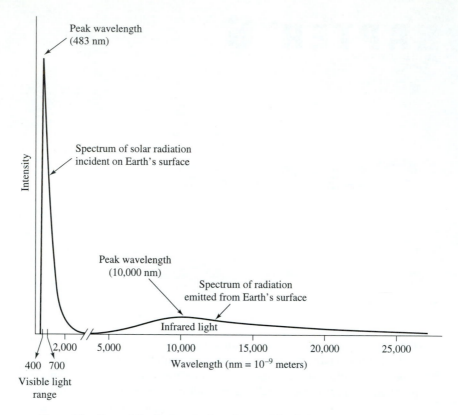

Figure 6.1   Spectral distribution of solar and terrestrial radiation.

then white as its temperature increases. The wavelengths of Earth's rays are too long to be detected by our eyes.

The spectral distribution of radiation from the sun and from Earth is shown in Figure 6.1. The curves are somewhat idealized; they are the expected "black body" radiation of objects with the temperature of the sun and of Earth. The spectrum of a black body is smooth, whereas the actual spectra of the sun and Earth are somewhat bumpy because specific atomic and molecular transitions contribute to the emissions. For a black body, the peak wavelength of the radiation is inversely proportional to the absolute temperature (Wein's law):

$$\lambda_{\text{peak}} \text{ (nm)} = 2.9 \times 10^6 \text{ (nm} \times \text{K)}/T(\text{K}) \tag{6.1}$$

The sun is a very hot body; its peak wavelength is 483 nm, corresponding to a temperature of 6,000 K. Most of its rays fall between 400 and 700 nm, in the region of visible light; these wavelengths are visible because our eyes have evolved in response to sunlight. Earth emits radiation with a peak wavelength of about 10,000 nm, corresponding to an average temperature of 288 K. Thus, while Earth absorbs radiation mainly in the visible region, characteristic of the high temperature at the surface of the sun, it gives off radiation in the infrared region, which corresponds to the much longer wavelengths characteristic of Earth's cooler surface temperature.

All the energy that Earth absorbs from the sun must eventually be re-emitted, so we can calculate Earth's steady-state temperature by setting its radiation rate equal to the rate at which Earth absorbs energy from the sun. The flux of solar energy directed at Earth, S_0, is 1,370 watts/m² (see Fundamentals 1.1).

It strikes Earth as if Earth were a disk of area πr^2, r being Earth's radius. Since Earth radiates from its entire surface, the area of which is $4\pi r^2$, we need to divide S_0 by 4. Moreover, not all of the sun's rays are absorbed; the fraction reflected back to space, the albedo (a), is close to 0.3. The rate at which solar radiation is absorbed and re-emitted is therefore

$$S = (1 - a)S_0/4 = 240 \text{ watts/m}^2 \qquad (6.2)$$

This rate can be used to calculate the temperature according to the Stefan-Boltzmann law, which states that the rate at which a black body radiates energy, S, is proportional to the fourth power of its absolute temperature:

$$S = kT^4 \qquad (6.3)$$

where k is the Stefan-Boltzmann constant, 5.67×10^{-8} watts/m² × K⁴. According to this calculation, the temperature of the surface of the Earth determined from the rate of solar emission is 255 K.

But 255 K is colder by 33 K than the average temperature at the surface of Earth. What is the source of this discrepancy? The atmosphere provides the answer: it traps much of the heat emanating from Earth's surface and radiates it back, raising the surface temperature.

This atmospheric trapping of infrared radiation is the greenhouse effect. Despite the name's negative associations, the greenhouse effect makes our planet habitable. At the much colder temperature that would prevail in its absence, all of Earth's water would freeze. The concern about the greenhouse effect is that it may become too much of a good thing, if the heat-trapping efficiency of the atmosphere increases further as a result of increasing concentrations of CO_2 and other greenhouse gases (see section 6.3, Greenhouse Effect, pp. 160–172). For now, we note that because of the heating from below, the atmosphere grows colder with increasing elevation above Earth's surface. The number calculated with equation (6.3), 255 K, is the average temperature that prevails at an altitude of about 5 km. The Earth-air system acts as if it radiates from somewhere in the middle of the atmosphere.

Can the energy balance be affected by human energy consumption? In principle, if we keep increasing the rate of fossil and nuclear fuel burning, the global heat load might become significant. From equation (6.3) we can calculate how much the energy input to Earth would need to rise in order to increase the average temperature by 1 K. For such a small change, we can differentiate equation (6.3) and divide by the total flux, to obtain

$$dS/S = 4dT/T \qquad (6.4)$$

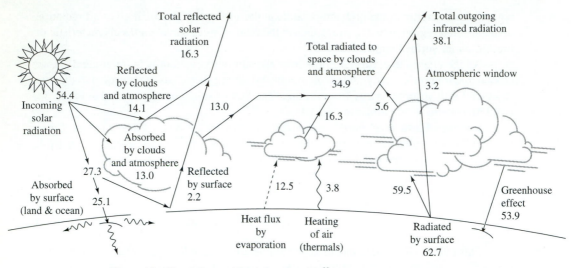

Figure 6.2   Heat balance of Earth (in units of 10^{20} kJ/yr).

In other words, the fractional increase in the energy budget is four times the fractional increase in the temperature. For a 1 K rise from 255 K, $dT/T = 0.00392$, and $dS/S = 0.0157$. Thus, human energy utilization would have to equal 1.57 percent of the solar input to produce a 1 K rise in the average temperature. Currently, the ratio of human energy consumption to solar energy flux on Earth is only 0.01 percent (see Table 1.1, p. 9), leaving a comfortable margin. How long until we reach the 1.57 percent ratio? Over the next 20 years, the Energy Information Agency of the U.S. Department of Energy forecasts an average annual global energy growth rate of 2.2 percent, which corresponds to a doubling time of 32 years; to reach 1.57 percent of the sun's energy would take about 240 years (see Worked Problem 1.2, p. 13). This rate of increase is not likely to be sustained due to resource and global pollution factors. Moreover, renewable energy does not count in the balance because renewable energy sources simply divert the current solar flux. To the extent that an increasing fraction of total human energy is extracted from renewable sources, Earth's heat load will be ameliorated.*

Thus, direct heating of the planet through our increasing use of energy is not likely to become a serious problem (although a "heat-island" effect can raise the temperature of urban areas by several degrees relative to the surrounding countryside). More serious is the potential for altering Earth's temperature indirectly through changes induced by human activity in either the albedo or the greenhouse effect. These matters are considered in the following sections.

The actual flows of energy through the atmosphere are quite complicated. Figure 6.2 shows the planet's inputs and outputs of energy in units of 10^{20} kJ per year. About 54.4 units

*For example, whether a given amount of biomass is burned as fuel or undergoes microbial degradation to CO_2 and H_2O, the amount of energy released is the same; rates of release, however, will differ, since the latter process is far slower than the former.

of solar energy impinge on Earth and its atmosphere, but about 16.3 units (30 percent) are reflected to space, exerting no influence on Earth's heat balance. Most of this light is reflected by clouds and the atmosphere; a smaller amount (2.2 units) is reflected by Earth's surface. The remaining 38.1 units (70 percent) are absorbed, 13.0 units (24 percent) by the atmosphere and clouds, and 25.1 units (46 percent) by Earth's surface.

All of the absorbed solar energy must be lost to space in order to maintain the planet's heat balance. However, a much larger amount of energy is in circulation as a result of the greenhouse effect. With an average temperature of 288 K, the Earth's surface radiates 62.7 units, consistent with the Stefan-Boltzmann law. Almost all of this radiation is absorbed by the atmosphere (including clouds); only 3.2 units escape to space through the *atmospheric window* (see p. 164). Thus the atmosphere absorbs 72.5 units of radiant energy, 59.5 from the Earth and 13.0 units from the sun. To this must be added 12.5 units of *latent heat,* the energy transferred to the atmosphere by the evaporation of water, and 3.8 units of *sensible heat,* the energy carried by updrafts of air. The grand total of energy gained by the atmosphere is 88.8 units. This is balanced by radiative losses, 53.9 units back to the Earth, and 34.9 units to space. Thus most of the planet's radiative cooling, 34.9 out of 38.1 units, is from the atmosphere.

The above discussion refers to global averages. In fact, the heat flow pattern is not uniform over Earth's surface. Most of the sun's rays are absorbed in the tropics, while the outgoing radiation of Earth is more uniform with latitude (see Figure 6.3). Thus, there is a constant movement of energy from the equator toward the poles, through the atmosphere and the oceans. These complexities do not alter the fundamental fact that the Earth-atmosphere

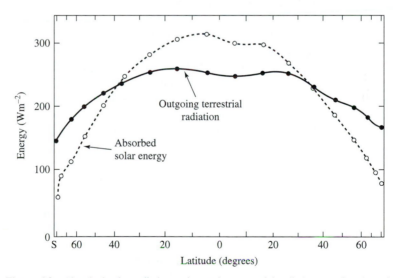

Figure 6.3 Absorbed solar radiation and outgoing terrestrial radiation as a function of latitude. *Source:* T. H. Von der Haar and V. E. Suomi (1971). Measurements for Earth's radiation budget from satellites during a five-year period. Part I. Extended time and space means. *Journal of Atmospheric Sciences* 28:305–314. Copyright © 1971, American Meteorological Society. Reprinted with permission.

system, taken as a whole, receives its energy from the sun and re-radiates it to space. The radiation balance has to be maintained.

WORKED PROBLEM 6.1: Volcanic Cooling

Q. *Large volcanic eruptions can cool the planet by increasing the albedo (see section 6.2b, p. 153). Calculate the expected temperature change if the albedo increases from 30 to 30.5 percent. Compare this estimate with the temperature record after the Mt. Pinatubo eruption (see Figure 6.7, p. 155)*

A. The albedo directly affects the radiation rate since

$$S = S_0(1 - a)/4$$

Differentiating this expression, and dividing by S, gives

$$dS/S = -da/(1 - a)$$

Since $a = 0.3$ and $a + da = 0.305$, then $1 - a = 0.7$, and $da = 0.005$, giving $dS/S = 0.00714$. According to eq. (6.4), the relative change in temperature is one-fourth this value:

$$dT/T = 0.00179$$

If T is 255 K, the average current atmospheric temperature, then $dT = 0.5$ K. This is about the maximum dip in the temperature record shortly (one year) following the Mt. Pinatubo eruption.

6.2   ALBEDO: PARTICLES AND CLOUDS

The albedo is a critical factor in the radiation balance because it directly determines the fraction of the solar radiation that is absorbed by the Earth-air system. Even a slight change in the average albedo can measurably affect global temperature (see Worked Problem 6.1). However, the reflectivity of solar radiation varies greatly from place to place. The effect on the overall planetary albedo for a specified surface type depends not only on its reflectivity but also on its spatial coverage. The albedo of Earth's surface varies considerably (see Figure 6.4). The darkest regions (with the lowest albedos) are the oceans, which constitute about 70 percent of Earth's total area. Albedos of the oceans range from 6–10 percent in the low latitudes to 15–20 percent near the poles, due to the low solar elevation. Sea ice with overlying snow has an albedo of 40–60 percent. The darkest land surfaces are the tropical forests and cultivated land (10–15 percent). The brightest parts of the globe are the snow-covered polar areas, with albedos as high as 80 percent. The major deserts have albedos of 25–40 percent.

The surface albedo can be affected by human activities. For example, the local albedo can be increased by clearing forest land for agriculture, followed by erosion and desertification.

a. Clouds.   However, the dominant factor in the global albedo is clouds. Peak cloud reflectivity occurs over the mid- and high-latitude oceans and in the tropical cirrus systems. As a global average, cloud albedo is around 35–40 percent. Since the annual

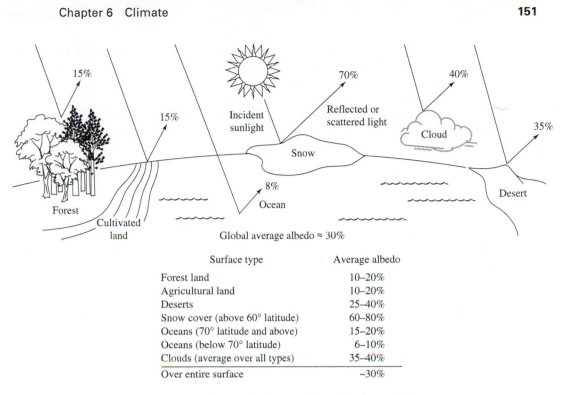

Figure 6.4   Variation in albedo with cloud cover and type of surface on Earth.

Surface type	Average albedo
Forest land	10–20%
Agricultural land	10–20%
Deserts	25–40%
Snow cover (above 60° latitude)	60–80%
Oceans (70° latitude and above)	15–20%
Oceans (below 70° latitude)	6–10%
Clouds (average over all types)	35–40%
Over entire surface	~30%

average global cloud cover is around 54 percent, the total sunlight reflected by clouds is around 20 percent, two thirds of the global albedo. The remaining third is divided between backscattering from air molecules (6 percent of incident sunlight) and by Earth's surface (only 4 percent).

As noted in the above discussion, clouds absorb both solar radiation and the long-wave radiation emitted from Earth's surface. The high albedo of clouds tends to cool Earth's surface by reflecting sunlight to space, but the absorption of outgoing radiation works to warm the surface by the greenhouse effect. The overall role of clouds in determining Earth's heat balance depends on the relative strengths of these two contrasting processes, which is currently the subject of intensive research. Even small shifts in global cloud coverage might contribute to significant changes in the heat balance.

Clouds are a natural part of the hydrological cycle, but the extent of cloudiness is extremely hard to predict. The sun's heat evaporates water at Earth's surface; as the moist air rises and cools, the water condenses out as droplets, forming the clouds. However, cooling is not the only factor in determining when cloud droplets form. Particles floating in the air are equally important, because they facilitate the coalescence of water molecules. These *condensation nuclei* provide a surface for the accumulation of water molecules. In the absence of such a surface it is difficult for the first few water molecules to stick to one another; the high *surface tension* of small aggregates of water molecules favors evaporation.

However, the film of water around a condensation nucleus has sufficiently low surface tension to permit droplet growth rather than evaporation. The principle of rainmaking by "cloud-seeding" is to inject particles that are effective in nucleating raindrops into supersaturated vapor.

There is a trade-off between droplet size and the number of condensation nuclei. A given amount of water vapor can form a small number of large drops or a large number of small ones. An excess of condensation nuclei can produce droplets that are too small to fall as rain. The fogs that often hover over cities probably reflect the large number of condensation nuclei in polluted air. An increase in the number of atmospheric particles is likely to increase the cloud cover, and therefore the albedo.

STRATEGIES 6.1 | RAINDROP FORMATION

In the presence of liquid water, there is an equilibrium with the vapor:

$$H_2O(g) = H_2O(l) \qquad (6.5)$$

Water condenses from the gas phase when its partial pressure, p, exceeds the equilibrium vapor pressure of liquid water, p_0. This is expressed in the free energy difference, ΔG, for the phase change:

$$\Delta G = -RT \ln p/p_0 \qquad (6.6)$$

where R is the gas constant (8.314 J/K) and T the absolute temperature. As the relative humidity, p/p_0, exceeds unity (100 percent), ΔG becomes negative, and liquid water forms spontaneously at equilibrium.

In pure air, however, the water molecules must first find a way to get together to form raindrops. Due to their large surface tension, very small drops evaporate quickly, even at relative humidities higher than 100 percent. In the condensation of a small number of molecules,

$$nH_2O = (H_2O)_n \qquad (6.7)$$

we have to include the surface free energy of the droplet as part of the free energy change,

$$\Delta G = -nRT \ln p/p_0 + 4\pi r^2 \gamma \qquad (6.8)$$

Here, γ is the surface tension (72.8 dynes*/cm at 20°C), r is the droplet radius, and n is the number of moles of water contained in the droplet.

[For a pool of water, n is very large, so that the surface tension term becomes negligible, and eq. (6.8) reduces to eq. (6.6) when the evaporation reaction is considered on a per mole basis.]

The number of molecules is also related to the radius of a droplet, via

$$n = (4\pi/3)r^3\rho/M \qquad (6.9)$$

where $(4\pi/3)r^3$ is the volume of the drop, ρ is its density (1 g/cm^3), and M is the gram molecular weight (18 g). Thus ΔG for a droplet is the result of two opposing terms that have a different dependence on r. Figure 6.5 is a plot of ΔG against r for a given value of p/p_0 (1.001, or 100.1 percent relative humidity). The curve goes through a maximum, which defines a critical radius, $r_c = 1 \mu m$. Droplets larger than this radius will accumulate more water molecules and become stable; droplets smaller than 1 μm will evaporate. A 1-μm drop contains 0.23×10^{-12} mole of water [see equation (6.9)] or 1.38×10^{11} molecules. It is very improbable that this many molecules can come together simultaneously to form a growing droplet; consequently with respect to precipitation, water vapor in pure air at 100.1 percent humidity is stable indefinitely. The critical droplet radius depends on the extent to which p/p_0 exceeds unity, that is, the extent to which the air is "supersaturated" with water. The dependence

---

*The *dyne* is a unit of force, which, multiplied by distance (in centimeters), yields work in *ergs;* 10^7 ergs = 1 joule.

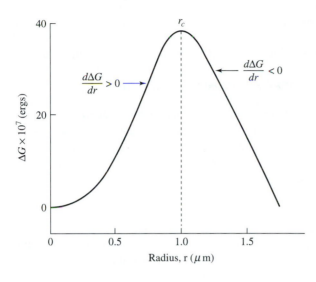

Figure 6.5 Variation of ΔG with drop size at $p/p_0 = 1.001$ (T = 20°C).

[obtained by differentiating equation (6.8) and setting $d(\Delta G)/dr$ equal to zero] is given by

$$r_c = 2M\gamma[\rho RT \ln (p/p_0)] \qquad (6.10)$$

Thus, r_c decreases slowly as p/p_0 increases. Pure air can be supersaturated to a high degree without precipitation.

In nature, however, condensation actually does occur in the range of 100.1 to 101 percent relative humidity, corresponding to $r_c = 1$ to 0.1 μm. This is because in natural air, droplets form around suspended particles, the condensation nuclei.

b. Aerosol particles. Total suspended particulate matter in air varies from less than 1 μg m^{-3} over polar ice caps and in mid-ocean, to as much as 30,000 μg m^{-3} in desert dust storms or forest fires. In a typical sample of urban air, mineral dust, sulfuric acid, ammonium sulfate, organic material, and soot may be found both as pure and as mixed particles (solid or liquid) in concentrations of around 100 μg m^{-3}. The effect of atmospheric particles on the heat flux of the atmosphere depends less on total concentration than on particle size and composition. Large, dark particles tend to absorb light, thus warming Earth's atmosphere. The most important of such particles is soot, arising from incomplete combustion of carbonaceous fuel and the burning of savannas and forests. In contrast, very small particles, regardless of color and composition, tend to scatter light, thus increasing the albedo of the atmosphere. The light-scattering effect seems to prevail at most latitudes, but at high latitudes, where snow and ice covered surfaces are highly reflective, absorption effects can dominate.

Two major natural sources of light-scattering aerosols appear to be 1) sulfate generated from biogenic gaseous sulfur emission in the deep ocean, and 2) organic carbon from partial oxidation of biogenic organic compounds, such as terpenes emitted from forests. In polluted atmospheres, light-scattering particles are produced by reactions with sulfur-, nitrogen-, and carbon-containing gases generated mostly from combustion processes. As a global average, anthropogenic sources are estimated to comprise between 25 percent to 50 percent of total aerosols.

The amount of light scattering by small particles or molecules is given by

$$s = (128\pi^5 r^6/3\lambda^4)(m^2 - 1)/(m^2 + 1) \qquad (6.11)$$

where r and m are the radius and refractive index of the particle, respectively, and λ is the wavelength of the incident light. Because of the $1/\lambda^4$ dependence, blue light is scattered more strongly than red. That is why the sky, which is seen in scattered light, is blue, while sunsets, which are seen in transmitted light, are red. The blue haze over the Great Smoky Mountains of the eastern United States is due to light scattering from small particles formed by the oxidation of volatile terpenes emitted by the sap of coniferous trees.

The particles in the atmosphere are known collectively as the atmospheric *aerosol*. Their size distribution is very wide, but as seen in Figure 6.6, the aerosol is dominated by the smallest particles. Dust and sea spray are blown about by the wind in great quantity, but these particles are large and settle out rapidly. The submicrometer particles are formed mainly by oxidation of sulfur-, nitrogen- and carbon-containing gases in the atmosphere. These particles stay aloft much longer, with lifetimes of days or weeks. Gradually, they

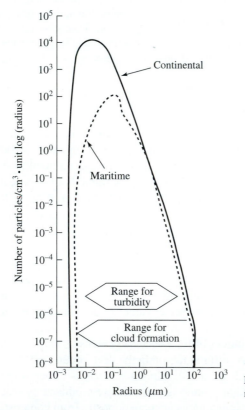

Figure 6.6   Size distribution of particulate matter in the lower atmosphere.

grow to larger ones via collisions and aggregation, and settle out. No long-term accumulation in the lower atmosphere or troposphere is thus possible.

The particulates in the atmosphere that are most important in influencing the global heat balance are the sulfate aerosols, due to their unique optical and chemical properties. Not only do they efficiently scatter the sun's rays directly, they also exert an indirect effect in their role as a major source of cloud condensation nuclei. They increase the concentration of cloud droplets, resulting in an increase in the scattering surface of clouds. Also, rain may be inhibited, since the mean droplet size decreases with increasing numbers of droplets, leading to a further increase in the cloud cover. On the other hand, as shown in Figure 6.2, clouds also absorb solar and terrestrial radiation; heat gained from this absorption will counter the heat lost by the scattering effect of clouds. Although the relative strengths of these two processes are not known with precision, evidence accumulated thus far indicates that the scattering effect dominates, with sulfates and cloud formation resulting in a net cooling of the planet.

Not all volcanic eruptions have an impact on climate, but those that emit copious quantities of sulfur and are forceful enough to inject the sulfur directly into the stratosphere provide dramatic evidence of the ability of sulfate aerosols to affect global climate. In the troposphere (see pp. 196–198 for discussion of troposphere and stratosphere), the lifetime of sulfate aerosol is a couple of days. In contrast, their lifetime in the stratosphere, which is a quiescent region with very low concentrations of constituent chemicals, can be a year or longer. The Mount Pinatubo eruption in the Philippines in June 1991, one of the largest blasts in the past 170 years, emitted an estimated 10 Tg (1 Tg $= 10^{12}$ g) of sulfur; it was followed by a perceptible decline in the mean global temperature over the next two years (see Figure 6.7).

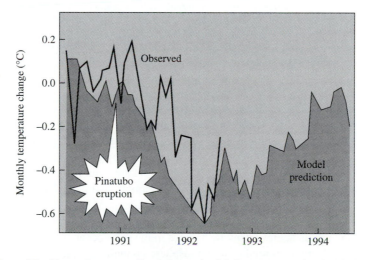

Figure 6.7   Observed versus predicted changes in global temperatures after the Mt. Pinatubo eruption.   *Source:* F. Pearce (1993). Pinatubo points to vulnerable climate. *New Scientist* 138(1878):7. (Based on model of J. E. Hansen, Goddard Institute of Space Studies, National Aeronautics and Space Administration (NASA), New York, New York). Copyright © 1993 by New Scientist. Reprinted with permission.

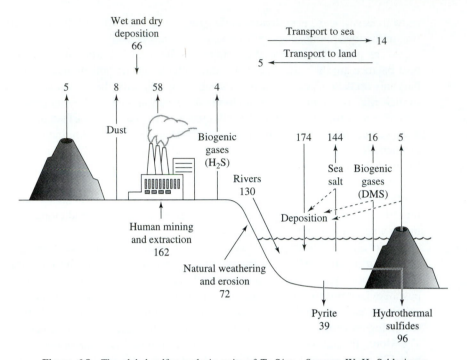

Figure 6.8 The global sulfur cycle in units of Tg S/yr. *Sources:* W. H. Schlesinger (1997). *Biogeochemistry, an Analysis of Global Change*, second edition (New York: Academic Press); U.S. Environmental Protection Agency (2000). *National Air Pollution Emission Trends, 1900–1998*. EPA-454/R-00-002; U.S. Energy Information Agency (2000). International Coal Consumption Data by Region with Most Countries and World, 1990–1999. (http://www.eia.doe.gov/emeu/international/coal.html#IntlConsumption) A. S. Lefohn et al. (1999). Estimating historical anthropogenic global sulfur emission patterns for the period 1850–1990. *Atmospheric Environment* 33:3435–3444.

c. Sulfur cycle.   The global sulfur cycle is diagrammed in Figure 6.8. Sulfur is abundant in Earth's crust in sulfide minerals and in calcium and magnesium sulfates. Considerable amounts of sulfur are processed throughout the microbial world. The main form of terrestrial sulfur emissions is H_2S produced by sulfate-reducing bacteria (see redox reaction 4a, Table 13.2, p. 310). These bacteria are responsible for the sulfur smell and black coloration, due to iron sulfides, of some aquatic habitats. Over the open ocean, the main biogenic sulfur compound is dimethylsulfide (DMS), CH_3SCH_3, which is produced in plankton by the enzymatic cleavage of *dimethylsulfonopropionate,* a compound that may help plankton achieve osmotic balance in the salty ocean water. It is estimated that some 20 Tg of sulfur are emitted to the atmosphere annually in volatile sulfur compounds from biogenic sources, of which DMS accounts for 16 Tg. The sulfur in these compounds is oxidized in the atmosphere to SO_2. In addition, as noted above, volcanoes occasionally spew large amounts of H_2S and SO_2 into the atmosphere.

The SO_2 from all sources is subsequently oxidized to sulfuric acid (under some conditions, the reduced sulfur compounds may be oxidized directly to H_2SO_4, without an intermediate SO_2 stage):

$$2SO_2(g) + O_2(g) + 2H_2O(1) = 2H_2SO_4(aq) \tag{6.12}$$

Direct reaction of SO_2 with O_2 is very slow, and the oxidation is actually carried out by more reactive species, particularly the hydroxyl radical and hydrogen peroxide (see p. 192). SO_2 lasts only a day or so before it is oxidized further. Some of the sulfuric acid in the atmosphere is neutralized (see discussion of global acidification, p. 305) by ammonia or calcium carbonate particles. The sulfuric acid and the sulfate salts are hygroscopic (absorb water), and form particles that serve as cloud condensation nuclei. Alternatively, the SO_2 is oxidized in the raindrops themselves. In either case, the sulfur cycle is completed by rainout (see Figure 6.8).

This natural cycle is now being strongly perturbed by human activity. Large quantities of SO_2 are emitted into the atmosphere by the burning of fossil fuels, especially coal, and by the smelting of sulfide minerals. In 1999, the amount of sulfur emitted from anthropogenic sources is estimated to have been 58 Tg/year, larger than the biogenic emission rate. Anthropogenic emissions had been higher a decade earlier, 69 Tg in 1990. Much of the subsequent decline occurred in the former Soviet Union (from 11 Tg of S emitted in 1990 to 5 Tg in 1999), due to the energy decline resulting from the collapse of the economy, and to a broad switch from coal to natural gas in Russia. Emissions also declined in Western Europe, Japan, and the U.S., as a result of regulation.

Implementation of the Clean Air Act Amendments of 1990 reduced U.S. emissions from 10.7 Tg in 1990 to 8.9 Tg in 1999, resulting in substantial reductions of sulfate deposition, especially in the eastern half of the country (see Figure 6.9). This was accomplished largely through an "emissions trading" program administered by the EPA. Each power plant of 25 MW or greater capacity is given a number of SO_2 allowances, and allowances in excess of actual emissions can be sold to other facilities. Each company can decide whether it is cheaper to install emission control technology, buy low-sulfur fuel, or buy additional allowances. This "market mechanism" allows the overall emissions target to be achieved at the least cost to the economy.

Since sulfuric acid is the main contributor to acid rain, the lower sulfate means less acid deposition. Before 1990, much of the northeastern part of the U.S. had sulfate deposition rates in excess of 8 kg (as S) per hectare per year, a level sufficient to acidify moderately vulnerable soils (see pp. 300–308 for discussion of soil buffering capacity), whereas in the late 1990s, the area subject to this deposition rate had shrunk to parts of Pennsylvania, West Virginia, Ohio, and Tennessee. Despite this progress, recovery of the soil and vegetation will require further reductions in acid deposition (see p. 303). Large areas of China, which is currently the largest sulfur emitter (13 Tg of S in 1999) are also subject to intensive sulfate deposition (see Figure 6.10), and the situation could grow much worse if coal use continues to expand without major sulfur abatement measures.

International shipping may play a significant role in the global sulfate aerosol. Satellite photographs show aerosol plumes hundreds of kilometers long in the wake of large ships. Their diesel engines burn the cheap "residual oil" that is left over at the refinery after the lighter hydrocarbons have been removed. This residual has a sulfur content of around 3 percent, and marine vessels are estimated to contribute 5 percent of the global sulfur emissions from all fuel combustion sources. The contribution can be much higher (up to 30 percent of

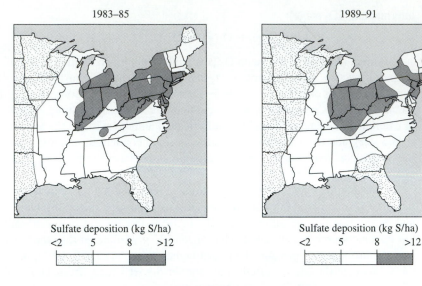

Figure 6.9   Mean annual wet sulfate deposition (kg S/ha-yr) in the eastern United States in 1983–1985, 1989–1991, and 1995–1997 based on measurements from the National Atmospheric Deposition Program.   *Source:* J. A. Lynch et al. (2000). Changes in sulfate deposition in eastern USA following implementation of Phase I of Title IV of the Clean Air Act Amendments of 1990. *Atmospheric Environment* 34:1665–1680. Copyright © 2000, Elsevier Science. Reprinted with permission.

ambient levels) in regions with heavy coastal traffic. In Northern Hemisphere oceans, the sulfur emission from ships is comparable to the biogenic flux from planktonic dimethylsulfoxide. In order to burn the residual oil, marine diesels operate at high temperatures, and also produce large quantities of NO_x, contributing further to the global aerosol.

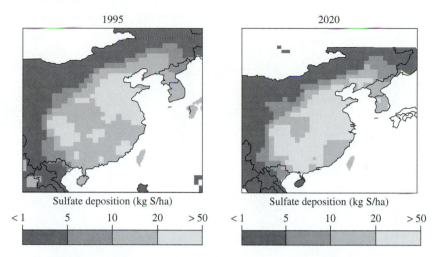

Figure 6.10   Sulfate deposition in China (kg S/ha-yr) in 1995 and projected deposition in 2020 if no policies are enacted to curb sulfur emissions.   *Source:* M. Amann et al. (2000). *RAINS-Asia Phase 2: An Integrated Assessment Model for Controlling SO_2 Emissions in Asia* (Laxenburg, Austria: International Institute for Applied Systems Analysis). Reprinted with permission.

How important is the sulfur contribution to the overall energy budget of Earth? This question is hard to answer, and scientists are grappling with large uncertainties, but the effect may be quite important. As shown in the bar graph in Figure 6.11, sulfate aerosol scatters enough sunlight to reduce the absorbed radiation by as much as 0.4 watts/m^2 (with an uncertainty range of between 0.2 and 0.8 watts/m^2). Sulfate's indirect effect of increasing cloudiness (and thus increasing the albedo) has an uncertainty range of 0.0 to 2.0 watts/m^2. The high end of the uncertainty range, a combined reduction of 2.8 watts/m^2, is over 1 percent of the absorbed radiation [see equation (6.2)]; on the basis of the Stefan-Boltzmann relation, it would cool Earth by 0.7 K [see equation (6.4)]. This much negative "radiative forcing"* would more than balance the positive forcing calculated for the current level of anthropogenic greenhouse gases (see next section). However, because of their complex characteristics, the uncertainties in the estimates for aerosols, including sulfate, are large, as indicated in Figure 6.11. Much effort is being directed toward reducing these uncertainties.

In any event, it seems that fossil fuel burning has two opposite effects on Earth's climate: greenhouse warming due to the emitted CO_2, and albedo cooling due to the emitted SO_2. Ironically, the international effort to improve air quality by reducing emissions of sulfur may have the unintended consequence of augmenting the greenhouse warming.

---

*Due to the complexity of the global climate system and its numerous feedback mechanisms, it is not currently possible to estimate precisely how each of these factors affects the climate. For this reason, scientists estimate "radiative forcing," rather than possible climatic responses. "Radiative forcing," expressed in Wm^{-2}, is the calculated change in the heat balance of Earth. Positive forcing warms the planet, while negative forcing cools it.

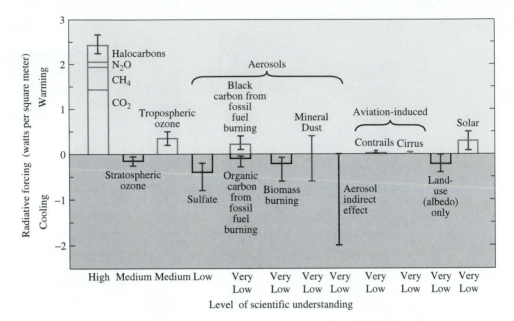

Figure 6.11   The global mean radiative forcing of the climate system for the year 2000, relative to 1750. The rectangular bars represent estimates of the contributions of these forcings, some of which yield warming (upper part) and some cooling (lower part). The indirect effect of aerosols shown is their effect on the size and number of cloud droplets. The vertical line about the rectangular bars indicates a range of estimates, guided by the spread in published values of the forcings and physical understanding. A vertical line without a rectangular bar denotes a forcing for which no best estimate can be given owing to large uncertainties.   *Source:* A Report of Working Group I of the Intergovernmental Panel on Climate Change (IPCC) (2000). *Summary for Policymakers* (Geneva, Switzerland: World Meteorological Organization/United Nations Environment Programme). (http://www.unep.ch/ipcc/pub/spm22-01.pdf)

However, reducing soot from fuel burning could ameliorate warming since the black carbon particles (see Figure 6.11) absorb sunlight.

6.3 GREENHOUSE EFFECT

a. Infrared absorption and molecular vibrations.   As mentioned above, the greenhouse effect is the trapping of heat from below by the atmosphere. Earth's atmosphere admits the visible rays from the sun, but traps the infrared rays emanating from Earth's surface. Some of Earth's heat is carried from the surface by air currents or by the evaporation of water (see Figure 6.2), but most of it is radiated to the atmosphere, which traps it and radiates much of it back, releasing the rest to space. But how is this trapping accomplished? Why do we worry about CO_2 and other minor constituents of the atmosphere, when it is made up almost entirely of N_2 and O_2 (see Table 6.1)?

TABLE 6.1   COMPOSITION OF DRY AIR AT GROUND LEVEL IN REMOTE
CONTINENTAL AREAS

Constituent	Formula	Concentration (by volume, ppm)
Nitrogen	N_2	780,900
Oxygen	O_2	209,400
Argon	Ar	9,300
Carbon dioxide	CO_2	370
Neon	Ne	18
Helium	He	5.2
Methane	CH_4	1.7
Krypton	Kr	1.1
Hydrogen	H_2	0.5
Nitrous oxide	N_2O	0.3
Xenon	Xe	0.08
Carbon monoxide	CO	0.04–0.08
Organic vapors		0.02
Ozone	O_3	0.01–0.04

The answer is that the major atmospheric gases are unable to absorb infrared light. They do not meet the two fundamental requirements for the absorption of electromagnetic radiation:

1) When radiation is absorbed by a molecule, the molecule undergoes a quantum transition, involving the movement either of its electrons or its nuclei; the energy of the radiation must therefore match the energy of the molecular transition. In the infrared region of the spectrum, the available transitions involve movement of the nuclei in molecular vibrations. That is why argon, the third most abundant atmospheric constituent (0.9 percent) is transparent to infrared radiation. Since argon is monatomic, it has no vibrations.

2) Because radiation is electromagnetic, its absorption requires that the transition change the electric field within the molecule, that is, the transition must alter the molecule's dipole moment (the vector sum of atomic charges times their distances from the molecule's center of mass). This second requirement is the reason that N_2 and O_2 are unable to absorb Earth's infrared radiation. Although their nuclei do vibrate along the bond joining them, and the energy of the vibration is in the infrared region, the vibration does not change the dipole moment. Because the molecule is symmetrical, the dipole moment remains zero no matter how much the bond is stretched. The vibration is infrared-*inactive*. This is true for all *homonuclear* diatomic molecules. The dipole moment is altered by vibrations of *heteronuclear* diatomic molecules, such as CO, NO, and HCl, since their atoms have different partial charges. However, these molecules do not contribute significantly to the greenhouse effect because their concentration in the atmosphere is too low and their infrared absorption too weak.

In contrast, polyatomic molecules have numerous vibrations ($3n - 6$ for nonlinear molecules, where n is the number of atoms, or $3n - 5$ for linear molecules); at least some of these vibrations change the dipole moment and are infrared-active. All the gases that

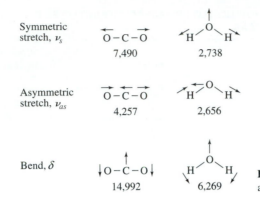

Figure 6.12   Molecular vibrations of CO_2 and H_2O in units of wavelengths (nm).

contribute significantly to the greenhouse effect are polyatomic. The two most important greenhouse molecules are water and carbon dioxide. Their vibrations are illustrated in Figure 6.12. All three vibrations of water change its dipole moment. For carbon dioxide, the symmetric stretching motion of the two O atoms leaves the dipole moment unchanged: the dipoles that each O atom generates relative to the carbon atom cancel one another due to the linear geometry. However, the net dipole moment is altered by the asymmetric stretch and the bending vibration.

Both water and carbon dioxide contribute to the greenhouse effect, but water is not listed in Table 6.1 because its content in the atmosphere varies greatly from place to place and time to time; on average, water molecules make up 0.4 percent of the atmosphere. The total amount of water is an order of magnitude greater than the amount of carbon dioxide. If Earth's temperature rises as a result of increasing carbon dioxide and other greenhouse gases, the water vapor pressure will also rise, and the amount of water in the atmosphere is expected to increase. This is an example of *positive feedback:* the greater the concentration of greenhouse gases, the higher the surface temperature; higher temperatures lead to greater amounts of atmospheric water, which in turn amplifies the greenhouse effect.

When greenhouse molecules absorb the terrestrial infrared radiation, they re-radiate it in all directions as illustrated in Figure 6.13. Absorption and re-radiation continue repeatedly with increasing altitude, leaving less than 10 percent of the absorbed infrared radiation available near the top of the atmosphere (see Figure 6.2). As a result, the lower atmosphere is warmer, but the higher atmosphere is colder than it would be in the absence of infrared absorbers.

It might appear that water and carbon dioxide would not be effective in heat trapping because Earth's emissions cover a wide spectrum of wavelengths, whereas the molecular vibrations correspond to specific energies. However, the molecules can undergo not only vibrations but also rotations. For each molecular vibration, infrared photons can induce transitions to many different rotational levels (rates of rotation). Consequently, each vibration has a broad absorption band. These bands are shown for water and carbon dioxide in Figure 6.14. In the top panel of the figure, the absorptions for these two molecules are added up and superimposed on Earth's emission spectrum. The combined absorption bands can be seen to block most of the terrestrial radiation. There is, however, a relatively

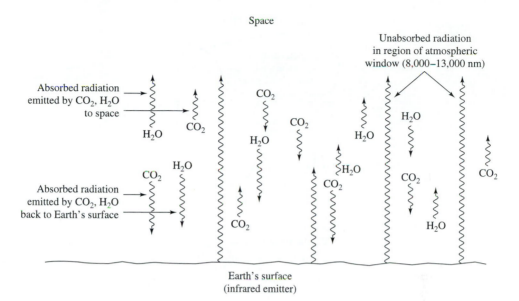

Figure 6.13   Re-radiation of terrestrial infrared radiation by greenhouse gases.

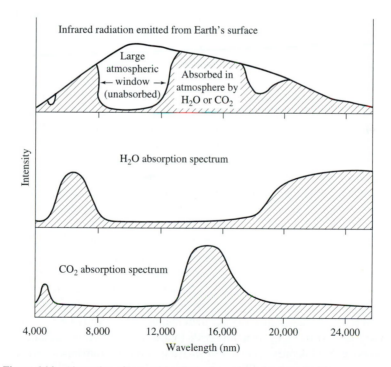

Figure 6.14   Absorption of terrestrial radiation by water and carbon dioxide.

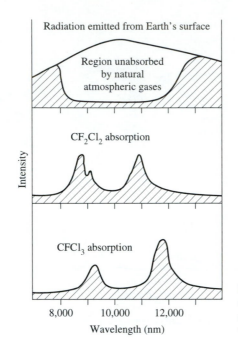

Intensity

Radiation emitted from Earth's surface

Region unabsorbed
by natural
atmospheric gases

CF_2Cl_2 absorption

$CFCl_3$ absorption

8,000   10,000   12,000

Wavelength (nm)

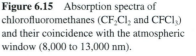

Figure 6.15   Absorption spectra of chlorofluoromethanes (CF_2Cl_2 and $CFCl_3$) and their coincidence with the atmospheric window (8,000 to 13,000 nm).

unobstructed region of the spectrum between 8,000 and 12,000 nm through which radiation can escape. This region is called the *atmospheric window.*

This window can be filled by other polyatomic molecules such as the chlorofluorocarbons (CFCs) (see Figure 6.15), methane (CH_4) and nitrous oxide (N_2O). The CFCs are of major concern as destroyers of stratospheric ozone (see p. 208), but they are also important greenhouse gases. Their impact is significant even though their concentration is about five orders of magnitude lower than that of carbon dioxide. The reason is that the CO_2 absorptions are nearly "saturated," that is, most of the radiation emitted within the absorption bands is already absorbed. As a result, each extra CO_2 molecule contributes only a relatively small amount to the total absorption. (The same is true for each additional water molecule.) In contrast, the contribution to overall absorption is relatively high for each extra CFC molecule precisely because the CFCs are very dilute and absorb only a small fraction of the radiation, but do so in the window region. One extra CFC molecule contributes thousands of times more to the greenhouse effect than one extra CO_2 molecule.

The relative *radiative forcing* (i.e., the relative contribution to the infrared absorption per molecule added to the atmosphere) is compared in Table 6.2 for methane, nitrous oxide, CFC-11 (CF_2Cl_2), historically one of the most widely produced CFCs, and HFC-23 (HCF_3), currently the major CFC substitute. The radiative forcing depends on the lifetime of the gas, as well as on its effectiveness as an infrared absorber. Thus, the N_2O contribution per molecule is much larger than that of CH_4, mainly because it is much longer lived. CH_4 is destroyed by reaction with hydroxyl radicals in the atmosphere (see p. 191), whereas N_2O is only removed by drifting into the stratosphere, where it is photolyzed by ultraviolet rays (see p. 208).

TABLE 6.2 SUMMARY OF PROPERTIES OF GREENHOUSE GASES AFFECTED BY HUMAN ACTIVITIES

	CO_2	CH_4	N_2O	CFC-11	HCF-23
Atmospheric concentration	ppmv	ppbv	ppbv	pptv	pptv
Preindustrial (1750–1800)	~280	~700	~270	zero	zero
Current	370	1745	314	268	14
Current rate of change/year*	1.5/yr[†]	7.0/yr[†]	0.8/yr	−1.4/yr	0.55
(% increase/year)	0.41	0.40	0.25	−0.52	3.92
Atmospheric lifetime (years)	5 to 200[‡]	12	114	45	260
Per molecule ratio of radiative forcing[§] $[\Delta F(\text{GHG})/\Delta F(\text{GHG } CO_2)]$	1	23	296	4,000	11,700
Major removal mechanism	I	II	III	III	III

ppmv, ppbv, and pptv are parts per million, parts per billion, and parts per trillion by volume, respectively.

*Rate is calculated over the period 1990–1999.

[†]Rate has fluctuated between 0.9 ppm/yr and 2.8 ppm/yr for CO_2, and between 0 and 13 ppb/yr for CH_4, over the period 1990–1999.

[‡]No single lifetime can be defined for CO_2 because of the different rates of uptake by different removal processes.

[§]Quoted forcings are for a 100-year time horizon.

I—slow exchange of carbon between surface waters and deeper layers of ocean; uptake into biomass.

II—reaction with hydroxyl radical in troposphere.

III—photolysis in stratosphere.

Sources: (a) Working Group I, IPCC (2001) *Technical Summary of the Working Group I Report.* Intergovernmental Panel on Climate Change: Geneva, Switzerland. (b) Carbon Dioxide Information Center (2000). *Current Greenhouse Gas Concentrations.* Oak Ridge National Laboratory: Oakridge, Tennessee.

b. Greenhouse gas trends. After the introduction of CFC compounds for a variety of products in the 1950s, the atmospheric CFC concentration increased rapidly until about 1990 (see Figure 6.16), but is now declining, thanks to international agreements to protect stratospheric ozone (see p. 195). Annual production of CFCs peaked at 1.1 Tg in the mid-1980s, and dropped to 0.08 Tg by the late 1990s. However, the decline in atmospheric concentration is slow because CFC lifetimes are long (see Table 6.2). Like N_2O, they are removed from the atmosphere only after transport to the stratosphere and encountering energetic ultraviolet radiation (see p. 208). Moreover, the ozone cure may actually worsen greenhouse warming, since CFC substitutes like HCF_3 (HCF-23) are themselves powerful greenhouse gases, and are still long-lived. Yearly production of HCF_3 stands at around 0.1 Tg and is increasing.

N_2O is a side-product of the microbial process of *denitrification,* in which NO_3^- is converted mainly to N_2 (see pp. 360–361); the ratio of N_2O to N_2 is about 1:16 but can vary depending on conditions (for example, pH, O_2 concentration). There are also indications that N_2O is a side-product of *nitrification,* in which ammonia is converted to nitrate. Natural sources are thought to account for 7–14 Tg of N per year as N_2O, while the anthropogenic contribution is 5–6 Tg/year (see Table 6.3A). Most of the anthropogenic contribution derives from nitrogen fertilizer use, which increases both nitrification and

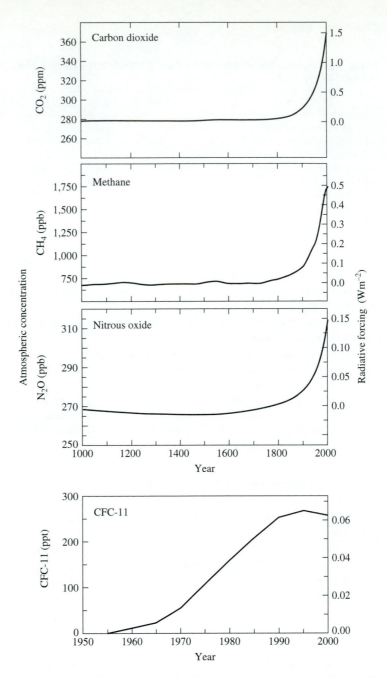

Figure 6.16 Estimated historical concentrations of major greenhouse gases. Carbon dioxide, methane, and nitrous oxide are natural components of the atmosphere, and their concentrations began to rise above background in the nineteenth century at the dawn of the Industrial Revolution. The chlorofluorocarbons (CFCs) are synthetic compounds without a natural component. Their atmospheric emissions began in the 1950s when they entered the world market in a variety of products. *Source:* A Report of Working Group I of the Intergovernmental Panel on Climate Change (IPCC) (2000). *Summary for Policymakers* (Geneva, Switzerland: World Meteorological Organization/United Nations Environment Programme). (http://www.unep.ch/ipcc/pub/spm22-01.pdf)

TABLE 6.3A   SOURCES OF ATMOSPHERIC EMISSIONS OF NITROUS OXIDE (2000)

Natural emissions	Emission (Tg N per year)
Terrestrial sources	
Tropical soils: wet forests, dry savannas (a)	2.7–5.7
Temperate soils: forests, grasslands (a)	0.6–4
Total soil	**3.3–9.7**
Aquatic sources	
Rivers, estuaries, continental shelf (b)	0.4
Deep ocean (b)	3.5
Total aqueous	**3.9**
Total natural emissions	**7.2–13.6**
Anthropogenic emissions	
Cultivated soils (c, d)	4.6*
Biomass burning (a)	0.2–1
Fossil fuel combustion (e)	0.3[†]
Adipic acid production (f)	0.07[‡]
Total anthropogenic emissions	**5.2–6.0**
Total natural and anthropogenic emissions	**12.4–19.6**

*Reference cited in (c) quantifies N_2O emissions in Asia (2.1 Tg/y) for all agricultural inputs that include synthetic fertilizer, animal wastes, biological N-fixation (in legumes), and crop residue burning. Since Asia's fertilizer inputs constitute about 48 percent of the global amount, an additional 52 percent was added to the Asian total. According to the analysis in reference (d), animal waste contributes 1.2 Tg of N_2O to the global total.

[†]Based on emission factors for burning coal, petroleum, and natural gas given in reference (e).

[‡]As described in reference (f), prior to 1999 emissions from industrial plants manufacturing adipic acid were about 0.37 Tg of N_2O. New abatement equipment added around 1999 has reduced emissions to the level quoted in the table.

Sources: (a) M. Prather et al. (1995). Other trace gases and atmospheric chemistry. In *Climate Change 1994: Radiative Forcing of Climate Change and an Evaluation of the IPCC IS92 Emission Scenarios,* J. T. Houghton et al. (eds.) (Cambridge, U.K.: Cambridge University Press).

(b) S. P. Seitzinger et al. (2000). Global distribution of N_2O emissions from aquatic systems: Natural emissions and anthropogenic effects. *Chemosphere–Global Change Science* 2: 267–279.

(c) A. R. Mosier and Z. Zhaoliang (2000). Changes in patterns of fertilizer nitrogen use in Asia and its consequence for N_2O emissions from agricultural systems. *Nutrient Cycling in Agro-ecosystems* 57:107–117.

(d) P. Czepiel et al. (1996). Measurements of N_2O from composted organic wastes. *Environmental Science and Technology* 30:2519–25.

(e) IPCC (1996). *Revised IPCC guidelines for national greenhouse gas inventories. Reference Manual, Vol. 3.* Intergovernmental Panel on Climate Change, Bracknell, U.K.

(f) A. Shimizu et al. (2000). Abatement technologies for N_2O emissions in the adipic acid industry. *Chemosphere—Global Change Science* 2:425–434.

TABLE 6.3B SOURCES OF ATMOSPHERIC EMISSIONS OF METHANE (2000)

Natural emissions	Emission (Tg CH_4 per year)
Wetlands (a)	120–175
Termites (b, c)	1.5–21
Wildfires (a, d)	5
Oceans (a)	5–25
Volcanoes (a)	3.5
Wild animals (a)	5*
Total natural emissions	**140–235**

Anthropogenic emissions	
Rice paddies (e)	20–100
Natural gas/drilling-transmission (f, g)	32–44[†]
Coal mining (h)	20–28
Enteric fermentation (domestic animals) (f, i, j)	40[‡]
Manure (f, i, j)	21[§]
Municipal solid waste disposal (f, j)	59[‖]
Biomass burning (j)	15[#]
Total anthropogenic emissions	**207–307**
Total natural and anthropogenic emissions	**347–542**

*Assumes emission from wild animals is one-third of emission in pre-industrial era as reported in reference (a).

[†]Emissions from the Soviet gas system were much higher up until the early 1990s, when emissions were estimated to range between 32 and 45 Tg per year. By the end of the 1990s, emissions are estimated to have declined to 10–23 Tg per year, with all sources outside of the former Soviet Union contributing about 22 Tg [see reference (g) for further information].

[‡]Calculated from emission factors for cattle per head per year gleaned from references (f) (developed countries) and (j) (developing countries), and world cattle populations given in reference (i).

[§]Includes manure from dairy and beef cattle, buffalo, sheep, goats, pigs, and poultry.

[‖]Different emission factors were calculated for developed and developing countries. The former was estimated from reference (f) and the latter from reference (j).

[#]Value for India provided in (j), and extrapolated for rest of developing countries.

(a) S. Houweling et al. (2000). Simulation of preindustrial atmospheric methane to constrain the global source strength of natural wetlands. *Journal of Geophysical Research* 105, No. D13:17,243–17,255.

(b) A. Sugimoto et al. (1998). Methane oxidation by termite mounds estimated by the carbon isotopic composition of methane. *Global Biogeochemical Cycles* 12:595–605.

(c) M. G. Sanderson (1996). Biomass of termites and their emissions of methane and carbon dioxide: A global database. *Global Biogeochemical Cycles* 10:543–557.

(d) E. J. Clugokencky et al. (2001). Measurements of an anomalous global methane increase during 1998. *Geophysical Research Letters* 28:499–502.

(e) R. L. Sass et al. (1999). Exchange of methane from rice fields: National, regional, and global budgets. *Journal of Geophysical Research* 104, No. D21:26,943–26,951.

(f) U.S. Environmental Protection Agency (1999). *U.S. Methane Emissions 1990–2000: Inventories, Projections, and Opportunities for Reductions.* EPA 430-R-013.

(g) A. I. Reshetnikov et al. (2000). An evaluation of historical methane emissions from the Soviet gas industry. *Journal of Geophysical Research* 105, No. D3:3517–3529.

(h) C. J. Bibler et al. (1998). Status of worldwide coal mine methane emissions and use. *International Journal of Coal Geology* 35:283–319.

(i) Food and Agriculture Organization (FAO) of the United Nations, FAOSTAT Agricultural Data (1990–2000) Agricultural Production, Live Animals, Livestock Primary, World, Developed Countries, Developing Countries. (http://www.apps.fao.org)

(j) A. Garg et al. (2001). Regional and sectoral assessment of greenhouse gas emissions in India. *Atmospheric Environment* 35:2679–2695.

denitrification; intensive agriculture is the likely cause of most of the observed increase in atmospheric N_2O (see Figure 6.16), currently estimated to be 0.25 percent annually. A significant industrial source ($\sim$0.4 Tg/y) had been identified* in the nitric acid oxidation of cyclohexanol to adipic acid, a precursor in nylon production, but process improvements have cut this rate by a factor of five.

Methane is the second most important greenhouse contributor after CO_2. Although it has a shorter lifetime (12 years) than CFCs or N_2O, its atmospheric concentration is much higher (see Table 6.2). Some of the methane comes from leaks in the gas distribution system, from coal mines, from biomass burning and wildfires, and from volcanoes (see Table 6.3B). Most of it, however, results from the action of anaerobic bacteria, the *methanogens,* which produce methane as an end-product of their metabolism. These bacteria are abundant in wetlands, rice paddies, manure piles, and municipal solid waste disposal sites. They are also found in the fore-stomachs of ruminant animals and in termites. Natural and anthropogenic sources are estimated to be of the same order of magnitude, roughly 190 and 260 Tg of CH_4 per year, respectively.

The natural contribution is dominated by wetlands, while the anthropogenic contribution is divided nearly equally among rice paddies, livestock, municipal solid wastes, and fossil fuel extraction and transmission (see Table 6.3B).

Global-scale measurements of atmospheric methane have revealed that although its concentration continues to increase (see Figure 6.16), the rate of increase has diminished over the last decade from about 0.9 percent to about 0.4 percent per year. The reasons for this slowdown are uncertain. A contributing factor may be improvements in the extensive natural gas distribution system of the former Soviet Union. Annual losses from this system were reported to be 32–45 Tg of CH_4 in 1990, but are now down to about 15 Tg. It has also been suggested[†] that the decreasing growth rate in atmospheric methane reflects its approach to a steady-state condition in which inputs to the atmosphere are balanced by removal via reaction with hydroxyl radical (see p. 191).

Although the relative share of the radiative forcing due to CFCs, N_2O, and CH_4 has been increasing, the largest effect is still due to CO_2 (see bar graph in Figure 6.11). Anthropogenic production of CO_2 from burning fossil fuels (plus a 3 percent contribution from cement production) is 6,300 Tg C/year, which dwarfs the other greenhouse gases. Deforestation is estimated to add another 1,600 Tg C/year. Of course, the natural flux of CO_2 to the atmosphere due to the constant respiration of the biosphere is much greater, but this flux is in balance with photosynthesis. The balance is nicely illustrated by the historical record of CO_2 concentration at the monitoring station in Mauna Loa, Hawaii (see Figure 6.17).

Every year the CO_2 declines to a minimum concentration in the summer when photosynthesis in the fields and forests of the Northern Hemisphere converts CO_2 to biomass; it rises to a maximum in winter when the dead vegetation decays, releasing its stored carbon as CO_2. The oscillatory pattern is regular from year to year, but it is superimposed on

---

*M. H. Tiemens and W. C. Trogler (1991). Nylon production: An unknown source of nitrous oxide. *Science* 251:932–934.

[†]E. J. Dlugokencky, K. A. Masarie, P. M. Lang, and P. P. Tans (1998). Continuing decline in the growth rate of the atmospheric methane burden. *Nature* 393:447–450.

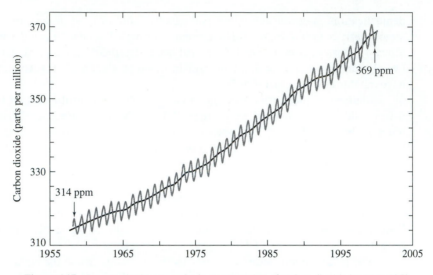

Figure 6.17   Increase in the atmospheric concentration of carbon dioxide between 1958 and 2000 at Mauna Loa, Hawaii. Carbon dioxide is taken up through photosynthesis in the summer producing the observed annual oscillations in concentration.   *Source:* C. D. Keeling et al., Scripps Institution of Oceanography, La Jolla, California (prior to 1974); National Oceanic and Atmospheric Administration, Washington, DC (since 1974).

a rising background of average CO_2 concentration, which increased from 314 ppm in 1958 to 369 ppm in 2000, a 17.5 percent increase in four decades.

Of the total 7,900 Tg C/year emitted by fossil fuels, cement production, and deforestation, only about 3,300 Tg C/year stays in the air. Where does the remaining CO_2 go? The oceans are an obvious possibility. Because seawater is alkaline and CO_2 is acidic, the oceans are a vast reservoir of CO_2 (see p. 285). However, only the surface layer of the ocean, the top 75 meters, is in equilibrium with the atmosphere, and its capacity for absorbing CO_2 is limited. Exchange of the surface layer with the deep oceans takes hundreds of years. It is estimated that the oceans are absorbing about 2,300 Tg of CO_2 carbon per year, half of the 4,600 Tg C/year not found in the atmosphere.

The location of the remaining carbon has been a subject of considerable debate, but it is now generally accepted that the biosphere itself is the reservoir. Studies of $^{13}C/^{12}C$ isotope ratios (which provide a measure of biosphere CO_2 flux, since plant photosynthesis discriminates against ^{13}C) have established a temperate latitude Northern Hemisphere CO_2 sink of about the right magnitude. Apparently vegetation in the Northern Hemisphere is absorbing a significant fraction of the emitted CO_2. However, much of this absorption can be attributed to forests that are regenerating on former agricultural land; as they reach maturity, their capacity for CO_2 storage will diminish.

Figure 6.18 illustrates the global carbon cycle with estimates for the pools of carbon residing in the air, soil, biomass, and oceans, as well as annual fluxes. (Here we switch to gigaton units to reduce the number of zeroes; 1 Gt $= 10^{15}$ g.) These fluxes are very large, making it hard to account fully for the fate of the relatively small anthropogenic contribution. Yet it is this contribution that is driving the accumulation of CO_2 in the atmosphere

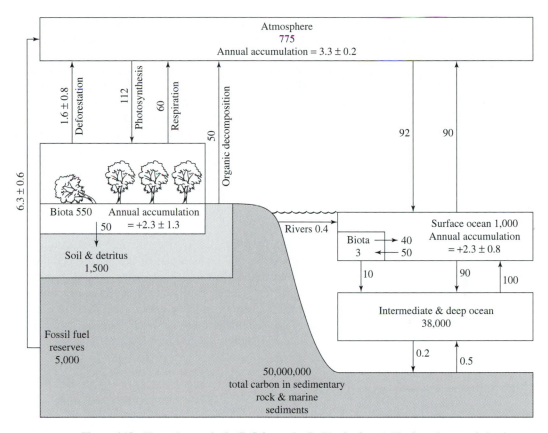

Figure 6.18 The carbon cycle (in Gt C for pools; Gt C/yr for fluxes). Net annual accumulation in biota is the difference between enhanced biomass accumulation (2.3 ± 1.3 Gt C/yr) and deforestation (1.6 ± 0.8 Gt C/yr), which equals about +0.7 Gt C/yr. *Sources:* Adapted from the Carbon Dioxide Information Analysis Center (2000). *Global Carbon Cycle (1992–1997)* (Oak Ridge National Laboratory, U.S. Department of Energy) (http://cdiac.esd.ornl.gov); Intergovernmental Panel on Climate Change (IPCC) (2000). *Summary for Policymakers, Land Use, Land-Use Change, and Forestry* (Geneva, Switzerland: World Meteorological Organization/United Nations Environment Programme).

and, with it, the greenhouse warming. When the estimates are integrated over the last century and a half (1850–1998), the picture is similar to the one we sketched above. The atmospheric CO_2 concentration has risen 30 percent, from 285 to 370 ppm, representing an excess carbon accumulation of 176 Gt. The estimate for total carbon emitted over this period is 270 ± 30 Gt from fossil fuels and cement production, and 136 ± 55 Gt from deforestation. The sum of these two contributions exceeds the atmospheric accumulation by about 230 Gt, of which half was taken up by the oceans, and the other half by the terrestrial ecosystem. Thus there appears to have been a small net loss of land vegetation (deforestation minus uptake) of perhaps 20 Gt.

Evaluating the future of biospheric carbon storage is critical for predicting atmospheric CO_2 levels, but the subject is riddled with uncertainties. Storage could increase because of the fertilizing effect of CO_2. In the laboratory, plants exposed to elevated CO_2 show enhanced growth. In the field, however, CO_2 may not be the limiting nutrient. The supply of nitrogen, phosphorus, or water may not support increased biomass production. Experiments in which small areas of natural forests are exposed to elevated CO_2 (from surrounding towers) have shown little change in biomass over several years. In addition, higher temperatures may increase the rate of respiration (through higher metabolic rates of microbes) faster than the rate of photosynthesis, so that global warming could shift the balance toward respiration, and net carbon loss. Finally, if global warming shifts the climate zones, forests may be unable to adapt, especially if the shift is rapid; again the result may be loss of biomass. These issues are currently under intensive scrutiny.

6.4 CLIMATE MODELING

The mean global temperature has risen by $0.6 \pm 0.2°C$ since the late 19th century (see Figure 6.19). Is this rise due to human activity? Or is it part of the natural variation? If Earth's temperature is tracked back over longer periods of time, using several geophysical techniques as well as fossil records, the temperature is found to have varied greatly. These variations are illustrated in Figure 6.20. The graph on the left shows that over the past several hundred years, the temperature has fluctuated more than 1°C. Temperatures dipped down from about 1400 to 1850, a period known as the Little Ice Age in Europe. Much

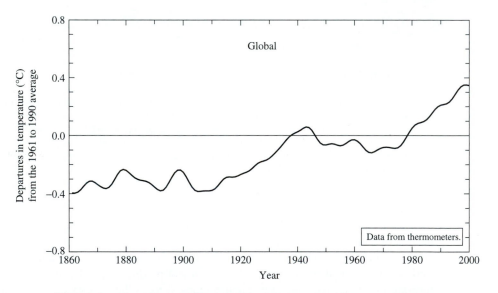

Figure 6.19   Mean global temperatures, 1860–2000. Trend line is filtered to smooth out annual fluctuations in temperature.   *Source:* A Report of Working Group I of the Intergovernmental Panel on Climate Change (IPCC) (2000). *Summary for Policymakers* (Geneva, Switzerland: World Meteorological Organization/United Nations Environment Programme). (http://www.unep.ch/ipcc/pub/spm22-01.pdf)

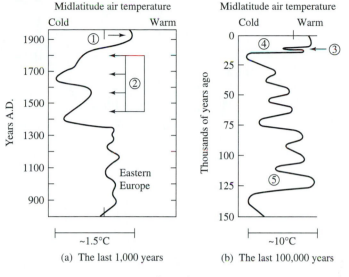

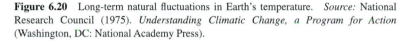

(a) The last 1,000 years   (b) The last 100,000 years

Legend
1. Thermal maximum of 1940s
2. Little Ice Age
3. Cold interval
4. Present interglacial (Holocene)
5. Last previous interglacial (Eemian)

Figure 6.20 Long-term natural fluctuations in Earth's temperature. *Source:* National Research Council (1975). *Understanding Climatic Change, a Program for Action* (Washington, DC: National Academy Press).

larger fluctuations occurred over a span of hundreds of thousands of years (see right graph). These reflect the global ice ages, the last one of which reached its lowest temperature about 20,000 years ago.

What caused these temperature swings? The matter is still under intense debate, but a role for the greenhouse effect is strongly suggested by the variation in CO_2 and CH_4 content found in the air bubbles trapped in the Antarctic ice sheet. The record now extends back over the past 150,000 years; as shown in Figure 6.21, the temperature is remarkably well-correlated with the CO_2 and CH_4 levels. Peaks and troughs in the records of temperature and greenhouse gases line up very well. Of course, a correlation does not establish cause and effect. It is possible that high temperatures induced higher CO_2 and CH_4 levels, rather than vice versa, or that some other factor induced similar variations in all three variables. However, it has been discovered that an intense warming period (4–8°C rise in high-latitude surface water temperatures, according to isotopic evidence) about 55.5 million years ago was most likely caused by methane escaping from hydrates (see pp. 22, 270) in deep ocean sediments.*

*M. E. Katz, D. K. Pak, G. R. Dickens, and K. G. Miller (1999). The source and fate of massive carbon input during the latest Paleocene thermal maximum. *Science* 286:1531–1533.

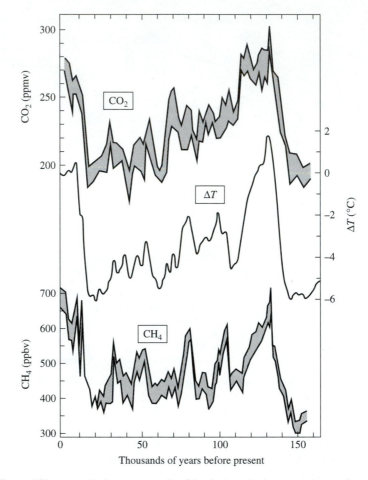

Figure 6.21 Antarctic ice-core records of local atmospheric temperature, and corresponding air concentrations of carbon dioxide and methane for the past 160,000 years. *Source:* J. T. Houghton et al., eds. (1990). *Climate Change: The IPCC Scientific Assessment* (Cambridge, U.K.: Cambridge University Press).

The $^{13}C/^{12}C$ ratio of microfossils in sediments dating from this period are unusually low. Since photosynthesis discriminates against ^{13}C (the heavier isotope reacts more slowly when CO_2 is processed by the plant enzymes), atmospheric CO_2 has a higher $^{13}C/^{12}C$ ratio than does plant matter or its products. Thus the low $^{13}C/^{12}C$ ratio in 55.5-million-year-old fossils indicates a massive injection of biogenic carbon into the atmosphere at that time. Evidence for massive slumping of sediments supports the view that this carbon came from sedimentary methane, released from methane hydrate as a result of a temperature rise of the deep ocean, possibly related to volcanic activity. The amount of methane required to produce the observed reduction in the $^{13}C/^{12}C$ ratio is very large, $\sim 10^{18}$ g; it would have been converted to CO_2 in the atmosphere, and the two gasses together would have produced a pronounced greenhouse warming. The fossil record indicates extinction of many deep-sea species at this time, and the rapid evolution on land of new animal species, including primates. (Methane release at low rates continues today at deep sea volcanic vents, but the methane is consumed by microorganisms and deposited as $CaCO_3$, often in limestone "chimneys.")

Given the variability of past climates, predicting the effects of human activities on future climates is a hazardous undertaking. It is also an important challenge, one that atmospheric scientists are trying to meet by calculating climate changes using global circulation models (GCMs). The general idea behind these models is to apply the laws of physics to the atmosphere and oceans, and to use the inputs of energy from the sun to predict the dynamics of air parcels around the globe. The system is extremely complicated and requires many approximations, even with the most powerful supercomputers. There are also conceptual difficulties such as determining the role of aerosols and clouds, including the myriad chemical reactions in the atmosphere, and describing chemical exchanges with the ocean and the biosphere.

In an effort to gain consensus based on the best scientific information, the *Intergovernmental Panel on Climate Change (IPCC)* was established in 1988 jointly by the World Meteorological Organization (WMO) and the United Nations Environment Programme (UNEP). Over 300 scientists from 26 nations have participated in the scientific assessment of climate change. The first report of the IPCC, *Climate Change* (1990), has been updated twice, the most recent edition published in 2001. Its projections estimate atmospheric CO_2 concentrations rising in a range between 540 and 940 ppm by the year 2100. The ranges for CH_4 and N_2O are 1,580–3,750 ppb and 350–460 ppb, respectively. The corresponding average global temperature increase is projected to be in the range 1.4 to 5.8°C (see Figure 6.22).

Such large temperature changes have not been experienced on Earth since the end of the last ice age. The effects would be profound. Due to both the expansion of water upon warming and the melting of mountain glaciers, the sea level would rise, flooding low-lying coastal zones where a substantial fraction of the human population currently lives. Computer models suggest a 0.1–0.9 meter rise over the next century. (It is worth noting that melting of Arctic ice would not affect sea level because most of it is floating on the water. The much thicker Antarctic ice sheet rests on land but is considered unlikely to melt unless the temperature rises even higher than currently predicted.)

The increasing temperature would also probably dry out the interior of continents, while increasing the precipitation rate in regions close to open water. The resulting gradients of temperature are likely to increase the severity of storms. The 2001 IPCC assessment considers it "very likely" that there will be higher maximum temperatures and more hot days over nearly all land areas, as well as higher minimum temperatures and fewer cold days and frost days. Intense precipitation events over many areas are also "very likely," while the prospect of increased risk of drought over most midlatitude continental interiors is in the "likely" category.

Additionally, as the average temperature rose, the climatic zones for flora and fauna would shift toward the poles. These shifts would force major readjustments of agriculture. Forests could be devastated if the rate of climate change outpaced the rate at which forest

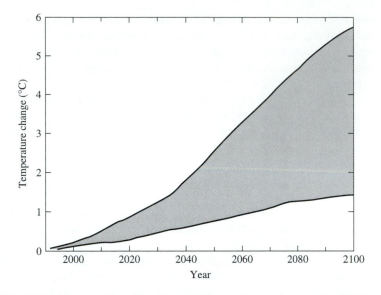

Figure 6.22   Shaded area shows range of global temperature increases from 2000 to 2100, based on different scenarios developed by the Intergovernmental Panel on Climate Change. The high end of the projections represents scenarios in which coal continues to provide a large share of the energy supply, energy efficiency implementation is slow, forests continue to be depleted, and alternative energy sources provide a relatively small share of total energy. The lower projections assume high energy efficiency, preservation of tropical forests, and a substantial shift to renewable and nuclear energy early in the 21st century. *Source:* A Report Accepted by Working Group I of the Intergovernmental Panel on Climate Change (IPCC) (2000). *Technical Summary* (Geneva, Switzerland: World Meteorological Organization/United Nations Environment Programme). (http://www.unep.ch/ipcc/pub/wg1TARtechsum.pdf)

species could migrate. There is also concern that tropical diseases would spread, as insect vectors increase their range. Not only is the magnitude of the projected change larger than experienced in Earth's recent history, but the rate of change is unprecedented.

Of course, the computer projections could be wrong, but confidence in their reliability is steadily increasing, because of advances in modeling climates of the recent and more distant past. For example, it has become possible to reproduce the temperature changes of the last 150 years quite accurately (see Figure 6.23a,b,c), and to separate out natural and anthropogenic radiative forcings in the model. There had been doubt about the human contribution to the current temperature rise because of a temperature peak in the mid-twentieth century, before anthropogenic greenhouse emissions were significant. In the modeling, this peak was found to be associated with decreased volcanic activity. The separate modeling of natural and anthropogenic forcings, and the quality of the fit to the data when they are combined, leaves little doubt that the current higher temperatures are indeed the result of human activity. Lingering questions about the adequacy of the computer models rest on their inability to reproduce the large temperature swings of the ice ages. The fundamental

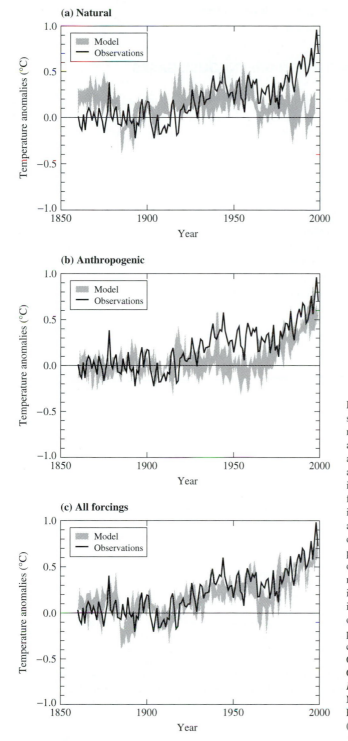

Figure 6.23 Simulated annual global mean surface temperatures. a) Model only includes natural forcings: solar variation and volcanic activity; b) model only accounts for anthropogenic forcings: greenhouse gases and an estimate of sulfate aerosols; c) model includes both natural and anthropogenic forcings. From (b) it can be seen that inclusion of anthropogenic forcings provides a plausible explanation for a substantial part of the observed temperature changes of the past century, but the best match with observations is obtained in (c) when both natural and anthropogenic factors are included. These results show that the forcings included are sufficient to explain the observed changes, but do not exclude the possibility that other forcings may also have contributed.   *Source:* A Report of Working Group I of the Intergovernmental Panel on Climate Change (IPCC) (2000). *Summary for Policymakers* (Geneva, Switzerland: World Meteorological Organization/United Nations Environment Programme). (http://www.unep.ch/ipcc/pub/spm22-01.pdf)

mechanisms of these large changes are still not understood, but there has been progress in simulating some aspects of ancient climates.

6.5   INTERNATIONAL AGREEMENTS ON GREENHOUSE GASES

Global warming is a problem of unprecedented dimension for human society, and it is not surprising that debate on what to do about it is lengthy and rancorous. To stem the tide of rising greenhouse gas (GHG) levels will require profound changes in many countries, with uncertain consequences for their economies. The scale of the problem is illustrated in Figure 6.24, which projects the trajectories of atmospheric CO_2 concentrations (which is the dominant anthropogenic greenhouse gas) for various levels of carbon emissions. Even if annual CO_2 emissions were capped in the next 20 years at 9 Gt, and then declined to 2 Gt (the level in the year 1950) over the next two centuries, the stabilized CO_2 concentration would still reach 450 ppm and stay there for hundreds of more years (mixing time of the oceans). Stabilizing CO_2 concentrations at 550 ppm, about double their pre-industrial level of 280 ppm and a common benchmark in discussions about global warming, would eventually require a reduction by all nations to 60 percent of the 1990 level. This is far more than cuts currently contemplated in international negotiations.

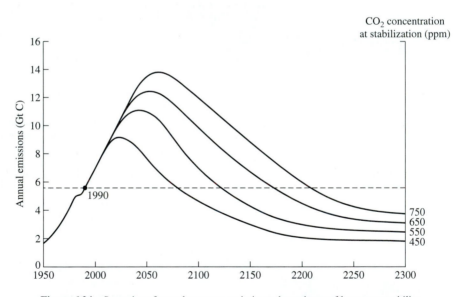

Figure 6.24   Scenarios of greenhouse gas emissions; dependence of long-term stabilization levels of atmospheric CO_2 upon timing and magnitude of peak emissions of carbon. Dashed line shows emissions in 1990, the benchmark year for emissions reductions in the Kyoto Protocol on Climate Change. *Source:* Adapted from Wigley et al. (1996). Economic and environmental choices in the stabilization of atmospheric CO_2 emissions. *Nature* 379:240–243. Copyright © 1996 by Nature, reprinted with permission from *Nature* and authors of journal article.

Recognizing the seriousness of the problem, the industrialized countries signed a Framework Convention on Climate Change at the 1992 Earth Summit in Rio de Janeiro. The Convention called for reducing emissions of GHGs to 1990 levels by the year 2000, but was non-binding and had no impact on subsequent emissions, which rose steadily through the 1990s (see Figure 6.24). It did, however, set the stage for subsequent international negotiations, culminating in the 1997 Kyoto Protocol. The goal was now to reduce GHG emissions to 5 percent below 1990 levels by 2008–2012, and specific targets were set for each country. The highest emitters—the U.S., Japan, and Europe—were assigned the greatest reductions, 6–8 percent—and less-developed countries (LDCs) were exempted.

The Protocol allowed that targets could to some extent be accomplished by emissions trading, although a trading system has yet to be specified. Emissions trading has worked well in lowering the cost of controlling SO_2 from U.S. power plants (p. 157), but GHG emissions trading across international borders will be a much more complex task. The Protocol also established a Clean Development Mechanism, by which developed countries could gain GHG credits by investing in GHG-reducing projects (e.g., tree planting) in LDCs. There remains significant disagreement on the extent to which targets can be achieved by these "market-based" mechanisms, as opposed to actual emission reductions.

The most contentious issue is the counting of biosphere uptake as a GHG credit. The U.S., with a large land area and evidence for substantial CO_2 uptake by its vegetation, claimed a credit worth about 20 percent of its target, a claim rejected by European countries, who want the Protocol to concentrate exclusively on emission reductions.

The fate of the Kyoto Protocol is currently in doubt, since the U.S. president, George W. Bush, has rejected it outright, claiming it would unduly hurt the U.S. economy. Nevertheless, there is considerable support for reigning in GHGs in the U.S., as well as elsewhere. For example, the city of Seattle has announced that it will exceed the Kyoto CO_2 reduction goals by improving energy efficiency, promoting renewable energies, reducing traffic, recycling industrial waste materials and heat, and planting trees.

The European Union is committed to maintaining its Kyoto Protocol reduction target (8 percent by 2008–2012). The U.K., historically a high CO_2 emitter, already reports a 7 percent reduction in 1999, relative to 1990, thanks largely to switching from coal to natural gas. A number of large corporations have established energy and/or GHG reduction targets. Even in the oil industry, which spearheaded political opposition to the Kyoto Protocol, the BP Corp. has broken ranks and announced its own GHG reduction target, implemented by an emissions trading program among units of the company. Ford has announced that it will voluntarily upgrade fuel economy of its SUVs.

Thus, momentum is growing for concerted action on global warming. However, the task is huge, and will be with us for a very long time.

CHAPTER 7

OXYGEN CHEMISTRY

The reactivity of oxygen is the controlling factor in the chemistry of the atmosphere and Earth's crust. Elemental oxygen is avid for electrons: among the elements, only fluorine has a higher electronegativity. Consequently, oxygen combines with most other elements to form stable oxides (see Figure 7.1). Solid oxides of iron, aluminum, magnesium, calcium, carbon, and silicon make up Earth's crust, while the oceans are filled with the oxide of hydrogen. The atmosphere contains volatile oxides, especially carbon dioxide and sulfur dioxide. The only common element stable in the presence of oxygen is nitrogen. The bond connecting the two nitrogen atoms is so strong that converting N_2 to any of the nitrogen oxides costs considerable energy. Nevertheless, nitrogen oxides, N_2O, NO, and NO_2, are found in the atmosphere because once formed, they are only slowly converted back to N_2 and O_2.

7.1 NITROGEN OXIDES: FREE ENERGY

Stability trends of atmospheric oxides can be gauged from the standard enthalpies and free energies of formation listed in Table 7.1. The standard enthalpy is the heat released or absorbed under standard conditions (25°C and 1 atmosphere of pressure) when the compound is formed from the elements, which define the zero of the standard enthalpy scale. Negative values mean that heat is released, while positive values mean that heat is absorbed. All the enthalpy values are strongly negative, except for NO, NO_2, and O_3, which have positive values. The free energy expresses the position of equilibrium for the formation reaction [eq. (7.7), p. 182], and is the true measure of stability. It differs from the enthalpy because of the entropy change of the reaction (see Fundamentals 5.2, p. 108).

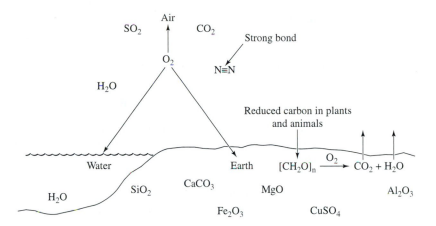

Figure 7.1   Most elements, except nitrogen, are stable as oxides.

TABLE 7.1   STANDARD (25°C) ENTHALPIES AND FREE
ENERGIES OF FORMATION FOR SOME ATMOSPHERIC
COMPOUNDS (kJ)

	ΔH_0	ΔG_0
O_3	142.2	163.4
CO_2	−393.4	−394.3
CO	−110.5	−137.2
NO	90.4	86.7
NO_2	33.8	51.8
$HNO_3(aq)$	−206.5	−110.5
SO_2	−296.8	−300.3
SO_3	−395.1	−370.3
$H_2SO_4(aq)$	−907.3	−741.8
$H_2O(g)$	−241.8	−228.6
$H_2O(l)$	−285.7	−237.2

However, the entropy contribution to the total energy is not large for these reactions, and the free energy trends parallel the enthalpy trends. All of the free energy changes are negative, meaning that formation of the compound is favored at equilibrium, except for NO, NO_2, and O_3. For these, the free energy change is positive, and consequently they are mostly decomposed back to the elements at equilibrium.

a. Free energy and the equilibrium constant.   In Fundamentals 5.2, the free energy change, ΔG, was introduced as the maximum amount of useful energy that can be extracted from a chemical reaction, while in Strategies 6.1, it was used to describe the energetics of raindrop formation. We now introduce a fundamental attribute of ΔG, namely that it determines the position of equilibrium of the reaction, i.e., the extent to which it proceeds toward the products or towards the reactants.

If we have a reaction

$$A + B = C + D \tag{7.1}$$

and mix the reactants together, the extent to which the reaction proceeds is governed by an *equilibrium constant,*

$$K = (C)(D)/(A)(B) \tag{7.2}$$

where (A), (B), (C), and (D) are the concentrations of the reactant and product molecules. Clearly the larger K is, the larger will be the concentrations of C and D, relative to A and B. The position of equilibrium lies toward products if K is large, and towards reactants if K is small. Equilibrium presupposes that the reaction can go in either direction, and that enough time has elapsed so that there is no further change in the concentrations.

How long it takes to reach equilibrium depends on reaction rates. The rate of the forward reaction

$$A + B \rightarrow C + D \tag{7.3}$$

can be expressed as

$$\text{rate}_f = k_f(A)(B) \tag{7.4}$$

where k_f is the forward *rate constant*. The concentrations of the reactant are multiplied together because the probability of an encounter between an A molecule and a B molecule increases as the product of the concentrations. k_f expresses the probability that a reaction will occur during an encounter. There is a corresponding rate for the reverse reaction

$$C + D \rightarrow A + B \tag{7.5}$$

$$\text{rate}_r = k_r(C)(D) \tag{7.6}$$

As soon as some C and D are formed, they can begin reacting back to give A and B. Equilibrium is established when the two rates become equal, after which there is no further change in the concentrations. We can then set the right-hand sides of eq. (7.4) and (7.6) equal to each other:

$$k_f(A)(B) = k_r(C)(D)$$

which, upon rearranging, is the same as eq. (7.2), and we recognize that the equilibrium constant is the ratio of forward and reverse rate constants:

$$K = k_f/k_r$$

The thermodynamic relationship between the equilibrium constant and the free energy change is:

$$\Delta G = -RT \ln K \tag{7.7}$$

where ln stands for natural logarithm, and R is a universal constant, (called the *gas constant*). Its value is 8.314 J/K. It is convenient to use base-ten instead of natural logarithms, and the factor between the two is 2.303. Thus:

$$\Delta G = -RT(2.303) \log K = -19.15T(\log K) \text{ joules}$$

At room temperature, 25°C (298 K), $\Delta G = -5706 \log K$ J or $-5.7 \log K$ kJ. Thus, every factor of ten increase in K corresponds to an additional 5.7 kJ of free energy change.

Once we know the magnitude of K, we have a quantitative measure of the position of the equilibrium. For given amounts of reactants we can calculate the amounts of products, and vice versa.

WORKED PROBLEM 7.1: Atmospheric NO Concentration

Q. *The concentration of NO in the atmosphere varies widely from place to place and from time to time; the average fraction is roughly 10^{-4} ppm. How does this compare with the expected equilibrium concentration at sea level, at the standard temperature, 25°C?*

A. NO is produced from atmospheric N_2 and O_2, via

$$N_2 + O_2 = 2NO \qquad (7.8)$$

To calculate the equilibrium NO concentration, we use the equilibrium constant equation [see eq. (7.2)]

$$K = (NO)^2/(N_2)(O_2) \qquad (7.9)$$

(The NO concentration is squared because two molecules of NO are produced in the reaction). The N_2 and O_2 concentrations in the atmosphere are known, so we can solve for (NO):

$$(NO) = [K(N_2)(O_2)]^{1/2} \qquad (7.10)$$

We express the concentrations of gases in terms of their partial pressure, in atmospheres (atm). The concentration of air molecules at sea level is 1 atm, and, since the atmosphere is 78 percent N_2 and 21 percent O_2, their concentrations are $(N_2) = 0.78$ atm and $(O_2) = 0.21$ atm.

We need only the value of K to complete the calculation. This is obtained from the free energy change of the reaction, which is twice the standard molar free energy of NO (see Table 7.1), because two moles of NO are produced:

$$2 \times 86.7 \text{ kJ} = 173.4 \text{ kJ}$$

Recalling eq. (7.7) of Fundamentals 7.1, $\Delta G = -RT \ln K$, and since $T = 298$ K, and $R = 8.314$ J/K, we have:

$$\ln K = 2.303 \log K = \frac{-173.4 \times 10^3 \text{ J}}{8.314 \text{ J/K} \times 298 \text{ K}}$$

or

$$\log K = -30.4, \text{ and } K = 10^{-30.4}$$

Now we substitute into eq. (7.10) to obtain

$$(NO) = (10^{-30.3} \times 0.78 \times 0.21)^{1/2}$$
$$= 10^{-15.5} \text{ atm}$$

(An easy way to obtain this result is to take logarithms of 0.78 and 0.21, add them to the exponent of K, and divide the total by 2.)

But the average concentration of NO at sea level is 1 atm $\times 10^{-4} \times 10^{-6} = 10^{-10}$ atm, a much larger number. Clearly NO is not in equilibrium with N_2 and O_2.

---

b. Free energy and temperature.   The free energy change of a reaction varies with temperature because of the entropy contribution. Recalling (see Fundamentals 5.2) that

$$\Delta G = \Delta H - T\Delta S \qquad (5.12)$$

we see that raising the temperature increases the difference between ΔH and ΔG. Recalling also that

$$\Delta G = -RT \ln K \qquad (7.7)$$

we can write

$$\ln K = -\Delta H/RT + \Delta S/R \qquad (7.11)$$

If ΔH is negative (*exothermic* reaction), $\ln K$ is positive (unless ΔH is overcome by a negative ΔS), meaning that products are favored over reactants. But as the temperature increases, $\ln K$ becomes less positive, meaning that products and reactants are more nearly balanced. This makes sense, since adding heat to a reaction that is already producing heat will tend to drive the reaction backward. Likewise, if ΔH is positive (*endothermic* reaction), $\ln K$ is negative, favoring reactants over products, but raising the temperature makes $\ln K$ less negative. Adding heat to a heat-absorbing reaction will tend to drive it towards products. In either case, the effect of raising the temperature is to produce a more even distribution of products and reactants.

If the equilibrium constant is known at a standard temperature, T_0 (usually 25°C), it can then be calculated at any other temperature from the enthalpy change,

$$\ln K - \ln K_0 = (1/T_0 - 1/T)\Delta H/R \qquad (7.12)$$

assuming that ΔH does not depend on the temperature. (This is a good assumption for gas-phase reactions, but not necessarily for reactions in solution, where interactions of the molecules with the solvent can change with temperature and affect ΔH.) The new value of K can be used to gauge the extent of the equilibrium reaction at the non-standard temperature.

WORKED PROBLEM 7.2: High-Temperature NO

Q. *If a fuel burns in air, some NO is produced. Calculate the equilibrium NO concentration (at sea level) at a temperature of 2,000 K.*

A. Again, NO is produced from atmospheric N_2 and O_2, via

$$N_2 + O_2 = 2NO \qquad (7.8)$$

for which the equilibrium constant at 25°C was calculated in Worked Problem 7.1 to be $10^{-30.3}$. ΔH at 25°C is obtained by doubling the molar enthalpy of NO formation in Table 7.1: $2 \times 90.4 = 180.8$ kJ. From eq. (7.12), we have

$$\ln K - \ln 10^{-30.3} = \left(\frac{1}{298} - \frac{1}{2{,}000}\right) \times \frac{180.8}{8.314}$$

which gives $K = 10^{-3.4}$. (Notice that the value of K is not very sensitive to how high the temperature is, as long as it is much higher than 298 K. 1/2,000 is already a small number compared to 1/298.)

This new value of K can now be used in eq. (7.10) (see Worked Problem 7.1)

$$(NO) = [K(N_2)(O_2)]^{1/2} \qquad (7.10)$$

using the atmospheric N_2 and O_2 concentrations, to give

$$(NO) = (10^{-3.4} \times 0.78 \times 0.21)^{1/2}$$
$$= 10^{-2.1} \text{ atm}$$

This value is 13 orders of magnitude larger than the value at 298! The high temperature has driven the equilibrium toward products to a very significant extent.

7.2   NITROGEN OXIDES: KINETICS

Because of its positive standard free energy, NO formation becomes significant only when air is heated sufficiently, as in combustion processes, or in the path of lightning bolts. Lightning and forest fires are major sources of atmospheric NO, and the anthropogenic contributions from fuel combustion are comparable to the natural sources (see Figure 15.1, p. 358).

When the temperature is lowered, the reverse of reaction (7.8) should proceed spontaneously. Once outside the combustion zone, NO is unstable with respect to conversion back to N_2 and O_2. But now kinetics intervenes. The rates of reactions are also strongly affected by temperature. In general, the higher the temperature, the higher the rate. Outside the combustion zone, the rate for NO decomposition slows down. It becomes so slow that another reaction intervenes, namely:

$$2NO + O_2 = 2NO_2 \qquad \Delta G_0 = -69.8 \text{ kJ} \qquad (7.13)$$

[This ΔG_0 is obtained by subtracting the standard free energy of formation (see Table 7.1) of NO from that of NO_2, and multiplying by 2.] Since ΔG_0 is negative, this reaction likewise proceeds spontaneously at low temperatures, and it is much faster than the reverse of reaction (7.8).

The NO_2 is in turn converted to HNO_3

$$4NO_2 + O_2 + 2H_2O = 4HNO_3 \qquad \Delta G = -239.6 \text{ kJ} \qquad (7.14)$$

The nitric acid molecules, being hygroscopic, are absorbed in raindrops and rained out of the atmosphere. This is the mechanism that removes nitrogen oxides from the atmosphere.

The relationship of thermodynamics to kinetics in the atmospheric chemistry of NO is diagrammed in Figure 7.2. On the left is the NO formation reaction. Because the free energy of formation is positive, energy must be provided to make the reaction go. But in addition, high temperatures are needed to achieve significant reaction rates. This requirement is captured in the concept of *activation energy*. There is a kinetic barrier to the

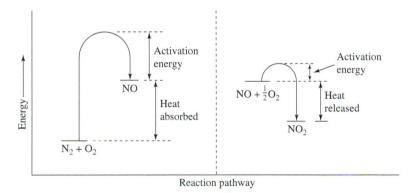

Figure 7.2   Energy profiles of nitrogen-oxygen reactions.

reaction, which must be surmounted by the input of additional energy. This barrier exists in the reverse as well as the forward direction. Consequently, even though the reverse reaction is highly exothermic, it still requires high temperature to surmount the barrier. The right side of the figure diagrams the further oxidation of NO to NO_2. The activation energy for this reaction is lower than for NO decomposition. Consequently it proceeds faster at low temperatures, even though the free energy difference driving the reaction is not as large as it is for NO decomposition.

What determines the activation energy? This question gets to the details of how the reaction works, its *mechanism*. The free energy change does not depend on the mechanism. It is a property inherent in the chemical nature of the reactants and products, whatever the mechanism of their interconversion. But the activation energy, and therefore the reaction rate, depends entirely on the mechanism. A given reaction may have more than one possible mechanism, each with a different activation energy. Naturally the reaction will go by the lowest activation energy path.

Generally speaking, gas phase reactions are inherently slow because the simplest available mechanisms have prohibitively high activation energies. The simplest mechanism would be to break the bonds of the reactant molecules and allow the atoms to form the new bonds of the product molecules. In reaction (7.13), for example, one can imagine breaking one O—O bond and two N—O bonds, and then recouping the energy by forming four N—O bonds in the two NO_2 molecules. But bond dissociation energies are large, and the fraction of reactant molecules with broken bonds is vanishingly small except at very high temperatures. Consequently, if a reaction does occur at ordinary temperatures, it must do so by a more complex mechanism.

For gas-phase reactions, the most important mechanisms involve *free radicals*.

7.3 FREE RADICAL CHAIN REACTIONS

Free radicals are atoms or molecules having an unpaired electron. We write the chemical symbol with a single dot. For example, the hydrogen atom, H•, is a free radical, as is the methyl radical, •CH_3, which results from the breaking of one of the C—H bonds in methane.

Most free radicals are highly reactive, and have only a fleeting existence. Since electron pairing is the basis of the covalent bond in chemistry, a free radical can always gain energy by pairing its electron with an electron on another atom. A few free radicals, notably the nitrogen oxides, are stable, because the unpaired electron resides in a relatively low-energy orbital. Even in these cases, however, reactivity is greater than in similar molecules in which all the electrons are paired.

A free radical can form a new bond at the expense of an already existing bond in another molecule, but the result is formation of another free radical, because there is still an unpaired electron. Thus:

$$A\bullet + B—C = A—B + C\bullet$$

The new radical can proceed to attack another molecule, generating still another radical, and so on. This chain of reactions can continue until a pair of radicals finds each other, forming a stable molecule:

$$A\cdot + C\cdot = A\text{---}C$$

Since no new radical appears, this is a *chain termination* reaction. The radicals, being highly reactive, are present at very low concentrations, and they are much more likely to encounter stable molecules in their vicinity than other radicals. Consequently, radical chains can be very long, leading to efficient overall reactions.

An example familiar in basic chemistry courses is the hydrogen-chlorine reaction:

$$H_2 + Cl_2 = 2HCl \qquad \Delta H = -184.6 \text{ kJ} \qquad (7.15)$$

The reaction is highly exothermic, but a mixture of H_2 and Cl_2 is stable indefinitely at room temperature, because the kinetic barrier is prohibitive. However, under certain conditions, the reaction occurs rapidly. For example, the mixture reacts explosively if it is irradiated with blue light. The blue light is absorbed by the Cl_2, which then dissociates into $Cl\cdot$ atoms:

$$Cl_2 + h\nu \text{ (blue)} = 2Cl\cdot \qquad (7.16)$$

The $Cl\cdot$ radicals react with H_2, forming an H—Cl bond at the expense of the H—H bond:

$$Cl\cdot + H_2 = HCl + H\cdot \qquad (7.17)$$

The $H\cdot$ atom left behind can then attack a Cl_2 molecule in the same way:

$$H\cdot + Cl_2 = HCl + Cl\cdot \qquad (7.18)$$

These two reactions add up to give the original reaction between H_2 and Cl_2 (7.15), but they provide a way of circumventing the kinetic barrier. We say that reaction (7.15) is *catalyzed* by the radical chain reactions, represented by (7.17) and (7.18).

Radical reactions require an *initiation* event, which produces the initial population of radicals. Somehow a bond must be broken in order to produce fragments with unpaired electrons. In the present illustration, the blue light provides the initiation, via reaction (7.16). This is a good model for what happens in the atmosphere, under the influence of sunlight. Atmospheric chemistry is largely governed by the free radicals that are produced by solar radiation.

a. Oxygen radicals.   Carbon dioxide has a large standard free energy of formation, providing a very strong driving force for oxygen to react with organic compounds (see the fuel energy calculations in Fundamentals 2.1, and Tables 2.1 and 2.2). Yet the biosphere houses a vast store of organic compounds in contact with an atmosphere that is 21 percent O_2. How is this possible?

While potentially very reactive, O_2 is also very slow to react at ordinary temperatures. In addition to the arguments in the preceding section for the inherent slowness of gas phase reactions, oxygen has another inhibitory factor, namely its electronic structure. Although O_2 has an even number of electrons, they are not all paired, as they are in other molecules. Two of them are left unpaired. The reason is that the last two electrons of O_2 have two orbitals available, of exactly the same energy. Each of these electrons enters a separate orbital.

FUNDAMENTALS 7.1: ELECTRONIC STRUCTURE OF DIATOMIC MOLECULES

The special electronic structure of O_2 emerges naturally from the *molecular orbital* picture of bonding. In this picture, the atomic orbitals on the two O atoms are allowed to combine into molecular orbitals, which have different energies, as illustrated in Figure 7.3. For an O atom,

the lowest valence orbital is the 2s orbital (the inner 1s orbital does not participate in bonding.) The 2s orbitals on the two atoms combine into a *bonding* and an *antibonding* molecular orbital, σ and σ^*. The bonding orbital results from positive overlap of the atomic orbitals,

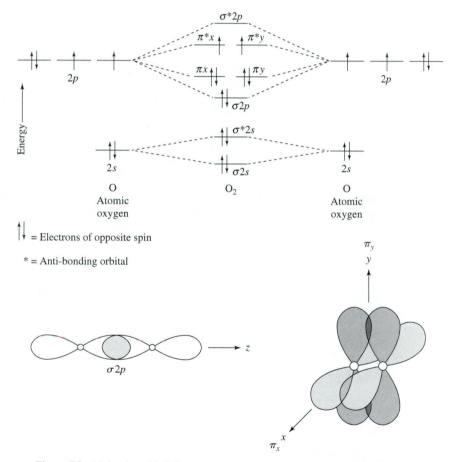

Figure 7.3   Molecular orbital diagram showing the electronic structure of O_2. The same diagram applies to other diatomic molecules of first row elements, for example, N_2, NO, and F_2, but the number of electrons available to fill the orbitals differs. (π^* is an anti-bonding orbital. Electrons occupying these orbitals weaken the O—O bond. See problem 16, end of Part II.) Overlap of the p atomic orbitals on the two O atoms to form σ and π molecular orbitals is illustrated.

while the antibonding orbital results from negative overlap. An electron in a bonding orbital has lower energy than if it were in a single atomic orbital, while an electron in an antibonding orbital has higher energy. Bonding occurs if the number of electrons in bonding orbitals exceeds the number in antibonding orbitals. The *bond order* (BO) is one-half the sum of the bonding electrons (n_b) minus the sum of the antibonding electrons (n_{ab}): BO = $(n_b - n_{ab})/2$. If all bonding and antibonding orbitals are filled, BO = 0, and there is no bond.

The higher-energy 2p orbitals on each O atom likewise form bonding and antibonding molecular orbitals, but these are of two kinds. One set combines the $2p_z$ orbitals, which point along the line (defined as z) between the two atoms. The $2p_z$-based molecular orbitals are also given the symbol σ, because, like the 2s-based orbitals, they are cylindrical about the internuclear axis. The other molecular orbitals combine the $2p_x$ and $2p_y$ atomic orbitals, which are perpendicular to p_z. The perpendicular molecular orbitals are called π. The overlap of atomic orbitals is less effective in the π than in the σ molecular orbitals (sideways versus head-on overlap, see Figure 7.3). Consequently, the energy of the π bonding orbitals is not as low as that of the σ bonding orbitals (the π bonds are weaker than the σ bonds), and likewise the π antibonding orbitals (π^*) are not as high as the σ antibonding orbitals.

Having constructed the energy level scheme, all we have to do is fill the orbitals with the available valence electrons and observe the consequences. Each oxygen atom has six valence electrons. Four electrons fill up the $\sigma 2s$ and $\sigma^* 2s$ orbitals, and the remaining eight arrange themselves among the 2p-derived orbitals. Two of these enter the $\sigma 2p$ orbital, creating a bond, and four more fill the π_x and π_y orbitals. The last two electrons enter the π^* orbitals, and reduce the bonding order to 2 [$(8 - 4)/2 = 2$].

However, because the π_x^* and π_y^* orbitals have exactly the same energy, they each acquire one of the two electrons, which are therefore unpaired. The two electrons occupy separate orbitals because their negative charges repel one another. This is why it costs a good deal of energy to pair them up in one of the two π^* orbitals, to make *singlet* oxygen (see next page).

The same molecular orbital diagram serves for other diatomic molecules of the first-row elements. In the case of N_2, we have ten electrons, which fill up the five lowest-energy molecular orbitals, producing a triple bond (one σ and two π bonds). NO has an additional electron, which enters a π^* orbital, reducing the bond order to 2.5 [$(8 - 3)/2 = 2.5$]. If we add an electron to O_2, it also enters a π^* orbital, and the resulting *superoxide* ion, O_2^- has a bond order of 1.5. The O=O bond is similarly weakened when O_2 combines with a radical (see next page) to form *peroxyl* radicals, ROO•. Finally, F_2 has all orbitals filled except $\sigma^* 2p$, and the bond order is 1.0, as it is in the *isoelectronic* peroxide ion, O_2^{2-}, and in H_2O_2, as well as in organic peroxides, RO_2H and RO_2R.

These orbitals are also the ones available for accepting electrons from other atoms or molecules. As noted above, oxygen combines with most other elements because it is avid for electrons. But as long as these electrons are paired up on the reacting atom or molecule (the electron *donor*) they cannot be transferred to O_2, since the available orbitals

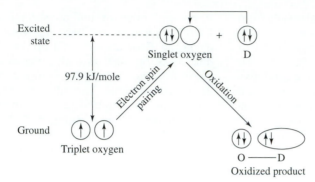

Figure 7.4   Slow reaction of oxygen with donor (D) containing two paired electrons.

already have one electron in them. For reaction to occur, the electron pairing on the donor must be disrupted, at a considerable energy cost, or else the unpaired electrons on O_2 must be paired up in order to leave one of the orbitals vacant (see Figure 7.4). This last process also costs energy, 97.9 kJ/mole, and results in an electronically excited form of O_2, called *singlet* oxygen. ("Singlet" refers to the fact that this form of oxygen has only a single magnetic state, in contrast to the ground state, called *triplet* oxygen, which has three magnetic states. The reason is that unpaired electrons *spin* on their axis, like gyroscopes, and generate magnetic fields; the rules of quantum mechanics allow three ways to align the two unpaired electrons, but only one when the electrons are paired.) Singlet oxygen can react rapidly with donor molecules, for example, with organic molecules having double bonds. However, because of its high energy, there is very little singlet oxygen available at normal temperatures. Once again there is a high activation barrier for the reaction. It is for this reason that organic molecules, even electron-rich ones, are stable in air. The special electronic structure of O_2 allowed life to evolve in the presence of an oxidizing atmosphere.

However, triplet O_2 *does* react rapidly if it encounters electron donors with unpaired electrons. These can be free radicals, or they can be transition metal ions with partially filled *d* orbitals. In general, free radicals control the gas-phase reactivity of O_2, while transition metal ions control its reactivity in aqueous solution and in biological tissues. The unpaired electron on the donor can readily enter one of the half-filled orbitals on O_2, forming a new electron pair bond with one of the O atoms. The result is a larger molecule, but one that still has an unpaired electron on the other O atom. Thus the product is itself a reactive-free radical.

b. Organic oxygen radicals.   An important example is the reaction of O_2 with an *alkyl* radical, $RCH_2\cdot$ (the symbol R is used here to indicate a generalized organic fragment):

$$RCH_2\cdot + O_2 = RCH_2O_2\cdot \tag{7.19}$$

In the product molecule, called an *alkylperoxyl* radical, the new C—O bond weakens the O—O bond relative to O_2 because the electron is added to an antibonding orbital (π^*), and leaves it vulnerable to fragmentation. Consequently, alkylperoxyl radicals are good O-atom donors to other molecules:

$$RCH_2O_2\cdot + X = XO + RCH_2O\cdot \tag{7.20}$$

X can be a variety of molecules, including NO and SO_2, or organic molecules with C=C double bonds; indeed, this reaction with peroxyl radicals is the primary mechanism for the oxidation of these three types of molecules in the atmosphere.

If R is itself an alkyl group, the $RCH_2O\cdot$ species remaining after the O-atom donation is an *alkoxyl* radical. In the atmosphere, alkoxyl radicals quickly react with O_2, donating an H atom:

$$RCH_2O\cdot + O_2 = RCHO + HO_2\cdot \tag{7.21}$$

By giving up H·, the alkoxyl radical becomes a stable molecule, an aldehyde. By receiving H·, O_2 becomes the *hydroperoxyl* radical. Like alkylperoxyl radicals, the hydroperoxyl radical readily donates an O atom to an acceptor, generating HO·, the *hydroxyl* radical:

$$HO_2\cdot + X = XO + HO\cdot \tag{7.22}$$

c. Hydroxyl radical.   The hydroxyl radical plays a special role in atmospheric chemistry. It readily abstracts an H atom from organic molecules, forming water as a product and leaving an organic radical:

$$HO\cdot + RCH_3 = RCH_2\cdot + H_2O \tag{7.23}$$

The O—H bond formed in this reaction is stronger than the C—H bonds of organic molecules. Reaction (7.23) initiates the oxygen radical cycle described in the preceding section. The organic reaction products of this cycle are subject to further attack by HO· and other radicals, and are eventually oxidized completely to CO_2.

Where do the hydroxyl radicals come from? They are continually produced in the atmosphere via the action of sunlight on ozone (O_3):

$$O_3 + h\nu(\lambda < 325\,nm) = O_2 + O \tag{7.24}$$

Ozone is a high-energy form of oxygen (positive free energy of formation, see Table 7.1). It exists mainly in the stratosphere where it provides a screen for UV light (see chapter on "Stratospheric Ozone," starting on p. 194). When O_3 absorbs a UV photon, it dissociates into O_2 and O atoms, a few of which (2 percent) are produced in electronically excited states. The excited O atoms, *O, have enough energy to react with water, producing hydroxyl radicals:

$$*O + H_2O = 2HO\cdot \tag{7.25}$$

There are not many O_3 molecules in the lower atmosphere, and they do not often encounter UV photons, most of which are filtered out by the stratospheric ozone; thus the number of hydroxyl radicals is limited. Nevertheless, hydroxyl radicals are so reactive that reactions (7.24) and (7.25) produce enough of them to ensure that radical chemistry consumes organic molecules in short order, even though the organic molecules in the presence of O_2 by itself would be stable indefinitely.

Hydroxyl radicals carry out other important atmospheric reactions. They convert NO_2 to nitric acid:

$$HO\cdot + NO_2 = HNO_3 \qquad (7.26)$$

and thereby provide the mechanism for washing nitrogen oxides out of the atmosphere. Likewise, they convert SO_2 to H_2SO_4 via the reactions

$$HO\cdot + SO_2 = HSO_3\cdot \qquad (7.27)$$

$$HSO_3\cdot + O_2 + H_2O = H_2SO_4 + HO_2\cdot \qquad (7.28)$$

and the resulting hydroperoxyl radicals produce additional hydroxyl radicals, via reaction (7.22).

Finally, hydroxyl radicals convert carbon monoxide to carbon dioxide:

$$HO\cdot + CO = CO_2 + H\cdot \qquad (7.29)$$

The great stability of CO_2 induces the rupture of the O—H bond, and H· atoms are left as a co-product. The H· atoms immediately attack O_2 molecules,

$$H\cdot + O_2 = HO_2\cdot \qquad (7.30)$$

producing hydroperoxyl radicals, which produce still more hydroxyl radicals. Carbon monoxide is a natural product of plant decay and an important component of combustion emissions. It is present in the atmosphere at levels exceeding those of organic compounds, and is therefore an important participant in atmospheric oxygen radical chemistry.

The reactions (7.19) to (7.30) constitute the main pathways for the oxidation and removal of most oxidizable molecules in the atmosphere. The hydroxyl radical is nature's vacuum cleaner.

d. Transition metal activation of O_2.   We complete our discussion of oxygen reactivity by considering the interactions of oxygen with transition metal ions in aqueous solution. Transition metal activation is important for atmospheric chemistry because a good deal of this chemistry actually occurs in raindrops. For example, raindrops are the sites of a substantial fraction of atmospheric H_2SO_4 production.

Transition metals activate oxygen for further reduction by donating single electrons and forming oxygen-containing free radicals. Because water is a neutral medium for radicals, they can persist in raindrops, where they are brought into close proximity to other reactive molecules.

Transition metal ions (M^{2+}) enter raindrops because dust particles are often included in condensation nuclei. Manganese and iron are particularly important, since they occur commonly in minerals. They are very reactive with O_2 in their divalent state, M^{2+}, because they have unpaired d electrons (five for Mn^{2+} and four for Fe^{2+}; see Periodic Table on inside front cover), which are readily donated to O_2. The first step in the reaction involves formation of a dioxygen adduct:

$$M^{2+} + O_2 = (MO_2)^{2+} \qquad (7.31)$$

Biology exploits this adduct formation by using the iron-containing protein hemoglobin to transport O_2 from the lungs to the tissue. The protein protects the bound O_2

from reacting further. However, when O_2 binds to free Fe^{2+} or Mn^{2+} ions in solution, it reacts with additional metal ions and protons to produce hydrogen peroxide. The overall reaction is:

$$2M^{2+} + 2H^+ + O_2 = 2M^{3+} + H_2O_2 \tag{7.32}$$

Additional H_2O_2 is produced in raindrops by the disproportionation of hydroperoxyl radicals from the atmosphere:

$$2HO_2\cdot = O_2 + H_2O_2 \tag{7.33}$$

In the presence of the same divalent metal ions, hydrogen peroxide is a potent source of hydroxyl radicals. Transfer of an additional electron from M^{2+} results in breakage of the O—O bond, leaving water and a hydroxyl radical:

$$M^{2+} + H_2O_2 + H^+ = M^{3+} + H_2O + HO\cdot \tag{7.34}$$

The hydroxyl radicals carry out rapid oxidations in solution, just as they do in the gas phase.

Since raindrops contain only traces of Fe^{2+} and Mn^{2+}, it might be thought that radical production would diminish when these traces are converted to M^{3+} ions and used up. However, the M^{3+} ions can be reduced back to M^{2+} by the action of sunlight. Under ambient pH conditions, the M^{3+} ions form hydroxide complexes, $M^{3+}OH^-$; when these complexes absorb a UV photon, the hydroxide transfers an electron to the metal ion, leaving the hydroxyl radical and M^{2+}, which is ready to react with O_2 to produce more radicals:

$$M^{3+}OH^- + h\nu = M^{2+} + HO\cdot \tag{7.35}$$

Thus a *photocatalytic* cycle is set up, in which trace concentrations of metal ions can induce sustained rates of radical production under the influence of sunlight and O_2.

CHAPTER **8**

STRATOSPHERIC OZONE

High in the atmosphere, a layer of ozone, O_3, acts as a solar filter, screening out the ultraviolet rays. These photons are energetic enough to induce photochemical damage in biological molecules. They are harmful to plants and animals alike, so much so that it is unlikely that complex life forms would have evolved without this ultraviolet shield.

The ozone layer developed as Earth's atmosphere grew rich in oxygen, through the advent of green plants. O_3 is a high-energy form of oxygen. The conversion of O_3 to O_2 releases energy; likewise, the conversion of O_2 to O_3 requires the input of energy. This energy is provided by ultraviolet photons that encounter O_2 high in the atmosphere. Thus, the ultraviolet shield itself is created by ultraviolet photons.

Although O_3 is inherently unstable, its decomposition is slow, as are other gas-phase molecular reactions in the stratosphere. It is subject, however, to *catalytic* destruction by chemicals that are avid for oxygen atoms. Scientists have long been aware of this possibility and have investigated potential threats to the ozone shield from gases emitted by human activity. In the early 1970s, attention focused on the potentially destructive effects of supersonic aircraft, which inject one of the ozone-destroying agents, NO, directly into the stratosphere.[*]

In a research article in 1974,[†] Mario Molina and F. Sherwood Rowland (who, along with Paul Crutzen, won the 1995 Nobel Prize in chemistry for their work on stratospheric

---

[*]This concern was a consideration in the cancellation at that time of the U.S. effort to develop a civilian supersonic transport plane. The issue resurfaced in the 1990s as the airline industry reconsidered supersonic jets to meet the expected increase in worldwide demand for air transport. New findings suggest that flights in the lower stratosphere may cause less damage to the ozone layer than initially expected (see p. 209), although more research is needed.

[†]M. J. Molina and F. S. Rowland (1974). Stratospheric sink for chlorofluoromethanes: chlorine-catalyzed destruction of ozone. *Nature* 249:810–812.

ozone) called attention to what has turned out to be a far greater danger. They pointed out that chlorine atoms, Cl•, are very efficient ozone-destroyers, and that chlorofluorocarbons (CFCs) deliver Cl• directly to the stratosphere.

These CFC molecules undergo no chemistry until they drift into the stratosphere, where the UV photons can break the C—Cl bonds. It is a great irony that CFCs came into widespread production precisely because their lack of reactivity in the troposphere makes them safe to use in many applications. It is the same lack of reactivity that preserves them until they migrate to the stratosphere, where they become potent ozone-destroyers. Moreover, because it takes about a century for the CFCs to complete their journey back to the troposphere in degraded form, the effect is felt long after their initial emissions.

Four years after Molina and Rowland's warning, the U.S. government banned the use of CFCs as propellants in aerosol cans, and a decade later international negotiations led to the signing of the 1987 Montreal Protocol. This was an unprecedented and ultimately successful effort to curb production of chemicals that science had shown to be a threat to the global environment. The Protocol set a future-year freeze and a subsequent 50 percent reduction in the production and use of CFCs and halons (brominated analogs of CFCs, used mainly to extinguish fires). Subsequent analysis showed that these curbs would not suffice to prevent the buildup of the gases in the atmosphere (see Figure 8.1), and further restrictions were negotiated. The 1990 London Amendments called for a complete phase-out by 2000, and the 1992 Copenhagen Amendments accelerated the phase-out dates to 1996 for CFCs and 1994 for halons (with a grace period to 2010 for developing countries).

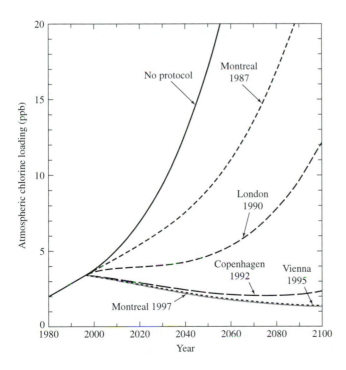

Figure 8.1   Projected abundances of chlorine in the atmosphere under the terms of the Montreal Protocol (1987), and its subsequent revisions. *Source:* Ozone Secretariat (1999). *A Decade of Assessments for Decision Makers Regarding the Protection of the Ozone Layer, 1988–1999; Synthesis of the Reports of the Assessment Panels of the Montreal Protocol* (Nairobi, Kenya: United Nations Environment Programme).

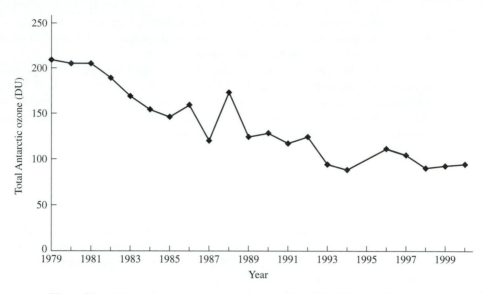

Figure 8.2   October mean ozone values in the Antarctic, 1979–2000, in Dobson units. *Source:* National Aeronautics and Space Administration, multimedia files (2001). (http://www.epa.gov/ozone/science/hole/sizedata.html#mintime)

It also controlled the *hydrochlorofluorocarbons* (HCFCs), which were being introduced as CFC substitutes (see p. 213) as "transitional substances," with a series of caps and reductions leading to a distant future phase-out. Final adjustments in Vienna (1995) and Montreal (1997) set caps on all 95 controlled substances, closing some previous loopholes. The final result of these agreements is expected to produce a gradual decline in the atmosphere of Cl- and Br-containing chemicals that destroy stratospheric ozone (see Figure 8.1).

A major impetus to action was the discovery of a dramatic "hole" in the ozone layer over the South Pole. Figure 8.2 illustrates the decline in the springtime ozone level over Antarctica, from 1980 onward. This effect had *not* been predicted by computer modeling at the time; the discrepancy revealed a role for an additional class of chemical reactions that had not been anticipated. These reactions, discussed on pp. 209–213, are now well understood, but the episode is a cautionary example of potential surprises that we might face, due to our incomplete knowledge of environmental chemistry.

8.1   ATMOSPHERIC STRUCTURE

Our gaseous envelope extends many miles from the surface of Earth. The atmosphere is quite uniform throughout its extent with respect to its major chemical constituents (see Table 6.1), except for water vapor, which is concentrated in the lower region. The air is far from uniform in other respects, however. It grows thinner with increasing altitude; the density falls off, roughly logarithmically, with increasing distance from the surface. Even

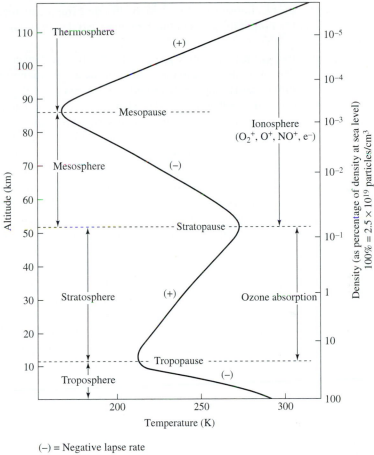

(−) = Negative lapse rate
(+) = Positive lapse rate

Figure 8.3   Layer structure of the atmosphere.

less uniform than the density is the temperature, which depends on altitude, as shown in Figure 8.3. The slope of the curve of temperature versus altitude is called the *lapse rate*. The lapse rate of the atmosphere reverses several times. Up to about 10 km, temperature decreases uniformly with increasing altitude. This reflects the fact that the atmosphere in this lower region is heated from below by convection and radiation from Earth's surface. Above 10 km, however, the temperature increases again with increasing altitude, reaching a maximum near 50 km; beyond this altitude, the temperature decreases once again. The maximum in the temperature profile reflects a heating process high in the atmosphere, which is due to the absorption of ultraviolet solar photons by the ozone layer. Beyond about 90 km, the temperature rises once more due to the absorption of solar rays in the far-ultraviolet region by atmospheric gases, principally oxygen. These far-ultraviolet rays are sufficiently energetic to ionize molecules and break them into their constituent atoms. Due to the thinness of the atmosphere at such high altitudes, fragments recombine only rarely; an appreciable fraction of the gases in this region, called the *thermosphere,* exists as atoms or ions.

The structure of the temperature profile reflects a physical layering of the atmosphere as well. A negative lapse rate leads to convection of the air; warm air rises from below and cool air sinks. But a positive lapse rate provides a region of stability with respect to convection, since warm air overlies cool air. The change from negative to positive lapse rate is called a *temperature inversion;* the point at which the lapse rate changes marks a stable boundary between two physically distinct layers of air. Local temperature inversions in the troposphere are quite common over many cities, particularly those nestled in surrounding mountains or, as in the case of Los Angeles, where prevailing winds from the ocean blow cool air into a region ringed on three sides by mountains. Warm air flowing over the mountaintops traps the cool air below, allowing pollutants to build up for considerable periods. Anyone flying into Los Angeles on a sunny day is likely to see the local temperature inversion, which produces a distinct boundary between the clear air above and the brown smoggy air below.

The temperature inversion shown in Figure 8.3 at an altitude of about 10 km is called the *tropopause.* It marks a global layering of the atmosphere into the troposphere below and the stratosphere above. The troposphere constitutes about 10 percent of the atmosphere's height but contains 80 percent of its mass. Due to its negative lapse rate, the air it contains is mixed rapidly by convection. It is also a region of much turbulence, due to the global energy flow that results from the imbalances of heating and cooling rates between the equator and the poles. The stratosphere, on the other hand, is a quiescent layer; due to its positive lapse rate, it mixes slowly. The tropopause itself is a stable boundary, and the flux of air across it is low. As we saw in the case of sulfate aerosols from volcanic explosions, residence times of molecules or particles in the stratosphere are measured on a scale of years. The air in the stratosphere is quite thin, so that pollutants in the stratosphere have a relatively greater global impact than they would in the much denser troposphere.

FUNDAMENTALS 8.1: Gas Volumes

Gas molecules are free to move about in space, and therefore fill up any container. The molecules exert pressure by colliding with each other and with the walls of the container. The collision rate depends on the number of molecules, on the volume they occupy, and on the temperature, which measures the average kinetic energy of the molecules. A simple equation brings all of these quantities together:

$$PV = nRT \qquad (8.1)$$

P is pressure, V is volume, T is temperature, N is the number of moles of the gas (which is proportional to the number of molecules), and R is the gas constant; we have encountered R previously in calculations involving free energy and the equilibrium constant. When P and V are expressed in units of atmospheres (atm) and liters (L), R has the value 0.0821.

This equation tells us that for a given temperature and pressure, the volume of a gas depends only on the number of molecules. For a mole of any gas, the volume is the same. Eq. (8.1) gives 22.4 liters for the molar volume when the pressure is 1 atm and the temperature is 0°C (273 K). By convention, these conditions are taken as the *standard* pressure and temperature.

The atmosphere is not in a closed container, of course. It is held in place by gravity, and the pressure falls continuously with increasing distance from the surface of the Earth. The major constituents of the atmosphere are uniformly mixed, but for reactive gases like ozone, there are spatial variations in the relative concentrations, depending on the location of sources and sinks. Ozone is found mainly in the stratosphere, spread over many kilometers of altitude (see Figure 8.3), but its concentration is low enough that it would occupy a band only 3 mm thick if all the molecules were brought together and spread evenly at sea level (1 atm pressure) at 0°C. The effective thickness of the ozone layer at a particular location is reported in *Dobson units* (DU), named after G. M. B. Dobson, who conducted pioneering measurements of the stratosphere in the 1920s and 1930s. One Dobson unit equals one-hundredth of a millimeter thickness of the ozone layer at standard temperature and pressure. Therefore, ozone columns typically have a thickness of 300 DU.

WORKED PROBLEM 8.1: How Much Ozone?

Q. *How many moles of ozone does the atmosphere hold? For a sense of scale, compare this amount to the total carbon dioxide emitted annually from fossil fuel burning, cement production, and deforestation (7,900 Tg/C, see p. 169).*

A. Since we know that the amount of ozone is equivalent to a 3-mm band around the Earth, at standard pressure and temperature, we can calculate the volume of this band and divide by the molar volume, 22.4 L, to obtain the number of moles.

For a thin band around a large sphere, we obtain the volume by multiplying the area by the thickness. The area of a sphere is given by:

$$A = 4\pi r^2$$

r is the radius, which is 6,378 km in the case of the Earth. Since we want to obtain the volume in L, we will convert length to cm and divide the resulting volume by 1,000 cm^3/L. Putting all this together we have:

moles ozone

$$= \frac{4\pi \times (6{,}378 \text{ km} \times 10^5 \text{ cm/km})^2 \times 0.3 \text{ cm}}{1{,}000 \text{ cm}^3/\text{L} \times 22.4 \text{ L/mole}}$$

$$= 6.85 \times 10^{13} \text{ moles}$$

To compare this with the 7,900 Tg/C annual emission of CO_2, we obtain the number of moles of C by dividing the weight by the atomic weight:

$$\text{moles C (or } CO_2) = \frac{7{,}900 \times 10^{12} \text{ g}}{12 \text{ g/mole}}$$

$$= 65.8 \times 10^{13} \text{ moles.}$$

Thus, humans emit nearly ten times as many moles of CO_2 per year as the moles of O_3 contained in the atmosphere.

8.2   ULTRAVIOLET PROTECTION BY OZONE

Concern about pollution of the stratosphere centers on possible threats to the ozone layer, which serves two essential functions: it protects living matter on Earth from the harmful effects of the sun's ultraviolet rays, and it provides the heat source for layering the atmosphere into a quiescent stratosphere and a turbulent troposphere.

Absorption of ultraviolet radiation by stratospheric ozone is strong enough to eliminate much of the ultraviolet tail of the solar radiation spectrum at the surface of the Earth (see Figure 8.4). Ozone absorbs light with wavelengths between 200 and 300 nm, in the UV range of the solar spectrum. These ultraviolet rays are harmful to life. They carry enough energy to break the bonds of organic molecules and produce reactive fragments. Ultraviolet radiation that passes through the ozone layer produces sunburn and also skin cancer, particularly in people with light pigmentation. The damage done to living tissue depends upon the solar wavelength, as shown in Figure 8.5. The solid curve in the figure shows the spectrum of the skin's sunburn sensitivity at constant light intensity. Also shown is the spectrum of the solar rays at ground level. The product of these two curves gives the action spectrum for

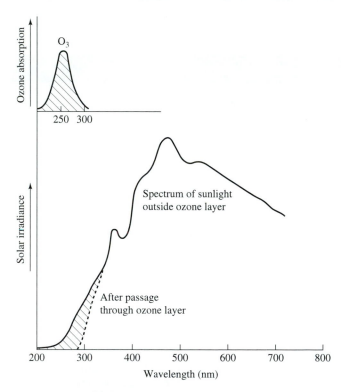

Visible light: 400–700 nm
Ultraviolet light: < 400 nm

Figure 8.4   Absorption of sun's ultraviolet light by ozone.

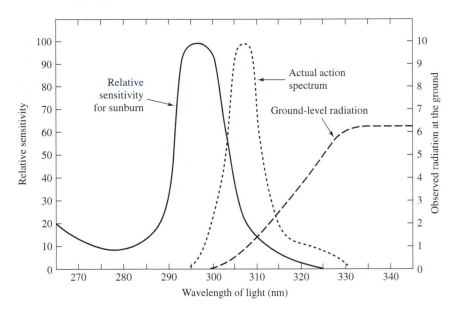

Figure 8.5   Action spectrum of ultraviolet radiation damage to living tissue.

sunburn, that is, the response of the skin to solar radiation. The overlap of ground-level radiation with the sunburn sensitivity curve would be much greater without the filtering effects of the ozone layer. Moreover, the effect of a given level of ozone reduction is greatest for the shortest, and most damaging, wavelengths. In assessing the damage potential of UV light, scientists measure the total amount of light exposure between 280 and 320 nm, a region called UV-B. It is estimated that in the absence of CFC and halon controls, surface exposure to UV-B radiation would have doubled by 2050 at midlatitudes of the Northern Hemisphere.

Exposure to solar ultraviolet radiation seems clearly linked to the incidence of skin cancer (see Figure 8.6). Both the incidence of skin cancer and the solar ultraviolet flux decrease with increasing distance from the equator. The UV flux at ground level depends upon latitude, because the sun's rays fall more directly at lower latitudes, and because the ozone layer is thinner over the equator than toward the poles. Between 30 and 46 degrees latitude north, the annual ultraviolet flux is reduced by a factor of three, and the incidence of malignant skin cancer falls roughly by a factor of two. Thus, the expected doubling from ozone loss in the absence of CFC and halon controls would have increased skin cancer incidence appreciably.

Ultraviolet rays also damage green plants. Their light-harvesting photosynthetic apparatus is tuned to visible radiation and can be destroyed by sufficiently intense ultraviolet radiation. Particularly susceptible are the light-harvesting phytoplankton that float at the surface of the ocean and provide the starting material for marine food chains. Antarctic phytoplankton populations have been shown to decrease in concert with the ozone depletion, although effects on the rest of the ecosystem have not yet been detected.

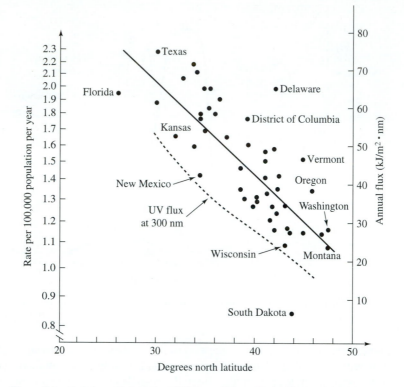

Figure 8.6 Variation with latitude of UV flux at 300 nm (dashed line) and death from skin cancer among white males in the United States (solid line) (excluding Alaska and Hawaii). *Source:* Adapted from F. S. Rowland (1982). Chlorofluorocarbons and stratospheric ozone. In *Light, Chemical Change, and Life.* J. D. Coyle et al., eds. (Washington, DC: Open University Press). Copyright © 1982 by Open University Press. Reprinted with permission.

FUNDAMENTALS 8.2: TRANSMISSION OF UV LIGHT THROUGH OZONE

How much does a given change in the ozone layer affect the amount of UV light reaching Earth's surface? The fraction of light transmitted by absorbing molecules falls off rapidly as the number of molecules in the path increases. In fact, the transmitted light falls off exponentially. We can work out the relationship in the following way. For an absorbing constituent of the atmosphere, the total number of molecules through which the light passes can be expressed as an equivalent thickness, l, of a global layer of the gas, brought to standard temperature (0°C) and pressure (1 atm). The fraction of light transmitted (T) through the layer decreases exponentially with increasing thickness:

$$T = I/I_0 = e^{-\varepsilon l} \qquad (8.2)$$

where ε is the *absorptivity* (also called the *extinction coefficient*) expressed in units of inverse length, and l is the path length of the light through a uniform sample. If the sample is not uniform, as in the case of the ozone layer, the l can be defined as the equivalent thickness, the thickness that would obtain if the sample were uniform. As we saw above, the equivalent thickness for the ozone layer, at standard pressure and temperature, is 0.3 cm.

Equation (8.2) is called the Beer-Lambert law. It holds for all wavelengths of light, and for all absorbing molecules. The exponential decay is the same as we have encountered for radioactive isotopes, except that it occurs along the light path, instead of in time. If the product εl is 1.0, then 37 percent of the incident light is transmitted, whereas if it is 10.0, only 0.005 percent is transmitted.

The value of ε depends on the wavelength because the absorption of light by ozone varies according to the wavelength (see Figure 8.4). The product εl goes from 1 to 10 between 310 and 290 nm, the critical region for sunburn and for skin cancer.

The effect of a small decrease in the ozone thickness can be estimated by differentiating equation (8.2):

$$dT = -\varepsilon e^{-\varepsilon l} \, dl \qquad (8.3)$$

Dividing through by equation (8.2) gives the relative response:

$$dT/T = -\varepsilon \, dl = -\varepsilon l \, dl/l \qquad (8.4)$$

Thus, a given fractional decrease in l produces a fractional increase in T, which is amplified by the product εl. A 1 percent decrease in the ozone layer gives a 1 percent increase in ultraviolet transmittance at 310 nm, a 3 percent increase at 300 nm, and 10 percent at 290 nm. Thus, ozone changes affect the UV flux most sensitively at the shortest wavelengths, where damage to biological molecules is greatest.

These calculations help to explain the concern over ozone depletion. During the 1980s, globally averaged total ozone was declining at a rate of 0.4 percent per year. Using the above calculations, the ozone loss per decade (4 percent) would cause an increase of 12 percent and 40 percent in the transmittance of UV light at 300 and 290 nm, respectively.

8.3   OZONE CHEMISTRY

a. Formation and destruction.   Ozone is formed in the stratosphere when O_2 molecules absorb solar radiation. The far-ultraviolet photons of the sun have enough energy to split oxygen molecules into oxygen atoms high in the atmosphere:

$$O_2 + h\nu \, (<242 \text{ nm}) \; \rightarrow \; 2O \qquad \text{(a)}$$

($h\nu$ is the symbol for a photon, and <242 nm indicates the wavelength range in which photons can induce the reaction.) The oxygen atoms produced in this reaction then combine with other oxygen molecules to form ozone:

$$O + O_2 + M \; \rightarrow \; O_3 + M \qquad \text{(b)}$$

In this reaction, M is a third molecule, which must be present at the encounter of O and O_2 in order to carry away some of the energy released in the reaction; otherwise the ozone molecules would fly apart as fast as they were formed. M can be any molecule that happens to be present. In the atmosphere, it is most likely to be nitrogen or another molecule of oxygen. (Likewise, many of the reactions we have previously considered involve a single product formed from two reactants. They all require a "third body" to carry off the excess energy. Usually, we don't bother to include M in the chemical equation because it appears on both sides and does not itself enter into the reaction. It does, however, influence the rate of the reaction, and is therefore included here.)

Ozone is also destroyed by solar radiation. When O_3 absorbs a solar UV photon, it dissociates into O_2 and O:

$$O_3 + h\nu\,(200\text{--}320 \text{ nm}) \quad \rightarrow \quad O + O_2 \tag{c}$$

Notice that longer wavelength photons can split O_3 [reaction (c)] because O_3 absorbs at longer wavelengths (see Figure 8.4). [We have already encountered reaction (c) in connection with the source of the atmospheric hydroxyl radical (see p. 207), noting that a small percent of the O atoms are electronically excited, and therefore energetic enough to react with H_2O.]

Also when O_3 encounters an O atom, the two can combine to form two O_2 molecules:

$$O_3 + O \quad \rightarrow \quad 2O_2 \tag{d}$$

[Note that no third body is required in reaction (d), in contrast to reaction (b), because the two O_2 molecules can carry off the energy of the reaction.] Reactions (c) and (d) limit the extent to which O_3 can build up.

The concentration of O_3 depends on the rates of the formation and destruction reactions. A *steady-state* concentration can be calculated by setting the total formation rate equal to the total destruction rate. As long as these rates do not change, the concentration of O_3 is constant. This would also be true at equilibrium, but the steady state is not an equilibrium condition. Because the free energy of O_3 is much higher than that of O_2, the equilibrium concentration of O_3 is extremely low. The steady-state concentration can be much higher because the formation reactions include an external energy contribution (the solar photons), which keeps the system out of equilibrium. When this energy source is removed, the system eventually returns to equilibrium (but may take a long time to do so).

b. Calculating the ozone steady state.   Reaction rates are given by the products of the reactant concentrations times the rate constants, k. For the four reactions involved in ozone formation and destruction:

$$\text{rate } a = k_a(O_2)$$
$$\text{rate } b = k_b(O)(O_2)(M)$$
$$\text{rate } c = k_c(O_3)$$
$$\text{rate } d = k_d(O)(O_3)$$

Ozone is a product only in reaction (b), so the ozone production is rate b, while the loss rate is rate c + rate d. The ozone steady-state condition is then

$$\text{rate } b = \text{rate } c + \text{rate } d$$

or

$$k_b(O)(O_2)(M) = k_c(O_3) + k_d(O)(O_3) \tag{8.5}$$

In this expression, (O_2) and (M) (the concentration of air molecules) are known, and the rate constants are known from experiments. There are two unknowns, (O_3) and (O). In order to eliminate one of them, we need an additional equation. This can be provided by assuming that the concentration of O atoms is also at steady state; the production and loss rates of O atoms must also balance. The rate of O production is twice the rate of reaction (a) (two oxygen atoms produced per O_2 molecule split) plus the rate of reaction (c), while the loss rate is rate b + rate d. At steady state,

$$2(\text{rate } a) + \text{rate } c = \text{rate } b + \text{rate } d$$

or

$$2k_a(O_2) + k_c(O_3) = k_b(O)(O_2)(M) + k_d(O)(O_3) \tag{8.6}$$

Equations (8.5) and (8.6) can now be solved for the steady-state concentrations of O and O_3. For example, subtraction of equation (8.5) from equation (8.6) gives:

$$2k_b(O)(O_2)(M) = 2k_a(O_2) + 2k_c(O_3) \tag{8.7}$$

which can be solved for (O). The expression is simplified if we assume that $k_a(O_2) \ll k_c(O_3)$, that is, that photolysis of O_2, reaction (a), produces far fewer O atoms than does photolysis of O_3, reaction (c). This assumption turns out to be valid for the lower reaches of the stratosphere, where O_3 is important because the flux of solar photons is much smaller in the far-ultraviolet region (O_2-splitting) than in the near ultraviolet (O_3-splitting). Dropping $k_a(O_2)$ from equation (8.7), we obtain

$$(O) = k_c(O_3)/k_b(M)(O_2) \tag{8.8}$$

Addition of equations (8.5) and (8.6) gives

$$2k_a(O_2) = 2k_d(O)(O_3) \tag{8.9}$$

After substituting equation (8.8) we can solve for the $(O_3)/(O_2)$ ratio:

$$(O_3)/(O_2) = [k_a k_b(M)/k_c k_d]^{1/2} \tag{8.10}$$

The value of this quantity depends on the altitude. As the altitude increases, several variables change: the concentration of air molecules (M) decreases; both k_a and k_c increase because the photon flux increases; and k_b and k_d increase slightly because the temperature increases (in the stratosphere). At an altitude of 30 km, the concentration of air molecules is $10^{17.7}$ molecules/cm^3, while the approximate values of the rate constants, expressed with this concentration unit (and averaged over Earth's surface), are $k_a = 10^{-11}$, $k_b = 10^{-32.7}$, $k_c = 10^{-3}$, and $k_d = 10^{-15}$. Inserting these numbers into equation (8.10) gives $(O_3)/(O_2) = 10^{-4}$. Thus, even in the ozone layer, the number of O_3 molecules is much smaller than the number of O_2 molecules.

At any given place in the atmosphere, the ozone concentration is subject to large variations because the solar flux varies throughout the day as well as seasonally. These variations are fairly regular, however, and the time-averaged ultraviolet flux depends only on the altitude, becoming smaller at lower altitudes as most of the rays are absorbed. Consequently, the steady-state calculation should give a reasonable estimate of the average ozone concentration at different altitudes. The calculation can be repeated at other altitudes,

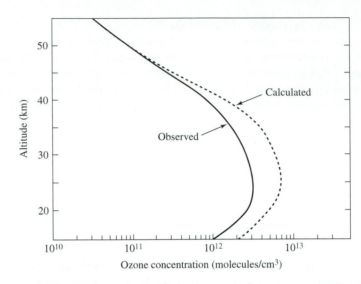

Figure 8.7 Plot of observed and calculated atmospheric ozone concentrations (not including reactions with OH, Cl, Br, and NO).

with appropriate modification of the constants, and the result is the dotted curve in Figure 8.7. The $(O_3)/(O_2)$ ratio is predicted to peak at about 30 km. At higher altitudes, ozone photolysis [reaction (c)] is increasingly rapid, whereas at lower altitudes, net ozone formation is limited by the decreasing supply of O atoms from reaction (a).

The measured ozone profile (solid line in Figure 8.7) has the same shape as the calculated curve, but the calculated values are about a factor or two too high. Since the rate of ozone production depends only on the UV flux, this discrepancy reflects the existence of mechanisms for ozone destruction other than the one considered so far.

WORKED PROBLEM 8.2: Steady State versus Equilibrium for O_3

Q. *Compare the steady-state $(O_3)/(O_2)$ ratio at 30 km altitude with the equilibrium value.*

A. The $(O_3)/(O_2)$ ratio at equilibrium can be obtained from K_{eq}, the equilibrium constant for the reaction

$$3O_2 = 2O_3$$

From the standard free-energy change at 298 K and 1 atm pressure, 2×163.4 kJ (see Table 7.1), K_{eq} is estimated to be $10^{-57.0}$. From the equilibrium expression

$$(O_3)^2/(O_2)^3 = K_{eq}$$

we obtain

$$(O_3)/(O_2) = [K_{eq}(O_2)]^{1/2}$$

Since at sea level the concentration of oxygen $[O_2]$ is 0.21 atm,

$$[K_{eq}(O_2)]^{1/2} = (10^{-57.0} \times 0.21)^{1/2}$$
$$= 10^{-28.8}$$

Because the (O_2) concentration and temperature are much lower in the stratosphere, this ratio in the stratosphere is much smaller. Thus, the steady-state ratio of $(O_3)/(O_2)$ exceeds the equilibrium value by many orders of magnitude.

8.4   CATALYTIC DESTRUCTION OF OZONE

Ozone is a meta-stable molecule; its equilibrium concentration is extremely low, but isolated molecules decompose very slowly. They can, however, be destroyed rapidly by catalytic chain reactions. In such reactions, the ozone is converted to dioxygen by a chain carrier, X, which is itself restored in the process. The general reactions are

$$X + O_3 = XO + O_2 \tag{e}$$

$$XO + O = X + O_2 \tag{f}$$

Adding these two reactions together produces reaction (d):

$$O_3 + O = 2O_2 \tag{d}$$

Reactions (e) and (f) provide an additional route to the reaction of ozone with oxygen atoms, thereby speeding up the destruction process. Since formation is unaltered, accelerating the destruction reaction lowers the steady-state concentration of ozone. Because X is regenerated, it is a catalyst; a single X molecule can destroy many ozone molecules. The chain reaction continues until X is removed by some side reaction that inactivates it.

There are many possible candidates for the chain carrier X, but four species have been identified as important for destruction of stratospheric ozone: the hydroxyl radical, chlorine and bromine atoms, and nitric oxide. They are considered in turn as follows.

a. Hydroxyl radical.   The hydroxyl radical accounts for nearly one-half of the total ozone destruction in the lower stratosphere (16 to 20 km). We have already seen (see p. 191) how the hydroxyl radical is formed in the troposphere, where it is produced by the reaction of water with excited O [reaction (7.25)], generated from ozone photolysis [reaction (7.24)]. Stratospheric HO• is formed in similar fashion. Excited O atoms react with a hydrogen source, supplied principally by H_2O and CH_4. Hydroxyl radical is capable of accepting an O atom from O_3 to produce H_2O [reaction (e)], which in turn can react with O atoms, regenerating HO• [reaction (f)]. The combined reactions lead to the ozone-destruction reaction [reaction (d)]. Since water and methane are components of the natural atmosphere, the HO• cycle of reactions constitutes a natural mechanism for ozone loss. There is some concern that this loss mechanism could be accelerated due to the rising atmospheric concentration of methane.

b. Chlorine and bromine.   Chlorine and bromine atoms are highly efficient chain carriers in reactions (e) and (f). However, Cl or Br in the stratosphere comes from few natural sources, because potential sources in the troposphere are unable to reach the stratosphere in significant quantities. For example, the oceans are abundantly endowed with chloride and bromide ions, but sea spray is efficiently cleared from the troposphere. Similarly, there are many organochlorine and organobromine natural products, and the more volatile molecules are emitted to the atmosphere, but they are destroyed in the troposphere by hydroxyl radicals via reaction (7.23) (see p. 191). However, the most abundant species, methylchloride and methylbromide, are long-lived enough to contribute a significant background level of halogen-induced ozone destruction.

Much larger sources of stratospheric chlorine and bromine have been inadvertently created by human manufacturing of organic compounds that contain one or two carbon atoms attached to only fluorine, chlorine, and/or bromine. These are the chlorofluorocarbons, CFCs, and the bromine-containing halons. CFCs have been widely used as refrigerants, blowing agents for plastic foams, propellants for aerosol sprays, and solvents for cleaning microelectronic components. The halons have been used as fire extinguishers; the heavy bromine-containing molecules provide a blanket of gas that effectively smothers flames. The CFCs and halons have been enormously useful in these applications because they are nontoxic and nonflammable. These desirable properties are directly related to the low chemical reactivity of these molecules in the troposphere. Because they lack H atoms, and thus contain no C—H bonds, CFCs and halons are not subject to oxidation, either in a flame or biochemically. Even hydroxyl radicals are unable to attack these molecules, which therefore escape the tropospheric fate of most organic species. Only in the stratosphere are CFCs and halons destroyed, by the action of UV photons. The result of absorbing UV photons is to break the weakest bond in the molecule, either C—Br or C—Cl:

$$RX + h\nu = R\cdot + X\cdot \tag{8.11}$$

Once released, chlorine and bromine atoms destroy ozone via the reactions (e) + (f).

Satellite measurements indicate that high levels of CFCs do reach the stratosphere; above about 20 km, their concentration decreases, while concentrations increase for HCl and HF, the ultimate products of CFC destruction. These molecules are formed by a variety of mechanisms, principally the slow attack of Cl· or F· on methane molecules, which drift up from the troposphere:

$$Cl\cdot (F\cdot) + CH_4 = HCl (HF) + CH_3\cdot \tag{8.12}$$

Since HF has no natural source, the data establish CFCs as the source of most of the stratospheric chlorine. It is for this reason that CFC production is being curtailed and phased out. Even with the ban on CFCs, CFC-catalyzed destruction of the ozone layer will persist for decades. It takes many years for these molecules to be transported up to the altitude of peak photoefficiency. The time lag between production, consumption, and eventual effects on the ozone layer is considerable.

c. Nitric oxide.   Although NO is produced abundantly in the lower atmosphere by combustion and lightning, almost all of it is oxidized to NO_2 and converted to nitric acid [see reaction (7.26), p. 192] in the troposphere, after which it is rained out before reaching the stratosphere. On the other hand, nitrous oxide, N_2O, although much less abundant, is also much less reactive; it does eventually reach the stratosphere. Above 30 km, most of the N_2O is photolyzed by UV photons to produce dinitrogen and excited oxygen atoms:

$$N_2O + h\nu (UV) = N_2 + {}^*O \tag{8.13}$$

A small percentage, 10 percent or less, of the N_2O molecules react with excited oxygen atoms formed via reaction (7.24), as well as (8.13), to produce NO:

$$N_2O + {}^*O = 2NO \tag{8.14}$$

This is the main source of NO in the stratosphere.

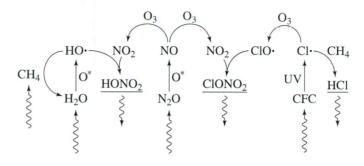

Figure 8.8   Diagram of stratospheric reaction linkages; reservoir molecules are underlined.

NO can act as X in ozone-destroying reactions (e) and (f), cycling through NO_2 in the process. However, NO_2 also reacts with other carriers in O_3 destruction chains, HO• and ClO•. The reaction with HO• produces nitric acid [reaction (7.26)], as we have already seen, while the reaction with ClO• produces an analogous molecule, chlorine nitrate:

$$ClO• + NO_2 = ClONO_2 \tag{8.15}$$

Nitric acid and chlorine nitrate do not participate in O_3 destruction directly. Rather, they are *reservoir* molecules; they sequester the chain-carrying species HO• and ClO• in less reactive forms, releasing them in response to UV light:

$$HONO_2 + h\nu = HO• + NO_2 \tag{8.16}$$

$$ClONO_2 + h\nu = ClO• + NO_2 \tag{8.17}$$

Although HO• and ClO• thereby remain available for O_3 destruction, binding to NO_2 reduces their concentration significantly and thereby reduces ozone destruction. Furthermore, the reservoir molecules can also be rained out when stratospheric and wet tropospheric air mix in the upper troposphere. These processes are diagrammed in Figure 8.8.

Thus, NO has a two-sided effect, on the one hand providing another catalytic chain mechanism for O_3 destruction, but on the other, inhibiting two other major mechanisms for O_3 destruction. Which of these effects dominates depends upon altitude. Above about 25 km, the net effect of nitrogen oxides is to lower O_3 concentrations via reactions (e) and (f). In fact, in the middle and upper stratosphere, nitrogen oxides account for over 50 percent of total ozone destruction. However, in the lower stratosphere, the overall effect of nitrogen oxides is to protect O_3 from destruction via reactions (7.26) and (8.15). These findings have given new impetus to the feasibility of supersonic jet transport in the lower stratosphere. Uncertainties persist, however, because jet emissions would not necessarily remain at low altitudes.

8.5   POLAR OZONE DESTRUCTION

Although the reactions described so far account for the observed average levels of stratospheric ozone, they are unable to account for the ozone hole over Antarctica or for the substantial ozone reduction that has also been detected in the Arctic region. Each spring the

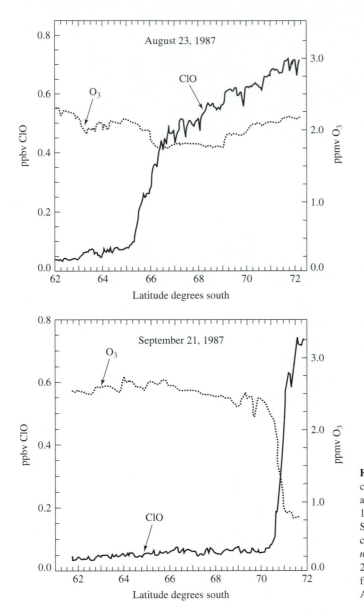

Figure 8.9   Chlorine oxide and ozone concentrations over Antarctica at 18-km altitude, August 23 and September 21, 1987.   *Source:* F. S. Rowland (1991). Stratospheric ozone in the 21st century: The chlorofluorocarbon problem. *Environmental Science and Technology* 25:622–628. Reprinted with permission from ES&T. Copyright © 1991 by the American Chemical Society.

ozone layer thins markedly over the poles. The phenomenon is limited both regionally and seasonally. Although several explanations have been considered, the cause was firmly established as catalytic destruction by Cl• when the O_3 loss was observed to coincide with a sharp rise in ClO• (see Figure 8.9). But the effect is too great and too sudden to be explained by the chain reactions (e) and (f), with X = Cl•.

Additional chemistry is involved, including an important role for reactions at the surface of cloud particles (see Figure 8.10). In the frigid winter temperatures over the poles,

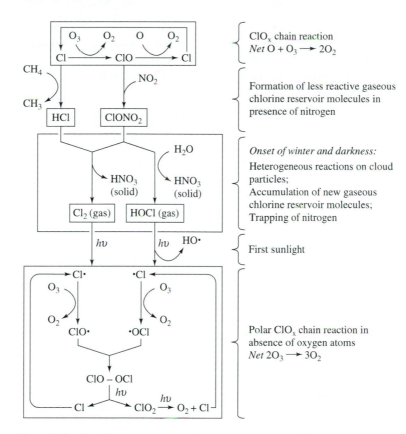

Figure 8.10   Reaction sequence responsible for Antarctic ozone hole.

the stratospheric air becomes trapped in a vortex in which clouds form, even though the air is very dry. These polar stratospheric clouds (PSCs) form first at 193 K, when particles of nitric acid trihydrate ($HNO_3 \cdot 3H_2O$) condense, and again at 187 K, when water ice particles condense. Under these conditions, the stage is set for the subsequent emergence of the ozone hole. First, the cloud particles efficiently absorb HNO_3 and $ClONO_2$, as well as HCl [the latter via reaction (8.12)]. Reactions on the surface of the particles then convert HCl and $ClONO_2$ to the more reactive Cl_2 and HOCl.

$$HCl + ClONO_2 = Cl_2 + HNO_3 \tag{8.18}$$

$$H_2O + ClONO_2 = HOCl + HNO_3 \tag{8.19}$$

These reactions are too slow in the gas phase to be of any importance, but they are greatly accelerated at the cloud-particle surfaces because 1) the reactant molecules are concentrated there, and 2) the formation of HNO_3 is assisted by hydrogen bonding to the water molecules in the particles. These reactions are further pushed to the right as the chlorine-containing gas products escape while the HNO_3 remains adsorbed on the ice particles. As the cloud particles grow over the winter, they sink to lower altitudes, thus physically separating the nitrogen from the active chlorine. Because nitrogen is unavailable, it cannot sequester the chlorine in the relatively unreactive molecule $ClONO_2$; the result is that, over the dark of winter, the vortex accumulates Cl_2 and HOCl.

As daylight returns in the springtime, Cl_2 and HOCl molecules are converted to Cl· by UV light:

$$Cl_2 + h\nu = 2Cl\cdot \qquad (8.20)$$

$$HOCl + h\nu = HO\cdot + Cl\cdot \qquad (8.21)$$

We can now understand why the ozone hole appears seasonally. During the dark polar winter, Cl_2 and HOCl accumulate; in spring, when the air is bathed in sunlight, the photolysis reactions produce a burst of Cl· atoms that react with the O_3 and produce ClO· (see Figure 8.9).

One more reaction is needed to complete the ozone-hole mechanism. In order to continue the catalytic cycle, the Cl· needs to be regenerated from ClO·. Normally this occurs via attack of O atoms in reaction (f). But in the lower stratosphere, where the polar ozone levels decline most steeply, there are not enough O atoms to keep the reaction going at the required rate. Instead, the ClO· builds up to a concentration (ppb range) sufficient to form dimers:

$$2ClO\cdot = ClO\!-\!OCl \qquad (8.22)$$

Radicals (like ClO·) form dimers readily; the two odd electrons pair up in making a new bond. Normally, dimerization terminates a radical chain reaction because the dimer is unreactive. Once again, however, photolysis intercedes, this time to split the O—Cl bond, regenerating the Cl· atoms:

$$ClOOCl + h\nu = ClOO\cdot + Cl\cdot \qquad (8.23)$$

$$ClOO\cdot + h\nu = O_2 + Cl\cdot \qquad (8.24)$$

This step completes the chain reaction, leading to the rapid destruction of a large quantity of O_3.

The chain reaction is broken when sunlight evaporates PSCs, leading to the conversion of HNO_3 to NO_2. The latter reacts with available ClO, converting it to $ClONO_2$, thus interrupting the chain sequence by eliminating reaction (8.22).

The extent of O_3 reduction depends on the extent of Cl_2 and HOCl formation during the dark of winter. This production rate in turn depends not only on the amount of chlorine in the stratosphere, but also on the number of cloud particles available to promote the reactions on the ice surface. Following the eruption of Mount Pinatubo in 1991, the polar ozone reduction became even more severe over the next two years because the sulfate aerosol from the volcano's SO_2 emission increased the surface available for reactions (8.18) and (8.19).

Although ozone depletion has been observed in both polar regions, the effect is more pronounced in Antarctica than in the Arctic region. This difference appears to be because temperatures in the Arctic do not fall low enough, or stay low long enough, to cause the removal of HNO_3 via the precipitation of large cloud particles. The temperature is higher in the Arctic than the Antarctic because there is more air movement into the stratosphere in the Northern Hemisphere. Although ClO concentrations in the Arctic region are elevated,

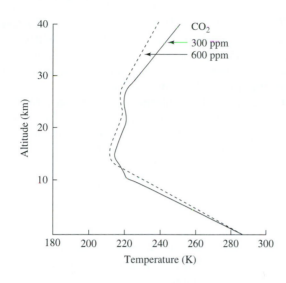

Figure 8.11   Effect of doubling the carbon dioxide concentration on the temperature profile of the atmosphere.   *Source:* S. Manabe and R. T. Wetherald (1967). Thermal equilibrium of the atmosphere with a given distribution of relative humidity. *Journal of Atmospheric Sciences* 24:241–259. Copyright © 1967, American Meteorology Society. Reprinted with permission.

just as they are in the Antarctic, HNO_3 provides a source of NO_2, which sequesters the reactive ClO. These findings suggest, however, that further cooling of the Arctic stratosphere could, in principle, cause an ozone hole comparable to that in Antarctica. This cooling could result from climate change, because Earth's surface temperature increases at the expense of decreasing temperature in the stratosphere. Figure 8.11 shows just such an effect in the temperature profile of the atmosphere projected for a two-fold increase in CO_2. Thus, the complex chemistry of polar clouds may link the dangers of climate change and those of ozone destruction.

8.6   OZONE PROJECTIONS

Now that the chemistry of ozone is understood in detail, atmospheric scientists are able to project changes in the ozone layer, with some confidence, using global atmospheric models. Figure 8.1 shows the projected trends in atmospheric chlorine levels, assuming implementation of international agreements. The figure indicates that by 2040, the atmospheric chlorine concentration is expected to decline to about 2 parts per billion (ppb), the level at which the Antarctic ozone hole began to appear in the late 1970s, and that it will further decline to about 1 ppb by the end of the twenty-first century. Although ozone losses are still large, they are expected to decline concomitant with the decrease in chlorine levels. Atmospheric data on $CFCl_3$ (CFC-11) (see Figure 6.16) show that the controls on emissions of CFCs appear to be working as planned.

8.7   CFC SUBSTITUTES

A great deal of effort has been applied to finding substitutes for CFCs. The main strategy has been to explore the suitability of hydrochlorofluorocarbons (HCFCs) and hydrofluorocarbons (HFCs). These molecules have hydrogen, as well as chlorine and/or fluorine substituents on the carbon. The presence of C—H bonds allows the HCFCs and HFCs to be attacked by hydroxyl radicals and thereby be destroyed in the troposphere. At the same time, the Cl and/or F substituents lend these chemicals some of the desirable properties of CFCs, such as low reactivity and fire suppression, good insulating and solvent characteristics, and boiling points suitable for use in refrigerator cycles.

Some of the CFCs and currently available substitutes are listed in Table 8.1. CHF_2Cl (HCFC-22) is a refrigerant that can substitute for CCl_2F_2 (CFC-12) in compressors of home air conditioners or refrigerators. The polyurethane foam insulation in refrigerator walls can be blown with CH_3CFCl_2 (HCFC-141b) or CF_3CHCl_2 (HCFC-123) instead of CCl_3F (CFC-11). They are good insulators and have low flammability.

However, while HCFCs have significantly shorter atmospheric lifetimes than CFCs, some of the molecules survive and drift into the stratosphere, where they contribute to ozone depletion. The three HCFCs shown in the table have ozone depletion potentials (ODP) ranging from 0.02 to 0.11 (relative to CFC-11). If enough of these HCFCs were to be produced, they could still deplete the ozone layer significantly and contribute to climate change. For this reason, HCFCs are viewed as transitional CFC substitutes; the amended Montreal Protocol calls for the eventual elimination of HCFCs and for their permanent replacement by substances without chlorine. HFCs, having no chlorine, are not ozone

TABLE 8.1 CHLOROFLUOROCARBONS (CFCs) AND THEIR SUBSTITUTES— HYDROCHLOROFLUOROCARBONS (HCFCs) AND HYDROFLUOROCARBONS (HFCs)

Trade name	Chemical formula	Market	Atmospheric lifetime (yr)	100 year GWP*	ODP[†]
CFC-11	CCl_3F	Blowing agent	50	4,000	1.0
CFC-12	CCl_2F_2	Refrigerant	102	8,500	1.0
CFC-113	CCl_2FCClF_2	Cleaning agent	85	5,000	0.8
HCFC-22	CHF_2Cl	Refrigerant Blowing agent	12.1	1,700	0.055
HCFC-141b	CH_3CFCl_2	Blowing agent	9.4	630	0.11
HCFC-123	CF_3CHCl_2	Blowing agent	1.4	93	0.02
HFC-134a	CH_2FCF_3	Refrigerant Blowing agent	14.6	650	0.0
HFC-23	CHF_3	Fire extinguisher	260	11,700	0.0
HFC-227ea	C_3HF_7	Fire extinguisher	36.5	2,900	0.0
HFC-245fa	$C_3H_3F_5$	Blowing agent	6.6	790	0.0

*GWP signifies "global warming potential." It is a measure of the degree of radiative forcing of a given molecule compared to a molecule of CO_2, which is assigned a GWP value of 1.

[†]ODP signifies "ozone depletion potential." It is the ratio of the impact on ozone of a chemical compared to the impact of a similar mass of $CFCl_3$ (CFC-11), which is assigned the value of 1.0.

Sources: IPCC *Special Report on Emission Scenarios,* Section 5.4.3. Halocarbons and other halogenated compounds (htpp://www.grida.no/climate/ipcc/emission/123.htm). Montreal Protocol on Substances that Deplete the Ozone Layer (http://www.unep.org/ozone/Montreal-Protocol/Montreal-Protocol2000.shtml).

destroyers, and are not covered by the Montreal Protocol. They have been used as blowing agents, refrigerants, and fire extinguishers.

However, both HCFCs and HFCs are potent greenhouse gases. Their global warming potentials (GWP) are many times larger on a per molecule basis than CO_2 (see Table 8.1). Although their lifetimes (with the exception of CHF_3) are shorter than those of the CFCs, they are longer than unsubstituted hydrocarbons, because the fluorine atoms stabilize the C—H bonds (see p. 230–231), making them slower to react with hydroxyl radicals. Concern for their potential impact on global climate has led to their inclusion on the list of greenhouse gases targeted for emissions reductions under the Kyoto Protocol. Current market trends indicate that relatively few CFCs will be replaced by HFCs. To date only HFC-134a has penetrated the market to any significant extent. The low percentage of replacement reflects the introduction of non-HFC substitutes, and increased efficiency, containment, and recycling of halocarbons currently in use.

New technologies that rely on neither HCFCs nor HFCs to replace CFCs are being developed. Aerosol propellants, for example, can be isobutane or dimethyl ether (mixed with water to suppress flammability). Similarly, CFCs have been replaced by hydrocarbons as blowing agents in styrofoam production. The foam insulation in refrigerator walls, first blown with CFC-11 and now with HCFC-141b, may soon be replaced with vacuum panels that are filled with a solid filler material and sealed under vacuum in a gas-tight envelope. The electronics industry, which relied heavily on CFC solvents for cleaning circuit boards, has switched to aqueous detergent cleaners and new imprinting methods that reduce the amount of cleaning needed.

The working fluids of refrigerators and air conditioners are the hardest to replace. Here, too, alternatives are coming into view. There has been much interest in older materials such as ammonia and hydrocarbons, but deterrents include the toxicity and corrosiveness of ammonia and the flammability of hydrocarbons. Flammability, at least, can be managed with proper engineering (we are exposed to flammability hazards all the time, for example, in handling automobiles and gas furnaces); on the market now is a German-made domestic refrigerator (called "Greenfreeze") that uses a propane-butane mixture. Moreover, there are air conditioners in development that do not have compressors, but that rely instead on evaporative cooling combined with desiccant drying of the cooled air. In the longer term, there is also interest in using sound waves for cooling.

There are currently no adequate substitutes for the halons, which are used to flood enclosed spaces such as offices, airplanes, and military tanks, for example, in case of fire. Since 1994 when production ceased, halons are being carefully banked, pending the development of substitutes. The halons exhibit a combination of low reactivity and effective fire suppression that is hard to match. The most promising candidate appears to be CF_3I, which, like CF_3Br (halon-1301) is heavy enough to blanket and smother fires. The C—I bond is rapidly photolyzed by solar photons, even at ground level, so the molecule's atmospheric lifetime is short. However, toxicity and corrosiveness have not been fully evaluated.

Overall, the shift away from massive reliance on CFCs is happening more quickly than anyone thought possible a few years ago. As usual, necessity is the mother of invention.

CHAPTER 9

AIR POLLUTION

While the greenhouse gases and the CFCs may be considered global pollutants (because they potentially harm the climate system and the stratospheric ozone layer worldwide), the term "air pollution" generally refers to substances that on local and regional scales directly harm animals, plants, and people and their artifacts. The phenomenon is hardly new. There have been complaints about air quality for centuries, especially in cities. But the steady expansion of population and industrial civilization has changed the nature of air pollution. The pervasive effects of emissions are increasingly manifest, and the need to control them is influencing to a greater degree the development of technology, particularly in the energy and transportation sectors.

9.1 POLLUTANTS AND THEIR EFFECTS

A wide range of chemicals can pollute the air, but the ones generally viewed as needing control measures are carbon monoxide, sulfur dioxide, toxic organics, particulates, nitrogen oxides, and volatile organic compounds. The first four directly harm human welfare, whereas the last two are ingredients of photochemical smog, whose harmful effects are due to the production of ozone and other "oxidant" molecules.

    a. Carbon monoxide.   As shown in Figure 9.1, carbon monoxide (CO) emissions in the United Sates peaked around 1970 at about 117 million metric tons per year, and have been declining at a rate of about 13 million metric tons per decade since then. By far the major source of CO has been transportation (on-road plus non-road). From 1940 to 1970, the increase in CO was directly proportional to the increase in vehicle miles traveled. Since

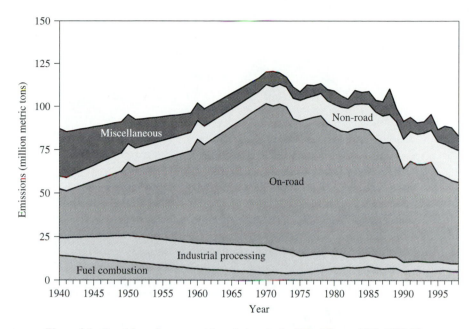

Figure 9.1 Trend in carbon monoxide emissions in the United States, 1940–1998. The category "fuel combustion" includes fuels used by electric utilities and industry, and residential wood burning; "industrial processing" includes chemical and allied product manufacturing, ferrous and nonferrous metal processing, and waste incineration; the "on-road" category includes automobiles, and light- and heavy-duty trucks; the "non-road" category includes lawn mowers, marine vessels, construction vehicles, farming machines, aircraft, and railroads; "miscellaneous" is primarily emissions from wildfires. *Source:* U.S. Environmental Protection Agency (2000). *National Air Pollutant Emission Trends, 1900–1998,* EPA-454/R-00-002 (March 2000). (http://www.epa.gov/ttn/chief/trends/trends98)

1970, emissions have declined, despite the continued increase in vehicle travel, due to increasingly stringent emission-control standards and improvements in energy efficiency.

Although carbon monoxide occurs naturally in the environment, it is an asphyxiating poison because it can displace the O_2 bound to hemoglobin (see Figure 9.2). The Fe binding sites in hemoglobin bind CO 320 times more tightly than O_2. Such high affinity means that in human blood, CO occupies about 1 percent of hemoglobin binding sites; in smokers, this percentage doubles, on average, due to the CO in the inhaled smoke. When the ambient concentration of CO reaches 100 ppm, the percentage occupancy of the hemoglobin binding sites rises to 16 percent. This concentration of CO can be encountered in dense traffic in enclosed spaces (tunnels, parking garages), and may result in headaches and shortness of breath. The severity of the effects depends on the duration of the exposure and level of exertion (see Figure 9.3), because it takes some time for the inhaled CO to equilibrate with the circulating blood. At concentrations higher than 750 ppm (0.1 percent of the air molecules), loss of consciousness and death occur quickly. At lower levels, the effects are reversed by breathing uncontaminated air, which allows O_2 to replace the CO bound to

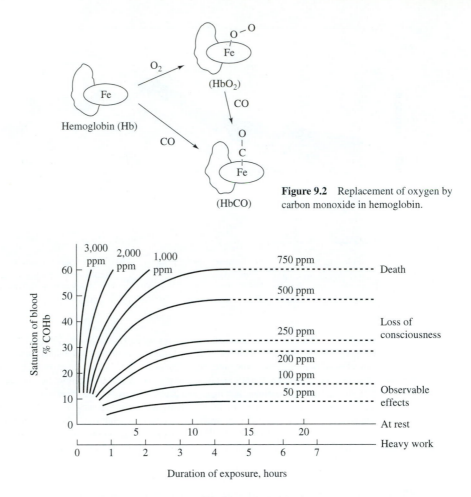

Figure 9.2   Replacement of oxygen by carbon monoxide in hemoglobin.

Figure 9.3   Dose-response curves of HbCO uptake in blood.

the hemoglobin molecules. However, individuals with heart problems are sensitive to even temporary oxygen insufficiency, and hospital admissions for congestive heart failure have been found to be influenced by CO levels in urban air.

While the main source of anthropogenic CO is automotive transport, individuals may be at greater risk of CO poisoning from malfunctioning stoves and space heaters in their homes. CO is produced whenever combustion is incomplete. The problem is particularly acute in poor countries where unvented and inefficient stoves are common.

b. Sulfur dioxide.   The main sources of anthropogenic sulfur dioxide air emissions have been stationary-source combustion of coal, and the smelting of ferrous and nonferrous metals, particularly copper. The sulfur content of refined petroleum is generally quite low, but the sulfur content of coal is quite high. Figure 9.4 shows sulfur dioxide

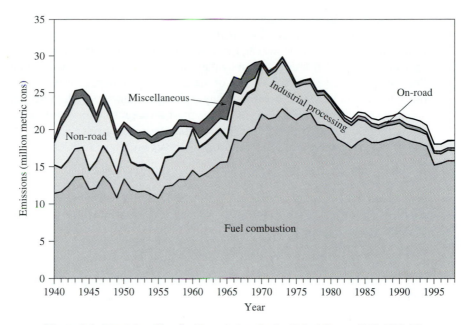

Figure 9.4   Trend in sulfur dioxide emissions in the United States, 1940–1998. The category "fuel combustion" is mostly coal consumption by electric utilities, industry, and commercial and residential users; "industrial processing" includes chemical and allied product manufacturing, ferrous and nonferrous metal processing, petroleum and related industries, pulp and paper production, and mineral products manufacturing; the "on-road" category includes automobiles, and light- and heavy-duty trucks; the "non-road" category includes lawn mowers, marine vessels, construction vehicles, farming machines, aircraft, and railroads.   *Source:* U.S. Environmental Protection Agency (2000). *National Air Pollutant Emission Trends, 1900–1998,* EPA-454/R-00-002 (March 2000). (http://www. epa.gov/ttn/chief/trends/trends98)

emissions in the United Sates over the period 1940 to 1998. The steep rise in emissions between 1940 and 1945, the decline in the late 1940s and early 1950s, and the sharp rise between the mid-1950s to 1973 reflect several opposing trends related to declining use of coal in the industrial, residential, and transport sectors on the one hand, and its expanded use for electricity generation on the other. In 1940, sulfur emissions were 11 percent from coal-fired power plants, 13 percent from residences, 15 percent from railroads, 17 percent from smelting, and 26 percent from industry. By 1970, railroads and houses no longer used significant amounts of coal, industry's share of sulfur emissions had fallen to 10 percent, smelting remained at 15 percent, while coal-fired power plants accounted for 51 percent. By 1998, coal combustion for electric power dominated all other sources at 63 percent of the total emissions. The overall decline in emissions since the peak year of 1973 is the result of continued fuel switching from coal to oil and gas, improvements in energy efficiency, and increasingly stringent regulations that have resulted in the substitution of high-sulfur coal by lower-sulfur coals, and implementation of emission-control devices, such as flue-gas desulfurization that strips sulfur out of the smokestack gases.

The sulfur in coal is converted to sulfur dioxide at the high temperatures of combustion. Sulfur dioxide itself is a lung irritant; it is known to be harmful to people suffering from respiratory disease. However, the most damaging health effects in urban atmospheres are caused not by sulfur dioxide but by the sulfuric acid aerosol formed from its oxidation. Sulfuric acid irritates the fine vessels of the pulmonary region, causing them to swell and block the vessel passages. Breathing may be severely impaired. The effect appears to be cumulative, with older people suffering the most severe respiratory problems.

Sulfuric acid aerosol is the major contributor to acid rain (see pp. 157–158, 302–303) and it corrodes human artifacts. It steadily dissolves limestone ($CaCO_3$),

$$CaCO_3 + 2H^+ = Ca^{2+} + CO_2 + H_2O \tag{9.1}$$

damaging outdoor monuments in cities around the world. Likewise, ancient stained glass windows are being eroded as the acid leaches out mineral constituents of the glass. Also, acid accelerates the corrosion of iron,

$$2Fe + O_2 + 4H^+ = 2Fe^{2+} + 2H_2O \tag{9.2}$$

Protection of iron and steel structures with corrosion-inhibiting paint costs billions of dollars annually. In heavily industrialized areas, zinc coatings on galvanized steel may last as little as 5–10 years.

c. Toxic organics.   Many organic compounds are toxic, but only a limited number are of concern as air pollutants. Several environmental toxins that are dispersed and transported through the air are not inhaled in significant quantities; rather they are deposited and delivered through the food chain. Examples of deposited toxins (discussed in Part IV) are dioxins, lead, and mercury. The toxic organic compounds that act as direct air pollutants are the small aldehydes, benzene, and polycyclic aromatic hydrocarbons (PAHs).

1) Formaldehyde and acetaldehyde.   Formaldehyde, CH_2O, is a reactive molecule that irritates the eyes and lungs at quite low concentrations, just over 0.1 ppm. The International Agency for Research on Cancer and the U.S. National Toxicology Program classify formaldehyde gas as a probable carcinogen, based on animal studies and limited evidence of carcinogenicity in humans. The evidence is strongest for nasal and nasopharyngeal cancer. In the U.S., industrial releases of formaldehyde to air in 1999 totaled 5,629 metric tons. Emissions have not decreased since 1988, when it was estimated that 5,651 tons were emitted, and it remains one of the largest emitters to air among suspected carcinogens.

Formaldehyde is also a source of indoor air pollution; it is released from formaldehyde resins used in construction materials such as plywood, particle board, and glass fiber insulation. Levels of formaldehyde can be quite high in mobile homes (greater than 1 ppm); exposed individuals develop symptoms that include drowsiness, nausea, headaches, and respiratory ailments. Because formaldehyde is potentially carcinogenic, chronic exposure, even at low doses, poses a public-health problem.

While outdoor levels of formaldehyde are low, there is concern that they may rise. Appreciable quantities of formaldehyde are produced by partial oxidation of methanol (CH_3OH). Likewise, the burning of ethanol releases significant amounts of acetaldehyde, which has similar toxic qualities. Hence, outdoor levels of formaldehyde and acetaldehyde may increase substantially if methanol and ethanol become future automotive fuels. In some parts of the United States, ethanol (C_2H_5OH) is already blended with gasoline, and its use as an additive is expected to grow significantly under current U.S. EPA regulations. In Brazil, a large fraction of the automotive fleet runs on ethanol from sugar cane.

2) Benzene.   Benzene is one of the top 20 chemicals produced in the U.S. by volume, with production of about 7.6 million metric tons annually. It is derived mainly from crude oil, and is widely used in the petroleum, chemical, and manufacturing industries. About 75 percent is used as a feedstock in production of styrene and phenol. Less than 2 percent of benzene is used in gasoline blending, and this amount is likely to decrease with continuing reformulation of gasoline (see p. 239). Air emissions of benzene from industry in 1999 are estimated to have been 3,466 metric tons, and have not declined significantly since 1995. However, they are much lower than in 1988 when they totaled 14,670 tons.

Benzene is one of the few chemicals classified as a known carcinogen, and is implicated as a causative agent in human leukemia. It is ranked in the top 20 list of hazardous substances by the U.S. Department of Health. According to the U.S. National Toxicology Program, general population exposures to benzene are not correlated with industrial or vehicular emissions, which account for 14 and 82 percent of the total. Instead, cigarettes account for 40 percent of benzene exposure, with an additional 5 percent from environmental tobacco smoke, while personal activities and car exhaust account for 18 percent each, and home sources for 16 percent. Industry accounts for only 3 percent of average benzene exposure, although individual workers may be exposed at much higher levels. Benzene use is strictly controlled in the workplace. In many applications, it is being replaced by alkylated benzenes such as toluene (methylbenzene), which are much less toxic than benzene; the alkyl groups are readily oxidized by enzymes in the liver, producing benzoic acid or related acids, which are excreted.

3) Polycyclic aromatic hydrocarbons (PAHs).   Compounds with four or more benzene rings fused together, such as benzo[a]pyrene (see Figure 9.5), are potent carcinogens. Interestingly, their carcinogenicity depends upon activation by the same class of liver enzymes, cytochromes P450 (see pp. 414, 417, 425), that metabolize toluene and other xenobiotics (molecules foreign to the organism). When these enzymes add oxygen to the

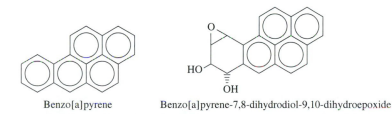

Benzo[a]pyrene   Benzo[a]pyrene-7,8-dihydrodiol-9,10-dihydroepoxide

Figure 9.5   Structures of benzo[a]pyrene and an oxygenated metabolite.

PAHs, they produce epoxide adducts (see Figure 9.5) that react strongly with the hetero-cyclic bases of DNA, altering genes.

PAHs are formed as side-products of carbon fuel combustion. Although they are present at low levels in automobile exhaust, the levels are much higher when large quantities of soot particles are produced, as in diesel exhaust or smoke from coal or wood fires. (The soot itself contains sheets of benzene rings, like graphite.) As long ago as 1775, soot exposure was linked to scrotal cancer in chimney sweeps in London. Coke oven workers have also been documented to have increased levels of lung and kidney cancer.

d. Particles.   Atmospheric particles are of concern for two major reasons: they significantly affect Earth's radiation balance, as discussed on pp. 153–155, and they are serious health hazards. Particles penetrate the lungs, blocking and irritating air passages, and can have toxic effects. Coal miners' black-lung disease, asbestos workers' pulmonary fibrosis, and the city dwellers' emphysema are all associated with the accumulation of particles in the lung. Small particles have the greatest health impact because they penetrate most deeply into the lung. Particles larger than several microns are trapped in the nose and throat, from which they are eliminated more readily.

Asbestos fibers are especially dangerous because they can cause *mesothelioma,* a cancer of the pleural cavity, even at quite low exposures. The most hazardous form of asbestos is *crocidolite,* in which the tiny fibers are rod-like and can penetrate deep into the lungs. The more common form of asbestos, *chrysotile,* has serpentine fibers that clump into bundles, most of which are intercepted in the upper airways, where they are less hazardous. Due to the risk of cancer, however, the use of all forms of asbestos has been sharply curtailed. Nevertheless, asbestos is still present as insulation and fireproofing material in many buildings. What to do about it is a contentious issue because, unless extraordinary precautions are taken, attempts to remove it can release large quantities of fibers into the air of a building. The released fibers settle on surfaces from which they can be re-suspended quite readily, thereby remaining much more hazardous than if they had remained locked and undisturbed in the building material.

Soot particles pose particular problems because they can adsorb significant amounts of toxic chemicals on their irregular surfaces. Soot particles are very prevalent in diesel exhaust and wood smoke. Coal fires release soot as well as SO_2; in foggy conditions, the resulting sulfate aerosol can combine with the soot to produce a toxic smog, with serious health consequences, especially for those with respiratory ailments. In London in December 1952, a heavy smog of this sort killed about 4,000 people within a few days. As a result of switching from coal to oil and gas for household heating, the London-type smog has largely disappeared in developed countries. But coal is still widely burned in developing countries, whose cities often have unhealthy air. In addition, indoor air pollution from cooking fires is a serious and widespread health problem in developing countries.

Epidemiological evidence associates particles more directly with disease and mortality than any gaseous pollutant. An influential study of six U.S. cities monitored a cohort of 8,111 adult subjects for 14–16 years from the mid-1970s and found that increased fine-particle concentrations were correlated with increased mortality (26 percent increase over an 18.6 μg/m^3 range of particle concentrations) from all causes, but especially from cardiopulmonary disease.* This correlation was confirmed in a much larger study under the aegis of the American Cancer Society.

In response to these findings, the U.S. EPA proposed an annual standard of 15 $\mu g/m^3$, and a daily standard of 65 $\mu g/m^3$, for fine particles, defined as those with a diameter of 2.5 μm or less ($PM_{2.5}$). These levels are currently exceeded regularly in many U.S. cities. Control measures will be difficult since fine particles derive from so many dispersed sources. Because much of the urban aerosol derives from SO_2, NO_x, and NH_3 emissions, as well as diesel exhaust, the proposed standards have serious implications for the transportation and industrial sectors. As a result, there has been intense opposition to the standards, and much questioning of the epidemiological studies on which they are based. However, reanalysis of the data by the industry-financed Health-Effects Institute has confirmed the original findings. Opponents of the standard successfully sued over the EPA's authority, but in 2000 the U.S. Supreme Court overturned the lower court ruling and confirmed the EPA's authority. The regulation currently awaits promulgation.

Trends in U.S. emissions of particles with a diameter of 10 μm or less (PM_{10}) from 1940 to 1998 are shown in Figure 9.6a. Historically, the largest source of particulates was the industrial processing sector, especially ferrous and nonferrous metals production, and minerals manufacturing, including cement. There was also a sizable input from stationary-source fuel combustion, from residential wood burning, and from the railroads. Emissions have dropped dramatically, beginning around 1970, thanks to the switch away from coal for heating and transport, the introduction of new, cleaner technologies, and the adoption of emission-control equipment such as electrostatic precipitators and bag filters to trap the larger particles. By the mid-1980s, emissions were almost equally distributed among fuel combustion, industrial processing, agriculture, and forestry (the latter two shown under the "miscellaneous" category in the figure), with smaller contributions from on-road and non-road transport vehicles.

Methods to measure fine particles were developed only recently. Estimates of $PM_{2.5}$ emissions (see Figure 9.6b) were fairly level during the 1990s at around 3 million metric tons, and account for about half of total particles. However, many of the sources are well removed from urban areas. These include agriculture and forestry (miscellaneous) that account for 32 percent of the total, and residential wood combustion (a category under fuel combustion) and non-road diesel vehicles that contributed 15 percent each. The totals, even in nonurban areas, do matter, however, when evaluating the global aerosol, and its influence on climate (pp. 153–155).

e. NO_x and volatile organics.   Nitrogen oxides (NO_x) and volatile organic compounds (VOCs) are not direct air pollutants, in that they rarely affect health directly. Rather, they are the main ingredients in the formation of *photochemical smog,* the brown haze that blankets many cities worldwide. Although most damage from smog results from the action of ozone and other oxidants, these oxidants cannot build up without the combined action of NO_x and VOCs. Controlling smog formation requires reducing emissions of NO_x and VOCs.

---

*D. W. Dockery, C. A. Pope, X. Xu, J. D. Spengler, J. H. Ware, M. E. Fay, B. G. Ferris, and F. E. Speizer (1993). An association between air pollution and mortality in six U.S. cities. *New England Journal of Medicine* 329:1753–1759.

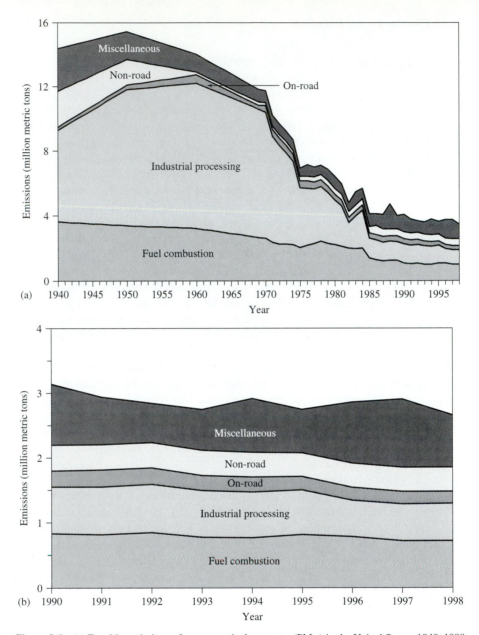

Figure 9.6 (a) Trend in emissions of coarse particulate matter (PM$_{10}$) in the United States, 1940–1998 (excluding fugitive dust sources). The category "fuel combustion" includes fuels used by electric utilities and industry, and residential wood burning; "industrial processing" includes ferrous and nonferrous metal processing, cement manufacturing, stone quarrying, agricultural county and terminal grain elevators, wood pulp and papermaking, and waste incineration; the "non-road" category, which was an important source in earlier years, was mainly from railroad emissions; "miscellaneous" includes dust and erosion from agricultural operations. (b) Trends in emissions of fine particulate matter (PM$_{2.5}$) in the United States, 1990–1998. The breakdown of categories is similar to that of the PM$_{10}$ emissions. *Source:* U.S. Environmental Protection Agency (2000). *National Air Pollutant Emission Trends, 1900–1998*, EPA-454/R-00-002 (March 2000). (http://www.epa.gov/ttn/chief/trends/trends98)

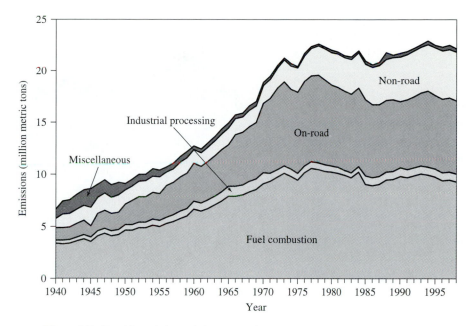

Figure 9.7   Trend in emissions of nitrogen oxides in the United States, 1940–1998. The category "fuel combustion" includes fuels used by electric utilities and industry, and residential and commercial space heating; the "on-road" category includes cars, light- and heavy-duty gasoline-powered trucks, and heavy-duty diesel vehicles; in the "non-road" category, the largest contributors are farming and construction equipment powered by diesel engines, and marine vessels.   *Source:* U.S. Environmental Protection Agency (2000). *National Air Pollutant Emission Trends, 1900–1998*, EPA-454/R-00-002 (March 2000). (http://www.epa.gov/ttn/chief/trends/trends98)

Trends in U.S. air emissions of NO_x compounds from 1940 to 1998 are shown in Figure 9.7. Almost all NO_x emissions are due to transportation (on-road and non-road sectors) and stationary-source fuel combustion (fuel combustion). In 1998, coal-fired power plants and heavy-duty diesel vehicles each accounted for 22 percent of NO_x emissions (4.9 million metric), while light-duty gasoline-powered cars and trucks accounted for 19 percent (4.3 million metric tons). As with all air pollutants generated from fossil fuel consumption, the increase prior to 1970 was due to increased fuel consumption with little attention paid to controlling air pollution. Over the last three decades, emissions have stabilized through implementation of emission controls and increased conservation of energy, particularly in transportation.

Emissions of VOCs are shown in Figure 9.8. The three largest source sectors are industrial processing, solvent utilization, and on-road and non-road vehicles. A host of industrial and consumer products and transportation vehicles add to the total emissions (16.3 million tons in 1998). Some of the larger specific emitters in 1998 were cars (16 percent) and light-duty trucks (11 percent), surface coatings (12 percent), and consumer solvents (6 percent).

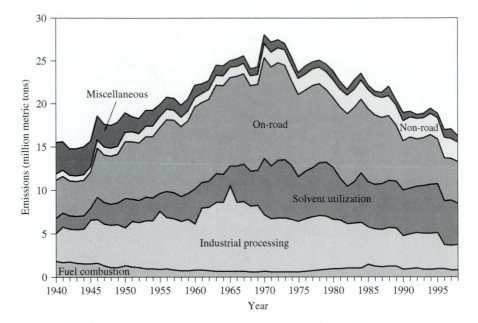

Figure 9.8 Trend in emissions of volatile organic compounds (VOCs) in the United States, 1940–1998. The category "industrial processing" includes chemical and petroleum industries, metals processing, and the storage of petroleum and petroleum products as industrial stocks and in service stations; the category "solvent utilization" includes degreasing, graphic arts, dry cleaning, surface coating, pesticide application, and consumer solvents; the most important "on-road" sources are gasoline-powered cars and light trucks; the most important "non-road" sources are gasoline-powered lawn mowers and recreational marine vessels. *Source:* U.S. Environmental Protection Agency (2000). *National Air Pollutant Emission Trends, 1900–1998*, EPA-454/R-00-002 (March 2000). (http://www.epa.gov/ttn/chief/trends/trends98)

f. Ozone and other oxidants.   While anthropogenic emissions are destroying ozone in the stratosphere, they are helping to create ozone in the troposphere via the phenomenon of photochemical smog (see next section). And while ozone in the stratosphere protects us from the harmful effects of UV rays, ozone at ground level is quite harmful, producing cracks in rubber, destroying plants, and causing respiratory distress and eye irritation in humans. These effects set in at quite low concentrations, around 100 ppb. As shown in Figure 9.9, large metropolitan areas around the U.S., particularly New York and surroundings, Chicago, Houston, Los Angeles, and San Francisco, are in nonattainment with the EPA's one-hour 120 ppb ozone standard. A revised, more stringent ruling, based on an eight-hour 80 ppb standard, was recently upheld by the U.S. Supreme Court. When fully implemented, the revised standard will likely broaden the areas of nonattainment.

These effects of ozone result from its being a strong oxidant and O-atom donor. Ozone reacts especially well with molecules containing C=C double bonds, forming epoxides. Such molecules are abundant in rubber, the photosynthetic apparatus of green plants, and the membranes lining the lung's air passages.

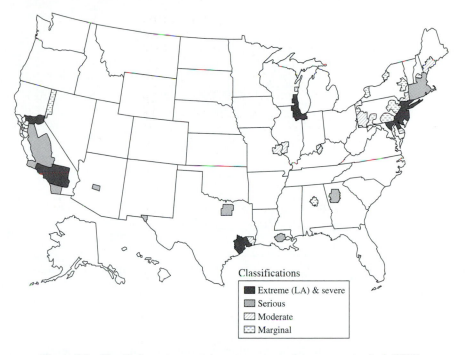

Figure 9.9   Classified ozone nonattainment areas, one-hour ozone standard (2001). *Source:* U.S. Environmental Protection Agency (2001). *Ozone, Green Book Home Page.* (http://www.epa.gov/oar/oaqps/greenbk/onmapc.html)

Other oxidant molecules are also formed in photochemical smog and produce similar damage. An example is peroxyacetyl nitrate (PAN), $CH_3C(O)OONO_2$, a potent eye irritant.

9.2   PHOTOCHEMICAL SMOG

Nitrogen oxides and volatile hydrocarbons are key ingredients in the formation of photochemical smog, a condition that afflicts an increasing number of cities and their surroundings. Photochemical smog can form whenever large quantities of automobile and industrial exhausts are trapped by an inversion layer over a locality that is, at the same time, exposed to sunshine. The classic location for smog is Los Angeles, with its dependence on the automobile, abundant sunshine, and frequent inversion layers, but automobile traffic has introduced the problem to many other cities. It is characterized by an accumulation of brown, hazy fumes, containing ozone and other oxidants, with the harmful effects described above.

Figure 9.10 shows the time-course of the key atmospheric ingredients for a classic smoggy day in Los Angeles. The concentration of hydrocarbons peaks during the early-morning rush hour. The concentration of nitric oxide reaches a peak at the same time, and then begins to decrease as the nitrogen dioxide concentration increases. Subsequently, the concentration of oxidants rises, and the hydrocarbon concentration falls.

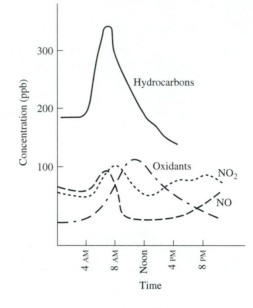

Figure 9.10 Example of a concentration-time profile of smog-forming chemicals in Los Angeles air.

Recall that the formation of ozone from oxygen requires energy, which is provided in the stratosphere by the absorption of UV light. What drives ozone formation near the ground level, where little UV arrives? The critical ingredient is NO_2, the only common atmospheric molecule capable of absorbing visible light. Its spectrum (see Figure 9.11) has maximum absorption at about 400 nm, in the blue region. It is this absorption that gives smoggy air its brown tint. The photoexcited NO_2 is unstable, and dissociates to NO and O atoms:

$$NO_2 + h\nu \ (<400 \ \text{nm}) = NO + O \tag{9.3}$$

The O atoms react immediately with the surrounding O_2 molecules to produce ozone, just as in the stratosphere [reaction (b), p. 203].

But since each molecule of ozone requires an oxygen atom from NO_2 dissociation, this mechanism cannot build up the ozone concentrations to levels greater than the NO_2 itself. Moreover, the NO produced in reaction (9.3) can react with the ozone to restore the NO_2 by the same reaction that destroys stratospheric ozone [reaction (e), p. 207, with X = NO]. Thus, nitrogen oxides and oxygen alone cannot lead to net ozone formation in sunlit air.

Hydrocarbons are needed as well. Hydrocarbons produce peroxyl radicals, which can react with the NO before the ozone does and regenerate NO_2 [reaction (7.20), p. 191, with X = NO]. If this happens, NO_2 can catalyze the formation of ozone as long as the supply of peroxyl radicals lasts, thereby allowing ozone to accumulate. The peroxyl radicals are formed when O_2 reacts with organic radicals [reaction (7.19), p. 190], which in turn are produced by the action of hydroxyl radicals on hydrocarbons [reaction (7.23), p. 191]. The hydroxyl radicals themselves are produced from ozone [reactions (7.24) and (7.25), p. 191],

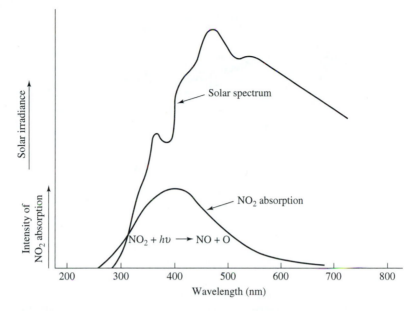

Figure 9.11   Absorption of solar light by nitrogen dioxide to form oxygen atoms. The reaction shown under the absorption curve occurs when nitrogen dioxide adsorbs light at wavelengths of less than 400 nm. For wavelengths greater than 400 nm, nitrogen dioxide is excited but does not decompose.

thereby completing the photochemical cycle. Since a single organic radical can produce many peroxyl radicals by successive rounds of O_2 combination and fragmentation, the ozone concentration can build up rapidly to levels exceeding the concentration of nitrogen oxides, and far exceeding the concentration of the hydroxyl radical. The interlocking NO_x, O_3, and hydrocarbon radical cycles are diagrammed in Figure 9.12.

The combination of reactive species in these cycles produces other oxidants. For example, the reaction of peroxyl radicals with NO_2 produces peroxyalkyl and peroxyacyl nitrates:

$$ROO\cdot + NO_2 = ROONO_2 \qquad (9.4)$$

[R signifies alkyl or acyl ($R{-}\overset{\overset{\displaystyle O}{\|}}{C}{-}$)]. An example is peroxyacetyl nitrate (PAN) formed from the peroxyacetyl radical, a relatively common smog constituent.

Because the initiation of the cycle depends on the formation of organic radicals, the extent of smog formation depends on the reactivity of hydrocarbons with the hydroxyl radical. Some hydrocarbons produce few radicals, others many more. As mentioned earlier, the H-atom abstraction proceeds spontaneously because the O—H bonds of the product water molecule are stronger than the C—H bond being attacked by the hydroxyl radical. But the rate of the reaction depends in some detail on the nature of the C—H bond being attacked.

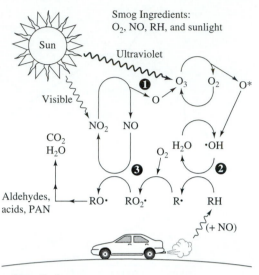

Smog Ingredients:
O_2, NO, RH, and sunlight

Smog Cycles:
❶ O_3 formed by O atoms from NO_2 photolysis
❷ HC radicals made by ·OH, from O_3 UV photolysis
❸ NO oxidation to NO_2 catalyzed by HC radicals

Figure 9.12   Smog formation from O_2, NO, hydrocarbon, and sunlight.

FUNDAMENTALS 9.1: C—H Bond Strengths

C—H bonds are not all the same. The bond dissociation energy listed in Table 2.1 (see p. 25) and used in combustion energy calculations, 410 kJ/mol, is an average value for hydrocarbons as a class. But individual bonds vary considerably in their dissociation energies, as illustrated in Table 9.1.

The reason for this variability is that the dissociation energy is the energy change for the reaction

$$R—H \rightarrow R· + ·H$$

which depends on the energies of both reaction products, relative to the reactant. The energy of the H atom is always the same, but the energy of the R· radical depends on its structure. An unpaired electron on a carbon atom is stabilized by additional carbon atom substituents; the electron can be delocalized to some extent over these substituents. For this reason a tertiary

C—H bond, one with three C-atom substituents (e.g., in tertiary butane, see Table 9.1) is more easily dissociated than a secondary C—H bond, one with two C-atom substituents (e.g., isopropane), because a tertiary C radical is

TABLE 9.1 DISSOCIATION ENERGY OF THE C—H BOND RELATED TO SUBSTITUENTS ON THE CARBON ATOM

Compound	Bond	Energy (kJ/mol)
Methane	$H_3C—H$	427
Ethane	$H_3CH_2C—H$	406
Isopropane	$[H_3C]_2HC—H$	393
Tertiary butane	$[H_3C]_3C—H$	381
Trifluoromethane	$F_3C—H$	446
Chloroform	$Cl_3C—H$	401
Methanol	$HOH_2C—H$	393
Ethylene	$H_2CHC—H$	444
Benzene	$H_5C_5C—H$	427
Toluene	$H_5C_6H_2C—H$	326

stabilized to a greater extent than a secondary radical. Likewise, a secondary C—H bond dissociates more readily than a primary C—H bond, one with a single C substituent (e.g., ethane). Hardest to dissociate is a methane C—H bond, because the methyl radical has no C substituent to stabilize the unpaired electron. Although hydroxyl radical attack is the mechanism for methane destruction in the atmosphere, the rate is too slow to make CH_4 a significant contributor to smog formation. Thus, switching automotive fuel to methane would help reduce smog.

While carbon substituents stabilize R·, fluorine atoms have the opposite effect (see trifluoromethane in Table 9.1). This is because fluorine has non-bonding electron pairs, which are brought close to the unpaired electron by the short C—F bond, and the electrostatic repulsion destabilizes the radical. The C—H bond is strengthened, leading to relatively long atmospheric lifetimes, and large global warming potentials, for HFCs and HCFCs (see Table 8.1, p. 214). (Chlorine atoms also have unpaired electrons, but the electrostatic effect is diminished because of the longer C—Cl bond. Cl atoms are only slightly less stabilizing than carbon substituents, see chloroform in Table 9.1.) Oxygen atoms have a similar strengthening effect on adjacent C—H bonds (see methanol in Table 9.1).

There are additional considerations for unsaturated hydrocarbons. Benzene and ethylene have high C—H dissociation energy because the orbital hybridization on carbon is sp^2, whereas it is sp^3 in alkanes. Bonds utilizing sp^2 orbitals are shorter and stronger because of the greater participation of the s orbital, which is concentrated near the nucleus. (However, hydroxyl radicals react rapidly with olefins, because their pi bonds are susceptible to attack; in the case of aromatic compounds, the pi bonds are stabilized by resonance and are not readily attacked, see Appendix). The bond energy for the methyl C—H bond on toluene is anomalously low, because the organic radical has a special stabilization mechanism. The unpaired electron on the C atom adjacent to the benzene ring can be delocalized over the entire benzene pi orbital system.

The reactivity of gasoline components with hydroxyl radicals (see PA column in Table 9.2) can be understood on the basis of the number of C—H bonds and their relative strengths. Reaction rates are lowest for benzene and compounds whose C—H bonds are mostly in methyl groups (methanol, ethanol, MTBE, methylpropane). As the number of secondary C—H bonds increases, so does the hydroxyl radical reactivity. Among the straight-chain alkanes, the rate increases from butane to pentane to hexane, and the rate is higher still for cyclohexane, in which the two methyl ends of *n*-hexane are replaced by methylene groups. High rates are also found for the methyl-substituted benzenes, toluene and xylene, because of the special stabilization of the methylene radicals by the benzene ring.

Even higher rates are seen for the alkenes (butene, methylpropene, pentene). However, for these molecules the mechanism is not H-atom abstraction, but rather attack by hydroxyl radical on the C=C bond to form a radical adduct:

$$R_2C{=}CR_2 + OH\cdot = R_2(OH)C{-}CR_2\cdot \qquad (9.5)$$

The electrons in the double bond are loosely held, and offer a favorable site for interaction with radicals, as they do with ozone. The steps in the propagation of the smog reaction

TABLE 9.2 PROPERTIES OF SOME COMPONENTS OF GASOLINE

Component	Research octane number (RON)	Motor octane number (MON)	Vapor pressure (psi @ 100°F)	PA*
butane			51	3.23
n-pentane	62	67	15.5	4.80
n-hexane	19	22	5.0	5.90
methyl propane			82	2.83
2-methylbutane	99	104	20	
2-methylpentane	83	79	6.6	5.82
2-methylhexane	41	42	2.2	6.85
iso-octane	100	110	1.65	3.15
1-butene	144	126	50	24.4
1-methylpropene	170	139	62	24.4
1-pentene	118	109	19	35.0
cyclohexane	110	97	3.3	8.50
methylcyclohexane	104	84	1.6	7.87
benzene	99	91	3.3	0.88
toluene	124	112	1.04	5.98
meta-xylene	145	124	0.33	22.8
ethanol	115[†]		17	3.3
methanol	123	93	60	1.0
methyl *tert*-butyl ether (MTBE)	123	97	8	2.6
ethyl *tert*-butyl ether (ETBE)	111[†]		4	8.1

*Photochemical activity measured as rate of reaction with OH radicals, units, cc/(molecule sec) $\times 10^{12}$.
[†]Average (RON + MON).

Source: D. Seddon (1992). Reformulated gasoline, opportunities for new catalyst technology. *Catalysis Today* 15:1–21.

cycle subsequent to reaction (9.5) are much the same as for the alkanes, depicted in Figure 9.12. Due to their high reactivity, however, alkenes are the most important hydrocarbon molecules in the dynamics of smog formation.

9.3 EMISSION CONTROL

Limiting atmospheric pollution depends upon two strategies: removing the pollutants before they are dispersed, and changing conditions to reduce the amount of pollutants produced in the first place. Both these strategies have been tried for most air pollutants, with mixed success.

a. Sulfur dioxide. In order to reduce the level of sulfuric acid aerosols in urban air, power plants are often built with tall smokestacks to disperse the plume over a wide area. This may alleviate the local problem, but at the expense of producing acid rain (see discussion, pp. 157–158, 278–279, 302–303) in areas that are downwind.

Actual abatement requires reducing the sulfur dioxide emissions or, alternatively, limiting the sulfur content of fuels. In coal-fired power plants, sulfur dioxide is currently removed from the stack gases by installing chemical scrubbers, in which the stack gas is passed through a slurry of limestone, converting it to calcium sulfite:

$$CaCO_3 + SO_2 = CaSO_3 + CO_2 \qquad (9.6)$$

Although limestone is relatively cheap, a great deal of it has to be used, and the resulting calcium sulfite sludge presents a significant waste-disposal problem (unless it can be used in gypsum for wallboard, see p. 127). An alternative is to use the more reactive (because more alkaline) $Ca(OH)_2$, which can be injected into the combustion gas, and the product collected on the fabric filter that nearly all power plants and industrial smokestacks employ for collecting particulates and other pollutants. This "dry sorbent injection" currently being tested lowers the volume of the $CaSO_3$ substantially. Another technology under development uses a regenerable amine salt as the scrubbing agent. Heating the resulting SO_2 adduct recovers the amine salt and drives off the SO_2, which can be converted to commercial-grade sulfuric acid. Alternatively, ammonia from fertilizer production can be diverted to the scrubber, and conditions adjusted to oxidize the SO_2 to ammonium sulfate, which can then be marketed as a fertilizer. Still another method, appropriate for coastal power plants, is to use seawater as the scrubber, returning the effluent to the ocean (which already has a substantial concentration of sulfate).

Another possibility is to remove the sulfur from the coal before or during combustion. The coal can be cleansed of the major sulfide mineral, FeS_2 (iron pyrite), by grinding the coal and floating the mineral particles away with a water/oil/surfactant emulsion. However, the coal still contains the organically bound sulfur. This sulfur can be removed by pulverizing the coal and mixing it with limestone in a fluidized bed combustor, a device in which air is passed through a screen from underneath, keeping the particles suspended until they burn. The limestone captures the SO_2 before it passes into the stack gas. However, the resulting calcium sulfite remains a disposal problem.

b. Nitrogen oxides, carbon monoxide, and hydrocarbons.   Combustion in air produces nitrogen oxides as inevitable byproducts. Their emission levels depend on the temperatures reached in the combustion process; the hotter the flame, the greater the NO production rate. Although all kinds of combustion contribute to NO_x emissions, the main contributors, at least in the developed world, are transportation and fuel combustion in stationary sources such as home furnaces, power plants, and industrial facilities. In the United States, transportation produces about 53 percent of the NO_x, while stationary sources account for most of the rest (see Figure 9.7).

Combustion also accounts for much of the atmospheric CO and hydrocarbons, at least in urban areas, because automotive exhaust contains substantial quantities of unburned gases. In addition, some of the volatile automotive fuel escapes before combustion, adding to the hydrocarbon levels. In the United States, transportation accounts for about 43 percent of volatile organic compound emissions. Industry, gasoline service stations, and commercial and consumer applications are significant sources, largely because

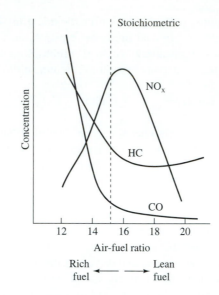

Figure 9.13   Composition of auto exhaust as air-fuel ratio varies.

of solvent evaporation (see Figure 9.8). In addition, substantial quantities are emitted by vegetation. Plants release numerous hydrocarbons, especially *terpenes,* which have C=C double bonds and therefore react rapidly with hydroxyl radicals. In some areas, the vegetation is a major contributor to the reactive hydrocarbons responsible for photochemical smog.

NO_x emissions are difficult to control because efficient energy conversion requires high combustion temperatures, whether in cars or power plants. Moreover, there is a trade-off between NO_x and unburned gases as the ratio of air to fuel in the combustion chamber is varied (illustrated for cars in Figure 9.13). The NO production rate is maximum near the *stoichiometric* ratio (just enough O_2 to completely oxidize the fuel), where the highest temperature is reached. If less air is admitted to the combustion zone ("fuel-rich"), the NO production rate falls along with the temperature, but the emission of CO and unburned hydrocarbons (HCs) increases.

It is possible to lower both NO and HC by carrying out the combustion in two stages, the first of which is rich in fuel and the second of which is rich in air. In this way the fuel is burned completely, but the temperature is never as high as it would be for a stoichiometric mixture. This two-stage approach is being incorporated into new power plants; it has been tried in cars via the "stratified-charge" engine, but with less success.

The other approach to reducing emissions is to remove the pollutant from the exhaust gases. In automobiles, this is accomplished with a three-way *catalytic converter,* so named because it reduces emissions of hydrocarbons (HCs), carbon monoxide (CO), and nitric oxide (NO). In order to deal with both NO and unburned gases, the converter has two chambers in succession (see Figure 9.14a). In the reduction chamber, NO is reduced to N_2 by hydrogen, which is generated at the surface of a rhodium catalyst by the action of water

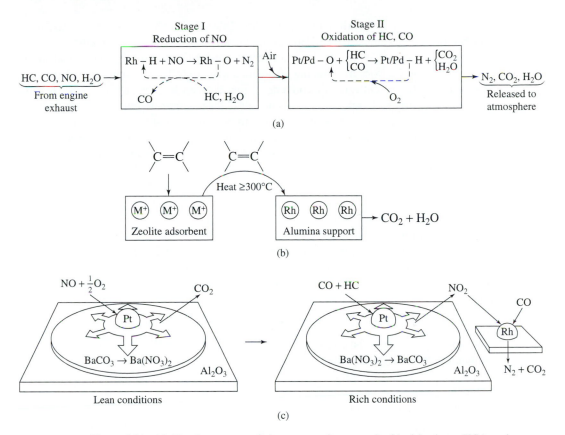

Figure 9.14 (a) The three-way catalytic converter for removal of hydrocarbons (HCs), carbon monoxide (CO), and nitric oxide (NO) from automobile exhaust. (b) Advanced methods of removal of HCs from exhausts during cold starts using zeolite adsorbents. (c) More efficient reduction of NO to N_2 using platinum embedded in barium carbonate.

on unburned fuel molecules (analogous to steam-reforming, see p. 41):

$$\text{hydrocarbons} + H_2O = H_2 + CO \tag{9.7}$$

$$2NO + 2H_2 = N_2 + 2H_2O \tag{9.8}$$

In the oxidation chamber, air is added, and the CO and unburned hydrocarbons are oxidized to CO_2 and H_2O at the surface of a platinum/palladium catalyst:

$$2CO + O_2 = 2CO_2 \tag{9.9}$$

$$\text{hydrocarbons} + 2O_2 = CO_2 + 2H_2O \tag{9.10}$$

The catalytic converter is quite effective in reducing automotive emissions. It is credited with significant reductions in ozone levels in some urban areas. In Los Angeles, peak ozone levels were cut nearly in half between 1970 and 1990, despite a 60 percent increase in the number of vehicle-miles driven. However, further improvements are required to meet increasingly stringent standards for hydrocarbons and NO_x stemming from the Clean Air Act of 1990 and California's ongoing efforts in smog control. Cold starts are a big part of the problem. The current catalytic converter is ineffective when the engine is cold, because the catalysts require a temperature of about 300°C for reactions to commence on their surface. Until this temperature is attained, relatively large amounts of volatile unburned hydrocarbons exit the tailpipe. One solution under development for this problem is to trap the hydrocarbons in a zeolite material, which absorbs them at low temperature and then releases them to the catalytic converter as the temperature rises above 300°C. The zeolite is impregnated with metal ions that form adducts with alkenes and aromatic molecules (see Figure 9.14b). These are the hydrocarbons with the highest smog-forming potentials (see Table 9.2, p. 232).

Another problem is the difficulty of reducing NO_x completely to N_2 in the three-way converter. To assist in this process, an additional stage is being developed, in which the NO formed during fuel-lean combustion is efficiently removed by a trap, which contains platinum embedded in barium carbonate (see Figure 9.14c). The platinum catalyzes the oxidation of NO to form barium nitrate salt, releasing CO_2:

$$4NO + 3O_2 + 2BaCO_3 = 2Ba(NO_3)_2 + 2CO_2 \qquad (9.11)$$

After the trapping step, the engine switches briefly to a fuel-rich mode. The hydrocarbon-containing exhaust gases reduce the nitrate to NO_2, which is swept out of the trap and reduced in the three-way converter. The combination of NO trapping and transient fuel-rich combustion increases the efficiency of NO_x removal.

In any event, it is increasingly recognized that the key to smog reduction is control of NO emissions; hydrocarbons are generally too abundant to be brought low enough to be the limiting factor. Even if the automotive contribution is reduced further, the contributions from other sources can often support substantial levels of smog production. But smog requires NO, whose only significant sources are automotive transport, electric power production, and industrial plants. These sources are therefore under increased scrutiny with respect to NO control.

In stationary power plants, it is possible to convert NO back to N_2 using catalytic converters, similar in concept to those developed for cars. Since unburned gases are generally a smaller fraction of the exhaust stream in power plants, some additional reductant is generally needed to effect substantial NO control. The reduction can be accomplished by injecting ammonia into the catalytic chamber:

$$6NO + 4NH_3 = 5N_2 + 6H_2O \qquad (9.12)$$

Careful control of conditions is required, however, to prevent oxidation of the ammonia directly to NO and NO_2 by residual O_2 in the exhaust stream. In another approach, the compound urea, $CO(NH_2)_2$, has been sprayed directly into the combustion flame in order to reduce the NO. The mechanism is complex, but the overall reaction is

$$2CO(NH_2)_2 + 6NO = 5N_2 + 2CO_2 + 4H_2O \qquad (9.13)$$

In the long run, the most effective NO_x abatement strategy will be to introduce fuel cells for electric power generation and transport (see "Fuel Cells," pp. 103–107), thereby eliminating the high temperatures that produce NO in the first place.

9.4   REFORMULATED GASOLINE: OXYGENATES

Because much of urban air pollution is produced by transportation, the liquid fuels used in transport have been scrutinized intensively, and pollution regulations have substantially changed the composition of gasoline. The challenge has been to design fuel formulations that perform well in terms of fuel efficiency, engine performance, and pollution reduction.

a. Knocking and octane.   Combustion is a radical chain process, similar to the oxygen-radical chemistry discussed in the preceding sections, that allows the oxygen to combine very rapidly with the fuel molecules. The spark-ignition engine of automobiles works by igniting a mixture of gasoline and air with a spark. The air/fuel mixture is first compressed by a piston in a cylinder, then ignited. The spark fragments the fuel molecules in its path, generating enough radicals to set off the chain reaction. The force of the explosion on the piston delivers power to the drive train. The greater the degree of compression, the greater the power. However, the compression stroke itself heats the fuel; at sufficiently high temperatures, the fuel molecules can react with thermally activated oxygen molecules to form radicals, thereby setting off the explosion prematurely. Pre-ignition lowers the power generated and produces extra wear on the engine. This is the *knocking* phenomenon, recognizable from the characteristic noise produced by the engine under acceleration.

The temperature required for radical generation depends on the structure of the fuel molecule. The reaction involves H-atom transfer to the hot oxygen molecules, and the rate depends primarily on the strength of the C—H bonds, just as it does for hydroxyl radical reactions. Branched-chain hydrocarbons are more resistant to radical formation than straight-chain hydrocarbons, because the branching increases the fraction of the hydrogen atoms that are on methyl groups, whose C—H bonds are stronger than for methylene groups (see Table 9.1, p. 230). The methylene groups are more susceptible to attack by the thermally activated oxygen molecules, which pull off a hydrogen atom, leaving a hydrocarbon radical. Since straight-chain hydrocarbons have more methylene groups than do branched-chain hydrocarbons, they are subject to pre-ignition at lower temperature.

The performance of gasoline is largely a function of its octane rating. Gasoline is a mixture of low-boiling hydrocarbons, most of them containing seven or eight carbon atoms. Among these, 2,2,4-trimethylpentane ("iso-octane") is particularly resistant to pre-ignition due to its highly branched structure. It is assigned an "octane" rating of 100. The zero of the octane scale is set by *n*-heptane, a straight-chain hydrocarbon with a strong knocking tendency. The octane rating of any gasoline is then set by comparison with these two hydrocarbons in standard engine tests. Table 9.2 lists octane ratings for a number of the compounds in gasoline. In some cases the rating exceeds 100, meaning that the fuel is even less prone to pre-ignition than iso-octane.

b. Diesels and cetane.   A diesel engine works very differently from a spark-ignition engine. In diesels, air in the piston is preheated by compression, and fuel is then sprayed into the hot chamber, burning on contact. Since pre-ignition is not an issue, the degree of compression can be very high, allowing for greater efficiency. The engine is built ruggedly to accommodate the higher compression forces, and since there are fewer moving parts than in a spark engine (no valves are needed), a diesel engine wears out more slowly. This is why diesel is generally the engine of choice for large trucks and buses.

Whereas easy fragmentation of fuel molecules is undesirable for spark engines, it is desirable for diesel engines, because it enhances the combustion of the injected fuel. Thus, straight-chain hydrocarbons are abundant in diesel fuel. The quality of diesel fuel is judged by its "cetane" number, which increases with fragmentation tendency, opposite to the octane number. Cetane (*n*-hexadecane, $C_{16}H_{34}$) is given a value of 100, while a highly branched isomer, heptamethylnonane is given a value of 15. Also, diesel engines work best with higher molecular weight molecules, in the $C_{11}-C_{16}$ range, whereas the optimum range for gasoline is C_6-C_{10}. Diesel fuel therefore takes a different cut of oil-refining production than gasoline (see p. 29).

Diesel emissions contain many more particles than spark-engine emissions because of the combustion characteristics of the injected fuel. Molecules at the air-fuel interface burn completely, but molecules at the center of the injected plume heat up before they have access to oxygen molecules, and therefore tend to decompose to solid carbon. In the spark engine, however, the fuel molecules are intimately mixed with air before combustion, and soot production is therefore much lower.

Because of the deleterious health effects of particles, there is pressure to abandon diesel engines, particularly in urban buses, whose slow progress along crowded streets is responsible for elevated particle concentrations in the street-level air. Many cities are trying natural gas-fueled bus fleets as an alternative. However, new diesel technology is also available, which greatly reduces particle emissions through the use of particle traps with catalysts, similar to the catalysts used in converters on cars, to oxidize the carbon particles in the exhaust. In Europe, new diesel cars are increasingly popular because of their fuel economy.

c. Lead in gasoline.   It was discovered in the 1920s that knocking could be diminished if organo-lead compounds, particularly tetraethyl- or tetramethyl-lead were added to the gasoline. In the succeeding decades, lead was added to virtually all gasolines in order to improve performance. The lead additives suppress the radical chain reactions in the pre-ignition phase. As the air/fuel mixture becomes compressed and heated, the weak alkyl-lead bonds break, releasing Pb atoms, which combine rapidly with oxygen to form particles of PbO and PbO_2. These particles provide attachment sites for hydrocarbon radicals, which recombine with one another, thereby terminating the chain reaction. To avoid building up lead deposits on the inside engine surfaces, gasoline containing lead usually also contains ethylene dichloride or ethylene dibromide. These organo-halogens act as scavengers for the lead, producing PbX_2 compounds (X = Cl or Br). Because these compounds are volatile at the high temperature of the exhaust gases, they remove the lead from inside the engine and release it into the atmosphere.

Beginning in the mid-1970s, unleaded gasoline was offered for sale in the U.S. and gradually displaced leaded gasoline. The reason was that lead compounds in the exhaust react with the rhodium and platinum catalysts in catalytic converters, "poisoning" their surfaces and rendering them inactive. As the fraction of the car fleet with catalysts increased, so did the lead-free fraction of the gasoline supply. This displacement had a salutary effect on lead pollution, as well as on air pollution in general, since human exposure to lead decreased dramatically when the lead was removed from gasoline, as discussed on pp. ??.

In Canada and some European countries, another organometallic compound has been introduced as an antiknock additive, methylcyclopentadienyl manganese tricarbonyl (MMT). This compound releases Mn atoms, which are converted to Mn_3O_4 particles in the combustion chamber, which, like PbO, capture hydrocarbon radicals and inhibit knocking. Unlike lead, the toxicity of manganese is low once released into the environment (indeed, it is a biologically essential element). Moreover, MMT does not significantly increase the human intake of Mn from natural sources. Nevertheless, MMT has not been approved as a gasoline additive by the U.S. EPA because the additive itself is toxic if ingested or inhaled. In addition, fouling of spark plugs and of onboard sensors has been reported. However, a federal appeals court has ruled against the EPA's ban on MMT.

d. Reformulated gasoline.   An alternative to adding free radical scavengers to reduce knocking is to alter the gasoline composition by reducing the fraction of low-octane components and increasing the fraction of high-octane components. A modern oil refinery has considerable latitude to alter the hydrocarbon molecules in oil through cracking, alkylation, and reforming reactions (see the discussion of petroleum composition and refining in Part I, pp. 29–31). In the United States, the removal of alkyl-lead additives was initially compensated for by increasing the content of aromatic compounds, principally benzene, toluene, and xylenes (sometimes called the BTX component). As seen in Table 9.2, benzene's octane rating is nearly as high as iso-octane's, and toluene's is even higher. Consequently, increasing the aromatic fraction of gasoline boosts its octane rating.

Despite the high-octane ratings of BTX gasoline, the BTX fraction is being decreased in the United States. The xylenes react rapidly with hydroxyl radicals (see value for PA in Table 9.2), and therefore have greater potential to form smog than do the alkanes. Benzene, although low in photochemical activity, is a carcinogen. The 1990 Clean Air Act Amendments require the benzene content in gasoline not to exceed 1 percent.

Beginning in the 1980s, the aromatics have been replaced by "oxygenates," fuel molecules that contain one or more oxygen atoms. The four oxygenates initially considered, methanol, ethanol, and the methyl and ethyl ethers of *tertiary*-butyl alcohol, MTBE and ETBE, all have octane ratings substantially higher than 100 (see Table 9.2). The oil industry soon settled on MTBE, which is made in refineries (see pp. 30–31), but ethanol emerged as a competitor in corn-growing regions, thanks to federal and local tax incentives. An added bonus is that MTBE and ethanol have relatively low hydroxyl-radical reaction rates and vapor pressures. For this reason, the 1990 Clean Air Act Amendments, which require reductions in ozone-forming volatile organic compounds, mandated oxygenates in reformulated gasoline (RFG), to a level of 2.0 percent oxygen, by weight. The RFG program applies to the nine major urban areas of the U.S.

A separate mandate of the same law required 2.0 percent oxygen by weight (amounting to 15.2 percent MTBE, and 7.6 percent ethanol, by volume) in 39 areas of the U.S., in order to bring them into compliance with standards for CO emissions, particularly during the winter months. CO emissions are especially high during cold engine starts and fuel-rich combustion. These emissions are reduced by adding oxygenates to the gasoline. Because the fuel molecules already contain oxygen atoms, their conversion to CO_2 during combustion is more complete when the combustion mixture is fuel-rich. As a result of this requirement, production has greatly increased for ethanol, and especially for MTBE; by 1997, MTBE production was 251,000 barrels per day.

However, MTBE has become highly unpopular because of its propensity to contaminate ground waters through gasoline leaks (see pp. 260, 344). No significant health hazards have been found for MTBE (toxicity is found at high doses in animal tests), but it has an unpleasant odor, and is unacceptable in water supplies. In 1999, the EPA formed a *Blue Ribbon Panel on Oxygenates in Gasoline,* which recommended that MTBE use be diminished, and that the oxygenate requirement be eliminated, allowing air quality standards to be met by other means. Many regions have taken action against MTBE, and California, which is a major part of the gasoline market, enacted an outright ban, by 2003. California also asked for exemption from the oxygenate rule, but this request has been rejected by the EPA.

Retention of the oxygenate rule, in the face of the likely demise of MTBE, sets the stage for a very large increase in ethanol production for gasoline, despite doubts about the wisdom of this course. Aside from the economic and environmental concerns about ethanol from corn (see p. 88), there are costs associated with the gasoline itself. Unlike MTBE, ethanol is very soluble in water, which is commonly found in pipelines and storage tanks associated with gasoline distribution. Once in contact with this water, ethanol will separate from the gasoline. Due to this potential for phase separation, ethanol is usually blended at the terminal, rather than the refinery. Also, ethanol's vapor pressure is twice that of MTBE (see Table 9.2), requiring additional adjustment in gasoline composition to lower volatile organics.*

This is accomplished by removing the lighter hydrocarbons from the blend. With MTBE, the vapor pressure requirement is met by removing the butane fraction at the refinery, but replacing MTBE with ethanol requires removing the pentanes as well, again increasing costs. Additional steps will have to be taken to maintain octane ratings, because the volume of ethanol, for a given oxygenate level, is only half that of MTBE.

---

*Pure ethanol is actually less volatile than gasoline because, although the molecular weight is low, ethanol is an associated liquid, with intermolecular hydrogen bonds. However, when ethanol is dispersed in gasoline, its hydrogen bonds are eliminated and its volatility increases greatly.

Summary

Looking back on our survey of atmospheric issues, we see that atmospheric balances can be upset both on a global scale through augmentation of the greenhouse effect and stratospheric ozone destruction, and on a local and regional scale through the buildup of fossil-fuel exhaust gases and their oxidation products. These problems are interrelated in complex and sometimes paradoxical ways. For example, hydroxyl radicals and nitrogen oxides catalyze ozone destruction in the stratosphere, but they are responsible (together with hydrocarbons) for ozone formation in polluted urban air. The chlorofluoromethane gases are harmless locally, but contribute globally to both the greenhouse effect and ozone destruction. Carbon monoxide and hydrocarbons are local pollutants whose noxious qualities can be eliminated by oxidizing them to carbon dioxide, but an increase in the global carbon-dioxide concentration adds to the greenhouse effect. The consequences of increasing the concentrations of greenhouse gases are uncertain because long-term trends in climate are currently unpredictable, but they could well be dire. While scientists have learned a great deal about the atmosphere, partly in response to recent environmental concerns, there is an urgent need to learn much more, in order to assess the human impact on it.

Despite the uncertainties, it is clear that overriding influences on air quality are the quantity of energy consumed, the kinds of fuels used, and the energy efficiencies of the applied technologies. At the end of Part I we discussed the benefits of increased energy efficiency and alternative sources of energy. In addition to conserving energy resources, these alternatives can provide a cleaner atmosphere.

PROBLEM SET

1. Sketch the heat balance of Earth in watts per square meter (Wm^{-2}), showing the following components: (1) short-wave solar radiation incident at the top of Earth's atmosphere; (2) short-wave radiation reflected by the atmosphere and Earth's surface; (3) short-wave radiation absorbed by the atmosphere and Earth's surface; (4) long-wave radiation emitted from Earth's surface; (5) the share of the long-wave radiation emitted directly to space (through the atmospheric "window") and the share absorbed by the atmosphere; (6) the long-wave radiation emitted downward from the atmosphere to Earth's surface, and outward from the atmosphere into space; and (7) the sensible and latent heat transfer from Earth's surface to the atmosphere. Compare your calculations with Figure 6.2.

   The following information (in units of Wm^{-2}) is sufficient to conduct this exercise:

   - $S_0 = 1{,}368\ Wm^{-2}$;
   - The total albedo is 30 percent, 86 percent of which is provided by the atmosphere, and the remainder by the Earth's surface;
   - 24 percent of the incident short-wave radiation is absorbed by the atmosphere, and 46 percent by the Earth's surface;
   - Earth's surface temperature is 288 K;
   - About 5 percent of the long-wave radiation emitted from Earth's surface radiates directly to space through the atmospheric window;
   - The radiative cooling of the atmosphere, and a corresponding radiative heating of Earth's surface, is about 106 Wm^{-2};
   - The Stefan-Boltzmann constant equals $0.567 \times 10^{-7}\ Wm^{-7}\ K^{-4}$.

2. Outgoing long-wave radiation from the top of Earth's atmosphere has been measured by satellites to be 237 Wm^{-2}. From this information, calculate the temperature at the top of the atmosphere. If S_0, the solar irradiance, is 1,368 Wm^{-2}, calculate Earth's albedo.

3. Calculate the long-wave emission by Earth's surface, given that its mean global temperature is 288 K.

4. The polar ice caps have an albedo of about 0.80, while the polar seas have a maximum albedo of about 0.20. How could this difference in albedo cause the spontaneous melting of much of the ice cap if a small rise in ambient temperature were to occur? How could large-scale soot deposition on the ice cap cause a similar melting?

5. Calculate the critical radius for water droplet formation in the atmosphere at 20°C and 101 percent relative humidity. [See Strategies 6.1, pp. 152–153 for information required to do the calculation.]

6. Calculate the temperature of Earth if there were no greenhouse effect. Assume S_0, the solar irradiance, equals 1,368 Wm^{-2}, and the total planetary albedo is 30 percent. (The Stefan-Boltzmann constant equals $0.567 \times 10^{-7} Wm^{-2} K^{-4}$.)

7. **(a)** What characteristic molecular properties of H_2O and CO_2 cause their absorption of infrared radiation?

   (b) Why are chlorofluorocarbons (CFCs) such as CF_2Cl_2 and $CFCl_3$ such effective greenhouse gases?

   (c) In addition to H_2O, CO_2, and CFCs, name two other greenhouse gases, and identify for each of these latter two, one human activity leading to atmospheric emissions.

8. Which of the molecules in the following table have the potential to cause a greenhouse effect similar to CFCs (see Figure 6.15)? For polyatomic molecules, identify the specific vibrations involved.

S—C—O (linear)	H—Cl	F—F	H_2S (non-linear)	Cl_2O (non-linear)
$\lambda_1 = 11{,}641$ nm	$\lambda = 3{,}465$ nm	$\lambda = 11{,}211$ nm	$\lambda_1 = 3{,}830$ nm	$\lambda_1 = 14{,}706$ nm
$\lambda_2 = 18{,}975$ nm			$\lambda_2 = 7{,}752$ nm	$\lambda_2 = 30{,}303$ nm
$\lambda_3 = 4{,}810$ nm			$\lambda_3 = 3{,}726$ nm	$\lambda_3 = 10{,}277$ nm

9. **(a)** Describe the two opposing effects on climate resulting from combustion of coal with high sulfur content.

   (b) Cite two processes by which sulfate particles in the troposphere affect climate.

10. **(a)** The total mass of carbon contained in fossil fuels that was burned in the world from 1750 to 2000 is estimated to be 2.77×10^{14} kg C. The amount of carbon released as CO_2 from agricultural expansion and deforestation over this period is estimated at 1.31×10^{14} kg C. The concentration of CO_2 in the atmosphere in 2000 was 360 ppm, corresponding to a total mass of 7.75×10^{14} kg C. If the concentration of CO_2 in 1750 was 280 ppm, calculate the percentage of CO_2 from these two sources that remained in the atmosphere over the two and one-half century period.

   (b) Plant studies indicate that the net primary production (NPP) of organic carbon by photosynthesis may increase with increasing CO_2 concentration in the atmosphere. Assume that the increase of NNP in the biosphere is 0.27 of the percentage increase of atmospheric CO_2. Given that the global NPP of the biosphere is currently estimated to be 1.10×10^{14} kg C/yr, estimate how much more carbon is being absorbed per year in NPP compared to the amount that would be absorbed if the atmospheric CO_2 concentration were 280 ppm.

   (c) Currently, about 6.3×10^{12} kg C/yr of fossil fuel carbon and 1.6×10^{12} kg C/yr of carbon from destruction of land vegetation are released to the atmosphere. Assuming the value obtained in part (b), what percentage of the emitted carbon might be taken away by increased absorption in the biosphere? The effect of deforestation may be reducing NPP. How does such a reduction affect the potential of biomass to serve as a sink for CO_2?

11. Name two strategies for reducing greenhouse gas emissions related to energy consumption.

12. The amount of CH_4 emitted annually is estimated to be 25 to 50 times greater than the amount of N_2O emitted annually, yet the atmospheric concentration of CH_4 is estimated to be only six times higher. Can you offer an explanation?

13. **(a)** Using the thermodynamic data in Table 7.1, calculate the equilibrium concentration ratio of NO_2 to NO at sea level and 25°C.
    (b) How does this ratio change at 40°C?
    (c) Consider the atmospheric concentrations on a typical smoggy day in Los Angeles in Figure 9.10. Estimate the $(NO_2)/(NO)$ ratio at 6 A.M. and at noon, and comment on the time variation of this ratio in relation to the equilibrium value. What role, if any, might the temperature variation play?

14. The rate constant (k) for the reaction

$$CH_4 + OH = CH_3 \cdot + H_2O$$

is 6.3×10^{-15} molecule^{-1} cm^3 second^{-1}. If the atmospheric mixing ratio of methane is 1,745 ppbv, and the concentration of OH and air molecules is 8.0×10^5 and 3.0×10^{19} molecules cm^{-3}, respectively, calculate the rate of reaction.

15. **(a)** From Figure 7.3, calculate the bond order of O_2 (defined as one-half the sum of electrons in the bonding orbitals, minus one-half the sum of electrons in the anti-bonding orbitals).
    (b) What property of the bonding in O_2 prevents rapid oxidation reactions with most molecules at ambient temperatures in the atmosphere and the biosphere?
    (c) Name the two classes of molecules (or atoms) that can bond effectively with O_2, even at ambient temperatures, and explain why they can?

16. When an electron donor such as an organic radical, $R \cdot$, or a heavy metal, $M \cdot$, reacts with O_2 to form $RO_2 \cdot$ or $MO_2 \cdot$, the O—O bond is weakened considerably. Use the molecular orbital diagram shown in Figure 7.3 to explain why this weakening occurs. Explain with at least two examples from atmospheric or biological chemistry how these electron donors can serve as catalysts for oxidation at ambient temperatures.

17. Biological compounds such as carbohydrates, fats, and proteins are thermodynamically unstable in the presence of O_2, since, upon combustion, they will undergo exothermic reactions with oxygen to form CO_2, H_2O, and other simple gases and compounds. On the other hand, aerobic life forms require O_2 for generating energy, via oxidation reactions, needed for myriad metabolic activities. Explain how the special electronic structure of O_2 allowed life on the planet to evolve in the presence of an oxidizing atmosphere, and how these life forms are able to harvest energy from oxidation reactions under controlled conditions within cells.

18. **(a)** Why is the hydroxyl radical such an important component of the atmosphere?
    (b) How is it generated?
    (c) Describe two of its reactions with air pollutants.

19. Isoprene (C_5H_8) is a natural atmospheric component emitted by coniferous forests, and is responsible for the "blue haze" of the Great Smoky Mountains in the southeastern U.S. It is quickly degraded in the atmosphere through reactions with OH or O_3, and the rate constants are known:

$$OH + \text{isoprene } (C_8H_8) = \text{radical product} + H_2O;\ k_{OH} = 1.0 \times 10^{-10} \text{ molecule}^{-1} \text{ cm}^3 \text{ second}^{-1}$$

$$O_3 + \text{isoprene } (C_8H_8) = \text{radical product} + H_2O;\ k_{O_3} = 1.4 \times 10^{-17} \text{ molecule}^{-1} \text{ cm}^3 \text{ second}^{-1}$$

Under atmospheric conditions and at a typical isoprene concentration of 5.9×10^{10} molecules cm^{-3}, the rates of these two reactions are found to be $(r_{OH}) = 4.7 \times 10^6$ molecule cm^{-3}

second^{-1}, and $(r_{O_3}) = 2.0 \times 10^6$ molecule cm^{-3} second^{-1}. What must the O_3 and OH concentrations have been?

Comment on why OH has been called the atmosphere's "vacuum cleaner."

20. **(a)** Write the equations for ozone formation and ozone destruction in the absence of catalytic chains, including the net reactions.
   (b) Write the equations of ozone destruction involving NO, OH, and Cl. For each of these, describe explicitly how they lead to ozone destruction, and describe the sources of these reactants.
   (c) Describe the two reactions by which NO_2 serves to protect the ozone layer.
   (d) How might stratospheric-flying aircraft perturb the ozone layer?

21. **(a)** How does stratospheric ozone shield Earth's surface from harmful UV radiation?
   (b) Calculate the fractional increase in transmission (dT/T) for a 1 percent decrease in the thickness (l) of the ozone layer for the following three wavelengths [in nanometers (nm)]: 310 nm, 295 nm, and 285 nm. (Assume $l = 0.34$ cm; $\varepsilon = 3$, 18, and 56 cm^{-1} for 310 nm, 295 nm, and 285 nm, respectively.)
   (c) As a result of the 1 percent decrease, describe what happens to the position of the *action spectrum* relative to the curve of *relative sensitivity to sunburn,* as shown in Figure 8.5.
   (d) Will the calculated increase in transmission at 285 nm cause more damage to biological tissue than the calculated increase in transmission at 295 nm? Explain your reasoning.

22. When the steady-state calculation of the O_3 concentration (see p. ?) is carried out for different altitudes, the value (O_3) is maximal at about 25 km altitude (see Figure 8.7). Consider the terms in the steady-state expression, and suggest why the calculated (O_3) falls at *both* higher and lower altitudes.

23. **(a)** In the ozone hole phenomenon over Antarctica, how is nitrogen removed from the atmosphere, and which molecules serve as "chlorine reservoirs" in the darkness of the Antarctic winter?
   (b) How does the sunlight at the beginning of spring initiate the polar ClO_x chain reaction?
   (c) Write the four equations involved in this chain reaction, and describe the equations that finally stop the chain reaction.

24. Give two reasons why the stratosphere is more susceptible to chemical pollution than the troposphere.

25. Assume that the concentration of suspended particulates in a polluted atmosphere is 170 μg/m^3. The particulates contain adsorbed sulfate and hydrocarbons comprising 14 percent and 9 percent of the weight, respectively. An average person respires 8,500 liters of air daily and retains 50 percent of the particles smaller than 1 μm in diameter in the lungs. How much sulfate and hydrocarbon are absorbed by the lungs in one year if 75 percent of the particulate mass is contained in particles smaller than 1 μm?

26. Choose one of the following four air pollutants, SO_2, NO_x, CO, or VOC, and provide the following information: (1) environmental effects; (2) human health effects; (3) major sources; (4) atmospheric reactions; and (5) atmospheric lifetime.

27. Describe the contribution of transportation to air pollution, including the four pollutants cited in problem 26, as well as particulates. Do the same for stationary-source fuel combustion.

28. Why is NO_2, unlike the higher oxides of carbon and sulfur (CO_2 and SO_3, respectively), unstable at 25°C in the presence of sunlight? Describe briefly how the reaction of NO_2 with sunlight plays a key role in smog formation and in the regeneration of NO_2 itself.

29. For the reaction of carbon monoxide to carbon dioxide:

$$CO + \tfrac{1}{2}O_2 \rightarrow CO_2$$

the equilibrium constant K_{eq} (at 25°C) $= 3 \times 10^{45}$. Given this enormous value, why doesn't carbon monoxide convert spontaneously to carbon dioxide in air? How does the use of platinum in the catalytic converter of automobiles facilitate the conversion to carbon dioxide? (Hint: platinum contains unpaired d electrons in its outer orbital.)

30. Why was lead added as a component of gasoline before the 1970s? Why was it removed with the advent of the catalytic converter? Why did the new unleaded gasolines contain higher concentrations of aromatics and liquid olefins? (Refer to Table 9.2; the aromatics are the fifth cluster of compounds, and the liquid olefins are the third cluster.) Among the properties listed in Table 9.2, in which are aromatics clearly superior to liquid olefins (discounting considerations of toxicity, particularly for benzene, which is a human carcinogen)? In addition to reducing CO, what other favorable advantages do ethanol and MTBE provide with respect to the properties listed in Table 9.2? Why do the producers of MTBE claim that their compound is superior to ethanol?

SUGGESTED READINGS

Chapter 6: Climate

Office of Air Quality Planning and Standards (1993). *The Plain English Guide to the Clean Air Act.* Report EPA-400-K-93-001 (Washington, DC: U.S. Environmental Protection Agency). (http://www.epa.gov/oar/oaqps/peg_caa/pegcaain.html)

G. P. Ayers and R. W. Gilbert (2000). DMS and its oxidation products in the remote marine atmosphere: Implications for climate and atmospheric chemistry. *Journal of Sea Research* 43:275–286.

J. J. Corbett, P. S. Fischbeck, and S. N. Pandis (1999). Global nitrogen and sulfur inventories for oceangoing ships. *Journal of Geophysical Research* 104:3457–3470.

A. S. Lefohn, J. D. Husar, and R. B. Husar (1999). Estimating historical anthropogenic global sulfur emission patterns for the period 1850–1990. *Atmospheric Environment* 33:3435–3444.

J. A. Lynch, V. C. Bowersox, and J. W. Grimm (2000). Changes in sulfate deposition in eastern USA following implementation of Phase I of Title IV of the Clean Air Act Amendments of 1990. *Atmospheric Environment* 34:1665–1680.

Office of Air and Radiation (1999). *Progress Report on the EPA Acid Rain Program.* Report EPA-430-R-99-011 (Washington, DC: U.S. Environmental Protection Agency). (http://www.epa.gov/acidrain)

J. Shah, T. Nagpal, T. Johnson, M. Amann, G. Carmichael, W. Foell, C. Green, J-P. Hettelingh, L. Hordijk, J. Li, C. Peng, Y. Pu, R. Ramankutty, and D. Streets (2000). Integrated analysis for acid rain in Asia: Policy implications and results of RAINS-ASIA model (2000). *Annual Reviews of Energy & Environment* 25:339–375.

M. E. Katz, D. K. Pak, G. R. Dickens, and K. G. Miller (1999). The source and fate of massive carbon input during the latest Paleocene thermal maximum. *Science* 286:1531–1533.

N. J. Shackleton (2000). The 100,000-year ice-age cycle identified and found to lag temperature, carbon dioxide, and orbital eccentricity. *Science* 289:1897–1902.

R. A. Kerr (2000). A North Atlantic climate pacemaker for the centuries. *Science* 288:1984–1986.

A. V. Fedorov and S. G. Philander (2000). Is El Niño changing? *Science* 288:1997–2002.

P. Czepiel, E. Douglas, R. Harriss, and P. Crill (1996). Measurements of N_2O from composted organic wastes. *Environmental Science and Technology* 30:2519–2525.

S. P. Seitzinger, C. Kroeze, and R. V. Styles (2000). Global distribution of N_2O emissions from aquatic systems: Natural emissions and anthropogenic effects. *Chemosphere—Global Change Science* 2:267–279.

A. I. Reshetnikov and N. N. Paramonova (2000). An evaluation of historical methane emissions from the Soviet gas industry. *Journal of Geophysical Research* 105(No. D3):3517–3529.

M. Cao, K. Gregson, and S. Marshall (1998). Global methane emission from wetlands and its sensitivity to climate change. *Atmospheric Environment* 32:3293–3299.

R. L. Sass, F. M. Fisher, and A. Ding (1999). Exchange of methane from rice fields: National, regional, and global budgets. *Journal of Geophysical Research* 104(No. D21):26,943–26,951.

Office of Air and Radiation (1999). *U.S. Methane Emissions 1990–2020: Inventories, Projections, and Opportunities for Reductions.* Report EPA-430-R-99-013 (Washington, DC: U.S. Environmental Protection Agency).

G. P. Robertson, E. A. Paul, and R. R. Harwood (2000). Greenhouse gases in intensive agriculture: Contributions of individual gases to the radiative forcing of the atmosphere. *Science* 289: 1922–1925.

H. Grassl (2000). Status and improvements of coupled general circulation models. *Science* 288: 1991–1997.

Working Group I of the Third Assessment Report of the Intergovernmental Panel on Climate Change (IPPC), J. T. Houghton, Y. Ding, D. J. Griggs, M. Noguer, P. J. van der Linden, and D. Xiaosu (eds.) (2001). *Climate Change 2001: The Scientific Basis* (Cambridge, U.K.: Cambridge University Press).

Working Group II of the Third Assessment Report of the Intergovernmental Panel on Climate Change (IPPC), J. J. McCarthy, O. F. Canziani, N. A. Leary, D. J. Dokken, and K. S. White (eds.) (2001). *Climate Change 2001: Impacts, Adaptation & Vulnerability* (Cambridge, U.K.: Cambridge University Press).

Working Group III of the Third Assessment Report of the Intergovernmental Panel on Climate Change (IPPC), B. Metz, O. Davidson, R. Swart, and J. Pan (eds.) (2001). *Climate Change 2001: Mitigation* (Cambridge, U.K.: Cambridge University Press).

Committee on the Science of Climate Change, National Research Council (2001). *Climate Change Science: An Analysis of Some Key Questions* (Washington, DC: National Academy Press).*

H. Herzog, B. Eliasson, and O. Kaarstad (2000). Capturing greenhouse gases. *Scientific American.* 282 (February):72–79.

R. A. Ney and J. L. Schnoor (2000). What course for carbon trading? *Environmental Science & Technology* 34(7):177A–182A.

L. Cifuentes, V. H. Borja-Aburto, N. Gouveia, G. Thurston, and D. L. Davis (2001). Hidden health benefits of greenhouse gas mitigation. *Science* 293:1257–1259.

Chapter 7: Oxygen Chemistry

R. G. Prinn et al. (2001). Evidence for substantial variations of atmospheric hydroxyl radicals in the past two decades. *Science* 292:1882–1888.

X. Liu, H. E. Jeffries, and K. G. Sexton (1999). Hydroxyl radical and ozone initiated photochemical reactions of 1,3-butadiene. *Atmospheric Environment* 33:3005–3022.

S. Oh and J. M. Andino (2000). Effects of ammonium sulfate aerosols on the gas-phase reactions of hydroxyl radical with organic compounds. *Atmospheric Environment* 34:2901–2908.

*This study originated from a White House request, in May of 2001, to help inform the Bush Administration's ongoing review of U.S. climate change policy.

A. Valavanidis, A. Salika, and A. Theodoropoulou (2000). Generation of hydroxyl radicals by urban suspended particulate air matter. The role of iron ions. *Atmospheric Environment* 34:2379–2386.

C. Anastasio and K. G. McGregor (2001). Chemistry of fog waters in California's Central Valley: 1. In situ photoformation of hydroxyl radical and singlet molecular oxygen. *Atmospheric Environment* 35:1079–1089.

Chapter 8: Stratospheric Ozone

Ozone Secretariat (2000). *The Montreal Protocol on Substances that Deplete the Ozone Layer* (Nairobi, Kenya: United Nations Environment Programme). (http://www.unep.ch/ozone/Montreal-Protocol2000.shtml)

Ozone Secretariat, D. L. Albritton, and L. Kuijpers (eds.) (1999). *Synthesis of the Reports of the Scientific, Environmental Effects, and Technology and Economic Assessment Panels of the Montreal Protocol—a Decade of Assessments for Decision Makers Regarding the Protection of the Ozone Layer: 1988–1999* (Nairobi, Kenya: United Nations Environment Programme). (http://www.unep.ch/ozone/pdf/Synthesis-Complete.pdf)

Chapter 9: Air Pollution

Office of Air Quality Planning and Standards (2000). *National Air Pollutant Emission Trends, 1900–1998*. Report EPA-454/R-00-002 (Washington, DC: U.S. Environmental Protection Agency). (http://www.epa.gov/ttn/chief/trends/trends98/contents.pdf)

Particle Epidemiology Reanalysis Project (2000). Reanalysis of the Harvard Six Cities Study and the American Cancer Society Study of Particulate Air Pollution and Mortality (Cambridge, Massachusetts: Health Effects Institute). (http://www.healtheffects.org/Pubs/Rean-ExecSumm.pdf)

A. Rabl and J. V. Spadaro (2000). Public health impact of air pollution and implications for the energy system. *Annual Reviews of Energy & Environment* 25:601–627.

Office of Integrated Analysis and Forecasting, Energy Information Administration (2000). *Analysis of Strategies for Reducing Multiple Emissions from Power Plants: Sulfur Dioxide, Nitrogen Oxides, and Carbon Dioxide*. Service Report SR/OIAF/2000-05 (Washington, DC: U.S. Department of Energy).

V. M. Thomas, R. H. Socolow, J. J. Fanelli, and T. G. Spiro (1999). Effects of reducing lead in gasoline: An analysis of the international experience. *Environmental Science & Technology* 33: 3942–3948.

Energy Information Agency (2000). *MTBE, Oxygenates, and Motor Gasoline* (Washington, DC: U.S. Department of Energy). (http://www.eia.doe.gov/emeu/steo/pub/special/mtbe.html)

PART III

HYDROSPHERE/ LITHOSPHERE

CHAPTER 10

WATER RESOURCES

10.1 GLOBAL PERSPECTIVE

We live, quite literally, in a watery world. All living things depend absolutely on a supply of water. The biochemical reactions of every living cell take place in aqueous solution, and water is the transport medium for the nutrients a cell requires and the waste products it excretes. Water is abundant on the planet's surface, but about 97 percent of Earth's supply of water is in the oceans, where it is too salty to be used by humans or other land creatures. Every day, however, the sun's rays distill a large quantity of water that falls back to the surface as rain. Proportionately more rain falls on land than on the oceans, providing a continual supply of freshwater. We treat water as if it was free, and, in a sense, it is free—a by-product of the enormous flux of solar energy on Earth. The hydrologic cycle accounts for about half of the solar energy absorbed by Earth's surface (see Figure 1.1, p. 5).

The global movement and storage of water is illustrated in Figure 10.1. Every year, 111,000 km^3 of water falls on land, and 70,000 km^3 returns to the atmosphere via evaporation from wet surfaces and transpiration from plants; these two processes are called collectively *evapotranspiration*. The remainder, 41,000 km^3, is the *runoff,* which eventually reaches the oceans. If the runoff were divided evenly, it could provide each person with 6,760 m^3 a year of freshwater (2000 population). But of course it is not divided evenly. Some continents are rainier than others, and the variation within continents is even greater (see Table 10.1). The runoff per km^2 of land area in South America, for example, is over four times as much as in Africa, while within Africa, the runoff per km^2 in the Congo is 12 times as much as in Kenya.

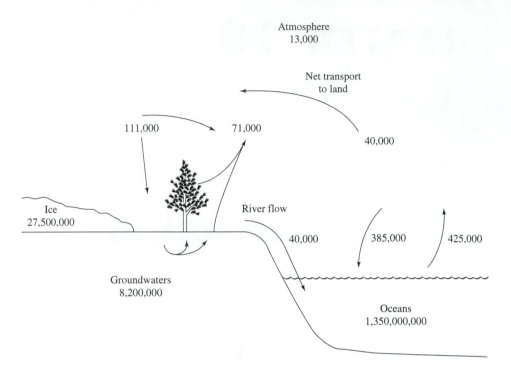

Figure 10.1   The global water cycle. The numbers are in km³ for the water reservoirs, and km³/yr for the flows. Figure from "The Global Water Cycle" in *Biogeochemistry: An Analysis of Global Change* by William H. Schlesinger. Copyright © 1991 by Academic Press. Reproduced by permission of the publisher.

The amount of water withdrawn for human uses is much less than the total runoff, averaging about 8 percent worldwide; but in some countries the fraction is considerably higher. Water is used for many purposes: drinking, cooking, washing, carrying wastes, cooling machines, and irrigating crops. As might be expected, utilization patterns depend on the level of economic development in a region (see Tables 10.1 and 10.2). Per capita water utilization varies widely; the continental averages range from 1,374 m³ per year in North America to 181 m³ per year in Africa. In Canada and Poland, three-quarters of the water utilization is for industry and power generation, while about one-tenth is for agriculture; in most of the developing world, these percentages are reversed. Interestingly, the European per capita utilization, 626 m³ per year, is less than half of that of North America, despite similar economic conditions.

Many people around the world are chronically short of water for personal needs. In many places, freshwater aquifers are being drained faster than they can be replenished. Local reservoirs might be insufficient, especially in times of drought. Water resources are further strained by increasing population; for example, unless the water supply in Africa and Asia increases significantly, the expected increases in population are predicted to

TABLE 10.1 ANNUAL WATER SUPPLY AND WITHDRAWAL FOR CONTINENTS AND VARIOUS COUNTRIES

Continents/ countries	Water supply			Water withdrawal		Per capita use/supply ratio (%)	Status[†]
	Total (km³)	Per km² (m³)	Per capita (m³)	Total* (km³)	Per capita (m³)		
World	41,022	314,386	6,761.5	3,240.0	534.0	7.9	Potential problems
Africa	3,996	134,842	4,995.0	145.1	181.4	3.6	Potential problems
Kenya	20	35,492	665.8	2.1	67.6	10.1	Scarcity
Congo, D.R.	935	412,430	17,992.9	0.4	6.9	0.0	Surplus
North America[‡]	6,365	302,698	14,373.2	608.4	1,373.9	9.6	Surplus
Mexico	357	187,249	3,586.9	77.6	779.0	21.7	Potential problems
Canada	2,850	309,024	92,624.5	45.1	1,466.0	1.6	Surplus
South America	9,526	543,435	27,628.7	106.2	308.0	1.1	Surplus
Peru	40	31,250	1,474.1	6.1	224.8	15.3	Stress
Brazil	5,190	613,728	30,508.8	36.5	214.4	0.7	Surplus
Asia	13,207	428,038	3,584.4	1,633.9	443.4	12.4	Potential problems
China	2,800	301,367	2,214.3	460.0	363.8	16.4	Potential problems
Indonesia	2,530	1,396,579	11,922.3	16.6	78.2	0.7	Surplus
Europe	6,235	275,826	8,570.3	455.3	625.9	7.3	Potential problems
Poland	49	162,276	1,278.2	12.3	317.7	24.9	Stress
Russia	4,313	255,363	29,695.5	77.1	530.9	1.8	Surplus
Oceania	1,614	190,105	52,072.6	16.7	539.7	1.0	Surplus
Australia	343	44,648	17,864.6	14.6	760.4	4.3	Surplus
Papua New Guinea	801	1,768,759	166,528.1	0.1	20.8	0.0	Surplus

*Total water withdrawals are for various years ranging from 1980 to 1995 as provided in WRI (1999).
[†]Refers to per capita water supply:
    Water surplus: >10,000 m³/capita
    Potential water management problems: >2,000 m³/capita <10,000 m³ capita
    Water stress: >1,000 m³/capita <2,000 m³ capita
[‡]Includes Central America

Sources: Population data is for mid-2000, as reported in the Population Reference Bureau (2000), *2000 World Population Data Sheet,* Washington, DC.

Other data from the World Resources Institute (in collaboration with the United Nations Environment Programme and the United Nations Development Programme) (1999). *World Resources 1998–1999* (Oxford, UK: University Press).

place both continents under "water stress" (see Table 10.1 notes) by 2025. In addition, water quality is just as important as quantity. The water in many places is contaminated because of failure to separate wastewater from the water supply. Waterborne diseases continue to be scourges of humankind. In many parts of the world, the critical need for sanitary water is all too often neglected.

TABLE 10.2   USES OF WATER FOR CONTINENTS AND COUNTRIES

Continents/countries	Domestic (%)	Industry/power (%)	Agriculture (%)
World	8	23	69
Africa	7	5	88
Kenya	20	4	76
Congo, D.R.	61	16	23
North America*	12	41	47
Mexico[†]	6	8	86
Canada[†]	18	70	12
South America	18	23	59
Peru[†]	19	9	72
Brazil[†]	22	19	59
Asia	6	9	85
China[†]	6	7	87
Indonesia[†]	13	11	76
Europe	14	55	31
Poland[†]	13	76	11
Russia	19	62	20
Oceania	64	2	34
Australia[†]	65	2	33
Papua New Guinea[†]	29	22	49

*Includes Central America

[†]Sectoral withdrawal estimates are for 1987

Source: World Resources Institute (in collaboration with the United Nations Environment Programme and the United Nations Development Programme) (1999). *World Resources 1998–1999* (Oxford, UK: University Press).

10.2 IRRIGATION

Worldwide, agriculture accounts for the lion's share of water use, 69 percent, and agricultural demand is growing as population continues to increase (see Figure 10.2). Large inputs of water are needed in agriculture because growing plants transpire rapidly. In order to capture atmospheric CO_2, their leaves are designed for efficient exchange of gases (see Figure 10.3). This exchange takes place in tiny pores called stomata; the cells in the stomata absorb CO_2 and release O_2 during photosynthesis. While the stomata are open, water vapor is released as well. Plants can conserve water by closing the stomata (in droughts, for example), but doing so curtails the CO_2 supply. Furthermore, as the seedlings develop, the leaf area expands until it is about three times the surface area of the soil below, and the amount of water lost by transpiration exceeds that lost by evaporation. Because the exchange of gases makes the evaporation of water from the moist cells unavoidable, photosynthesis is always accompanied by copious transpiration. The amount of water required to produce a bushel (25 kg) of corn, for example, is 5,400 gallons (20 m^3). Moreover, agriculture uses more water than that required by the growing plants, because much of the water runs off or evaporates before it even reaches the plants. The global average irrigation efficiency is estimated to be only 37 percent.

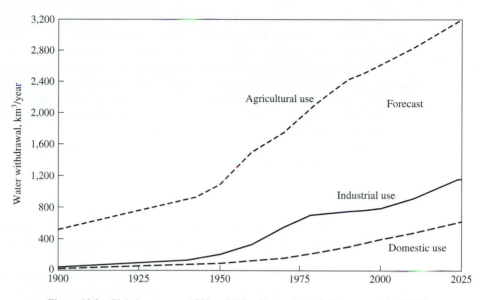

Figure 10.2 Global water use, 1900 to 2025. *Source:* I. A. Shiklomanov (ed.) (1999). *World Water Resources at the Beginning of the 21ˢᵗ Century* (St. Petersburg, Russia: State Hydrological Institute/UNESCO).

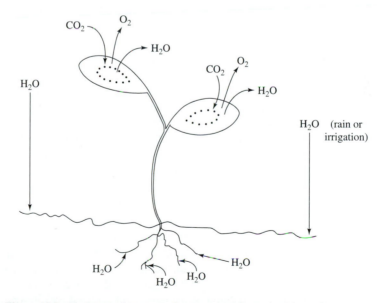

Figure 10.3 Exchange of gases on the stomata of leaves, and loss of water through transpiration.

New irrigation technologies can increase this percentage substantially. Thus drip irrigation, involving a network of perforated plastic tubing installed on or below the surface, can achieve up to 95 percent efficiency, because the water is delivered directly to the roots of the plants. Moreover, drip irrigation can increase crop yields, because it avoids saturating the soil with water, which can leach nutrients from the soil, and damage the roots during the subsequent drying out of the soil. An alternative to drip irrigation is a newly designed sprinkler system, which delivers water in small doses just above the soil surface. This system is particularly effective for grain crops.

An important development in water conservation is the accurate leveling of fields, using laser-guided graders to scrape the soil. Water runs off much more slowly when the field is level. It is estimated that the water saved by laser-leveling farmland in central Arizona during the 1980s was equivalent to the entire water supply of the half-million people living in the city of Tucson. Finally, detailed weather monitoring, combined with new computer technologies, can provide farmers with continual estimates of how much water their crops actually need, further improving the efficiency of the irrigation system.

Most of the world's farmers lack the needed capital and maintenance facilities to implement advanced irrigation systems. However, continued technological improvements are bringing costs down. For example, drip irrigation can cost up to $2,500 per hectare to install, but a Denver-based firm has developed drip systems appropriate for small farms that cost as little as $250 per hectare. Field tests are underway in Nepal and India. Weather information is not inherently expensive, and the International Water Management institute in Sri Lanka has developed a computerized tool for irrigation and crop planning, which integrates information from a worldwide network of weather stations.

Improving irrigation efficiency is important not only for water conservation per se, but for ameliorating problems caused by overuse of irrigation water. A variety of ecological disruptions are associated with construction of the dams, dikes, and levees that are required to channel surface waters into irrigation. These diversions of the water flow reduce the ability of rivers to support migrating fish, filter pollutants through wetlands (see p. 328), and deliver fertile silt to agricultural lands in the flood plain. The extent of these constructions can be diminished if irrigation water is used more efficiently.

Increased efficiency also reduces *salinization,* a condition in which salts build up in the soil, eventually inhibit plant growth, and diminish crop yields. Salts are a natural consequence of neutralization of soil minerals by rainwater (see p. 297–298) and accompany the rainwater as it drains away. When this water is applied to crops, the salts are left behind as the water is transpired and evaporated. Unless there is good drainage, the salts build up in the soil over time. Drainage is slowed when irrigation water is applied in excessive amounts, causing the water table to rise. Eliminating the excess by providing only what the plants need can eliminate salinization. The problem is widespread, especially in arid areas, where evaporation is rapid. Salinization affects 23 percent of U.S. irrigated land, including two-thirds of the land irrigated in the lower basin of the Colorado river. Likewise, the worldwide fraction of irrigated land affected by salinization is estimated to be 21 percent, with much higher percentages in irrigated dry lands.

10.3 GROUNDWATER

Some of the runoff is carried to the ocean directly in surface waters, but much of the water falling on land percolates into permeable rock layers and is stored as groundwater in aquifers. Here the hydrosphere comes into intimate contact with the lithosphere. Wells dug into these aquifers supply a substantial fraction of the water used by humans.

There are special problems associated with groundwater utilization. One is simply overpumping: the rate at which water is extracted from wells is often greater than the rate at which the aquifer is recharged from the surface water. This practice can continue for some time if the aquifer is large, but eventually the storage capacity is exceeded, and the wells run dry. Deepening the wells, as the water table is lowered, can extend the water supply, but pumping costs also increase. In parts of the High Plains region of the U.S. (a band of states reaching from Montana and North Dakota south to Texas), farms are being abandoned as pumping costs for irrigation escalate. The area is underlain by the Ogallala aquifer, the largest groundwater reservoir in the world, which provides irrigation water for wheat, corn, sorghum, and cotton crops. However, the water table has fallen by over 30 meters, and in some areas water is being drawn as much as 40 times faster than it is being recharged.

Additional difficulties can arise when aquifers are overdrawn. In some cases, the loss in fluid pressure can lead to subsidence of the overlying soils. Occasionally sinkholes suddenly appear, when underlying limestone caves collapse after the water is drawn out of them. During a 1981 drought in Florida, a sinkhole appeared in the town of Winter Park and swallowed a house, two businesses, a street with several cars, and a community swimming pool.

In coastal areas, aquifers are in contact with the oceans. The flow of freshwater keeps the saltwater out, but when pumping depletes the freshwater, the saltwater can intrude and contaminate the wells. Saltwater intrusion has occurred in coastal areas of Florida, Louisiana, Texas, California, Washington, and in the Northeast and Mid-Atlantic States. In dry areas like the southwestern states, the drop in the water table can dry out rivers during much of the year, with loss of native animals and plants.

Another pervasive problem is well-water contamination from the surrounding soils. Soils have a large capacity for absorbing contaminants, and can actually serve as an effective filter. But when this capacity is exceeded, the contaminants enter the groundwater. Pathogenic microbes are the main cause for concern, but a variety of chemical contaminants can also affect well water. In farm areas, pesticides and herbicides applied to the soil can sometimes reach drinking wells. The herbicide atrazine (see p. 398), which is moderately soluble in water, is a common contaminant in well water and ground water drinking supplies. In addition, wells in agricultural regions often have elevated levels of nitrate ion, resulting from the leaching of fertilizers (see pp. 344–345).

In urban areas, groundwater is affected by runoff from streets, and by contaminant spills and leaks, particularly leaks from underground storage tanks. Underground tanks for oil and gasoline storage are numerous, and corrosion produces leaks over time. The low solubility of petroleum in water limits the extent of contamination, but soluble additives can pose significant problems. The most notorious recent instance is the widespread contamination of urban wells in the U.S. with the gasoline additive methyl tertiary-butyl ether (MTBE). MTBE was added to gasoline in order to combat air pollution (see pp. 239–240) but is being removed in some states because it has shown up in significant concentrations in many wells. In Santa Monica, California, wells supplying 50 percent of the city's drinking water had to be taken out of service because of MTBE contamination. MTBE has high solubility in water and it is only slowly degraded by soil microbes, so it can travel quite far from the site of a leak.

Overpumping exacerbates the contamination problem, because some of the contaminant is swept past any given well by the natural flow of the groundwater. The higher the pumping rate, the larger the fraction of the contaminant that is captured by the well. In the case of Santa Monica, the pumping rate at the time the MTBE contamination was discovered was about twice the natural flow through the 2-km-wide aquifer drained by the city's wells, so that all leaked material within 1 km of the wells would have been captured.

Finally, we note that not all well contaminants are anthropogenic. In some cases the rock formation holding the aquifer can leach toxic minerals into the water table. Arsenic can reach toxic levels in some areas. In Bangladesh, thousands of villagers suffer from arsenic poisoning as a tragic consequence of the drilling of tube wells, intended, ironically, to provide them with clean water (see p. 444).

A less serious, but widespread problem, is the presence of high iron levels in well water. Iron is a major constituent of rocks and soils, but its solubility is very low, provided that it is in its oxidized (ferric) state. The solubility is much higher in its reduced (ferrous) state, and iron levels are high in wells drawing water with little dissolved oxygen. Iron is not very toxic (see p. 433), but the water tastes metallic, and precipitates brown rust as the iron becomes oxidized in air.

10.4 U.S. WATER RESOURCES

To gain a more detailed perspective on water resources, we examine the pattern of distribution and use in the continental United States. A schematic of the water flows is given in Figure 10.4. As in the global water cycle, about two-thirds of the precipitation is returned to the atmosphere via evapotranspiration; the balance goes to runoff, amounting to about 2,000 km^3/yr. Most of the water is in the east of the country, and flows to the Atlantic and Gulf coasts. A quarter of the annual runoff (468 km^3) is withdrawn for various uses, of which three-quarters (338 km^3) is subsequently returned to the streamflow and the remainder is "consumed," mostly through evapotranspiration after or during use; eventually, of course, all the water is returned to the global cycle.

The pattern of water utilization is graphed in Figure 10.5. Of the 470 km^3 of freshwater used in 1995, 77 percent was from surface supplies and the rest was from groundwater, which is eventually recharged from the surface. Water usage is dominated by agriculture and by the cooling of electric generators, each of which requires about 40 percent of the supply. The remaining 20 percent is divided between industry and mining on the one hand (8 percent), and domestic and commercial uses on the other (12 percent). Major

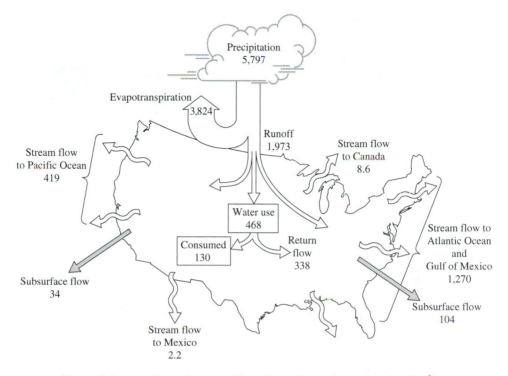

Figure 10.4   Annual water flows in the United States. The numbers are in units of km³/yr. *Source:* Adapted from U.S. Water Resources Council (1978). *The Nation's Water Resources, 1975–2000: Second National Water Assessment, Vol. 1: Summary* (Stock number 052-045-00051-7) (Washington, DC: Superintendent of Documents).

industrial users are the steel, chemical, and petroleum industries; the mining uses include water extraction of minerals and fossil fuels, and milling and related activities. Most of the industrial and mining water is obtained directly from surface and groundwater, but about 20 percent of it comes from public supplies, which also provide most of the domestic and commercial water. Domestic uses (drinking, food preparation, bathing, washing clothes and dishes, flushing toilets, and watering lawns and gardens) amounted to about 380 liters (100 gallons) per day per person. The total was about three times higher than for commercial uses (hotels, restaurants, office buildings, and so on).

While agriculture is a heavy user of water throughout the country, irrigation is especially important in the arid West, where extensive diversion of the major waterways has permitted large-scale crop production. California and Idaho are the largest users of irrigation water, together accounting for 34 percent of the national total.

The influx of city-dwellers also strains the water resources of arid regions. The Southwestern metropolises of Los Angeles, San Diego, Las Vegas, Tucson, and Phoenix all experience serious water management problems and conflicts over access to distant supplies of water.

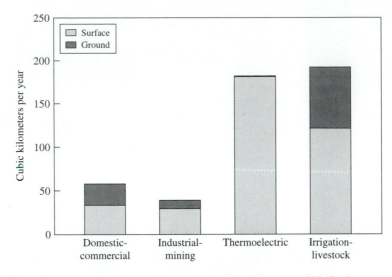

Figure 10.5 Sources and uses of water in the United States in 1995. Total use was 470 km³/yr. *Source:* Data from W. B. Solley et al. (1998). *Estimated Use of Water in the United States in 1995,* U.S. Geological Survey, Circular 1200 (Washington, DC: U.S. Department of Interior).

10.5 THE OCEANS

Not only do the oceans contain most of the world's water, they also cover most of Earth's surface. The oceans form an enormous solar collector and the solar heating of the oceans drives the hydrological cycle. Energy exchanges between ocean waters and the atmosphere are major determinants of the climate, and changes in ocean current can have major climatic impacts. Thus, a periodic northward shift of the main tropical current in the Pacific Ocean, called *El Niño,* temporarily alters weather patterns around the globe.

Currents at the surface of the oceans follow a circulatory pattern (see Figure 10.6) driven by the trade winds, which blow from east to west in an equatorial band. When the currents reach the continents, they divide, moving north and south in the respective hemispheres. As a result of Earth's rotation, the circulation around each ocean basin is clockwise in the Northern Hemisphere and counterclockwise in the Southern Hemisphere. Because of the open water around Antarctica, there is also an Antarctic current that flows continuously from west to east.

Solar heating divides the surface waters from the deep ocean by a temperature gradient, called the *thermocline.* As in the case of a temperature inversion in the atmosphere, there is little mixing between the warm layer above and the cold layer below the thermocline, which is found from 75 to 200 meters below the surface. Above the thermocline, the waters are well mixed by the action of wind and waves. The average temperature is 18°C above the thermocline, and 3°C below it. (Recall that this temperature difference is the basis for schemes to extract energy via OTEC devices, see p. 96.)

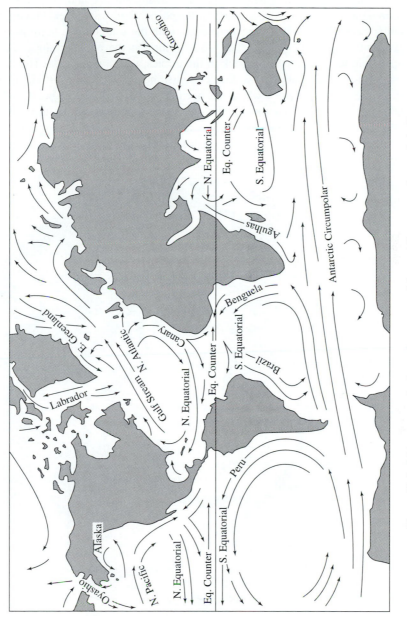

Figure 10.6 Major currents in the surface waters of the world's oceans. *Source:* J. A. Knauss (1978). *Introduction to Physical Oceanography* (Saddle River, NJ: Prentice Hall). Copyright © 1978. Reprinted by permission of Pearson Education, Inc., Upper Saddle River, NJ.

The surface waters cool as they move away from the equator along the currents, and the thermocline disappears near the poles, allowing surface and deep waters to mix. In addition, the surface becomes saltier near the poles because salt is excluded from ice when water freezes. Because the density increases with the salt content, there is net downward movement of the polar waters. This downwelling at the poles produces currents in the deep ocean, which carry water back to the tropics, where it upwells. Thus in addition to the wind-driven circulation of surface currents, there is a vertical circulation between the surface and the deep waters, which has been likened to a conveyor belt (see Figure 10.7). Because of its dependence on salt (*halide*) as well as temperature, this conveyor belt is called the *thermohaline* circulation. The deep waters are rich in nutrients, thanks to the constant rain of decomposing organisms from above. Consequently there is high

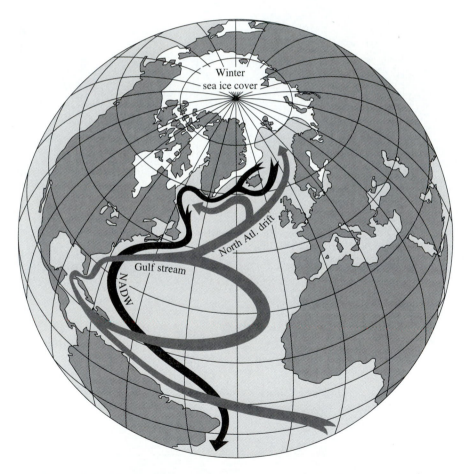

Figure 10.7   Pathway of the Gulf Stream, and transfer of heat to Northern Europe. NADW signifies "North Atlantic Deep Water." *Source:* S. Rahmstorf (1997). Risk of sea-change in the Atlantic. *Nature* 388: 825–826. Copyright © 1997 by *Nature*. Reprinted with permission from *Nature* and author of the journal article.

biological productivity in regions of upwelling, giving rise, for example, to rich fisheries along many tropical shores.

There is serious concern that if global warming continues long enough, it could disrupt the thermohaline circulation by melting enough polar ice to diminish, or even abolish, the salt gradient. The consequences for ocean ecosystems, and indeed for terrestrial ones, could be severe. Because of the importance of the ocean currents to global heat flows, there would be major alterations in weather patterns. For example, the temperate climate enjoyed by Northern Europe, despite its high latitude, is due to the heat provided by the Gulf Stream.* If this heat supply were to diminish, the continent would be considerably colder.

10.6   WATER AS SOLVENT AND AS A BIOLOGICAL MEDIUM

In considering issues surrounding water use and quality, it can be helpful to look at the two distinct roles water plays in Earth's environment. On the one hand, water is a remarkable solvent; its unusual molecular properties allow it to dissolve and transport a wide range of materials. On the other hand, water houses ecosystems; a large percentage of the biosphere lives in some form of aqueous environment. We often define water quality in terms of the ability of the aqueous environment to support the normal range of biological species with their accompanying biochemical processes. These two perspectives on water are interrelated, of course: the materials dissolved in water affect its ability to support life, and the biological processes of aqueous environments often play critical roles in clearing dissolved substances from contaminated waters.

We shall follow these two perspectives on water by tracing the course of water through the hydrological cycle. From atmosphere to soil, the dominant aspect of water is its role as a solvent. Water is almost pure when it enters the atmosphere, but immediately begins to interact and form solutions with other substances. As discussed in Part II, water condenses in clouds through the nucleation of atmospheric particles, including nitrates and sulfates. Atmospheric water, whether in clouds or rain, is in equilibrium with other constituents in the atmosphere, particularly carbon dioxide.

When rain falls to Earth, the number of interactions increases as water enters the soil, where it absorbs increasing numbers of ions. Eventually, water and its load of dissolved materials accumulate in surface waters and the oceans; at this point, the role of water as a solvent is augmented by its role as ecosystem. Most critical to considerations of the aqueous biosphere is the *redox potential* (see pp. 307–312), determined in large part by the amount of dissolved oxygen in the water relative to the nutrients necessary for life. The balance changes with time and place, leading to distinct ecological niches in freshwater lakes, rivers, wetlands, and oceans.

---

*Norway and Sweden are as far north as Greenland and Alaska. Trondheim, Norway (latitude 63° N) has an average January temperature of –3.1°C (26.4°F) and an average annual temperature of 4.8°C (40.6°F); in contrast, Fairbanks, Alaska (latitude 64° N) has an average January temperature of –18.1°C (–0.5°F), and an average annual temperature of –1.5°C (29.3°F).

CHAPTER 11

FROM CLOUDS TO RUNOFF: WATER AS SOLVENT

11.1 UNIQUE PROPERTIES OF WATER

Water is the most commonplace of substances, and yet its properties are unique among chemical compounds. Perhaps most remarkable are its high melting and boiling temperatures. Table 11.1 lists the melting and boiling points for hydrides of the elements from carbon to fluorine in the first row of the periodic table. Compared to the neighboring hydrides, water has by far the highest melting and boiling points. One characteristic that makes Earth uniquely suitable for the evolution of life is its surface temperature, which, over most of the planet, lies within the liquid range of water.

a. Hydrogen bonding.   The high temperature required for boiling implies that liquid water has a high cohesive energy; the molecules associate strongly with one another. One source of this cohesion is water's high dipole moment (1.85 Debye*). The O—H bonds are highly polar; negative charge builds up on the oxygen end while positive charge builds up on the hydrogen end. The dipoles tend to align with one another, increasing the cohesive

*The dipole moment, μ, is defined as the charge times the distance separating the charge. 1 Debye (D) = 3.336×10^{-30} coulomb (C) $\times$ meters (m). For example, for one electron (1.60×10^{-19} C) separated from a proton by 1 Å (10^{-10} m), $\mu = 1.6 \times 10^{-29}$ C $\times$ m = 4.80 D.

TABLE 11.1   LIQUID TEMPERATURE RANGE FOR WATER AND NEIGHBORING FIRST-ROW ELEMENT HYDRIDES

	CH_4	NH_3	H_2O	HF
Melting point (°C)	−182	−78	0	−83
Boiling point (°C)	−164	−33	100	20

energy. But if the dipole-dipole interaction were the only factor, then HF would have a higher boiling point than water, because it has a higher dipole moment (1.91 Debye). The greater cohesion of water derives, in addition, from the molecule's bent structure, with hydrogen at each end. Water is able to donate two hydrogen bonds, one from each of its H atoms, while simultaneously accepting two hydrogen bonds, one to each of the electron lone pairs on the oxygen atom. Thus water forms a three-dimensional network of H-bonds, whereas HF, having only one donor site, is limited to forming linear H-bond chains (see Figure 11.1).

The H-bond network of water is demonstrated strikingly by the crystal structure of ice, in which each water molecule is H-bonded to four other molecules in connected six-membered rings (see Figure 11.2). These six-membered rings produce a very open structure, which accounts for another unusual property of water, its expansion upon freezing. Most substances contract upon solidification because the molecules take up more room in the chaotic liquid state than they do in the close-packed solid. The liquid continues to expand as it is heated due to the increased motion of the molecules. When ice melts, however, the open lattice structure collapses on itself and the density *increases*. H-bonded networks still exist, but they fluctuate very rapidly, allowing the individual water molecules to be mobile and closer together. As liquid water is heated from 0°C, the H-bonded network is further disrupted and the liquid water continues to contract. When the temperature reaches 4°C, the density of liquid water reaches a maximum. As the temperature rises above 4°C, the water slowly expands, reflecting the increasing thermal motion. The lower density of ice relative to water is of great ecological significance to the world's temperate-zone lakes and rivers. Because winter ice floats on top of the water, it protects the aquatic life below the surface from the inhospitable climate above. If ice were denser than water, the lakes and rivers would freeze from the bottom up, creating arctic conditions.

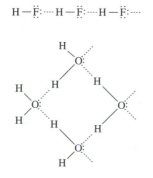

Figure 11.1   Linear H-bonding in HF versus network H-bonding in H_2O.

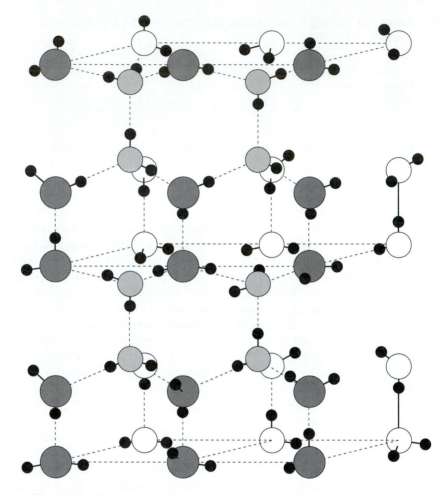

Figure 11.2   The crystal structure of ice. The larger circles and smaller circles represent oxygen atoms and hydrogen atoms, respectively. Oxygens with the same color shading indicate they lie in the same plane.

FUNDAMENTALS 11.1: HYDROGEN BONDS

It is important to remember that hydrogen bonds (often called H-bonds for short) are not conventional electron pair bonds. The two H atoms in the water molecule meet their electronic requirement by pairing their single valence electrons with the two available O atom electrons. The H-bond is an electrostatic interaction between the partial positive charge on a water H atom and the partial negative charge on the O atom of a neighboring water molecule. The interaction is stronger than that of ordinary dipoles because the bare proton of the

H atom can be right next to the lone pair of the O atom of the water neighbor. Still, this distance is not as short as the O—H covalent bond distance, and the H-bond is much weaker than is the covalent bond.

In general, H-bonds form between an H atom on an *electronegative* atom, X (one that has a strong attraction for electrons), and a lone pair on another atom, Y. We use a dotted line to represent the H-bond:

$$X—H \cdots \cdots \cdots Y—$$

The electronegativity of X is important, since it determines the extent to which the bonding electrons are pulled toward X and away from H. Thus, hydrocarbons do not form H-bonds to a significant extent, because C and H have similar electronegativities, resulting in an essentially neutral H atom. The elements to the right of carbon in the periodic table—N, O, and F—are increasingly electronegative, because their increasing nuclear charge draws the valence electrons closer to the nucleus. They are the ones that can donate H-bonds, and they are also the elements that retain lone pairs when they form compounds.

The same principles apply to elements of the second and higher rows of the periodic table, but H-bonding is weaker for these elements (e.g., P, S, and Cl) since they are less electronegative than those in the first row (because the nucleus is farther from the valence electrons). The most important H-bonds in nature are those in which X and Y are N and/or O.

H-bonding is very important in biological molecules, for which N and O atoms occur at key points in the structure. The H-bonds help maintain the three-dimensional shape of these molecules, and are responsible for determining how they interact with one another. Perhaps the most important H-bonds are those that establish complementarity of the bases in the DNA double helix (see p. 411), and thereby determine the genetic code.

b. Clathrates and water miscibility. The ability of water molecules to hydrogen-bond to one another accounts for the existence of crystalline *clathrates* (also called *hydrates*), compounds in which water molecules can pack themselves around nonpolar molecules by forming ordered arrays of H-bonds. The structure of xenon hydrate, $8Xe.46H_2O$, is shown in Figure 11.3. The crystals consist of arrays of dodecahedra formed from 20 water molecules H-bonded together. Within the dodecahedra are the Xe atoms, or nonpolar molecules of similar size, such as CH_4 or CO_2. By filling the voids, the guest molecules stabilize the clathrate H-bond network. The clathrates are stable to temperatures appreciably higher than the melting point of ice. For example, methane hydrate melts at 18°C at atmospheric pressure. Clogging of natural gas pipelines by methane hydrate was once a problem for gas distribution. It was solved by removing water vapor before the gas was fed into the lines.

It has recently come to be appreciated that methane hydrates occur widely in nature, principally in ocean sediments or in the arctic permafrost. Anaerobic microbes in sediments or swamps decompose organic matter to methane (see pp. 310–312). When methane is released into the surrounding water, it forms methane hydrate, provided the temperature is low enough, and/or the pressure is high enough. As discussed elsewhere (see p. 22), the

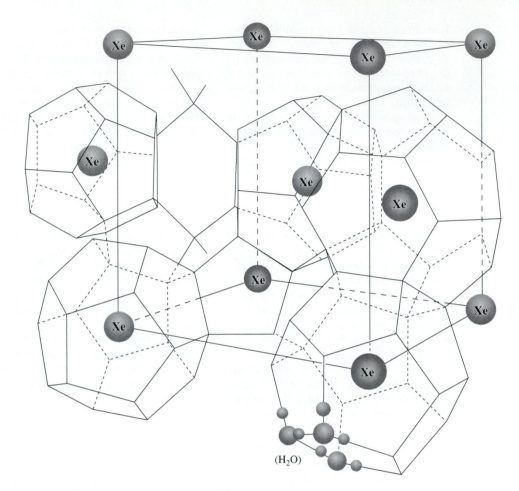

(H$_2$O)

Figure 11.3   The structure of a clathrate crystal, xenon hydrate; the xenon atoms occupy cavities (eight per unit cube) in a hydrogen-bonded three-dimensional network formed by water molecules (46 per unit cube).

amount of methane held in hydrate deposits is enormous, and there is much interest in capturing the methane for fuel. From the perspective of climate change, however, there is concern that large amounts of methane could quickly be released into the atmosphere, either as a result of mining operations, or simply because of seawater warming, thereby amplifying the greenhouse effect, as apparently happened earlier in Earth's history (see pp. 173–175).

The ability of water to form extended H-bond networks around nonpolar molecules also explains why oil and water do not mix. Water molecules actually attract hydrocarbon molecules; as shown in Table 11.2, the enthalpy of dissolution in water is negative for simple hydrocarbons, meaning that the molecular contact releases heat. The heat release

TABLE 11.2   FREE ENERGY, ENTHALPY, AND ENTROPY OF SOLUTION IN LIQUID
WATER AT 298 K

Process	ΔG (joules/mol)	ΔH (joules/mol)	ΔS (joules/K/mol)
CH_4 in benzene $\rightarrow$ CH_4 in water	+10,878	−11,715	−75
C_2H_6 in benzene $\rightarrow$ C_2H_6 in water	+15,899	−9,205	−84
C_2H_4 in benzene $\rightarrow$ C_2H_4 in water	+12,217	−6,736	−63
C_2H_2 in benzene $\rightarrow$ C_2H_2 in water	+7,824	−795	−29
Liquid propane $\rightarrow$ C_3H_8 in water	+21,129	−7,531	−96
Liquid n-butane $\rightarrow$ C_3H_{10} in water	+24,476	−4,184	−96
Liquid benzene $\rightarrow$ C_6H_6 in water	+17,029	0	−59
Liquid toluene $\rightarrow$ C_7H_8 in water	+19,456	0	−67
Liquid ethyl benzene $\rightarrow$ C_8H_{10} in water	+23,012	0	−79

Source: W. Kauzmann (1959). Some factors in the interpretation of protein denaturation. *Advances in Protein Chemistry* 14:1–63.

reflects an attractive force created by the interaction between water and the guest molecule. Nevertheless, the free energy values are positive; the reaction is disfavored by large negative entropy changes ($\Delta G = \Delta H - T\Delta S$). The negative entropy implies that mixing increases molecular order, consistent with the water molecules packing around the nonpolar solute, as they do in the clathrate structures. Thus, oil and water do not mix because the packing tendency of the water inhibits it.

This tendency of water and nonpolar molecules to separate is responsible for many important structures in biology. For example, proteins keep their shape in water by folding in such a way that *hydrophobic* ("water-hating") alkyl and aryl groups of the amino acids are in the interior, away from the water, while *hydrophilic* ("water-loving") polar and charged groups face the exterior, where they contact the water. Likewise, biological membranes are bilayers of lipid molecules (see Figure 11.4), with the hydrocarbon chains of the lipids packed together away from the water in the interior of the bilayer.

The solubility of organic molecules in water increases when they have functional groups capable of forming H-bonds with water. Thus, alcohols, ethers, acids, amines, and amides are all significantly soluble in water, the extent of solubility depending on the size of the hydrocarbon portion of the molecule. The smaller the hydrocarbon portion, the higher the solubility. Small alcohols, like methanol and ethanol, are completely miscible with water. The —OH group can both donate and accept H-bonds with water molecules, providing a strong impetus for mixing. Ethers are not completely miscible, since the O atom can accept but not donate an H-bond. However, the solubility is considerably enhanced by the H-bond acceptance, relative to the solubility of a comparable hydrocarbon. Thus the solubility of the gasoline additive MTBE (methyl tertiary-butyl ether, see pp. 30–31) is 4,700 mg/L, compared to only 24 mg/L for the equivalent hydrocarbon, 2,2-dimethylbutane. This is why MTBE is a far more serious groundwater contaminant than the gasoline to which it has been added.

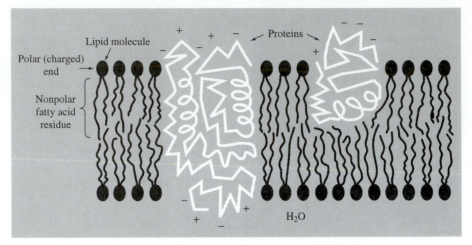

(a)

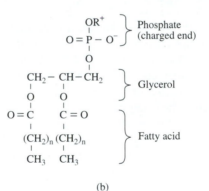

(b)

Figure 11.4 (a) The hydrophilic-hydrophobic bilayer structure of a biological membrane; (b) molecular structure of a lipid molecule.

11.2   ACIDS, BASES, AND SALTS

a. Ions, autoionization, and pH.   Another unique property of water is the ease with which it dissolves ionic compounds. The forces between ions of opposite charge in a crystal are very strong, and cost a good deal of energy to disrupt. This can be seen in the high melting point of an ionic crystal, such as ordinary table salt. Yet salt dissolves readily in water, because the water molecules strongly *solvate* both the positive sodium ions and the negative chloride ions. The strong interionic forces are replaced by equally strong solvation forces.

The strong solvation forces stem from water's large dipole moment and H-bonding ability. Anions interact with the positive end of the dipole via H-bonds, while cations interact with the negative end via coordinate bonds from the oxygen lone-pair electrons (see Figure 11.5).

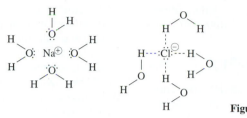

Figure 11.5   Solvation of ions in water.

Among the ions that are stabilized in water are the hydrogen and hydroxide ions. Strong acids such as HCl, HNO_3, and H_2SO_4 release H^+ ions when dissolved in water, while strong bases such as NaOH or KOH release OH^-. These ions are special because they react together to form water:

$$H^+ + OH^- = H_2O \qquad (11.1)$$

Reaction (11.1) is a *neutralization* reaction; the strong base neutralizes the strong acid and vice versa. The reverse of reaction (11.1) is *autoionization;* the water can ionize itself because the ions are stabilized by other water molecules. This reverse reaction does not proceed to any great extent; the position of equilibrium for reaction (11.1) lies far to the right. Nevertheless, autoionization is a key factor in acid-base chemistry because it sets the range of available acid and base strengths in water.

In contrast to reactions in the gas phase, reactions in aqueous solution are usually rapid. The water molecules are in continuous contact with one another, and the H-bonds and coordinate bonds are broken and reformed rapidly. There are some metal ions for which the coordinate bonds to water are long-lived for special electronic reasons, as in the complex ion $Cr(H_2O)_6^{3+}$. But for most ions, the bonds are broken and reformed very rapidly, producing an averaged solvation environment for the ion. Consequently, the extent of reaction is generally set by the equilibrium constant, rather than the kinetics, even if the position of equilibrium lies far to the right or to the left. We can use the experimentally de-termined equilibrium constant to calculate the extent of reactions from various starting conditions. This is a particularly useful capability in acid-base reactions, which are funda-mental to the chemistry of the aqueous environment.

FUNDAMENTALS 11.2: THE pH SCALE

For solutions of a strong acid, HX, the concentration of H^+ ions is the same as the *analytical* concentration (C_{HX}), i.e., the number of moles per liter of dis-solved HX. We write:

$$(H^+) = C_{HX} \, M* \qquad (11.2)$$

Likewise, for solutions of a strong base, MOH:

$$(OH^-) = C_{MOH} \, M \qquad (11.3)$$

Because (H^+) is generally less than 1 M, and frequently much less, it has

*M signifies molar, i.e., moles per liter, the standard concentration units in solution.

become standard to express the concentration as the negative logarithm, symbolized by "p":

$$pH = -\log(H^+) \qquad (11.4)$$

For example, when the HX concentration is $10^{-3}\ M$, the pH is 3, and when it is twice this much, the pH is 2.7 [because $-\log(2 \times 10^{-3}) = 0.3 - 3 = 2.7$]. In the same way:

$$pOH = -\log(OH^-) \qquad (11.5)$$

If we now consider the autoionization reaction [reverse of equation (11.1)]:

$$H_2O = H^+ + OH^- \qquad (11.6)$$

we can write the equilibrium expression as

$$K = (H^+)(OH^-)/(H_2O) \qquad (11.7)$$

We see that (H^+) and (OH^-) are reciprocally related:

$$(H^+) = K(H_2O)/(OH^-) \qquad (11.8)$$

or

$$pH = -\log K(H_2O) - pOH \qquad (11.9)$$

In an aqueous solution, the concentration of the water molecules is essentially constant; a liter of water weighs 1,000 g, and contains $1{,}000/18 = 55.5$ moles. (The added weight from the dissolved ions is a small fraction of 1,000 g, unless the solution is very concentrated). Therefore it is customary to include (H_2O) in the "effective" equilibrium constant. For the autoionization equilibrium, the effective constant, $K(H_2O)$, is called K_w; its experimental value is $10^{-14}\ M^2$. Substituting this value in equation (11.9), we see that

$$pH = 14 - pOH \qquad (11.10)$$

In other words, the autoionization equilibrium requires that the pH and pOH add up to 14.

In pure water there is no source of H^+ or OH^- other than the autoionization reaction itself, which requires that one H^+ be produced for every OH^-. In that case, $(H^+) = (OH^-)$, and plugging this condition into the equilibrium expression gives

$$(H^+)^2 = 10^{-14}\ M^2 \quad \text{or} \quad (H^+) = 10^{-7}\ M$$

Thus the autoionization reaction requires that pure water have a pH of 7. This is also the pH of a solution in which a strong acid has been exactly neutralized by a strong base. A pH of 7 defines neutrality; pH values below 7 are acidic, while pH values above 7 are basic, or *alkaline.*

---

b. Weak acids and bases. Because water dissolves so many different substances, the only place in the environment where pure water is found is at the start of the hydrological cycle, in water vapor. Even raindrops include other substances; as detailed in Part II, the formation of raindrops in saturated air is nucleated by atmospheric particles, especially nitrates and sulfates. In addition, the water in cloud vapor and raindrops is in equilibrium with other constituents of the atmosphere, especially carbon dioxide. These substances tend to lower the pH of rainwater significantly below neutrality.

Strong acids transfer a proton completely to water, but many acidic substances hold the proton more or less tenaciously, and the transfer to water may be incomplete. For a generalized acid, HA, the extent of the transfer depends on the equilibrium constant, K_a, for the *acid dissociation* reaction:

$$HA = H^+ + A^-$$ (11.11)

$$K_a = (H^+)(A^-)/(HA)$$ (11.12)

K_a is often referred to as the *acidity constant*. If K_a is small, then only a small fraction of HA will be dissociated, creating a considerable difference between the concentration of acid molecules and the concentration of hydrogen ions.

WORKED PROBLEM 11.1: pH of a Weak Acid

Q. *What is the pH of a 0.1 M solution of acetic acid, which has a K_a of $10^{-4.75}$ M^2?*

A. Dissociation of HA produces an equal number of H^+ and A^- ions, and therefore $(H^+) = (A^-)$. Substituting this equality in equation (11.12) and rearranging gives

$$(H^+)^2 = K_a(HA)$$ (11.13)

If only a small fraction of HA is dissociated, then (HA) is essentially the same as the analytical concentration, C_{HA} (this will be the case provided $K_a \ll C_{HA}$), and therefore:

$$(H^+)^2 \sim K_a C_{HA}$$ (11.14)

In this problem, $C_{HA} = 0.1\ M$, and $K_a = 10^{-4.75}\ M^2$ (which is much smaller than C_{HA}) so

$$(H^+)^2 = 10^{-4.75} \times 10^{-1.0} = 10^{-5.75}$$

$$(H^+) = 10^{-2.88}\ M \quad \text{or} \quad pH = 2.88$$

Therefore, a 0.1 M solution of acetic acid has a pH close to 3.0.*

If HA holds its proton tenaciously, then the anion, A^-, will have some tendency to remove a proton from water. Thus, if a salt of A^- is added to water, the reaction

$$A^- + H_2O = HA + OH^-$$ (11.15)

which is called a *hydrolysis*, or a *base dissociation* reaction, will proceed to some extent. Since OH^- is produced, A^- is acting as a base. Assuming that the equilibrium does not lie completely to the right, A^- is a weak base. The equilibrium constant, called the *basicity constant,* is

$$K_b = (HA)(OH^-)/(A^-)$$ (11.16)

[As in the case of the autoionization reaction, the water concentration, (H_2O), being constant, is lumped into the equilibrium constant.]

The acid and base dissociation reactions are linked by the autoionization reaction. This can be seen by adding reaction (11.11) to reaction (11.15) to yield reaction (11.6). When reactions are added, the equilibrium constants multiply, i.e., $K_a K_b = K_w$. [This result can be checked by substituting the equilibrium expressions, eqs. (11.12), (11.16), and (11.7).] Thus, determining K_a also specifies K_b.

---

*Since the value of (H^+) is only 1 percent of the value of C_{HA}, the approximation that $(HA) = C_{HA}$ is a good one in this case. When this approximation is poor, then allowance must be made for the lowering of HA due to the ionization reaction. Since every molecule of HA that ionizes produces one H^+ ion, $C_{HA} = (HA) + (H^+)$, or $(HA) = C_{HA} - (H^+)$. Substituting this into equation (11.12) and rearranging, we have:

$$(H^+)^2 + K_a(H^+) - K_a C_{HA} = 0$$

which may be solved with the quadratic formula:

$$(H^+) = (-K_a \pm \sqrt{K_a^2 + 4K_a})/2$$

WORKED PROBLEM 11.2: pH of a Weak Base

Q. *What is the pH of a 0.1 M solution of sodium acetate?*

A. Since base dissociation produces one OH^- for every HA [eq. (11.15)], then $(OH^-) = (HA)$ for a salt of A^- dissolved in water. Then (OH^-) can be calculated by substitution in equation (11.16) and rearranging,

$$(OH^-)^2 = K_b(A^-) \approx K_b C_{A^-} \qquad (11.17)$$

provided that $K_b \ll C_{A^-}$, so that the difference between (A^-) and C_{A^-} is negligible.

In this problem, $C_{A^-} = 0.1\ M$, and $K_b = K_w/K_a = 10^{-14}/10^{-4.75} = 10^{-9.25}\ M$ (which is much less than C_{A^-}), so that:

$$(OH^-)^2 = 10^{-9.25} \times 10^{-1.0} = 10^{-10.25}$$

or

$$(OH^-) = 10^{-5.12}\ M$$

Since $(H^+) = K_w(OH^-)$ then $pH = 14 - 5.12 = 8.88$. Thus, a 0.1 M sodium acetate solution has a pH close to 9, a fairly alkaline value.

11.3 CONJUGATE ACIDS AND BASES; BUFFERS

A weak acid, HA, and its anion, A^-, constitute a *conjugate acid-base pair.* The two are the same molecular entity, differing only by one proton. The actual charge is unimportant in this context. The base could equally well be neutral, and then its conjugate acid would have a positive charge. For example, the ammonium ion and ammonia, NH_4^+ and NH_3, constitute a conjugate acid-base pair.

An important characteristic of a conjugate acid-base pair is that a mixture of the two partners has a pH that is close to the negative logarithm of the acidity constant, called the pK_a. This can be seen by rearranging the equilibrium expression, equation (11.12):

$$(H^+) = K_a(HA)/(A^-)$$

or

$$pH = pK_a - \log(HA)/(A^-) \qquad (11.18)$$

If $(HA) = (A^-)$, then the pH is exactly pK_a. Moreover, the pH will be close to the pK_a, as long as $(HA)/(A^-)$ is not far from unity. Even if this ratio reaches a value of 10 or 0.1, the pH will deviate from pK_a by only one unit. Thus the pH is *buffered* against large changes. A mixture of HA and A^- constitutes a *buffer* solution, because it resists large pH changes when small amounts of other acids or bases are added to the solution.

The buffering effect is best illustrated by a *titration* curve of the acid. Figure 11.6 shows a titration curve for a 0.1 M acetic acid solution, to which successive amounts of a strong base (e.g., NaOH) are added. At the start of the titration the pH is close to 3 (see Worked Problem 11.1), while at the endpoint, when all of the HAc has been converted to NaAc, the pH is close to 9 (see Worked Problem 11.2). At the halfway point, $(HAc) = (Ac^-)$ and $pH = pK_a = 4.74$. From the 10 percent to the 90 percent points of the titration, the pH varies only by minus and plus one unit from this value. But as the endpoint is approached, the *buffer capacity* of the HAc/Ac$^-$ mixture is used up, and the pH shoots

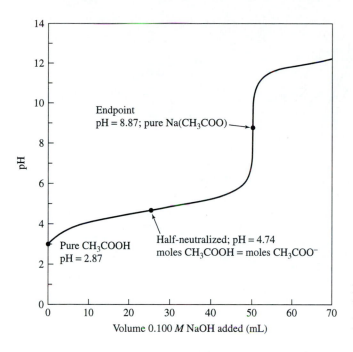

Endpoint
pH = 8.87; pure Na(CH$_3$COO)

Half-neutralized; pH = 4.74
moles CH$_3$COOH = moles CH$_3$COO$^-$

Pure CH$_3$COOH
pH = 2.87

Volume 0.100 M NaOH added (mL)

Figure 11.6   The change in pH during the titration of a weak acid with a strong base; 50.0 mL of 0.100 M acetic acid is titrated with 0.100 M NaOH.

up to 9 and beyond. With only a 10 percent excess of the base titrant [(OH$^-$) = 10^{-2} M], the pH reaches 12.

11.4   WATER IN THE ATMOSPHERE: ACID RAIN

Although pure water is neutral, and has a pH of 7.0, rainwater is naturally acidic because it is in equilibrium with carbon dioxide. When dissolved in water, carbon dioxide forms carbonic acid, a weak acid:

$$CO_2 + H_2O = H_2CO_3 \tag{11.19}$$

We can gauge the amount of carbonic acid from the equilibrium constant for reaction (11.19):

$$K_s = (H_2CO_3)/P_{CO_2} = 10^{-1.5} \ M/\text{atm} \tag{11.20}$$

where P_{CO_2} is the partial pressure of CO$_2$, in atmospheres.*

---

*Actually, dissolved CO$_2$ is mostly unhydrated; only a small fraction exists as H$_2$CO$_3$ molecules:

$$CO_2(aq) + H_2O(l) = H_2CO_3(aq) \qquad K_h = (H_2CO_3)/(CO_2) = 10^{-2.81}$$

The total concentration of dissolved CO$_2$ is:

$$(CO_2)_T = (CO_2) + (H_2CO_3) = (H_2CO_3)(1 + K_h^{-1}) = 10^{2.81}(H_2CO_3)$$

Most equilibrium measurements are made with respect to total dissolved CO$_2$, and (H$_2$CO$_3$) is to be understood as (CO$_2$)$_T$.

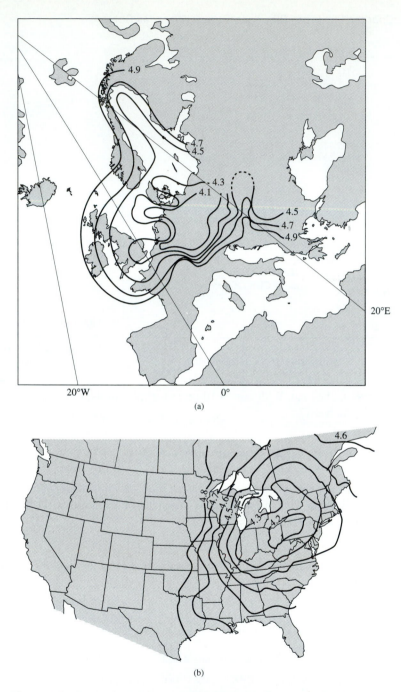

Figure 11.7 Contour lines of long-term, volume-weighted averages of pH in precipitation in (a) Europe, 1978–1982, and (b) eastern North America, 1980–1984. *Sources:* EMEP/CCC (1984). *Summary Report,* Report 2/84 (Lillestrøm, Norway: Norwegian Institute for Air Research); National Acid Precipitation Assessment Program (NAPAP) (1987). *NAPAP Interim Assessment* (Washington, DC: U.S. Government Printing Office).

Since the atmospheric concentration of CO_2 is currently 370 ppm, $P_{CO_2} = 370 \times 10^{-6}$ or $10^{-3.4}$ atm (at sea level), and therefore $(H_2CO_3) = K_h P_{CO_2} = 10^{-1.5} \times 10^{-3.4} = 10^{-4.9}$ M.

Rainwater is thus a dilute solution of carbonic acid. The acid dissociates partially to hydrogen and bicarbonate ions:

$$H_2CO_3 = H^+ + HCO_3^- \qquad K_a = 10^{-6.40} \ M \qquad (11.21)$$

From K_a we can calculate (H^+), as in Worked Problem 11.1

$$(H^+)^2 = K_a(H_2CO_3) = 10^{-6.4} \times 10^{-4.9} = 10^{-11.3}$$

$$(H^+) = 10^{-5.7} \quad \text{and} \quad pH = 5.7$$

Thus, atmospheric CO_2 depresses the pH of rainwater by 1.3 units from neutrality. Even in the absence of anthropogenic emissions, rainwater is naturally acidic (although it can occasionally be neutral or alkaline as a result of contact with alkaline minerals in wind-blown dust, or with gaseous ammonia from soils or from industrial emissions).

To the acidifying effect of carbon dioxide must be added the contributions from other acidic constituents of the atmosphere, particularly HNO_3 and H_2SO_4. These acids can both be formed naturally: HNO_3 derives from NO produced in lightning and forest fires, while H_2SO_4 derives from volcanoes and from biogenic sulfur compounds. At the natural background concentrations, these acids rarely influence rainwater pH significantly.

In polluted areas, however, the concentrations of these acids can be much higher and can reduce the pH of rainwater substantially over extended regions, producing what is known as *acid rain*. It is not uncommon in polluted areas to find the pH of rainwater in the pH 5–3.5 range. Some fogs, which can be in contact with pollutants over many hours, have had pH readings as low as 2.0, a degree of acidity equivalent to a 0.01 M solution of a strong acid. Moreover, the acid rain can fall quite far from the sources of pollution, due to long-range atmospheric transport. In particular, acid rain is a pressing problem for areas downwind of coal-fired power plants, whose tall smokestacks ameliorate local pollution by lofting SO_2 and NO emissions high into the air. Thus, power plants in western and central Europe affect the rain falling on Scandinavia, and power plants in the U.S. midwestern industrial belt similarly affect rain falling on northeastern regions of the United States and on southeastern Canada (see Figure 11.7a,b), although emissions of SO_2 have been reduced in the 1990s (see Figure 9.4, p. 219). While acid rain is a phenomenon in the hydrosphere, it depends upon atmospheric conditions—the extent of acidic emissions and the prevailing weather patterns.

FUNDAMENTALS 11.3: POLYPROTIC ACIDS

Carbonic acid actually has two ionizable protons, but they dissociate successively, and with very different K_a values:

$$H_2CO_3 = H^+ + HCO_3^- \qquad K_{a1} = 10^{-6.40} \qquad (11.22)$$

and

$$HCO_3^- = H^+ + CO_3^{2-} \qquad K_{a2} = 10^{-10.33} \qquad (11.23)$$

Because the bicarbonate ion is negatively charged, it holds its proton more tenaciously than does carbonic acid, and the second dissociation reaction makes no significant contribution to the H^+ concentration of a carbonic acid solution, or to that of a H_2CO_3/HCO_3^- buffer. However, we can neutralize both protons on H_2CO_3 with a strong base, or equivalently start with a solution of a carbonate salt, and then the basic properties of CO_3^{2-} come into play:

$$CO_3^{2-} + H_2O = HCO_3^- + OH^-$$

$$K_b = K_w/K_a = 10^{-14}/10^{-10.33} = 10^{-3.67}$$

$$(11.24)$$

For a solution which is 0.1 M in a carbonate salt (just as in Worked Problem 11.2 for acetate, see p. 276):

$$(OH^-)^2 = K_b(CO_3^{2-})$$

$$= 10^{-3.67} \times 10^{-1} = 10^{-4.67}$$

and

$$(OH^-) = 10^{-2.34}$$

giving pH = 11.66, a highly alkaline value.

Also, a buffer having equimolar HCO_3^- and CO_3^{2-} has

$$pH = pK_a - \log(HCO_3^-)/(CO_3^{2-})$$

$$= 10.33$$

However, if only one of the two H_2CO_3 protons is neutralized, or, equivalently, if we start with a solution of a bicarbonate salt, then things are slightly more complicated, since HCO_3^- is both a conjugate base (of H_2CO_3) and a conjugate acid (of CO_3^{2-}). To the extent that HCO_3^- dissociates a proton [via reaction (11.23)], the proton is taken up by another HCO_3^- ion [via the reverse of reaction (11.22)]. Consequently, the main acid-base reaction in a bicarbonate solution is:

$$2HCO_3^- = H_2CO_3 + CO_3^{2-} \quad (11.25)$$

This reaction is obtained by subtracting reaction (11.22) from reaction (11.23), and consequently the equilibrium constant is given by:

$$K_{a2}/K_{a1} = (H_2CO_3)(CO_3^{2-})/(HCO_3^-)^2$$

From equation (11.25), $(H_2CO_3) = (CO_3^{2-})$ for a solution containing only bicarbonate initially. Then:

$$K_{a2}/K_{a1} = (H_2CO_3)^2/(HCO_3^-)^2$$

or

$$(HCO_3^-)/(H_2CO_3) = (K_{a1}/K_{a2})^{1/2}$$

To calculate the pH, we can insert this ratio into the equilibrium expression for reaction (11.22):

$$(H^+) = K_{a1}(HCO_3^-)/(H_2CO_3)$$

$$= K_{a1}(K_{a2}/K_{a1})^{1/2} = (K_{a1}K_{a2})^{1/2}$$

or

$$pH = (pK_{a2} + pK_{a1})/2$$

$$= (10.33 + 6.40)/2 = 8.36$$

$$(11.26)$$

Equation (11.26) can be generalized to any diprotic acid: the pH of a solution of the monoprotic form is just the geometric mean of the two pK_a values.

There can also be more than two ionizable protons. For example, phosphoric acid has three:

$$H_3PO_4 = H^+ + H_2PO_4^- \quad pK_{a1} = 2.17$$

$$(11.27)$$

$$H_2PO_4^- = H^+ + HPO_4^{2-} \quad pK_{a2} = 7.31$$

$$(11.28)$$

$$HPO_4^{2-} = H^+ + PO_4^{3-} \quad pK_{a3} = 12.36$$

$$(11.29)$$

As in the case of carbonic acid, the pH of a solution of H_3PO_4 or of PO_4^{3-} can be calculated just as it would be for a monoprotic acid or base, while for solutions of the diprotic and monoprotic forms, the pH is the geometric mean of the bracketing pK_a values, i.e., for $H_2PO_4^-$:

$$pH = (pK_{a1} + pK_{a2})/2 = 4.74$$

while for HPO_4^{2-}:

$$pH = (pK_{a2} + pK_{a3})/2 = 9.83$$

The phosphoric acid system has three buffer zones, corresponding to the three pK_a values, i.e., solutions having comparable concentrations of H_3PO_4 and $H_2PO_4^-$, or of $H_2PO_4^-$ and HPO_4^{2-}, or of HPO_4^{2-} and PO_4^{3-}.

WORKED PROBLEM 11.3: Phosphate Protonation

Q. *Given the pK_a values in Fundamentals 11.3, what is the dominant form of the phosphate ion in a) a lake with pH = 5.0, and b) seawater with pH = 8.0, and what is the ratio of this form to the next most abundant form?*

A. The lake pH is higher than pK_{a1} but lower than pK_{a2}, so most of the phosphate will be present as $H_2PO_4^-$. For the seawater, HPO_4^{2-} dominates, because the pH is between pK_{a2} and pK_{a3}. Since the lake pH is closer to pK_{a2} than pK_{a1}, the next most abundant form is HPO_4^{2-},

while for the seawater, $H_2PO_4^-$ is the next most abundant form, since the pH is closer to pK_{a2} than to pK_{a3}. The ratios can be obtained from the equilibrium expression for the second ionization, (11.28):

$$(HPO_4^{2-})/(H_2PO_4^-) = K_{a2}/(H^+)$$

At pH = 5.0, this ratio is $10^{-7.31}/10^{-5.0} = 10^{-2.31} = 0.0049$, while at pH = 8.0, the ratio is $10^{-7.31}/10^{-8.0} = 10^{0.69} = 4.9$ [or $(H_2PO_4^-)/(HPO_4^{2-}) = 0.20$].

CHAPTER 12

WATER AND THE LITHOSPHERE

12.1 EARTH AS ACID-BASE REACTOR

Earth is believed to have formed some 4.5 billion years ago from the coalescence of meteorites that circled the early sun. Heating from gravitational forces and nuclear decay melted the interior of the evolving planet, allowing the minerals to separate according to their density. The result is a molten core, mainly of iron, and a surrounding mantle, formed mainly of silicate-based rock, silicon and oxygen being by far the most abundant elements in the mantle. The heating of Earth led to plate tectonics, the thermally induced circulation of segments of the mantle, which leads to the spreading of crustal plates at mid-ocean ridges, and subduction and uplift of crustal material at the plate boundaries, which are found at continental margins.

Volatile compounds, such as H_2O, HCl, CO_2, SO_2, and N_2, were expelled from the interior in volcanic eruptions, forming the oceans and the atmosphere. This natural separation of volatile and non-volatile compounds was also a separation into acidic and basic compartments, because CO_2, SO_2, and especially HCl are acidic, while the minerals left behind tend to be basic. The basicity of the lithosphere arises principally from the alkali metals and alkaline earths, especially Na, K, Mg, and Ca, which are relatively common in Earth's crust (see Table 12.1). These elements form basic oxides, which are incorporated into the dominant silicate framework of the major mineral phases. In addition, calcium carbonate (sometimes containing magnesium, as well), or *limestone,* is abundant in Earth's crust, and carbonate is a basic anion (see Fundamentals 11.3).

The fundamental reason for this global acid-base separation is that the base-forming elements at the left side of the periodic table form exclusively ionic bonds, because of their

TABLE 12.1:   MAJOR ELEMENTS IN EARTH'S CRUST
AND THEIR ABUNDANCE IN SEAWATER

Element	Crustal average (percent)	Seawater (ppm)
O	46.6	(88%)
Si	27.7	3
Al	8.1	1×10^{-3}
Fe	5.0	3×10^{-3}
Ca	3.6	0.041
Mg	2.1	1.3×10^{3}
Na	2.8	1×10^{4}
K	2.4	3.9×10^{-2}
Ti	0.44	1×10^{-3}
H	0.14	(10%)
P	0.11	0.09
Mn	0.10	2×10^{-3}
F	0.06	1.3
Cl	0.01	1.9×10^{4}
Ba	0.04	0.02
Sr	0.04	8
S	0.03	0.09
C	0.02	28

relatively low effective nuclear charge; consequently their compounds are non-volatile ionic solids. However, the acid-forming elements toward the right of the periodic table have a stronger tendency to form covalent bonds, because of their higher effective nuclear charge. Consequently, they more readily produce isolated (volatile) molecules.

Earth's volatile acids react back with the non-volatile bases through the aqueous medium of the hydrosphere. Thus, the seas are salty because the outgassed HCl dissolved in the condensed H_2O and reacted with the most basic constituents of Earth's crust, mainly the sodium oxide equivalents in silicate minerals, in a gigantic neutralization reaction. Although this happened a long time ago, global neutralization continues to this day in the form of rock *weathering*. Earth's tectonic processes expose new crustal material to the atmosphere, where rainfall mediates erosion of the rock through neutralization reactions.

The principal acidic constituent of today's atmosphere is CO_2. It dissolves limestone by neutralizing the carbonate ions, converting limestone to soluble calcium bicarbonate:

$$CO_2 + H_2O + CaCO_3 = Ca^{2+} + 2HCO_3^- \qquad (12.1)$$

The acid-base reactions of silicate rocks are harder to capture in chemical equations because of the complexity of silicate chemistry. Silicon dioxide, or *silica,* is a polymeric solid with a three-dimensional network of silicon atoms bound tetrahedrally to four oxygen atoms, each of which is bound in turn to two silicon atoms (see Figure 12.1). In silicate minerals, this network is rearranged to accommodate other metal oxides. When these oxides are neutralized through weathering, the network rearranges further, to produce a

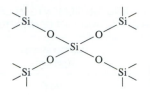

Figure 12.1 Polymeric structure of silicon dioxide.

secondary mineral. For example, weathering of the mineral *feldspar,* $NaAlSi_3O_8$, can be represented as

$$2NaAlSi_3O_8 + 2CO_2 + 11H_2O = 2Na^+ + 2HCO_3^- + 4Si(OH)_4 + Al_2Si_2O_5(OH)_4$$

$$(12.2)$$

$Al_2Si_2O_5(OH)_4$ is the formula for the secondary mineral *kaolinite,* a member of the clay mineral class. Although it has a specific structure (see p. 294), it can formally be considered as being made up of a mole of Al_2O_3, two moles of SiO_2, and two moles of H_2O. Likewise, a mole of feldspar can be considered as a mole of Al_2O_3, three moles of SiO_2, and one mole of Na_2O. The acid-base part of the reaction involves the neutralization of Na_2O by H_2CO_3 to form two moles of sodium bicarbonate. But when this happens, the feldspar is converted to kaolinate, with the release of two moles of SiO_2 in the hydrated form, $Si(OH)_4$. This form, *silicic acid,* exists as free molecules in very dilute solutions. As rain falls on feldspar, the sodium bicarbonate and silicic acid gradually leach away, leaving kaolinite behind. This is a very slow process, but the accumulation of kaolinite, and of other clays from silicate minerals, is a key to the formation of soils.

The weathering of limestone [reaction (12.1)] is much faster than the weathering of silicate minerals, since no rearrangement of the crystal lattice is required. Both types of weathering are greatly accelerated if the rain contains strong acids, such as sulfuric or nitric acids. In this case, protons work directly on the minerals, instead of through the weakly acidic H_2CO_3 molecules. This is why statues and stone buildings erode rapidly when exposed to acid pollution.

12.2 ORGANIC AND INORGANIC CARBON CYCLES

In Parts I and II, we discussed the carbon cycle (see pp. 19–21, 170–171), which results from the biological processes of photosynthesis and respiration, and the disturbance of this cycle by the burning of fossil fuels. In reality, the cycle of photosynthesis and respiration is imbedded in a much more complex biogeochemical cycle, involving sedimentation and burial, and the tectonic processes of Earth's crust.

This larger cycle (see Figure 12.2) is actually two cycles, linked by the CO_2 molecule. One cycle is organic, and involves the photosynthesis/respiration and combustion balance, and also the burial of reduced carbon compounds, together with their reoxidation when the rock bearing them is exposed to the atmosphere through geological uplift. The other cycle is inorganic, and involves the weathering of silicate rocks, the precipitation of calcium carbonate in the ocean, and finally the conversion of the carbonate to calcium

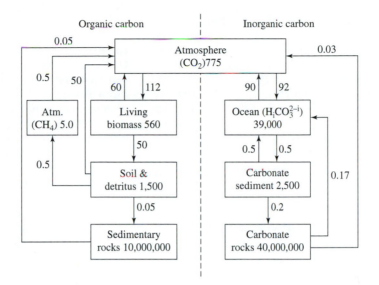

Figure 12.2   Reservoirs (boxes) and annual flows (arrows) of carbon (in units of 10^{15} g) through the organic (left) and inorganic (right) cycles. Outputs from the atmosphere to living biomass and the ocean are roughly 2×10^{15} g larger than inputs to the atmosphere from the living biomass and the ocean. This is attributed to net accumulations of carbon in these reservoirs owing to anthropogenic emissions of CO_2 from the burning of fossil fuels (see Figure 6.18, and ensuing discussion, p. 171).

silicates plus CO_2 through tectonic activity. The geological phases of these cycles operate very slowly, on time scales of millions of years, but they ultimately control the level of CO_2 in the atmosphere.

FUNDAMENTALS 12.1: RESERVOIRS, FLOWS, AND RESIDENCE TIMES

It is often helpful to represent environmental processes in terms of *reservoirs* and *flows*, as Figure 12.2 does for the carbon cycle. A reservoir is a region of the environment that holds a significant amount of the material under consideration, and is physically or chemically separated from other reservoirs. These amounts can be estimated from the concentration of the material, obtained by sampling, multiplied by the computed size of the reservoir.

To understand the dynamics of the system, one needs to understand how much material is transferred from one reservoir to another per unit of time, i.e., the flow. These numbers are generally much harder to determine, and a great deal of ingenuity has gone into making flow estimates.

If the flows and reservoirs are known, one can calculate *residence times*, assuming that the system is in a steady state, i.e., that the amounts do not change with time (or at least not rapidly). A steady state implies that inflows and outflows are in balance for each reservoir. The residence time is the average length of time the material spends in a reservoir, and is just the ratio of the reservoir size to the inflow or outflow rate. For example, the atmosphere contains 775×10^{15} g of C as CO_2, and delivers 112×10^{15} g/yr to green plants (nearly all of which is returned through the various re-oxidation routes, at steady state). Considering only the organic side of the cycle, the residence time for atmospheric CO_2 is $775/112 = 6.9$ yr; in other words, each CO_2 molecule cycles through the biosphere once every 6.9 years, on average. On the other hand, the organic carbon in sedimentary rocks turns over much more slowly. Since this carbon is estimated to accumulate (and erode) at a rate of 0.05×10^{15} g/yr, and the total size of the reservoir is estimated to be 10 million $\times 10^{15}$ g, the residence time is 200 million years.

---

a. The carbonate control.   The chemistry of the organic carbon cycle was laid out in Part I. Green plants harness solar photons and store chemical energy by reducing CO_2 to carbohydrates via photosynthesis:

$$CO_2 + H_2O = CH_2O + O_2 \tag{12.3}$$

Energy is released through the reverse reaction, during respiration or combustion. The annual exchange of carbon between the living biosphere and the atmosphere is large, 110×10^{15} g (see Figure 12.2), amounting to flows that are about one-fifth and one-seventh of their respective reservoirs. [The regular seasonal variations in this flow can be seen in the saw-tooth pattern of the atmospheric CO_2 record (see Figure 6.17, p. 170)]. About half of the return flow of C to the atmosphere from the organic carbon cycle goes by way of decomposition in soils, a reservoir of organic carbon that is nearly three times the size of the living biosphere. (A small part of this latter flow occurs by way of methane release to the atmosphere and subsequent oxidation to CO_2.) A tiny fraction of the organic carbon, 0.05×10^{15} g/yr, is buried in sedimentary rock, and is eventually oxidized back to CO_2 when the rock is recycled in the crust and exposed to the atmosphere. Although the burial is low, the amount of reduced C accumulated in rock over the eons is enormous, 10^{22} g; the turnover time (see Fundamentals 12.1) is 200 million years. It is this part of the cycle that we are accelerating by burning fossil fuels. Even though fossil fuel deposits account for less than 0.1 percent of buried organic carbon (see Figure 2.1, p. 20), the rate of oxidation is so much faster than the natural rate that the CO_2 reservoir size is being significantly increased.

The chemistry of the inorganic carbon cycle is determined by the solubility of calcium carbonate. The pH of the ocean is about 8, reflecting the basic character of the minerals with which it is in contact. CO_2 dissolves readily at this pH, the carbonic acid being converted to bicarbonate. The ocean is therefore a vast store of inorganic carbon ($39,000 \times 10^{15}$ g; see Figure 12.2).

A small fraction of this carbon is present as CO_3^{2-}. [Since the pK_a of HCO_3^- is 10.33, the $(CO_3^{2-})/(HCO_3^-)$ ratio at pH 8 is $10^{-2.33}$, or 0.005, see Fundamentals 11.3]. However, its concentration is sufficient that when Ca^{2+} enters the ocean as a result of the weathering of terrestrial rocks, calcium carbonate precipitates (mostly as shells of sea creatures). As a result, more carbonate is formed via the disproportionation of bicarbonate ion:

$$Ca^{2+} + CO_3^{2-} = CaCO_3$$

$$2HCO_3^- = CO_3^{2-} + H_2CO_3$$

The overall process is:

$$Ca^{2+} + 2HCO_3^- = CaCO_3 + H_2CO_3 \qquad (12.4)$$

The precipitation of each mole of calcium carbonate produces a mole of H_2CO_3, which outgasses CO_2 to the atmospheric reservoir. However, if the source of the calcium (or magnesium) is weathering of terrestrial carbonate mineral, then reaction (12.4) has been run in reverse on land, and the net effect on global CO_2 is zero. Weathering of terrestrial carbonates consumes the same amount of CO_2 as is released by precipitation of carbonates in the ocean.

The situation is quite different when silicate minerals are weathered, because silicate acts as a base, but contributes no bicarbonate to the pool. Silicate weathering reactions are complex [see, e.g., the feldspar weathering reaction (12.2), p.284], but the essence of the acid/base process is that the equivalent of one SiO_3^{2-} unit consumes two protons and is converted to $Si(OH)_4$:

$$SiO_3^{2-} + 2H^+ + H_2O = Si(OH)_4$$

If the protons are provided by CO_2, and if the silicate is the source of the Ca^{2+} ions that later precipitate carbonate in the ocean, then the silicate weathering contribution to the carbon cycle can be represented as:

$$CaSiO_3 + 2H_2CO_3 + H_2O = Si(OH)_4 + Ca^{2+} + 2HCO_3^- \qquad (12.5)$$

When this reaction is added to the carbonate precipitation reaction (12.4), the resultant is:

$$CaSiO_3 + H_2CO_3 + H_2O = Si(OH)_4 + CaCO_3 \qquad (12.6)$$

Thus if silicate, rather than carbonate, is the source of the Ca^{2+} that subsequently produces carbonate precipitation, then one mole of H_2CO_3 (or equivalently of CO_2) is consumed for every mole of carbonate precipitated. Consequently, silicate weathering draws down the atmospheric reservoir of CO_2.

In fact, weathering provides negative feedback on fluctuations in the CO_2 level, because of the operation of the greenhouse effect. The weathering rate increases with temperature and moisture. If the CO_2 rises, so does Earth's surface temperature and the flow of water through the geological cycle. The weathering rate increases, and so does the flow of CO_2 to the carbonate sediments, thereby diminishing the atmospheric CO_2 concentration. This mechanism acts to stabilize CO_2 levels over geologic time.

If reaction (12.6) operated only in the forward direction, then CO_2 would slowly but surely disappear from the atmosphere. However, the reaction is eventually reversed by the operation of plate tectonics. As Earth's crust is subducted at the plate boundaries, the increase in temperature and pressure reverses reaction (12.6); the volatile components, H_2O

and CO_2, escape through volcanoes, leaving the silicate rock behind. This process completes the inorganic carbon cycle.

As seen in Figure 12.2, most of Earth's carbon is trapped in sedimentary carbonate rock. This reservoir is 40×10^{21} g, four times larger than the reduced carbon reservoir in rocks. The rate of carbonate rock formation from sediment is 0.2×10^{15} g/yr, but most of this flow is reversed by redissolution of carbonate in the ocean. (Calcium carbonate dissolves in the deep ocean because of increasing levels of CO_2 that result from the decomposition of organic matter as it sinks toward the bottom, and from the increase in CO_2 solubility at low temperatures and high pressures.) The rate at which the carbonate rock is recycled to silicate and CO_2 through tectonic activity is only 0.03×10^{15} g/yr. Dividing this into the carbonate rock reservoir gives a residence time of 1.3 billion years. Thus, the inorganic carbon cycle operates over a very long time indeed.

b. Carbonate sequestration.   One of the ideas for sequestering the CO_2 currently emitted to the atmosphere by burning fossil fuels is to react the CO_2 with basic minerals; this idea amounts to speeding up the weathering process. If limestone were used as the reactant [reaction (12.1)], soluble bicarbonate would be the product, and would require long-term disposal as a liquid, possibly in the deep ocean. However, if calcium or magnesium silicates are used as the reactant, then the product would be insoluble $CaCO_3$ or $MgCO_3$ [reaction (12.6)], which could simply be buried in the ground.

Calcium and magnesium silicates are abundant in Earth's crust, and concentrated deposits of magnesium silicates are common, especially near the continental margins, where people are also concentrated. It would be technically feasible to collect CO_2 from power plants, or from reforming plants where fossil fuels are converted to H_2 and CO_2, and ship it to magnesium silicate mines. Here the CO_2 could be reacted with the silicate, and the resulting $MgCO_3$ and silica could be deposited back in the mine. The main problem is the slowness of mineral phase reactions. They can be speeded up by grinding the mineral (to increase surface area) and by elevating the pressure and temperature. Additional rate enhancement could be obtained by using a secondary volatile reactant that is a stronger acid than CO_2 (e.g., HCl); the secondary acid could be recovered in a separate heating step. However, the additional reaction loop would add complexity and a corrosive chemical to the process.

Even if the reaction rate problem can be overcome, it is not clear whether mineral carbonation is superior to direct injection of CO_2 into the oceans or deep aquifers (see Figure 2.12, p. 43). Solid carbonates offer the advantage of thermodynamic stability, whereas injected gases could leak out over time (the leakage rate is a major issue to be evaluated in various injection schemes). However, the costs of mineral carbonation may be higher than direct CO_2 injection, and public reaction to the required mining operations may be problematic.

c. CO_2, H_2O and Earth's planetary neighbors.   The inorganic carbon cycle is an instructive vantage from which to compare Earth's atmosphere with the very different atmospheres of its neighboring planets, Venus and Mars. Venus has a runaway greenhouse effect, as a result of a massive atmosphere (100 times Earth's atmospheric mass) that is 98 percent CO_2. Venus is closer to the sun and has twice as high a solar flux, but dense clouds reflect 75 percent of the sunlight, producing a radiative temperature, 229 K, which is actually lower than that of Earth (255 K, see p. 147). However, the greenhouse effect raises the surface temperature to 750 K. There is no liquid water, and water vapor amounts to only 0.1 percent (by weight). It is likely that the surface of Venus was never cool enough to allow water to liquify, and that most of the water outgassed from the planet's interior was eventually lost via ultraviolet photolysis and loss of the hydrogen to space. Without water there was no trapping of the CO_2 as carbonate, and the outgassed CO_2 stayed in the atmosphere.

In contrast to Venus, Mars is much colder than Earth. The solar flux is less than half that of Earth, and the radiative temperature is only 210 K. Moreover, there is little greenhouse warming (8 K), because although, like Venus, the atmosphere is mostly (95 percent) CO_2, it is very thin, less than 0.01 percent as massive as that of Venus. Part of Mars' CO_2 and essentially all of its water is locked in its ice caps. However, the Martian terrain bears clear evidence of liquid water in the past. Evidently Mars was tectonically active, and outgassed its volatile compounds, just as Earth did, but that tectonic activity ceased, perhaps a billion years ago. With a diameter only half that of Earth, Mars would have lost internal heat much faster; internal heat is the driving force for the circulation of the planet's crust. Without tectonic activity, outgassing would have stopped, and weathering would have drawn down the atmospheric CO_2, and its greenhouse warming, until surface temperatures no longer permitted water to remain liquid. Thus Earth had two key advantages over its neighbors: a sufficient distance from the sun to permit water to remain liquid at the surface, and sufficient size and internal heat to maintain tectonic activity. These attributes produced a planet that is neither too hot nor too cold (this has been called the "Goldilocks effect"), but just right for the evolution and maintenance of life.

12.3   WEATHERING AND SOLUBILIZATION MECHANISMS

a. Ionic solids and the solubility product.   Although water is generally an excellent solvent for ions, many ionic compounds are only sparingly soluble because the ion-water forces are outweighed by the forces that hold the ions together, particularly when the ions can arrange themselves in an energetically favorable way in a crystalline lattice. The energies stabilizing a lattice are generally maximal when the positive and negative ions have equal size and/or charge. For example, lithium fluoride is less soluble than lithium iodide because fluoride is closer in size to the small lithium ion than is iodide; in contrast, cesium iodide is less soluble than cesium fluoride because the cesium ion is closer in size to iodide. Similarly, both sodium carbonate and calcium nitrate are highly soluble because the positive and negative ions differ in charge, but calcium carbonate is sparingly soluble because both the cation and anion are doubly charged and therefore have a large lattice energy. Magnesium carbonate is more soluble than calcium carbonate, while sodium silicates are more soluble than potassium silicates, in both cases because the smaller cations fit the lattice less well, and interact more strongly with water. These differing solubilities account for magnesium and sodium being much more abundant in seawater than calcium and potassium, respectively, despite the similar abundance of these pairs of metals in Earth's crust (see Table 12.1, p. 283).

The solubility of a sparingly soluble salt is governed by the equilibrium constant for the dissolution reaction, called the solubility product, K_{sp}. For example, barium sulfate dissolves with a solubility product K_{sp} of 10^{-10}:

$$BaSO_4 \text{ (s)} = Ba^{2+v} \text{ (aq)} + SO_4^{2-} \text{ (aq)}$$

$$K_{sp} = (Ba^{2+})(SO_4^{2-}) = 10^{-10} \, M^2 \qquad (12.7)$$

[$BaSO_4$ does not appear in equation (12.7) because the effective concentration of a solid phase is constant.] Equation (12.7) is valid as long as there is solid barium sulfate in equilibrium with the solution. If the concentration product of Ba^{2+} and SO_4^{2-} exceeds the value of K_{sp}, barium sulfate will precipitate. It actually does precipitate on dead phytoplankton in the ocean, because their decomposition produces enough sulfate that the barium ions in the ocean are sufficient to exceed the solubility product.

WORKED PROBLEM 12.1: BaSO₄ Solubility

Q. *What is the solubility of barium sulfate in a) pure water, or b) one millimolar sodium sulfate in water?*

A. a) When pure water is equilibrated with barium sulfate, the concentration of barium matches the concentration of sulfate in solution:

$$(Ba^{2+}) = (SO_4^{2-}) = K_{sp}^{1/2} = 10^{-5} \, M$$

Thus the $BaSO_4$ solubility is $10^{-5} \, M$.

b) When either Ba^{2+} or SO_4^{2-} is present in excess, the solubility is the concentration of its partner, which decreases in inverse proportion to the excess concentration, via equation (12.7). Thus when $(SO_4^{2-}) = 1 \times 10^{-3} \, M$, the solubility is $(Ba^{2+}) = K_{sp}/(SO_4^{2-}) = 10^{-10}/10^{-3} = 10^{-7} \, M$.

---

b. Solubility and basicity.   If one of the ions of a sparingly soluble salt is an acid or a base, then the solubility increases, because the ion is partially converted to its basic or acid form, thereby pulling the solubility equilibrium toward more dissolution. For example, the K_{sp} of calcium carbonate is $10^{-8.34}$, and if the only reaction were

$$CaCO_3 \text{ (s)} = Ca^{2+}(aq) + CO_3^{2-} \text{ (aq)} \qquad (12.8)$$

the solubility would be $10^{-4.17} M$. However, the solubility equilibrium is shifted to the right because the basic carbonate ion reacts further with water (see Fundamentals 11.3, p. 279):

$$CO_3^{2-} + H_2O = HCO_3^- + OH^- \qquad K_b = 10^{-3.67} \qquad (12.9)$$

and is largely converted to bicarbonate.

The solubility is increased further when the solution is exposed to the atmosphere, because the carbonate ion then reacts with the acidic CO_2:

$$CO_3^{2-} + H_2CO_3 = 2HCO_3^- \qquad K = K_{a1}/K_{a2} = 10^{3.93} \qquad (12.10)$$

The equilibrium constant is large, so the equilibrium lies far to the right, and the $CaCO_3$ solubility is increased considerably. [Reaction (12.10) is obtained by subtracting the second dissociation reaction (11.23) from the first (11.22) (see Fundamentals 11.3), so the equilibrium constant is obtained by dividing the two acidity constants.]

An ominous consequence of this simple chemistry is that sea creatures may have difficulty making their calcium carbonate shells as the atmospheric CO_2 rises. Experiments in controlled environments have demonstrated that adding CO_2 to seawater slows the rate at which coral and reef-building algae secrete $CaCO_3$. Based on these experiments, the rise in the CO_2 level to the present 370 ppm, from the pre-industrial value of 280 ppm, is calculated to have slowed by 6–11 percent the $CaCO_3$ production by the coral and algae.

WORKED PROBLEM 12.2: Calculating the CaCO₃ Solubility

The combination of solubility and acid-base equilibria can make for fairly complicated calculations, and it is important to keep one's eye on the main reactions going on. Their *stoichiometry* (the number of moles of reactants and products consumed and produced) provide the key equalities.

In the case of calcium carbonate in contact with water (but not with the atmosphere), there are two reactions to consider:

$$CaCO_3\text{ (s)} = Ca^{2+}\text{ (aq)} + CO_3^{2-}\text{ (aq)}$$
$$K_{sp} = 10^{-8.34} \qquad (12.8)$$

and

$$CaCO_3\text{ (s)} + H_2O =$$
$$Ca^{2+}\text{ (aq)} + HCO_3^-\text{ (aq)} + OH^-$$
$$(12.11)$$

To obtain the equilibrium constant for the second reaction (which we will label $K_{12.11}$), we recognize that the equation is obtained by adding equations (12.8) and (12.9), so:

$$K_{12.11} = K_{sp}K_b = 10^{-12.01}$$
$$= (Ca^{2+})(HCO_3^-)(OH^-)$$

If (12.8) is the main reaction, then $(Ca^{2+}) = (CO_3^{2-}) = K_{sp}^{1/2} = 10^{-4.17}$, or $0.67 \times 10^{-4}\,M$. On the other hand, if (12.11) is the main reaction, then its stoichiometry requires that the concentrations of all three product ions be the same: $(Ca^{2+}) = (HCO_3^-) = (OH^-)$. Consequently, $(Ca^{2+})^3 = K_{12.11}$, giving $(Ca^{2+}) = 10^{-4.00}\,M$, which is about 50 percent higher than the value, obtained by considering reaction (12.8) alone.

To take both reactions into account, we recognize that the total Ca^{2+} concentration must be the sum of the concentrations of carbonate and bicarbonate:

$$(Ca^{2+}) = (CO_3^{2-}) + (HCO_3^-) \qquad (12.12)$$

(This is an example of a *mass balance*.) But also, from reaction (12.11), $(HCO_3^-) = (OH^-)$. Substituting this equality into the equilibrium expression for reaction (12.11) allows us to express (HCO_3^-) in terms of (Ca^{2+}):

$$(HCO_3^-)^2 = K_{12.11}/(Ca^{2+})$$

or

$$(HCO_3^-) = [K_{12.11}/(Ca^{2+})]^{1/2}$$

Likewise, (CO_3^{2-}) can be expressed in terms of (Ca^{2+}) using the equilibrium expression for reaction (12.8):

$$(CO_3^{2-}) = K_{sp}/(Ca^{2+})$$

Substituting these two expressions into the mass balance (12.12), and multiplying through by (Ca^{2+}) gives:

$$(Ca^{2+})^2 = K_{sp} + [K_{12.11}(Ca^{2+})]^{1/2} \quad (12.13)$$

The first term on the right-hand side of equation (12.13) represents the contribution from reaction (12.8), and the second term, the contribution from reaction (12.11). The equation can be solved (with difficulty) algebraically, or (more simply) by successive approximations. Recognizing that (Ca^{2+}) must be at least $10^{-4.0}\,M$ (the higher of our two approximate answers above), we can plug this into the right-hand side of equation (12.13), and take the square root to obtain a better estimate of (Ca^{2+}). Repeating this procedure a few times produces a convergent answer, 1.30×10^{-4} (or $10^{-3.89}$) M. Thus, the solubility of $CaCO_3$ is twice what it would be if carbonate were not a base [i.e., considering only reaction (12.8)].

Having solved for (Ca^{2+}), we can use this value to calculate the pH of the solution, using the equilibrium expression for reaction (12.11),

and remembering that $(HCO_3^-) = (OH^-)$:

$$K_{12.11} = (Ca^{2+})(OH^-)^2,$$

or

$$(OH^-) = K_{12.11}/(Ca^{2+})^{1/2}$$
$$= (10^{-12.01}/10^{-3.89})^{1/2} = 10^{-4.06}$$

Thus, pOH = 4.06 and pH = 9.94. Dissolution of $CaCO_3$ produces a basic solution.

WORKED PROBLEM 12.3: CaCO₃ Solubility and CO₂

The solubility calculation is actually simpler for a solution exposed to the atmosphere, because then the only reaction of importance is:

$$CaCO_3 + H_2CO_3 = Ca^{2+} + 2HCO_3^-$$
$$(12.14)$$

The equilibrium constant (which we will call $K_{12.14}$) is:

$$K_{12.14} = (Ca^{2+})(HCO_3^-)^2/(H_2CO_3)$$
$$= K_{sp}K_{a1}/K_{a2} = 10^{-4.41} \qquad (12.15)$$

[Equation (12.14) is obtained by adding equations (12.8) and (12.10)]. From the stoichiometry of reaction (12.14), we see that $(HCO_3^-) = 2(Ca^{2+})$, and consequently, $K_{12.14} = 4(Ca^{2+})^3/(H_2CO_3)$. (H_2CO_3) is fixed by the

atmospheric CO_2 concentration at $10^{-4.9}$ M (see section 11.4). Therefore:

$$(Ca^{2+}) = (10^{-4.41}\, 10^{-4.9}/4)^{1/3} = 10^{-3.30}\ M$$

The pH is obtained from the H_2CO_3 acidity constant, recognizing that $(HCO_3^-) = 2(Ca^{2+}) = 10^{-3.00}$ M; then:

$$pH = pK_{a1} - \log(H_2CO_3)/(HCO_3^-)$$
$$= 6.40 - \log(10^{-4.9}/10^{-3.00}) = 8.30$$

Comparing these values with those obtained in Worked Problem 12.2, we see that equilibration with atmospheric CO_2 increases the calcium carbonate solubility by a factor of four, and lowers the pH by 1.6 units.

---

c. Ion exchange; clays and humic substances.   Many solids have ions that are loosely held at fixed-charge sites. These can be exchanged with ions that are free in solution. The exchanging ions may be positively charged (cations) or negatively charged (anions):

$$R^-M^+ + M'^+ = R^-M'^+ + M^+ \quad \text{cation exchange} \qquad (12.16)$$

$$R^+X^- + X'^- = R^+X'^- + X^- \quad \text{anion exchange} \qquad (12.17)$$

In these reactions, R represents a fixed-charge site. It attracts ions of the opposite charge, the strength of attraction being proportional to the charge/radius ratio. (Multiply-charged ions can occupy more than one ion exchange site.) This is the underlying mechanism of *ion exchange resins,* organic polymers having numerous covalently attached charged groups.

Cationic resins can be strong or weak acid exchangers, depending on the proton affinity of the fixed-anionic sites. Strong acid resins generally contain sulfonate groups, $-OSO_3^-$. Like sulfuric acid, protonated sulfonate groups readily give up their proton when another cation is available for exchange. Weak acid resins generally contain carboxylate groups, whose pKa's, when protonated, are similar to acetic acid (~4.5). Because of the relatively high proton affinity of the carboxylate group, other cations are readily displaced by the protons. Likewise, anionic resins can be strong or weak base exchangers. Strong base resins have quaternary ammonium groups, $-N(CH_3)_3^+$, while weak base resins have protonated amine groups, $-NH_3^+$, which have high affinity for hydroxide. Cation and anion exchange resins are widely used in tandem to deionize water:

$$R^-H^+ + M^+ + X^- = R^-M^+ + H^+ + X^- \tag{12.18}$$
$$R^+OH^- + H^+ + X^- = R^+X^- + H_2O \tag{12.19}$$

The spent resins can be regenerated by washing them with strong acids and bases.

Because of their abundance in Earth's crust, silicates are the major component of soils (63 percent on average), and are responsible for most of the ion exchange capacity. Because of its closed three-dimensional network of oxygen atoms (see Figure 12.1), silica itself (as found naturally in quartz) has no ion exchange sites. However, substitution of metal ions in the silica framework leaves uncompensated negative charges, because the charge on the metal ions ($+3$, $+2$, or $+1$) is less than the charge on silicon ($+4$) (counting the oxide ions as -2). These uncompensated charges are balanced by mobile cations, giving the silicate mineral cation exchange character.

Of special importance are the clays, which result from the weathering of primary silicate minerals (see section 12.1), and are therefore abundant in soils. Clays contain sheets of polymerized silicate tetrahedra arranged in layers (see Figure 12.3). Three of the oxygen

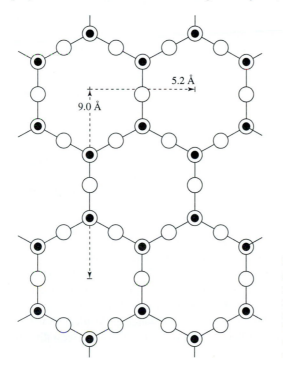

Figure 12.3   Layer structure of polymerized silicate tetrahedral. (The black circles represent silicon atoms and the open circles represent oxygen atoms. Each silicon atom is tetrahedrally bound to four oxygen atoms. The oxygen atoms shown superimposed on the silicons are directed upward and bound to a second parallel layer.)

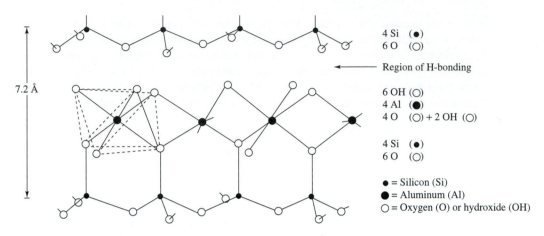

Figure 12.4   Structure of kaolinite, $Al_4Si_4O_{10}(OH)_8$. The plates contain phyllosilicate sheets bound to octahedral aluminum layers. (Dashed lines show six-fold coordinate positions in octahedral layer.) The distance between two successive plates is 7.2 Å.

atoms around each silicon atom are linked to neighboring silicon atoms in the sheet, while the fourth oxygen atom extends upward out of the sheet, bound to a second parallel layer. In the common clay minerals kaolinite and pyrophyllite, the fourth oxygen atom is bound to an aluminum ion, Al^{3+} (see Figure 12.4). Aluminum prefers octahedral coordination and is surrounded by six oxygen atoms. In kaolinite, two of these oxygen atoms are provided by neighboring silicate groups, while the remaining oxygens come from hydroxyl groups; the aluminum hydroxyl groups form hydrogen bonds with adjacent silicate oxygen atoms to hold the layers together. In pyrophyllite, the aluminum octahedra are sandwiched between two silicate sheets (see Figure 12.5); the next triple layer is held only weakly to the first one, since the facing silicate oxygen atoms lack protons with which to form hydrogen bonds. The interlayer space can be filled with water molecules, and pyrophyllite swells considerably in water. In many clays, some of the Al^{3+} ions are replaced by Fe^{3+} ions.

Other aluminosilicates have the kaolinite and pyrophyllite structures, but with some of the aluminum or silicon ions substituted by metal ions of lower charge. Thus, the common clay mineral montmorillonite has the pyrophyllite structure, but about one-sixth of the Al^{3+} ions are replaced with Mg^{2+}. Likewise, the illite clays share this structure, but Al^{3+} ions replace some of the Si^{4+} ions in the silicate sheet.

All of these substitutions of cations with lower charge produce excess negative charge, which is balanced by the adsorption of other cations, commonly Na^+, K^+, Mg^{2+}, or Ca^{2+}, in the medium between the aluminosilicate layers. These are the *base cations,* so called because their oxides are strong bases. They are readily exchanged for other cations in solution. The order of exchange depends upon the affinity of the cations for anionic sites on the clay compared to their attraction for water molecules. Generally, aluminum cations are the most difficult to exchange, and sodium cations the least difficult, with other

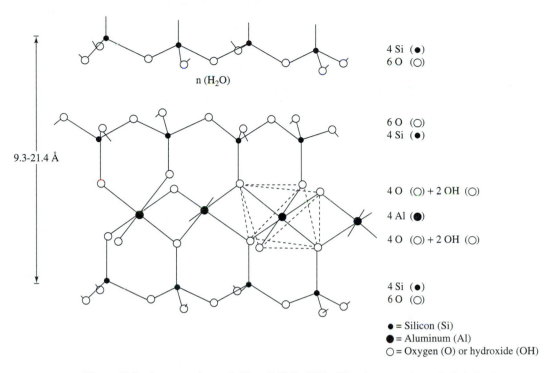

Figure 12.5   Structure of pyrophyllite, $Al_2Si_4O_{10}(OH)_2$. The plates contain octahedral aluminum layers sandwiched between two phyllosilicate sheets. The distance between the plates can vary up to 21 Å., depending on the amount of water present between plates.

ions between, in the order $Al^{3+} > H^+ > Ca^{2+} > Mg^{2+} > K^+ > NH_4^+ > Na^+$. Since protons are more tightly held than most other cations, if the soil solution is acidic, the protons will exchange with the adsorbed cations. The exchange of protons for other ions increases both the solution's pH and its base cation concentration. Thus, like limestone, clays neutralize the acids in soil water while increasing the concentration of base cations.

Soil also contains organic matter, the humus, consisting of the remains of decomposed vegetation. The remains that are most resistant to degradation, and therefore most abundant in soils, are complex polymeric materials, which have a high content of aromatic groups (as does lignin, see p. 24). Partial oxidation in the soil introduces many carboxylate and phenolic OH groups (see Figure 12.6a,b for schematic structures). Collectively these are called *humic substances,* and are sometimes divided into *humin, humic acid,* and *fulvic acid,* based on their response to extraction by strong base and subsequent acidification. Humin is the non-extractable fraction, while humic and fulvic acid both dissolve in strong base. Humic acid precipitates upon subsequent acidification, while fulvic acid remains in solution. In the soil, the numerous carboxylate groups provide cation exchange sites, complementing those of the clays.

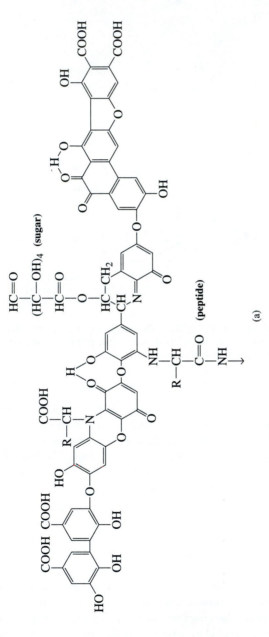

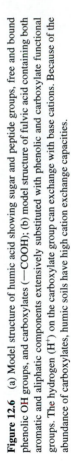

Figure 12.6 (a) Model structure of humic acid showing sugar and peptide groups, free and bound phenolic OH groups, and carboxylates (—COOH); (b) model structure of fulvic acid containing both aromatic and aliphatic components extensively substituted with phenolic and carboxylate functional groups. The hydrogen (H^+) on the carboxylate group can exchange with base cations. Because of the abundance of carboxylates, humic soils have high cation exchange capacities.

12.4   EFFECTS OF ACIDIFICATION

a. Soil neutralization.   Figure 12.7 diagrams the acid-base reactions that occur after rain falls to the ground. Initially the pH declines, because topsoil contains large amounts of CO_2 due to bacterial decomposition of the organic matter, up to 100 times the concentration of CO_2 in the atmosphere. In addition, plants exude a variety of organic acids, and the decay of plant matter produces other acids en route to the eventual conversion of the organic molecules to CO_2. pH values in the topsoil are frequently below 5.

As the water percolates into the mineral layers, neutralization reactions come into play. In soils containing limestone (*calcerous* soils), neutralization via reaction (12.14) raises the pH toward 8.3 (see Worked Problem 12.3). The effectiveness of this reaction is

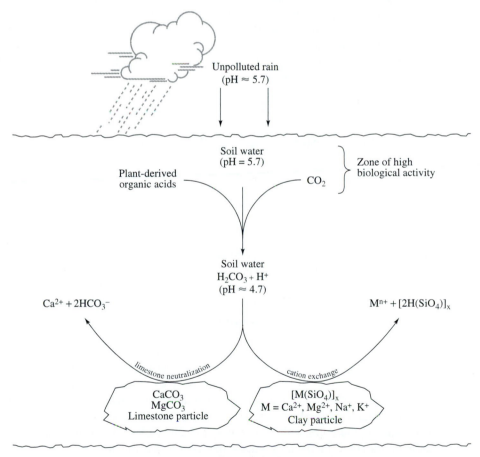

Figure 12.7   Percolation of rainwater through soil and neutralization by limestone and clays.

reflected in the widespread practice of *liming* (mixing crushed limestone into the soil) agricultural fields, as well as lawns and gardens, in order to raise the pH of acid soils.

When limestone is absent, neutralization occurs via exchange of protons for the base cations on ion exchange sites of clay particles and the humus. The number of exchangeable cation sites is the *cation exchange capacity* (*CEC*), measured in units of acid equivalents (*eq*) per square meter of soil. The fraction of the sites occupied by base cations, rather than by H^+, is called the *base saturation* (β). These ion exchange sites consist of anionic O atoms bound to the silicate lattice (or organic acid anions, in the case of humus) and are much less basic than is the carbonate ion. Consequently, the pH rises much less in non-calcerous than in calcerous soils; a typical pH in non-calcerous soils is about 5.5.

Because the ion exchange sites are on the particle surface, the CEC is limited, and is much less than the neutralizing capacity of calcerous soils. However, the CEC is slowly replenished by silicate weathering reactions, such as reaction (12.2) which release additional base cations, and provide new ion exchange sites. In some of the weathering reactions, Al^{3+} is also released from the aluminosilicate framework of clays. Because $Al(OH)_3$ has a very low solubility, it precipitates at pH values above about 4.2 and stays bound to the soil particles. If soil acidification exceeds the CEC, so that nearly all the base cations are replaced by protons (β declines toward zero), then the pH drops below 4.2 and aluminum is solubilized.

$$Al(OH)_3 + 3H^+ = Al^{3+} + 3H_2O \qquad (12.20)$$

b. Hardness and detergents.   The neutralization reactions that raise the pH of natural water as it percolates through the soil (see Figure 12.7) also bring into solution appreciable quantities of calcium and magnesium ions. Water with relatively high concentrations of Ca^{2+} and Mg^{2+} is considered "hard," while water with low concentrations is "soft." Soft water also has lower pH because the low Ca^{2+} and Mg^{2+} concentrations reflect poor availability of limestone or clays for neutralization.

The hard and soft appellations reflect the fact that the doubly charged Ca^{2+} and Mg^{2+} ions can precipitate detergents. Detergents are molecules with long hydrocarbon chains and ionic or polar head groups (see Figure 12.8a). When added to soft water, the detergent molecules aggregate into micelles (see Figure 12.8b), with the hydrocarbon tails pointing to the inside and the polar groups pointing out toward the water. The hydrophobic interior of the micelles solubilizes grease and dirt particles and removes them from clothing, dishes, or other items being washed. In soft water, the micelles float freely and are prevented from aggregating with one another by the mutual repulsions of the polar hydrophilic head groups.

This solubilization process is prevented by hard water. The polar head groups of most detergents are negatively charged; they interact with divalent cations such as Ca^{2+} and Mg^{2+} and precipitate. The micelles are then unavailable for solubilizing dirt; worse still, the precipitates themselves tend to be scummy and stick to the items being washed. Consequently, in areas with hard water, a clean wash requires the addition of agents— called *builders* in the laundry trade—that tie up polyvalent cations and thereby prevent them from precipitating the detergent.

Detergent products contain a variety of kinds of builders. One class comprises *chelating agents,* molecules with several donor groups that can bind the cations through

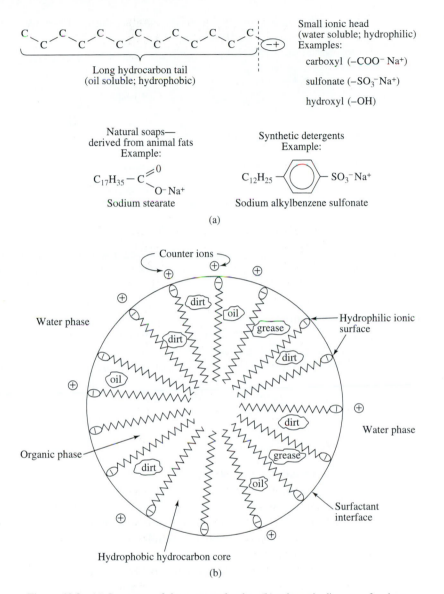

Figure 12.8   (a) Structures of detergent molecules; (b) schematic diagram of a detergent micelle.

multiple coordinate bonds. A particularly effective chelating agent is sodium tripolyphosphate (STP), which is shown bound to a Ca^{2+} ion in Figure 12.9a. This compound is relatively inexpensive and has the advantage that it rapidly breaks down in the environment to sodium phosphate, a naturally occurring mineral and nutrient for plants. However, STP-containing detergents release sodium phosphate into natural bodies of water and can

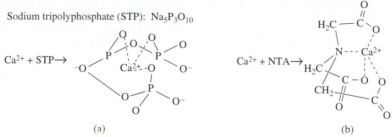

Sodium nitrilotriacetate (NTA): $N(C_2H_2O_2)_3Na_3$

Sodium tripolyphosphate (STP): $Na_5P_3O_{10}$

$Ca^{2+} + STP \rightarrow$

$Ca^{2+} + NTA \rightarrow$

(a)   (b)

Figure 12.9   Chelating agents: chemicals that tie up positive ions in solution so they can no longer react with detergents (a) sodium tripolyphosphate (STP); (b) sodium nitrilotriacetate (NTA).

stimulate plant growth, leading to eutrophication (see discussion beginning on p. 319). In response to warnings from environmentalists, phosphates have been banned in the U.S. for use in laundry detergents. They are still used, however, in automatic dishwashing detergents and detergents used by institutions and industry. They have not been banned in most other countries. In Europe, detergents are considered to be a small source of environmental phosphate, compared to runoff from farms and animal feedlots.

A variety of alternative chelating agents have been explored, such as sodium nitrilotriacetate, NTA (see Figure 12.9b). NTA is not widely used currently, partly because of cost, and partly because of concern that NTA does not break down as readily as STP does and therefore might mobilize metals other than calcium and magnesium. The most successful alternative builders are *zeolites,* synthetic aluminosilicate minerals that can trap Ca^{2+} or Mg^{2+} by ion exchange, releasing Na^+ ions in their place. Zeolites are used in place of STP in U.S. laundry detergents.

c. Acid deposition and watershed buffering.   As described in Part II, the atmosphere receives substantial inputs of SO_2 and NO from both natural and anthropogenic sources. These emissions are cleared from the air within a few days by oxidation reactions, and subsequently transfer to the soil, either directly by dry deposition in aerosols, or indirectly by wet deposition in rainfall. Such reactions are vital to the health of the biosphere because they cleanse the air of noxious fumes. If SO_2 and NO accumulated in the atmosphere as CO_2 does, the air would quickly become toxic.

However, the cleansing of the atmosphere transfers sulfuric and nitric acids to soils. Hence, soils can be described as *sinks* or *reservoirs* for atmospheric pollutants. Pollutants emitted to the air flow through the environment, mediated by a series of physical and chemical processes, as illustrated schematically in Figure 12.10. The first step is air transport and then deposition onto soil (1). Soils, in their capacity as *chemical filters,* may adsorb, neutralize, or otherwise retain and store the pollutant. When buffering capacities are diminished, the soil may release the pollutant to rivers and lakes (2a) or to the groundwater (2b). Eventually, pollutants are discharged to the oceans through stream (3a) and subsurface (3b) flow, and deposited in the ocean sediment (4), their ultimate repository.

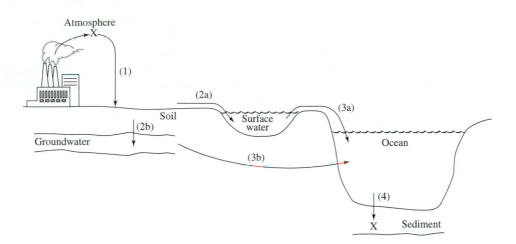

Figure 12.10 Flow of pollutant X from sources to sinks.

The effectiveness of the soil as a chemical filter of acidic inputs depends upon its buffering capacity and the rate of acid deposition. Although the buffering capacity of most soils is sufficient to neutralize naturally occurring acids, over time the capacity can be overwhelmed by high inputs of acid deposition. Figure 12.11 illustrates schematically

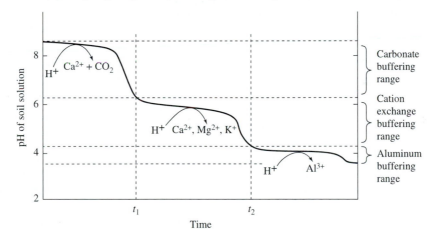

Figure 12.11 Scheme showing progression of decline in soil solution pH over time in response to atmospheric acid inputs. For a given soil, the time-scales over which the soil solution passes from one buffering range to the next depends on the intensity of acid deposition and the concentration of exchangeable cations. Arrows indicate the chemistry by which soils retain H^+ from the soil solution by exchanging it for cations. The period from t_1 to t_2 is the time over which soils lose 90–95 percent of their original exchangeable base cations. *Source:* W. M. Stigliani and R. W. Shaw (1990). Energy use and acid deposition: The view from Europe. *Annual Review of Energy* 15:201–216. Copyright © 1990 by Annual Reviews, (http://www.AnnualReview.org). Reprinted with permission, from the *Annual Review of Energy.*

the course of events as soil is continuously acidified. If limestone is present, the pH is initially maintained at around 8, through the dissolution of carbonate. When the limestone is dissolved away, then the pH drops to about 5.5 as acid displaces the base cations in the soil. When these are all displaced, the pH falls to around 4, where the $Al(OH)_3$ is gradually dissolved. Thus there are three distinct buffer regions, as soil is titrated by protons.

Figure 12.11 can also represent the fate of a watershed that is subject to continuing inputs of acid. The pH can stay fairly constant, at either the carbonate or the silicate buffer pH, for extended periods of time, and then drop rapidly when the buffer capacity is exceeded. It is hard to predict when this will happen. The length of time depends upon many factors, including the rate of deposition, the nature of the soil, the size of the watershed, and the flow characteristics of the lake or groundwater.

One case where the historical record of acidification has been well documented, however, is the watershed of Big Moose Lake in the Adirondack Mountains of New York state. This area has received some of the highest inputs of acid deposition in North America because it is downwind from western Pennsylvania and the Ohio Valley, historically the industrial heartland of the United States. Pollutants carried by the westerly winds are trapped in the mountains and deposited via wet and dry deposition.

The difficulties of recognizing acidification while it is occurring are illustrated in Figure 12.12, which shows the historical trends in Big Moose Lake water pH (dashed line), SO_2 emissions upwind from the lake (solid line), and the extinction of different fish species. The Adirondack watersheds lie on granitic rock, and are without limestone. The pH of the lake remained nearly constant at around 5.6, the clay buffer value, over the entire period from 1760 to 1950. Then, within the space of 30 years, from 1950 to 1980, the pH declined more than one whole pH unit to about 4.5. The decline in pH lagged behind the rise in SO_2 emissions by some 70 years, and the peak years of sulfur emissions preceded the decline in pH by 30 years. The deposition rate is estimated to have been about 2.5 grams of sulfur $m^{-2} yr^{-1}$ during the peak period. These quantities, deposited year after year, were large enough to deplete the capacity for base cation exchange in the watershed. Thus, beginning around 1950, atmospheric acid deposition moved through the buffer-depleted soils of the watershed and percolated into the lake with diminished neutralization (see problem 12, Part III). At that point, acid-sensitive fish species such as smallmouth bass, whitefish, and longnose sucker began to disappear, followed in the late 1960s by the more acid-resistant lake trout.

From the shape of the historical pH trend, we can see that Big Moose Lake was the subject of an inadvertent titration experiment conducted over four generations of industrial activity. The coal-driven industrialization of the Ohio Valley, upwind from the lake, supplied the acid inputs, mostly as sulfuric acid formed from SO_2 released during coal combustion. The soils of the watershed provided the supply of buffering chemicals. Thus, the watershed's natural buffering capacity delayed recognition of the deleterious effects of coal burning for about three generations. During this time, there was no direct evidence of how pollution was affecting the pH of lake water or fish mortality. As this example suggests, polluting activities may be far displaced in time from their environmental effects.

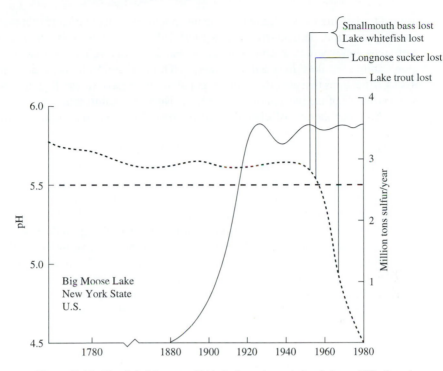

Figure 12.12   Trends in lakewater pH (dashed curve), upwind emissions of SO_2 from the U.S. industrial Midwest (solid curve), and fish extinctions for the period from 1760 to 1980. *Source:* W. M. Stigliani (1988). Changes in valued capacities of soils and sediments as indicators of nonlinear and time-delayed environmental effects. *Environmental Monitoring and Assessment* 10:245–307. Copyright © 1988. Reprinted with permission of Kluwer Academic Publishers.

d. Ecosystem effects of acid rain.   Although loss of fish in acidified lakes is a dramatic indicator, ecosystem effects of acid rain are much more widespread. Painstaking studies at Hubbard Brook Experimental Forest, in New Hampshire, have revealed that concentrations of calcium and magnesium have been lowered to half their historic values and that the overall growth of vegetation is at a standstill. Studies at many other sites in the northeastern states likewise show reductions in nutrient cation levels, and also the release of aluminum, which is sometimes found precipitated on the rootlets of trees, blocking nutrient uptake. In Vermont forests, acid fogs and rain have been found to leach calcium directly from spruce needles, leaving the trees susceptible to drought and insects. Similar effects are now being reported in the Southeast of the U.S., some 20 years after they appeared in the Northeast. The delay is attributed to the buffering effect of thicker southern soils, which are now becoming saturated with acid.

Over the past three decades, progressively tougher pollution rules have reduced U.S. emissions of SO_2 by about 40 percent, but ecologists estimate that an additional 80 percent reduction will be required to permit affected soils to regenerate the base cation levels needed for healthy trees. Meanwhile, the nitric acid component of acid rain is receiving increasing attention, particularly in the western U.S., where low-sulfur coals minimize SO_2 emissions, but where increasing population and traffic (as well as the manure from the region's animal feedlots) have led to increased NO_x in the atmosphere. Ancient spruces in the Rocky Mountains that are downwind of populous areas have shown high levels of nitrogen and low levels of magnesium in their needles. Nearby streams, which are usually nutrient-poor, show populations of diatoms shifting toward species that do better in nitrate-rich waters.

     Damage from acidification is magnified when soils are also polluted by toxic metals such as cadmium, copper, nickel, lead, and zinc. As cations, these metals compete with hydrogen and the base cations for cation exchange sites. At high pH, metal ions in well-buffered soils are generally retained at the exchange sites; their concentrations in soil solution are low. However, as the pH declines from 7 to 4, the leaching velocity at which an ion migrates through the soil may increase by an order of magnitude as shown by the example of cadmium (see Figure 12.13). Thus, at near neutral pH, the soil will accumulate heavy metals such as cadmium, only to release them as the soil acidifies. Once in the aqueous phase, cadmium ions are mobile and biologically active. They can be transported to lakes via surface or subsurface flow, transferred to groundwaters, or taken up by vegetation, with toxic effects. Al^{3+} is also toxic to plants and aquatic organisms. Some of the

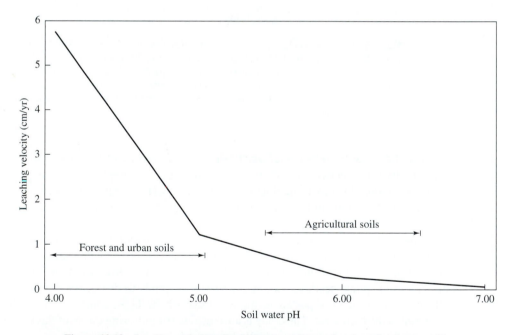

Figure 12.13   Leaching velocity of cadmium in soil as a function of soil water pH. *Source:* W. M. Stigliani and P. R. Jaffe (1992), private communication (Laxenburg, Austria: International Institute for Applied Systems Analysis).

deleterious effects of strong acidification are probably due to the Al^{3+} leached from clay particles in the soil.

e. Acid mine drainage.   A problem related to acid rain is acid mine drainage. Coal mines, especially those that have been abandoned, are known to release substantial quantities of sulfuric acid and iron hydroxide into local streams. The first step in the process is the oxidation of pyrite (FeS_2), which is common in underground coal seams:

$$FeS_2 + \tfrac{7}{2}O_2 + H_2O = Fe^{2+} + 2HSO_4^- \qquad (12.21)$$

This reaction is analogous to the first step in the generation of acid rain, in which sulfur is oxidized during coal combustion. In acid mine drainage, however, this step is mediated under aerobic conditions by the bacterium *thiobacillus ferrooxidans,* which oxidizes FeS_2 as an energy source much the same way that other aerobic bacteria oxidize organic carbon (CH_2O) in respiration reactions (see p. 307). The oxidation step occurs spontaneously at ambient temperatures once iron sulfide, which is stable in the absence of air, is exposed to the atmosphere. In the second step, the ferrous iron (Fe^{2+}) formed from reaction (12.21) combines with oxygen and water in the overall reactions:

$$Fe^{2+} + \tfrac{1}{4}O_2 + \tfrac{1}{2}H_2O = Fe^{3+} + OH^- \qquad (12.22)$$

$$Fe^{3+} + 3H_2O = Fe(OH)_3 + 3H^+ \qquad (12.23)$$

The sum of reactions (12.21), (12.22), and (12.23) yields the following reaction:

$$FeS_2 + \tfrac{15}{4}O_2 + \tfrac{7}{2}H_2O = Fe(OH)_3 + 2H^+ + 2HSO_4^- \qquad (12.24)$$

Thus, one mole of pyrite produces two moles of sulfuric acid and one mole of ferric hydroxide, which is removed from solution as a brown precipitate. The pH of streams receiving this drainage can be as low as 3.0.

These reactions continue long after coal mining operations have ceased. The resulting pollution problem can be quite severe locally; in coal mining areas, streams are often highly polluted by sulfuric acid. The problem is difficult to prevent because sealing up mines effectively is arduous and expensive. Since coal mining is a major activity on most continents, the problem of acid mine drainage has global implications.

f. Global acidification.   Thus far we have been discussing acidification on local and regional scales. Acid mine drainage is tied to particular watersheds; acid rain depends upon industrial activity and prevailing weather patterns. But acid deposition on a global level from industrial sources is of the same order of magnitude as deposition from natural sources (see Table 12.2; see problem 13, Part III). Acid deposition is not the whole story, because there are also natural and anthropogenic sources of alkaline chemicals in the atmosphere. These include ammonia (NH_3) and alkaline particles derived from ash, as well as windblown alkaline minerals. These chemicals have been crudely estimated to neutralize between 20 and 50 percent of the generated acidity. The atmosphere has acted

TABLE 12.2   ESTIMATED NATURAL AND ANTHROPOGENIC SOURCES OF GLOBAL ATOMOSPHERIC ACIDITY

Source	10^{12} moles of H^+ generated per year
NATURAL	
Unpolluted rainwater	1.0
Lightning*	1.4
Volcanoes[†]	1.3
Biogenic sulfur	4.1
TOTAL NATURAL	**7.8**
POLLUTION	
Coal combustion/metal smelting[†]	5.8
Combustion processes*	1.4
TOTAL FROM POLLUTION	**7.2**

*Refers to acidity generated from NO_x emissions.

[†]Refers to acidity generated from SO_2 emissions.

Source: Adapted from W. H. Schlesinger (1991). *Biogeochemistry, an Analysis of Global Change* (San Diego, California: Academic Press, Inc.).

as an acidic medium throughout geologic time, but natural sources of acidity, although of the same order of magnitude as anthropogenic sources, are spread evenly across the globe. Polluting sources are concentrated near industrial and urban centers, with levels of acidity exceeding 50 to 100 times the natural background. It is the excessive concentration of acidity in particular regions that causes problems for the biosphere.

CHAPTER 13

OXYGEN AND LIFE

We now turn to water as the medium that supports life. All organisms require water, and a large fraction of them make their home in rivers, lakes, and the oceans. Life started in the ocean, and occupied dry land only later. Moreover, biological processes have a profound influence on the chemistry of natural waters, and indeed of the entire globe. Were it not for the evolution of photosynthetic organisms, first in the ocean, and then on land, the atmosphere would be devoid of oxygen. The profound influence of oxygen on the chemistry of the atmosphere was considered at length in Part II. O_2 is also the dominant actor in the chemistry and biochemistry of the hydrosphere. The limited availability of O_2 in water sets the boundary between aerobic and anaerobic life, with crucial consequences for water quality and the health of ecosystems.

13.1 REDOX REACTIONS AND ENERGY

Life is powered by redox reactions, chemical processes in which electrons are transferred from one molecule to another, with the release of energy. Organisms have evolved machinery, made up of proteins and membranes, which channels this energy into the biochemical pathways that support vital functions.

In an aerobic environment, the most important biological redox process is respiration,

$$(CH_2O) + O_2 = CO_2 + H_2O \tag{13.1}$$

which we encountered previously as part of the global carbon cycle (see pp. 19–21). In this case carbohydrate molecules provide electrons for the reduction of dioxygen. All higher life forms obtain their energy via respiration. However, many other redox processes are utilized by bacteria. Indeed, bacteria have evolved to exploit just about any redox process that is available in nature. Anyplace where a supply of oxidizable molecules coexists with molecules capable of oxidizing them, it is a good bet that bacteria are present that can utilize the potential redox reaction. The oxidation of FeS_2 by *thiobacillus ferrooxidans* in the discussion of acid mine drainage (see p. 305) is a good example.

FUNDAMENTALS 13.1: OXIDATION LEVELS AND WATER

Many elements can exist in multiple oxidation states, depending on the number of electrons added to or removed from the valence shell of the atoms. In an aqueous world, the stability of these different oxidation states depends on the properties of water. Thus we are familiar with Na^+ and Mg^{2+} ions, because sodium and magnesium have one and two electrons, respectively, in their valence shells, which are easily removed when water molecules are available to stabilize the resulting ions (see Figure 11.5). All metals form positive ions in water, and in the case of transition metals, multiple oxidation states are available; for example, iron can exist in water as Fe^{3+} or Fe^{2+}.

Nonmetals, being electronegative elements, readily attain negative oxidation levels, depending on the number of electrons that their valence shells can accommodate. Thus the lowest oxidation levels attainable by F, O, N, and C are −I, −II, −III, and −IV; we use Roman numerals to denote the oxidation number, to distinguish them from the actual charge. Thus, although Cl^- ions exist as such in water, O^{2-} ions do not. Their proton affinities are high enough that they are completely converted to OH^- (or to H_2O, depending on the pH). The lowest oxidation levels for N and C are represented by NH_3 (or NH_4^+) and CH_4.

Positive oxidation levels are also accessible to the nonmetals because of the stabilization available through bonding to oxide ions. Thus C, N, S, and Cl are in their maximum oxidation states, +IV, +V, +VI, and +VII, when surrounded by oxide: CO_2 (or CO_3^{2-}), NO_3^{2-}, SO_4^{2-}, and ClO_4^-. The actual charges on the central atoms in these molecules are much less than +4, +5, +6, or +7, since electrons are shared in the polar but covalent bonds with the O atoms. Nevertheless, the oxidation state is crucial in determining the possibilities for redox chemistry. For example, eight electrons must be removed from N in order to convert NH_3 to NO_3^{2-}. In the case of the respiration reaction, (13.1), carbon in (CH_2O) is in the oxidation state 0 (the rules are that O counts for −2 and H counts for +1 in determining the "effective" charge, i.e., the oxidation state, of the remaining atoms); four electrons are transferred to O_2 in converting (CH_2O) to CO_2.

WORKED PROBLEM 13.1: Calculating the Oxidation State and Balancing Redox Equations

Q. *What is the oxidation state of N in the nitrite ion, NO_2^-?*

A. Since O counts as −2, and there is an overall −1 charge, N must have an effective charge of +3. The oxidation level is III.

Q. *Write a balanced chemical equation for the reduction of NO_2^- to NH_3 by H_2.*

A. $NO_2^- + 3H_2 + H^+ = NH_3 + 2H_2O$

First balance the number of electrons transferred from oxidant to reductant. Since N changes from III to −III, six electrons are transferred. H changes from 0 to I, so six H atoms, or three H_2 molecules, are required to receive the electrons. Since the reaction is in water, it is permissible to add H_2O, H^+, or OH^- to either side of the reaction, as needed. Seeing that nitrite had two O atoms, we balance these by adding two water molecules to the right-hand side. The total H count on the right-hand side is now seven, which we balance by adding one H^+ to the left-hand side. This also balances the charge.

a. Biological oxygen demand.   Wherever oxygen is present, respiration provides life-supporting redox energy, but in liquid water, oxygen can easily become depleted. The solubility of O_2 in water is only 9 mg/L (about 0.3 millimolar) at 20°C, and less at higher temperatures. The oxygen supply can be replenished by contact with the air, as in rapidly flowing streams. But in standing water or in waterlogged soils, the diffusion of oxygen from the atmosphere is slow relative to the speed of microbial metabolism, and the oxygen is used up.

Given the centrality of oxygen to metabolism, a parameter called *biological oxygen demand* (BOD) has been defined to measure the reducing power of water that contains organic carbon. BOD is the number of milligrams of O_2 required to carry out the oxidation of organic carbon in one liter of water. Values for various industrial wastes and municipal sewage are given in Table 13.1.

TABLE 13.1   TYPICAL BODs FOR VARIOUS PROCESSES

Type of discharge	BOD (mg O_2/liter wastewater)
Domestic sewage	165
All manufacturing	200
Chemicals and allied products	314
Paper	372
Food	747
Metals	13

WORKED PROBLEM 13.2: BOD

Q. *What is the BOD of water in which 10 mg of sugar (empirical formula CH_2O) is dissolved in a liter? How does this compare with the O_2 solubility at 20°C?*

A. Since each mole of CH_2O requires one mole of O_2 [equation (13.1)], we divide 10 mg by the molecular weight of CH_2O (30 g) to obtain the required number of moles of O_2, and then multiply by the molecular weight of O_2 (32) to obtain the number of mg:

$$BOD = 10 \times 32/30 = 10.7 \text{ mg/L}$$

This exceeds the O_2 solubility (9 mg/L) by about 20 percent.

TABLE 13.2   REDOX REACTIONS, PRODUCTS, AND CONSEQUENCES

Redox reaction	Reaction products/consequences
1. $O_2 + CH_2O \rightarrow CO_2 + H_2O$	The aerobic condition, characterized by the highest redox potential, occurs when there is an abundance of O_2, and the relative absence of organic matter owing to oxic decomposition by aerobic microorganisms. Two examples are the aerobic digestion of sewage wastes, and the decomposition of organic matter near the surface of well-aerated soils. The end products, CO_2 and water, are nontoxic.
2. $\frac{4}{5}NO_3^- + CH_2O + \frac{4}{5}H^+$ $\rightarrow CO_2 + \frac{2}{5}N_2 + \frac{7}{5}H_2O$	When molecular oxygen is depleted from the soil or water column, as would be the case, for example, in waterlogged soils and wetlands, available nitrate is the most efficient oxidant. Denitrifying bacteria consume nitrate and release N_2. N_2O, a greenhouse gas, is also released as a side-product. In agricultural soils, denitrification can lead to losses of nitrogen fertilizer amounting to as much as 20 percent of inputs. Denitrifying bacteria are also very active in heavily polluted rivers, or in stratified estuaries where organic matter accumulates. In some estuary systems, denitrification may significantly affect the transfer of nitrogen to the adjacent coastal waters and atmosphere.
3a. $2MnO_2 + CH_2O + 4H^+$ $\rightarrow 2Mn^{2+} + 3H_2O + CO_2$ **3b.** $4Fe(OH)_3 + CH_2O + 8H^+$ $\rightarrow 4Fe^{2+} + 11H_2O + CO_2$	In anaerobic environments where nitrates are in low concentration and manganese and ferric oxides are abundant, the metal oxides are a source of oxidant for microbial oxidation. This may be the case in natural soils, and in the sediments of lakes and rivers. The environmental significance of these metal oxides is that they serve a dual role. Not only are they a source of oxidants to microorganisms, they are also important for their capacity to bind toxic heavy metals, deleterious organic compounds, phosphates, and gases. When the metal oxides are reduced, they become water-soluble and lose their binding ability. This loss may result in the release of toxic materials.
4a. $\frac{1}{2}SO_4^{2-} + CH_2O + H^+$ $\rightarrow \frac{1}{2}H_2S + H_2O + CO_2$	Sulfidic conditions are brought about almost entirely by the bacterial reduction of sulfate to H_2S and HS^- accompanying organic matter decomposition. Sulfate reduction is very common in marine sediments because of the ubiquity of organic matter and the abundance of dissolved sulfate in seawater. In freshwater, such reactions are important in areas affected by acidic deposition in the form of sulfuric acid. H_2S is an extremely toxic gas. Sulfides are also important in scavenging heavy metals in bottom sediments.
4b. $MS_2 + \frac{7}{2}O_2 + H_2O$ $\rightarrow M^{2+} + 2SO_4^{2-} + 2H^+$	Conversion of a heavy-metal sulfide (MS_2) to sulfate may also occur when anaerobic sediments are exposed to the atmosphere, as in the case of the raising of dredge spoils. It may also occur when wetlands containing pyrites (FeS_2) are drained for agriculture, or in coal-mining areas as acid mine drainage. One consequence may be an increase in acidification from the generation of sulfuric acid; another might be the release of toxic metals.
5. $CH_2O + CH_2O$ $\rightarrow CH_4 + CO_2$	Under anaerobic conditions at a redox potential of about -200 mV, and in the presence of methogenic bacteria as may be found in swamps, flooded areas, rice paddies, and the sediments of enclosed bays and lakes, partially reduced carbon compounds can disproportionate to produce methane as well as CO_2. This reaction is more typical in freshwater systems because sulfate concentrations are much lower than in marine environments, averaging about one one-hundredth the concentration in seawater. Methane is a critical gas in the determination of global climate. Since the early 1970s, global atmospheric methane levels have been increasing at a rate of 1 percent per year. Although the reasons for this increase are still under investigation, the expansion of rice paddy cultivation in southeast Asia has been cited as a contributing cause. See discussion, Part II, pp. 123–125.

Source: W. M. Stigliani (1988). Changes in valued capacities of soils and sediments as indicators of nonlinear and time-delayed environmental effects. *Environmental Monitoring and Assessment* 10:245–307. Copyright © 1988. Reprinted with permission of Kluwer Academic Publishers.

b. Natural sequence of biological reductions. When water is depleted of oxygen, organisms that depend upon aerobic respiration cannot survive, and anaerobic bacteria take over. These bacteria utilize oxidants other than O_2. These alternative oxidants are less powerful than O_2, and cannot produce as much energy. Nevertheless, bacteria are quite capable of surviving on lower energy processes; in doing so, they can fill ecological niches that are not available to aerobic organisms. The oxidizing power of anaerobic environments in the biosphere is mainly controlled by five molecules. In decreasing order of energy produced, they are nitrate (NO_3^-), manganese dioxide (MnO_2), ferric hydroxide ($Fe(OH)_3$), sulfate (SO_4^{2-}), and, under extreme conditions, carbon dioxide (CO_2) itself. The biological oxidation processes supported by these oxidants are described in Table 13.2.

The oxidizing power of a molecule depends on the specific reaction being carried out, and is measured as the *reduction potential* associated with the reduction of the oxidant. These are listed in Table 13.3 for the environmental oxidants we are considering. Microbial populations first use the oxidant that produces the most energy until it is depleted; only then does another agent become the dominant oxidant. Thus, the redox potential of a body of water tends to fall in a stepwise pattern as BOD increases (see Figure 13.1).

As oxidants are consumed in the conversion of reduced carbon to CO_2, the reduction potential falls to successively lower plateaus, corresponding to the successively lower potential redox couples O_2/H_2O, NO_3^-/N_2, MnO_2/Mn^{2+}, $Fe(OH)_3/Fe^{2+}$, SO_4^{2-}/HS^-, and CO_2/CH_4. These couples do not give reversible potentials at electrodes, but the metabolic activity of the vast array of microbes in soils and water ensure that electron transfer does occur on a time-scale of hours or days. Consequently, all redox-active materials respond to the reduction potential established by the microbial activity.

TABLE 13.3 THERMODYNAMIC SEQUENCE FOR REDUCTION OF IMPORTANT ENVIRONMENTAL OXIDANTS AT pH 7.0 AND 25°C

Reaction	Eh (V)*
Disappearance of O_2 $O_2 + 4H^+ + 4e^- \rightleftarrows 2H_2O$	0.812
Reduction of NO_3^- to N_2 $NO_3^- + 6H^+ + 5e^- \rightleftarrows \frac{1}{2}N_2 + 3H_2O$	0.747
Reduction of MnO_2 to Mn^{2+} $MnO_2 + 4H^+ + 2e^- \rightleftarrows Mn^{2+} + 2H_2O$	0.526
Reduction of Fe^{3+} to Fe^{2+} $Fe(OH)_3 + 3H^+ + e^- \rightleftarrows Fe^{2+} + 3H_2O$	−0.047
Formation of H_2S $SO_4^{2-} + 10H^+ + 8e^- \rightleftarrows H_2S + 4H_2O$	−0.221
Formation of CH_4 $CO_2 + 8H^+ + 8e^- \rightleftarrows CH_4 + 2H_2O$	−0.244

*Eh(V) is the $E°$ value recalculated for pH7 (see p. 314)

Source: W. H. Schlesinger (1997). *Biochemistry: An Analysis of Global Change* (2nd ed.) (San Diego: Academic Press).

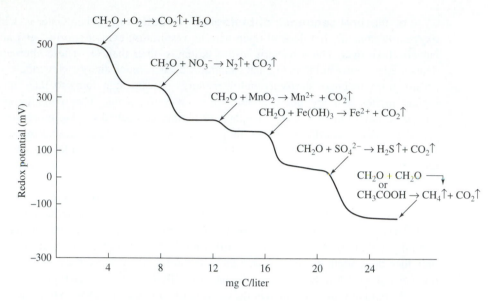

Figure 13.1   Sequence of redox reactions in aqueous environments. O_2 in natural waters at 20°C is sufficient to oxidize about 3.4 mg of organic carbon (shown here as CH_2O) per liter of water. When the rate of replenishment of O_2 from the atmosphere is slower than the rate of oxidation of CH_2O, oxygen is depleted and microbes will select the next most energetic oxidant in the sequence shown. For simplicity, only major products and their valence states are shown. See Table 13.2 for balanced equations.   *Source:* W. M. Stigliani (1988). Changes in valued capacities of soils and sediments as indicators of nonlinear and time-delayed environmental effects. *Environmental Monitoring and Assessment* 10:245–307. Copyright © 1988. Reprinted with permission of Kluwer Academic Publishers.

Note, however, that while there is a general correspondence with the *Eh* values of the half-reactions, the plateau potentials in Figure 13.1 deviate substantially from the numbers listed in Table 13.3. This is because conditions in the environment are far from the standard conditions that establish the *Eh* values. While the pH may be close to 7, the concentrations of other reactants and products are unlikely to be 1.0 *M* (or 1 atm, in the case of a gas).

FUNDAMENTALS 13.2: REDUCTION POTENTIALS

All redox reactions can be divided, at least conceptually, into two reduction *half-reactions,* one proceeding forward and the other in reverse. For example, the oxidation of hydrogen by oxygen,

$$2H_2 + O_2 = 2H_2O \qquad (13.2)$$

can be divided into

$$O_2 + 4e^- + 4H^+ = 2H_2O \qquad (13.3)$$

and

$$4H^+ + 4e^- = 2H_2 \qquad (13.4)$$

Subtracting half-reaction (13.4) from (13.3) gives the overall reaction (13.2). These half-reactions can actually be carried out at the electrodes of a hydrogen-oxygen fuel cell, as discussed in Part I (see pp. 103–107). A potential difference is developed between the oxygen electrode and the hydrogen electrode, allowing a current to flow through the external circuit. For the hydrogen-oxygen fuel cell, this potential difference approaches 1.24 volts (V) at the standard temperature of 25°C, when the gases are at 1 atmosphere pressure, and the electrodes are behaving reversibly, i.e., when the reactants and products are at equilibrium with the electrodes (implying rapid electron transfer rates).

The potential difference, ΔE, is the energy of the electrochemical cell per unit of charge delivered. [Specifically, $1\text{ V} = 1\text{ J/C}$, where V = volt, J = joule, and C (coulomb)* is the unit of charge.] ΔE is related to the free energy of the cell reaction by the relation

$$\Delta G = -nF\,\Delta E \qquad (13.5)$$

where F (Faraday) is the amount of charge in a mole of electrons (96,500 C) and n is the number of electrons transferred in the reaction. Thus, in reaction (13.2), four electrons are transferred from $2H_2$ to O_2, and $\Delta G = -4 \times 96{,}500 \times 1.24 = -479{,}000$ J, or -479 kJ. (Recall that this value, in combination with the entropy of the reaction, gives a theoretical energy conversion efficiency of 80 percent for the H_2/O_2 fuel cell, p. 103.)

Numerous electrode combinations are possible in electrochemical cells, and it is convenient to specify a *standard potential*, E^0, for each electrode by referencing it to the hydrogen electrode, the standard potential of which is defined as zero. Thus $E^0 = 1.24$ V for the oxygen electrode, represented by half-reaction (13.3). The standard conditions for E^0 are unit activities (partial pressure or molar concentration) of the reactants and products at 25°C.

There are many half-reactions for which the electrode potential cannot actually be measured, because the electron transfer reaction at an electrode is too slow. These potentials can nevertheless be calculated from the free energy of appropriate redox reactions. For example, the formation of NO from N_2 and O_2, the thermodynamics of which was considered in Part II, is a redox reaction:

$$O_2 + N_2 = 2NO \qquad (13.6)$$

which can be divided into the half-reactions

$$O_2 + 4e^- + 4H^+ = 2H_2O \qquad (13.7)$$

and

$$2NO + 4e^- + 4H^+ = N_2 + 2H_2O \qquad (13.8)$$

From the free energy of the overall reaction (see p. 180), 173.4 kJ, we obtain a cell potential of -0.45 V [using equation (13.5)]. Then, knowing that the standard potential of the oxygen electrode is 1.24 V, we can readily calculate that the standard potential for half-reaction (13.8) is 1.69 V, even though it is impossible to measure this potential directly, because the electron transfer between the electrode and the NO and N_2 molecules is too slow to establish a reversible potential.

---

*A coulomb is the quantity of charge that passes a fixed point in an electric circuit when a current of one ampere flows for one second. It takes 6.24×10^{18} electrons to produce one coulomb of charge.

FUNDAMENTALS 13.3: CONCENTRATION DEPENDENCE OF THE POTENTIAL; pH AND E^0(w)

What happens to the reduction potential when conditions are not standard? As in all chemical reactions, the driving force for electrochemical processes depends on the concentrations of reactants and products. This dependence is given by the *Nernst* equation

$$E = E^0 - (RT/nF)\ln Q \qquad (13.9)$$

where E^0 is the standard potential, R is the gas constant, n is the number of electrons transferred in the reaction, and Q is the equilibrium quotient, i.e., the concentration expression for the equilibrium constant. In the fuel cell reaction (13.2), for example, $Q = 1/P_{O_2}P_{H_2}^2$ (the water activity being defined as unity), and $n = 4$. Therefore:

$$E = 1.24 - (RT/4F)$$
$$\times (-\ln P_{O_2} - 2\ln P_{H_2})$$

A convenient form of the Nernst equation is

$$E = E^0 - (0.059/n)\log Q \qquad (13.10)$$

where 0.059 is the value of RT/F at 25°C, multiplied by the conversion factor from natural to base-ten logarithms (ln 10 = 2.303). For temperatures other than 25°C, the factor 0.059 must be raised or lowered accordingly.

The Nernst equation applies equally to whole cell reactions or half-reactions. Thus the potential of the hydrogen electrode [half-reaction (13.4)] at 25°C is (after dividing through by $n = 4$):

$$E = 0 - 0.059[\log P_{H_2}^{1/2}/(H^+)] \qquad (13.11)$$

From this we see that the hydrogen electrode potential becomes more negative as (H^+) diminishes. Thus H_2 gas is a

more powerful reductant in alkaline solution than in acid. E falls by -0.059 V for every unit rise in pH. At pH 7, the hydrogen electrode potential is -0.42 (when all other conditions are standard).

Likewise, O_2 is a less powerful oxidant in alkali than in acid, because protons are consumed in the reduction half-reaction (13.3). The oxygen potential (again after dividing by $n = 4$) is:

$$E = 1.24 - 0.059 \log[1/P_{O_2}^{1/4}(H^+)] \qquad (13.12)$$

Again the potential drops 0.059 V for every unit rise in pH and is 0.82 V at pH 7. Because pH 7 is closer to most biologically and environmentally relevant conditions than is pH 0, electrode potentials are often cited for pH 7, as they are in Table 13.3. The Eh values are E^0 values recalculated for pH 7.

Even if no protons appear explicitly in a half-reaction, the potential may be pH-dependent because of secondary acid-base reactions. For example, the potential of the Fe^{3+} reduction half-reaction

$$Fe^{3+} + e^- = Fe^{2+} \qquad (13.13)$$

has no proton dependence per se, but the equilibrium quotient, $(Fe^{2+})/(Fe^{3+})$, is highly dependent on pH because of the acidic character of Fe^{3+}. At quite low pH, it forms a series of hydroxide complexes, and precipitates as the highly insoluble $Fe(OH)_3$ ($K_{sp} = 10^{-37}$). In contrast, Fe^{2+} forms hydroxide complexes only at high pH, and $Fe(OH)_2$ ($K_{sp} = 10^{-15}$) is soluble. Consequently, the reduction potential falls with increasing pH, because (Fe^{3+}) declines more rapidly than (Fe^{2+}).

WORKED PROBLEM 13.3: $E^0(w)$ and K_{sp} of Fe(OH)$_3$

Q. *The $Fe^{3+/2+}$ standard potential [equation (13.13)] is 0.77 V. From this value and the K_{sp}, calculate $E^0(w)$ for the reduction of $Fe(OH)_3$ to Fe^{2+} (see Table 13.3).*

A. $E^0(w)$ is the $Fe^{3+/2+}$ potential at pH 7. This potential can be calculated from the Nernst equation:

$$E = 0.77 - 0.059[\log(Fe^{2+})/(Fe^{3+})]$$

and (Fe^{3+}) can be calculated from $K_{sp} = (Fe^{3+})(OH^-)^3$, i.e., $(Fe^{3+}) = K_{sp}(OH^-)^3$. Substi-

tution gives

$$E = 0.77 - 0.059[\log(Fe^{2+})$$
$$+ \log K_{sp} - 3\log(OH^-)]$$

At pH 7,

$$E = 0.77 - 0.059[\log(Fe^{2+}) - 37 + 21]$$
$$= -0.18 - 0.059[\log(Fe^{2+})]$$

which is the Nernst equation for Fe(OH)$_3$ reduction, with $E^0(w) = -0.18$.

WORKED PROBLEM 13.4: Effective Oxygen Potential

Q. *The first plateau in Figure 13.1, corresponding to O_2 reduction, is at 0.5 V, whereas the $E^0(w)$ value (see Table 13.3) is 0.816 V. What might account for this difference?*

A. Assuming that the environmental pH is 7, the difference must arise from the O_2 concentration dependence. The potential diminishes with decreasing O_2 concentration. Recall that

$$E = 1.24 - 0.059 \log[1/P_{O_2}^{1/4}(H^+)] \quad (13.12)$$

or, at pH 7, $E = 0.816 - 0.059 \log(1/P_{O_2}^{1/4})$. If $E = 0.50$, then

$$\log(1/P_{O_2}^{1/4}) = (-\log P_{O_2})/4$$
$$= -0.316/(-0.059) = 5.36$$

or $P_{O_2} = 10^{-21.4}$ atm. This may seem a bizarrely low value, but it reflects the fact that when microbes are actively respiring in an aqueous medium, they draw down the O_2 to very low levels in their immediate vicinity.

c. Biological oxidations. Bacteria also catalyze oxidation of reduced substances by molecular oxygen, even though such reactions can occur spontaneously in an aerobic environment. Thus HS^- oxidation to sulfate is catalyzed by sulfide oxidizers. These bacteria manage to extract energy from the HS^-/SO_4^{2-} and O_2/H_2O redox couples. Another important oxidation process is *nitrification,* the conversion of NH_4^+ to nitrate ion. Since plants take up and utilize nitrogen mainly in the form of nitrate, this is a key reaction in nature, especially in connection with the use of ammonium salts in fertilizers (see p. 361). The process actually occurs in two steps, ammonium to nitrite, NO_2^-, and nitrite to nitrate:

$$NH_4^+ + 2H_2O = NO_2^- + 8H^+ + 6e^- \quad (13.14)$$
$$NO_2^- + H_2O = NO_3^- + 2H^+ + 2e^- \quad (13.15)$$

These half reactions are catalyzed by two separate groups of bacteria, *Nitrosomonas* and *Nitrobacter,* each utilizing the oxidizing power of O_2 to extract energy from the process.

In summary, redox potential can be considered as a kind of chemical switch in the aqueous environment, one that determines the sequence by which oxidants and reductants are utilized by microorganisms. Changes in redox potential can have important consequences for environmental pollution (see Table 13.2).

13.2 AEROBIC EARTH

O_2 was not always a constituent of the atmosphere; it arose from the evolution of life itself. The primitive Earth had an atmosphere derived from outgassing of the minerals in the interior. Once the surface cooled sufficiently to condense water, and with it acidic gases like HCl and SO_2, the main atmospheric constituents would have been N_2 and CO_2.

Life arose quite early in Earth's history; microfossils resembling modern *cyanobacteria* have been found in 3.5 billion-year-old rocks. How life started is unknown, and remains one of the great scientific issues of our time. It is known that simple organic molecules are common in the universe, and are present in meteorites, which would have bombarded the young Earth. Laboratory experiments show that they could also have been formed from inorganic precursors when subjected to electric discharges from lightning, or to ultraviolet irradiation. The ultraviolet flux would have been intense, since, in the absence of an oxygen atmosphere, Earth would have lacked an ozone shield. Many of the organic building blocks of organisms could have been produced in this way. Alternatively, the building blocks might have been formed on the surfaces of sulfide minerals under the high pressures and temperatures found in hydrothermal vents on the sea floor. (These vents are found in regions where the crustal plates are being formed through upwelling from Earth's mantle.) Recent experiments show that complex organic molecules can be formed in this way. How the organic building blocks were assembled into the first self-replicating organisms remains an unanswered question, although many ingenious proposals have been put forward.

The very first organisms must have been *heterotrophic,* assimilating organic compounds from their environment. Since there was no O_2, they must have obtained their energy from redox reactions other than respiration, similar to the modern anaerobic processes discussed in the preceding section. The splitting of simple organic molecules, such as acetic acid,

$$CH_3COOH = CH_4 + CO_2 \qquad (13.16)$$

may have been the first of such processes; this reaction still provides energy for modern *acetogenic* bacteria.

However, photosynthesis evolved quite early on, probably in the cyanobacteria mentioned above, which survive today as photosynthetic organisms in the oceans. Photosynthesis made these organisms *autotrophic,* capable of synthesizing their own organic molecules from CO_2. They had a strong selective advantage over heterotrophs. In addition to the fossil evidence mentioned above, carbon isotope measurements on the fossil organic carbon show photosynthesis to be at least 3.5 billion years old. The fossil carbon is found to be depleted in the stable ^{13}C isotope, relative to ^{12}C, as a result of the slightly slower diffusion of $^{13}CO_2$ and its slower rate of capture by the CO_2-fixing enzyme *ribulose bis-phosphate carboxylase.*

O_2 was a byproduct of the rise of autotrophic organisms. Because of the reactivity of O_2, it would have been a toxic by-product; most anaerobes are very sensitive to O_2, and cannot survive in an aerobic environment. However, O_2 did not become a significant constituent of the atmosphere for a long time after the advent of photosynthesis, because it was first consumed by oxidizable elements in the ocean and in Earth's crust, particularly iron and sulfur. The early ocean would have had a high concentration of Fe^{2+}, which is abundant in silicate minerals of the mantle, and is quite soluble, in contrast to Fe^{3+}. Photosynthetic O_2 would initially have been used up by reaction with Fe^{2+} to produce precipitates of $Fe(OH)_3$. Indeed, ferric oxide begins to be seen in sedimentary rock that is about 3.5 billion years old, occurring in *Banded Iron Formations,* in which Fe_2O_3 deposits are interleaved with siliceous sediment. These formations reach a peak occurrence in rock which is 2.5–3 billion years old.

Once the oceanic Fe^{2+} was used up, the accumulating O_2 attacked oxidizable minerals on land, principally FeS_2 (pyrite), producing $Fe(OH)_3$ and H_2SO_4 (the same chemistry that still produces acid mine drainage, see p. ?). Evidence for this transition is found in the occurrence of *Red Beds,* deposits of Fe_2O_3 found in geologic layers of terrestrial origin, starting about 2 billion years ago, after the last of the Banded Iron Formations were formed.

Finally, when the rate of O_2 production exceeded its rate of consumption by exposed oxidizable material, the O_2 concentration in the atmosphere began to rise, permitting the evolution of respiring organisms. Fossil evidence of *eukaryotic* organisms has been found in rocks that are 1.3–2 billion years old. Eukaryotes (in contrast to the more primitive *prokaryotes*) have mitochondria, organelles that are specialized for respiration. Some eukaryotes can survive on O_2 at only 1 percent of the present concentration, suggesting that this level was attained over 1 billion years ago. O_2 production would have accelerated with the evolution of *chloroplasts* in the eukaryotes, organelles which are specialized for photosynthesis. The rising O_2 was also accompanied by the production of stratospheric ozone, which permitted life to colonize the continents, freed from the destructive effects of UV radiation. Fossils of multicellular organisms have been found in sedimentary rocks that are 680 million years old, but the rise of green plants, and with them the modern O_2 atmosphere, dates to 400 million years ago.

The time-line for the course of O_2 production is shown schematically in Figure 13.2. The present atmospheric reservoir accounts for only about 2 percent of the estimated cumulative production of O_2, the rest having been used up in the oxidation of minerals. Interestingly, the O_2 concentration seems to have stayed at about 20 percent of the atmospheric gases over the last 400 million years; this constancy suggests some sort of feedback control. As with any reservoir (see Fundamentals 2.1, p. 26), the amount of O_2 reflects the balance between the rate of production and the rate of consumption. Over geologic time, O_2 consumption results from exposure and weathering of reduced carbon-bearing rock; this rate is set largely by Earth's tectonic movements. O_2 production results from the burial of reduced carbon, whose rate depends (among other things) on the total biomass. The biomass is limited, at least in part, by forest fires, and it is possible that feedback

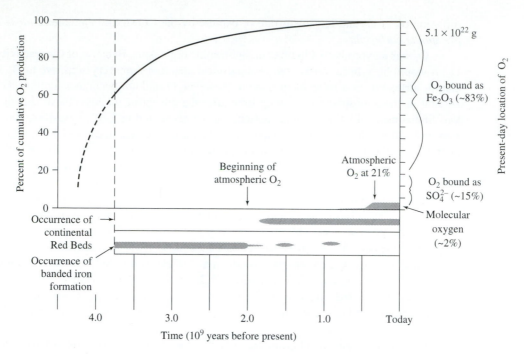

Figure 13.2   Cumulative history of O_2 released by photosynthesis through geologic time. Of more than 5.1×10^{22} g of O_2 released, about 98 percent is contained in seawater and sedimentary rocks, beginning with the occurrence of *Banded Iron Formations* at least 3.5 billion years ago (bya). Although O_2 was released to the atmosphere beginning about 2.0 bya, it was consumed in terrestrial weathering processes to form *Red Beds,* so that the accumulation of O_2 to present levels in the atmosphere was delayed to 400 mya. Figure from "Origins," p. 37 in *Biogeochemistry: An Analysis of Global Change,* 2nd ed., by William H. Schlesinger. Copyright © 1997 by Academic Press. Reprinted by permission of the publisher.

control arises from the dependence of fires on the O_2 concentration. It is known that fires cannot be maintained when the O_2 concentration is less than 15 percent, while even wet organic matter burns freely at a concentration greater than 25 percent.*

If carbon burial has balanced O_2 accumulation for the last 400 million years, what accounts for the rising O_2 level starting 4 billion years ago? A much larger carbon burial rate seems unlikely. It has recently been suggested[†] that UV photolysis of CH_4 could have provided the driving force. Methane production would have been much higher when O_2 levels were low; methane-producing anaerobes would have been abundant, and the methane would have escaped to the atmosphere without oxidation. In the absence of the ozone UV shield, the methane would have been exposed to photons energetic enough to

*See J. E. Lovelock (1974). *Gaia: A New Look at Life on Earth.* (Oxford University Press: Oxford, U.K.)

[†]D. C. Catling et al. Biogenic methane, hydrogen escape, and the irreversible oxidation of early Earth (2001). *Science* 293:839–843.

break the C—H bonds. At the top of the atmosphere, the light H atoms would have escaped Earth's gravitational field, and would have been lost to space. This removal of oxidizable H atoms from the earth-atmosphere system would provide a mechanism for O_2 accumulation.

13.3   WATER AS ECOLOGICAL MEDIUM

a. The euphotic zone and the biological pump.   Biological productivity depends on *primary producers,* organisms that fix carbon via photosynthesis, and provide the food for the animal food chain. In water, the primary produces are cyanobacteria, phytoplankton, and algae. Because of their dependence on sunlight, they are limited to the region near the surface, where sunlight can penetrate. This is the *euphotic zone.* Its depth depends on the clarity of the water.

Most biological activity takes place in the euphotic zone. The primary producers are eaten by animals or decomposed by bacteria, in a continuing cycle of photosynthesis and respiration. However, because of gravity, some dead organisms fall below the euphotic zone. In the deeper layers, bacterial decomposition continues and the waters are enriched in carbon and the other elements of life. Because of thermal stratification, there is little physical mixing between the warmer surface layer and the cold deep layer. Consequently, there is a kind of "biological pump," which transfers carbon and other nutrients from the surface to the deep layers and sediments. Figure 13.3 shows the effect of biological production on oceanic depth profiles of nitrate and iron, as well as oxygen. O_2 is high at the surface and diminishes sharply over the first few hundred meters. Nitrate and iron are drawn down at the surface, due to uptake by organisms, but increase sharply with depth as the organisms are decomposed; below the surface layer their concentrations remain at elevated levels.

In the oceans, the biological pump is responsible for increasing the carbonate concentration of the deep layers with respect to the surface layers. This drawing down of carbonate from the surface increases the rate of transfer of CO_2 from the atmosphere. This is an important contribution to the global carbon cycle. It has been calculated that the atmospheric CO_2 level would double in the absence of the biological pump.

b. Eutrophication in freshwater lakes.   Because the supply of oxygen is restricted, the species that inhabit an aquatic ecosystem are in a dynamic balance, one that is easily disturbed by humans. In water, the O_2 concentration falls with increasing distance from the air-water interface. Thus, aerated soils support oxygen-utilizing microbes as well as higher life forms, while deeper in the soil, in the *saturated zone* where the soil pores are filled with water, anaerobic bacteria dominate and utilize progressively lower $E^0(w)$ redox couples. Likewise in lakes, the sediments are generally oxygen-starved and rich in anaerobic microorganisms, while in the water column above, the O_2 concentration rises toward the surface. The concentration of O_2 at the surface is increased not only because the surface is in contact with air, but because the surface waters support the growth of vegetation and algae, which release O_2 as a product of photosynthesis.

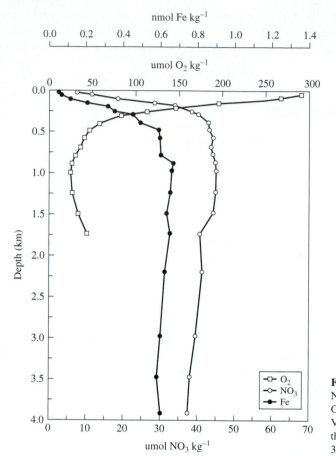

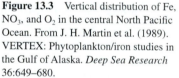

Figure 13.3   Vertical distribution of Fe, NO_3, and O_2 in the central North Pacific Ocean. From J. H. Martin et al. (1989). VERTEX: Phytoplankton/iron studies in the Gulf of Alaska. *Deep Sea Research* 36:649–680.

The biological productivity of a temperate lake varies annually in a cycle (see Figures 13.4 and 13.5). The onset of winter diminishes the solar heating of the surface. The thermal stratification disappears and the water's density becomes uniform, allowing easy mixing by wind and waves, which brings nutrient-rich waters to the surface. In winter, the nutrient supply is high, but productivity is inhibited by low temperatures and light levels. Spring brings sunlight and warming, leading to a bloom of phytoplankton and other water plants. As plant growth increases, the nutrient supply diminishes and phytoplankton activity falls. Bacteria decompose the dead plant matter, gradually replenishing the nutrient supply, and a secondary peak of phytoplankton activity is observed in the autumn. Because the nutrient supply is limited in unpolluted waters, the BOD in the surface waters rarely outstrips the available oxygen.

This natural cycle can be disrupted, however, by excessive nutrient loading from human sources such as wastewaters or agricultural runoff. The added nutrients can support a higher population of phytoplankton, producing "algal blooms." When masses of algae

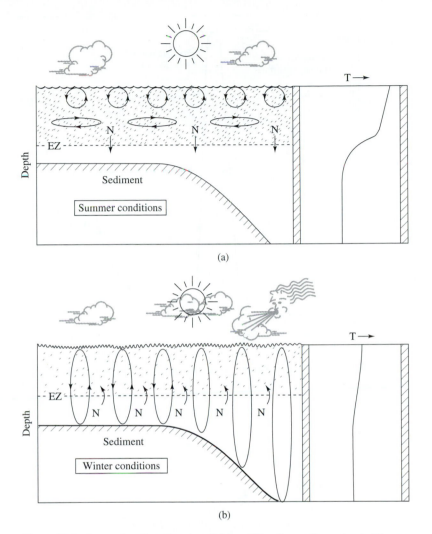

Figure 13.4   Seasonal cycling of nutrients in lakes. EZ = thermocline and end of the euphotic zone; stipple represents phytoplankton growth; N → signifies direction of nutrient
flow; enclosed arrows indicate circulation of waters. The solid line at the right is the temperature profile of the water column.

die off, their decomposition can deplete the oxygen supply, killing fish and other life
forms. If the oxygen supply is exhausted, the bacterial population may switch from
predominantly aerobic bacteria to mainly anaerobic microorganisms that generate the noxious products (NH_3, CH_4, H_2S) of anaerobic metabolism. This process is called *eutrophication,* or, more accurately, *cultural eutrophication.* Eutrophication is the natural process
whereby lakes are gradually filled in (see Figure 13.6). Over time, an initially clear
(*oligotrophic*) lake eutrophies gradually, filling with sediment and becoming a marsh

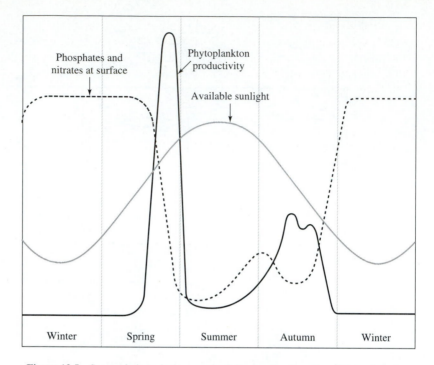

Figure 13.5   Seasonal phytoplankton productivity as a function of sunlight and nutrient concentration.   *Source:* Adapted for W. D. Russel-Hunter (1970). *Aquatic Productivity* (New York: Macmillan Publishing Co., Inc.). Reprinted with permission from W. D. Russel-Hunter.

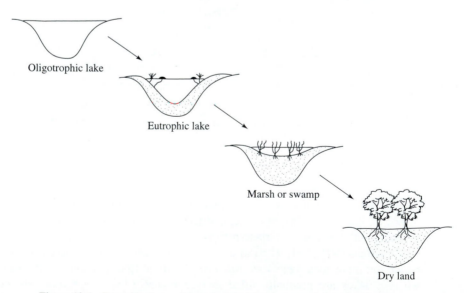

Figure 13.6   Eutrophication and the aging of a lake by accumulation of sediment.

eventually, and then dry land. This process normally proceeds over thousands of years because biological growth and decomposition in the euphotic zone are closely balanced—
the surface layers remain well-oxygenated, and only a small fraction of biological production is deposited as sediment. When this balance is upset by overfertilization of the water,
the eutrophication process accelerates greatly.

c. Nitrogen and phosphorus: The limiting nutrients.   The slow pace of
natural eutrophication reflects the nutrient dynamics of an aquatic ecosystem (see Figure 13.7). The nutrients are assimilated from the environment by the primary producers,
which serve as food for *secondary producers,* including fish. Dead plant and animal tissues
are decomposed by bacteria, which restore the nutrients to the water. The growth of the primary producers is controlled by the *limiting* nutrient, the element that is least available in
relation to its required abundance in the tissues. If the supply of the limiting nutrient
increases through overfertilization, the water can produce algal blooms, but not otherwise;
conversely, management of the aquatic ecosystem requires that the supply of the limiting
nutrient be restricted.

The major nutrient elements are carbon, nitrogen, and phosphorus, which are
required in the atomic ratios 106:16:1, reflecting the average composition of the molecules
in biological tissues. Numerous other elements are also required, including sulfur, silicon,
chlorine, iodine, and many metallic elements. Because the minor elements are required in

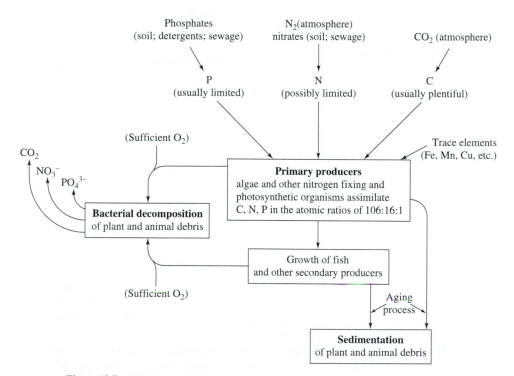

Figure 13.7   Nutrient cycling in an aquatic ecosystem.

small amounts, they can usually be supplied at adequate rates in natural waters. On the other hand, carbon, the element required in the largest amounts, is plentifully supplied to phytoplankton from CO_2 in the atmosphere. Phytoplankton outrun the supply of CO_2 only under conditions of very rapid growth such as in some algal blooms. In these cases, the pH of the water can be driven as high as 9 or 10 through the required shift of the carbonate equilibrium:

$$HCO_3^- + H_2O = H_2CO_3 + OH^-$$
$$H_2CO_3 = CO_2 + H_2O$$

The increase in pH can in turn alter the nature of the algal growth, selecting for varieties that are resistant to high pH.

Normally, the limiting nutrient element is either N or P. Although nitrogen makes up 80 percent of the atmosphere, it is unavailable except through the agency of N_2-fixing bacteria, living in symbiotic association with certain species of plants. On land, these species are rare enough to make nitrogen the limiting nutrient under most conditions. In water, however, N_2-fixing algal species are common, and nitrate ions are often abundant because of runoff from the land. Consequently, nitrogen is not usually limiting, although it may be in some regions, especially the oceans, where nitrate concentrations are low.

This leaves phosphorus as the element that is usually limiting to growth. Phosphorus has no atmospheric supply because there is no naturally occurring gaseous phosphorus compound. Moreover, the input of phosphorus in runoff from unfertilized lands is usually low because phosphate ions, having multiple negative charges, are bound strongly to mineral particles in soils. In surface waters, most of the phosphorus is contained in the plankton biomass; the phosphorus availability depends on recycling of the biomass by bacteria.

Some of the phosphorus is lost to the deeper water and the sediments when dead organisms sink. When a lake turns over in winter, the phosphorus in the deep waters is carried to the surface and supports the plankton bloom in the spring. Whether this phosphorus is available to the surface waters depends on conditions in the lake. At the bottom, phosphate ions may be adsorbed onto particles of iron and manganese oxide. However, when the sediment becomes anoxic, the metal ions are reduced to the divalent forms, the oxides dissolve, and the phosphate ions are released into solution (see notes on manganese and iron oxides in Table 13.2). Phosphate solubility is also increased through acidification, since at successively lower pH values, HPO_4^{2-}, $H_2PO_4^-$, and H_3PO_4 are formed (see Fundamentals 11.3, p. 279).

Under conditions of phosphorus limitation, human inputs of phosphate lead to enhanced biological production and the possibility of oxygen depletion. These inputs can arise from sewage, from agricultural runoff, especially where synthetic fertilizers and manure (both of which contain phosphate) are applied intensively, and from polyphosphates in detergents. When phosphorus is added to lakes and rivers where the availability of phosphorus limits biochemical productivity, biomass production will increase in proportion to the amount of excess phosphorus added. The increased biomass raises the BOD of the water; as the BOD increases, oxygen is depleted, leading to anoxia and anaerobic conditions. The most notorious instance of phosphate-induced eutrophication was in Lake Erie, which "died" in the 1960s. Excessive algal growth and decay killed most of the fish and fouled the shoreline. A concerted effort by the United States and Canada to reduce phosphate inputs was put into effect in the 1970s. Over $8 billion was spent building sewage treatment plants to remove phosphates from wastewater, and the levels of phosphate in detergents were restricted. These efforts, along with other pollution-control measures, succeeded in bringing the lake back to life. Commercial fisheries have revived, and the beaches are once again in use.

d. Anoxia and its effects on coastal marine waters.

Not only is anoxia a problem in freshwater lakes, it can be a problem in the intermediate or deep waters of an enclosed estuary, gulf, or fjord with restricted circulation between deep and surface waters. If biomass productivity is high, dead biomass sinks into deeper waters where aerobic bacteria progressively consume the oxygen; if the deep layers fail to mix with surface layers, the oxygen is not replenished and anoxia sets in. Because seawater is rich in sulfate salts, the favored reaction under anaerobic conditions is sulfate reduction to hydrogen sulfide (H_2S), a chemical that is extremely toxic to fish and humans. Although H_2S is generally confined to the lower layers of seawater, during storms, the deeper, anoxic layers can mix with surface layers, exposing aquatic life to the deadly gas.

Such events occurred in 1981 and 1983 in the enclosed marine areas off the east coast of Denmark. These events killed unprecedented numbers of fish through suffocation or poisoning by H_2S gas (see Figure 13.8). The two episodes were triggered by sea storms, but the underlying cause was nutrient enrichment of the coastal waters. The source of the nutrient inputs was found to be primarily nitrogen in runoff from agricultural lands. The extra nitrogen increased the production of biomass (and hence organic carbon), which, upon bacterial decomposition, outstripped the oxygen available for aerobic degradation; anaerobic conditions produced H_2S, setting the stage for disaster.

Similar mechanisms have been at work in the United States, causing severe pollution of, for example, Chesapeake Bay. Because massive amounts of nutrients enter the bay from sewage, runoff, and atmospheric deposition, phytoplankton grow much faster than they can be foraged by consumer organisms such as oysters. The phytoplankton mass clouds the waters; when these organisms sink to lower depths, they die from lack of light. In the depths, the dead plankton are consumed by bacteria that outstrip the supply of dissolved oxygen, making the bottom anoxic and allowing the production of H_2S. Without oxygen, the *benthic* (bottom-dwelling) organisms such as oysters and rooted plants cannot survive, and fish are displaced from their habitats.

The pollution of the Chesapeake appears to involve both nitrogen and phosphorus. Levels of these nutrients in the estuary rise and fall annually in seasonal patterns that are fairly complex (see Figure 13.9). In winter, cold temperatures and lack of biochemical activity allow the concentration of O_2 to reach its annual maximum. At the same time, nitrogen enters in large amounts because winter is the period of maximum freshwater flow, with accompanying transport of sediment and runoff. Simultaneously, sedimentation is removing phosphorus from the water column, mainly through the precipitation of manganese and iron oxides, which absorb phosphorus efficiently and are insoluble under aerobic conditions. (Phosphorus is also removed during the settling of organic debris.) Beginning in the late spring and early summer, the oxygen levels decline due to increased biological activity. Nitrogen concentrations also decline for these reasons: 1) nitrogen is

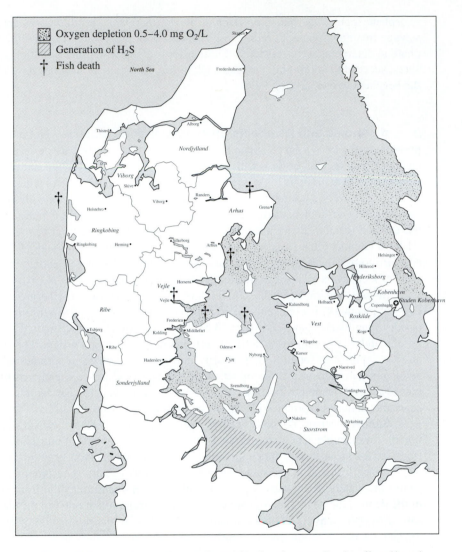

Figure 13.8   Coastal areas of eastern Denmark and southwestern Sweden affected by oxidation depletion, fish suffocation, and generation of H_2S.   *Source:* Miljostyrelsen (1984). *Oxygen Depletion and Fish Kill in 1981: Extent and Causes* (in Danish) (Copenhagen: Miljostyrelsen).

incorporated into biomass and sinks as the organisms die; 2) little new nitrogen is introduced in runoff; and 3) nitrogen is depleted as increasingly anoxic conditions force a switch from oxygen to nitrate as oxidant.

The opposite situation prevails for phosphorus. Under anaerobic conditions, phosphorus is liberated from the sediments, in large part due to the reduction of manganese and ferric oxides to Mn^{2+} and Fe^{2+}. In the II valence states, the metals are soluble and release

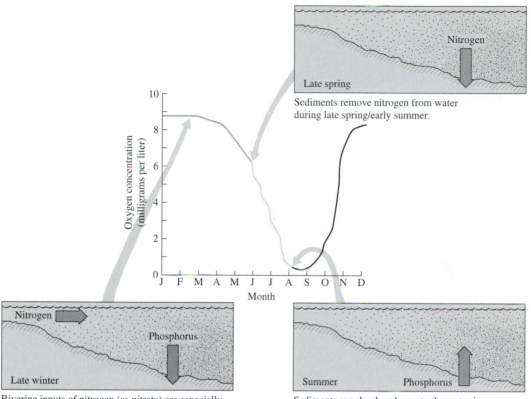

Figure 13.9   Oxygen concentration in water overlying the sediments with major seasonal net fluxes of nitrogen and phosphorus (insets) in the Patuxent River at the estuary of Chesapeake Bay. *Source:* C. F. D'Elia (1987). Too much of a good thing: Nutrient enrichments of the Chesapeake Bay. *Environment* 29(2):6–11, 30–33. Reprinted with permission of the Helen Dwight Reid Educational Foundation. Published by Heldref Publications, 1319 Eighteenth St., NW, Washington, DC 20036-1802. Copyright © 1987.

the bound phosphorus formerly adsorbed to the insoluble oxides of the metals. The phosphorus is readily mixed with the surface layers given the mechanical turbulence of estuarine environments. Thus, as conditions cycle from aerobic to anaerobic and back, the phosphorus is continuously recycled between the surface waters and the sediments. During anaerobic periods, phosphates are released to the water column, to be taken up by microorganisms; during aerobic periods, phosphates are returned to the sediments. The amount of phosphate trapped in this cycle is vast, much greater than the annual quantities entering the estuaries from sewage effluents or other sources; it represents the cumulative inputs of many years. Thus, even though Maryland and Virginia banned detergents with phosphates in the 1980s, phytoplankton productivity is still excessive. Now the limiting

Figure 13.10   The "dead zone" in the Gulf of Mexico due to nutrient enrichment from land use activities in the drainage basin of the Mississippi River.

nutrient may well be nitrogen, but nitrogen inputs are very difficult to control. Chesapeake Bay receives some of the highest atmospheric NO_x emissions in the world, mainly due to the density of traffic in the adjacent areas. Part of the strategy for cleaning up Chesapeake Bay might include reducing NO_x from vehicle exhausts, demonstrating once again the link between the atmosphere and the hydrosphere.

Still another example of marine anoxia is in the Gulf of Mexico, where a vast "dead zone" has come to light in recent years (see Figure 13.10). In this 18,000 km^2 area, the O_2 concentration is too low to support aquatic life during the spring and summer. These are the seasons of great algal blooms, resulting from overfertilization of the Gulf by the nutrients in the outflow of the Mississippi River. The Mississippi drains the vast mid-continental farmlands of the United States. The algal blooms are attributed to the 1.5 million tons of dissolved nitrogen discharged annually by the river. Agriculture accounts for 80 percent of this total, 25 percent from animal manure, and 55 percent from synthetic fertilizer. More than 40 percent of U.S. commercial fisheries are located in the Gulf of Mexico, and these have been hard hit by the annual appearance of the dead zone.

e. Wetlands as chemical sinks.   The episodes of fish-kills along the Danish coast depicted in Figure 13.8 might have been avoided if the original coastal wetlands had not been drained. Wetlands are typically anoxic and have large amounts of organic carbon; they create a natural buffer zone for nearby fresh or marine waters by trapping nitrates. The nitrates enter the wetlands in runoff, but are utilized by bacteria to oxidize stored carbon via the reduction of nitrate to N_2 or N_2O, which are vented to the atmosphere

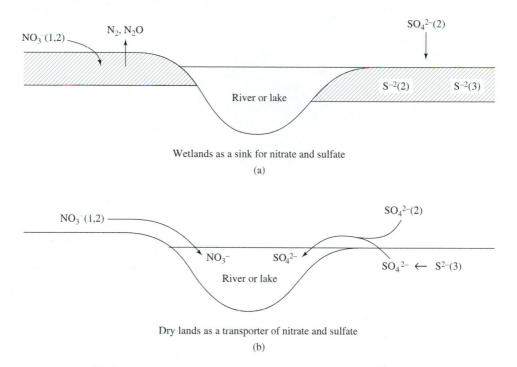

Wetlands as a sink for nitrate and sulfate

(a)

Dry lands as a transporter of nitrate and sulfate

(b)

1. Runoff of nitrogenous fertilizer
2. Input from acid deposition
3. Sulfide minerals from former marine sediments

Figure 13.11   (a) Ability of wetlands to buffer against nitrate and sulfate inputs to water bodies; (b) under conditions where wetlands become dry, none of the protective reducing reactions occur. In addition, accumulated sulfides may oxidize to sulfate as sulfuric acid, and leach into adjacent rivers or lakes.   *Source:* W. M. Stigliani (1988). Changes in valued capacities of soils and sediments as indicators of nonlinear and time-delayed environmental effects. *Environmental Monitoring and Assessment* 10:245–307. Copyright © 1988. Reprinted with permission of Kluwer Academic Publishers.

(see Figure 13.11a). By depleting the nitrates before they can enter the estuary, the surrounding wetlands limit the excessive growth of biomass and subsequent anoxic conditions in the estuary. Restoration of wetlands has been proposed in many areas as a means of reducing overfertilization from runoff.

If wetlands are of marine origin, they are likely to contain high concentrations of sulfur in the form of reduced sulfide minerals such as pyrite. Under the redox/pH conditions prevalent in wetlands, these sulfides are highly insoluble and immobilized (see Figure 13.11a). Draining the wetlands (see Figure 13.11b) exposes these compounds to oxidizing conditions, producing a situation similar to acid mine drainage (see p. 305).

One example of this phenomenon occurred in a coastal area of Sweden near the Gulf of Bothnia, where wetlands were drained in the early 1900s for use as agricultural lands.

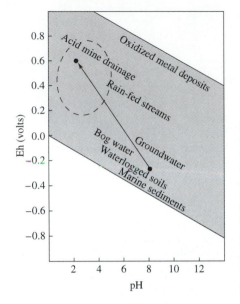

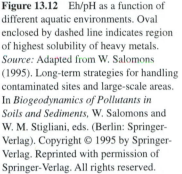

Figure 13.12 Eh/pH as a function of different aquatic environments. Oval enclosed by dashed line indicates region of highest solubility of heavy metals. *Source:* Adapted from W. Salomons (1995). Long-term strategies for handling contaminated sites and large-scale areas. In *Biogeodynamics of Pollutants in Soils and Sediments,* W. Salomons and W. M. Stigliani, eds. (Berlin: Springer-Verlag). Copyright © 1995 by Springer-Verlag. Reprinted with permission of Springer-Verlag. All rights reserved.

As shown in Figure 13.12, draining the wetland shifted the *Eh*/pH conditions diagonally to the upper left, from the values typical of waterlogged soils to conditions close to those of acid mine drainage. The draining exposed sulfides to the atmosphere, and their oxidation to sulfuric acid acidified the soil and nearby lakes. The pH in one of these lakes, Lake Blamissusjon, dropped from 5.5 or higher in the last century to a current value of 3. Even though agricultural activities ceased in the 1960s, the lake has not recovered; it is widely known as the most acidic lake in Sweden.

f. Redox effects on metals pollution. Changes in the redox potential can have important consequences for environmental pollution, especially with respect to metal ions such as cadmium, lead, and nickel. In general, the solubility of heavy metals is highest in oxidizing and acidic environments (see Figure 13.12). At neutral-to-alkaline pHs in oxidizing environments, these metals often adsorb onto the surface of insoluble $Fe(OH)_3$ and MnO_2 particles, especially when phosphate is present to act as a bridging ion. When the redox potential shifts to only slightly oxidizing or slightly reducing conditions as a result of microbial action, and the pH shifts toward the acidic range, $Fe(OH)_3$ and MnO_2 in soils and sediments are reduced and solubilized. The adsorbed metal ions likewise become solubilized and move into groundwater [or into the water column of lakes when there is $Fe(OH)_3$ or MnO_2 in the sediment]. However, if sulfate is reduced microbially to HS^-, metal ions are immobilized as insoluble sulfides. But as we have seen, if sulfide-rich sediments are exposed to air through drainage or dredging operations, then HS^- is oxidized back to sulfate, and the heavy metal ions are released.

A particularly important instance of biological redox mediation of heavy-metal pollution occurs in the case of mercury. Inorganic mercury, in any of its common valence states, Hg^0, Hg_2^{2+}, and Hg^{2+}, is not toxic when ingested; it tends to pass through the digestive system, although Hg^0 is highly toxic when inhaled. But the methylmercury ion, $(CH_3)Hg^+$, is very toxic, regardless of the route of exposure. The environmental route to toxicity involves sulfate-reducing bacteria that live in anaerobic sediments. As part of their metabolism, these bacteria use methyl groups to produce acetate. When exposed to Hg^{2+}, the bacteria transfer the methyl groups to the mercury, producing $(CH_3)Hg^+$; because methylmercury is soluble, it enters the aquatic food chain, where it is bio-accumulated in the protein-laden tissue of fish (see discussion in Part IV, pp. 437–441).

g. Fertilizing the ocean with iron.   Although nitrogen and phosphorus are the limiting aquatic nutrients near land, it has become evident that in large areas of open ocean, it is actually iron that limits biological production. Among the "trace metals" essential for life, iron is required in the largest amount. Iron is utilized in many enzymes involved in electron transport, and in processing O_2 and N_2, as well as their reduction and (for N_2) oxidation products. Thus all organisms require a steady supply of iron. Since iron is abundant in Earth's crust, iron limitation is not a problem for land plants, or for phytoplankton growing near land. However, the concentration of iron in the ocean is extremely low (see Table 12.1, p. 283), because of the low solubility of $Fe(OH)_3$ (see p. 317) in the alkaline (pH = 8) seawater.

In much of the oceans the settling of dust from the land provides phytoplankton with sufficient iron for growth. Prevailing winds blow sands from the Sahara and Gobi Deserts far out over the Atlantic and Pacific Oceans. Recent satellite measurements show a fairly good correlation between patterns of dust in the air and phytoplankton growth in the oceans below. However, there are large areas that are relatively dust-free, especially in the equatorial Pacific Ocean, and the waters ringing Antarctica at greater than 60° south latitude, called the Southern Ocean. These areas have less phytoplankton than could be supported by the available nitrogen and phosphorus. It has been known for some time that adding iron to samples of these waters stimulates phytoplankton growth in the laboratory, and a series of field experiments in the 1990s showed that spreading iron over areas of nutrient-rich ocean produced phytoplankton blooms.

Iron limitation on biological productivity is an important ingredient in the carbon cycle, because phytoplankton take up CO_2 and transport some of it to the deep ocean when they die. This is the mechanism of the "biological pump" for CO_2, discussed on p. 319. In iron-limited areas, adding iron to the oceans could increase the speed of the biological pump, drawing down the atmospheric CO_2. Indeed, it has been suggested that iron supplementation could offer a "geoengineering" solution to the problem of rising atmospheric CO_2. However, this solution has been set aside for several reasons:

a. The remedy would be very expensive, because the iron stimulation of phytoplankton blooms is a transient effect. The blooms quickly fade as the excess iron precipitates out of the photic zone. [The duration depends somewhat on the form of the added iron. Ferrous salts are soluble, but rapidly oxidize to insoluble $Fe(OH)_3$. Ferric chelates are longer lived, but the chelating agents (see pp. 298–299, 450–451) would add to the expense.] Consequently, iron would have to be added continuously to have a permanent effect.

   b. Modeling indicates that the maximum effect on the atmospheric CO_2 concentration
   would be a ~60 ppm lowering, making a relatively small difference in the rising
   level.
   c. There could be unforeseeable consequences to the biology of the oceans from such
   an intervention.
   d. There would have to be international agreement on ocean alteration, particularly in
   the region of Antarctica, which is protected by international law.

The evidence that iron can fertilize the oceans, and that dust is an important source
of iron, raises the possibility that changes in global dustiness may have contributed to the
temperature changes that produced the Ice Age. Data from ice cores and deep sea sediments indicate that there was much more iron in ocean water during the ice ages. Thus, the
biological pump would have been stimulated; the ~60 ppm lowering in the CO_2 level that
might have been available from this mechanism corresponds approximately to the CO_2
lowering which is also detected in ice cores (see p. 174). The increased iron might have resulted from dust due to drying of the continents and expansion of deserts. However, as is
usual in reconstructing the past, it is difficult to decide which factor is cause and which is
effect.

CHAPTER 14

WATER POLLUTION AND WATER TREATMENT

The quality of surface and groundwater is of concern from two overlapping but distinct points of view: 1) human health and welfare; and 2) the health of aquatic ecosystems. Both aspects of water quality are enhanced by minimizing the impacts of human activities, but the specific issues and control measures are different.

14.1 WATER USE AND WATER QUALITY: POINT AND NONPOINT SOURCES OF POLLUTION

Water quality is as important an issue as water quantity. Although most of the water supply is returned to the stream flow after use, its quality is inevitably degraded. The effects are summarized in Table 14.1. The cooling of power plants by circulating water raises the temperature (thermal pollution), with adverse effects on the biota of the receiving waters. Discharge of sewage from homes and commercial establishments reduces the dissolved oxygen content, again upsetting the biological balance of surface waters. Industrial and mining activities contaminate water with a variety of toxic materials. Agriculture can foul surface and groundwaters with excess nutrients, and can lead to salinization of soil when irrigation waters evaporate, leaving salts behind.

TABLE 14.1   EFFECTS ON WATER QUALITY FROM WATER USE

Water use	Effects on water quality
Domestic/commercial	Decreases dissolved oxygen
Industrial/mining	Decreases dissolved oxygen; pollutes water with toxic heavy metals and organics; causes acid mine drainage
Thermoelectric	Increases water temperature (thermal pollution)
Irrigation/livestock	Causes salinization of surface and groundwaters; decreases dissolved oxygen (near feedlots)

In considering effects on water quality, it is useful to distinguish *point sources* and *nonpoint sources* of pollution. Point sources are factories and other industrial and commercial installations that release toxic substances into the water. In recent years toxic releases from point sources have been substantially reduced, especially in developed countries. Large amounts continue to be discharged (see Table 14.2a,b), but they are far

TABLE 14.2A   TOP TEN INDUSTRIES FOR SURFACE WATER DISCHARGE (1999)

Industry*	Pounds	Metric tons	Percent of total
Chemicals	77,097,472	35,002	29.8
Primary metals	62,513,740	28,381	24.2
Food	50,225,853	22,803	19.4
Paper	19,118,393	8,680	7.4
Petroleum	15,655,884	7,108	6.1
Electric utilities	4,510,038	2,048	1.7
Electrical equipment	4,393,066	1,995	1.7
Fabricated metals	2,429,536	1,103	0.9
Measurement/photo	1,320,125	599	0.5
Metal mining	447,029	203	0.2
Total	**237,711,136**	**107,921**	**91.8**

*Note: List does not include industries with multiple codes that discharge 8,680 metric tons.

TABLE 14.2B   TOP TEN CHEMICALS IN SURFACE WATER DISCHARGE (1999)

Chemical	Pounds	Metric tons	Percent of total
Nitrate compounds	231,367,165	105,041	89.4
Ammonia	7,917,711	3,595	3.1
Manganese compounds	5,398,239	2,451	2.1
Methanol	3,873,380	1,759	1.5
Barium compounds	2,182,327	991	0.8
Sodium nitrite	1,593,212	723	0.6
Zinc compounds	1,373,162	623	0.5
Ethylene glycol	544,047	247	0.2
Chlorine	391,583	178	0.2
Copper compounds	361,467	164	0.1
Total	**254,640,826**	**115,607**	**98.4**

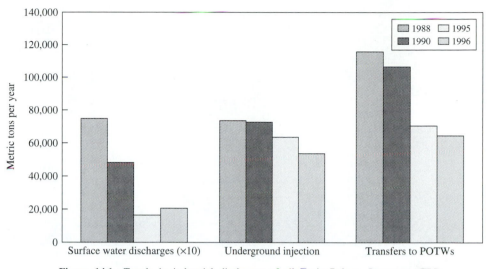

Figure 14.1   Trends in industrial discharges of all Toxic Release Inventory (TRI) chemicals with potential impact on water quality.   *Source:* U.S. Environmental Protection Agency, Office of Environmental Information (1999). *1999 Toxic Release Inventory, Public Data Release* (Washington, DC: U.S. EPA).

less than in former years. Figure 14.1 shows the reduction of U.S. discharges from 1988 to 1996 of chemicals included in the Toxic Release Inventory (TRI) that is maintained by the EPA. Discharges to surface waters were reduced by nearly a factor of four, while transfers to publicly owned treatment works (POTWs) were nearly halved. The POTW transfers are particularly significant because treatment works are designed to handle domestic sewage, primarily, and have often been disabled by industrial discharges in the past. Figure 14.1 actually understates the improvement in pollutant discharges because the TRI is an aggregate of all toxic chemicals, the most abundant of which are the least toxic (see Table 14.2b). Releases of the most toxic chemicals have been cut more sharply. Figures 14.2 and 14.3 show that POTW transfers have been cut by factors of five and nine for benzene and chromium, respectively, two chemicals that are known to cause cancer. (The discharges to surface waters have been cut less, because they were relatively small to begin with.)

Nonpoint sources represent a much harder problem. They include emissions from transport vehicles, agricultural runoff, which can carry excess nutrients, pesticides, and silt into streams and groundwaters, and urban runoff, which can carry toxic metals and organics through storm drains into sewage treatment plants, or directly into rivers and lakes (see Figure 14.4). The progress made in controlling point sources of pollution has drawn attention to nonpoint sources, which account for an increasing fraction of the total pollutant load. For example, the relative contribution to the cadmium (Cd) load to the Rhine River from point and nonpoint sources changed dramatically between the mid 1970s and mid 1980s (see Figure 14.5). While most of the Cd came from point sources in the 1970s, this is no longer the case, thanks to industrial controls. Now the greater share of Cd derives

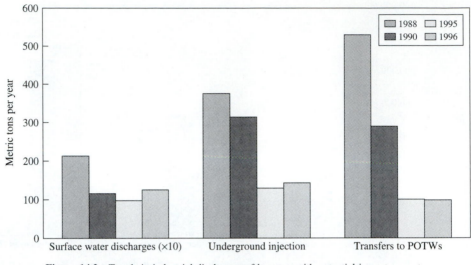

Figure 14.2   Trends in industrial discharges of benzene with potential impact on water quality.   *Source:* U.S. Environmental Protection Agency, Office of Environmental Information (1999). *1999 Toxic Release Inventory, Public Data Release* (Washington, DC: U.S. EPA).

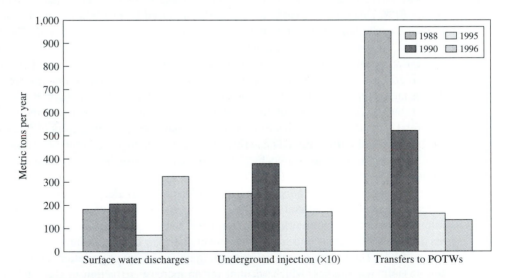

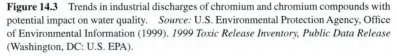

Figure 14.3   Trends in industrial discharges of chromium and chromium compounds with potential impact on water quality.   *Source:* U.S. Environmental Protection Agency, Office of Environmental Information (1999). *1999 Toxic Release Inventory, Public Data Release* (Washington, DC: U.S. EPA).

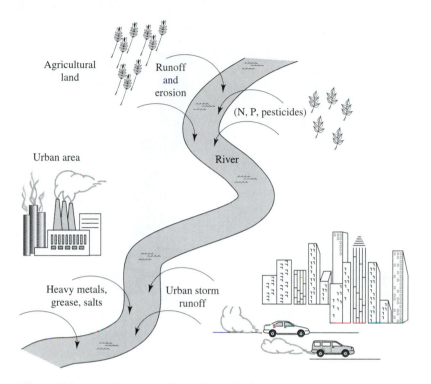

Figure 14.4   Nonpoint-source pollution from agricultural and urban areas.

from nonpoint sources from urban and agricultural runoff (Cd is a contaminant in urban dust as well as in phosphate fertilizer). As discussed above, nonpoint sources are responsible for the overfertilization of lakes and bays.

14.2   REGULATION OF WATER QUALITY

Governments around the world try to regulate water quality in the interests of public health and environmental protection. In the U.S., the legal instruments for regulation are two basic laws, the *Clean Water Act* (CWA) of 1972 and the *Safe Drinking Water Act* (SDWA) of 1974, both of which have been amended several times. Under these acts, the EPA is required to set standards that protect from harmful effects of contaminants—discharges of point-source pollutants are subject to a permitting process, industries are advised with respect to the best available pollution control technologies, and resources are provided for the construction of municipal treatment plants. Similar initiatives have been undertaken in most other industrialized countries. These measures have helped to produce marked reductions in pollutant discharges to water bodies, as noted in the preceding section.

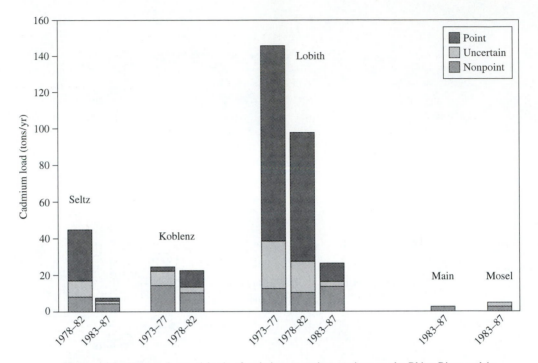

Figure 14.5   Estimated annual loads of cadmium at various stations on the Rhine River and its tributaries. The station at Lobith is at the German-Dutch border; the cadmium load there represents the sum of all upstream inputs in the Rhine Basin. The basin includes most of Switzerland, northeastern France, Luxembourg, a large part of southwestern Germany, and most of The Netherlands.   *Source:* H. Behrendt (1993). Point and Diffuse Loads of Selected Pollutants in the River Rhine and Its Main Tributaries, Report RR-93-1 (Laxenburg, Austria: International Institute for Applied Systems Analysis). Copyright © 1993, International Institute for Applied Systems Analysis. Reprinted with permission.

Under the SDWA, standards have been set for microorganisms, radionuclides, and 54 organic and 14 inorganic chemicals, which can sometimes be found in drinking water. The standards (which may be found on the EPA web site: http://www.epa.gov/safewater/mcl.html) are set as *maximum contaminant levels* (MCL), which are as close as is technologically feasible to the *maximum contaminant level goal* (MCLG). The MCLG is the level of a contaminant in drinking water below which there is no known or expected risks to health. Although sensible-sounding, the MCLG is a difficult quantity to establish. As more data on health effects accumulate, and more sensitive measures are used, the MCLGs tend to be readjusted downward. In many cases there may be no threshold below which some health effect could be expected. Indeed, several MCLGs have been set at zero (e.g., benzene, dichloromethane, carbon tetrachloride, dioxin, PCBs, lead), although zero is an unattainable goal.

The legally enforceable MCLs are set by considering the best available technology that is economically feasible. The MCLs range from a high of 10 mg/L (nitrate ion) to a low of 0.0002 mg/L (heptachlor epoxide and lindane). For microorganisms, the rule is that filtration and disinfection must remove or inactivate 99.9 percent of *Giardia* and 99.99 percent of viruses, and no more than 5 percent of samples in any month can test positively for coliform bacteria. In the case of copper and lead, the source of drinking water contamination is generally the plumbing system, and water utilities are required to take treatment steps if 10 percent of tap water samples exceed 1.3 mg/L of copper or 0.015 mg/L of lead.

For surface waters generally, the quality issue comes down to the ability to support aquatic ecosystems, and human uses such as swimming and fishing. The Clean Water Act of 1972 established a mechanism for protecting these waters by requiring states to set a *total maximum daily load* (TMDL) of pollutants at a level that ensures water quality standards in a given waterway. If the TMDL is exceeded, required pollution reductions are to be allocated among the pollution sources. This provision of the CWA was not enforced for many years, until lawsuits brought by environmental groups in more than 30 states led the EPA to propose an implementation rule in 1999. The need for action was justified by the results of a 1998 survey that indicated that some 40 percent of assessed streams, lakes, and estuaries in the U.S. were not clean enough to support uses such as swimming or fishing. The EPA's rule specifies procedures for states to follow in identifying impaired waterways, establishing a budget of pollution reductions, and producing an implementation plan. Although the rule is intended to implement a 30-year-old law, and gives states up to 15 more years to do so, it has run into strong opposition because of the difficulty of controlling nonpoint sources of pollution, which have become the new battleground of environmental protection. Some of the provisions of the EPA's rule are intended to bring pollution generated by forestry and agriculture under the CWA regulations, although these sectors have previously been exempted. Much of the opposition centers on these provisions.

14.3   WATER AND SEWAGE TREATMENT

Municipalities treat their water supplies for domestic and commercial uses to ensure freedom from disease and to eliminate odors and turbidity; they treat their sewage to reduce water pollution and eutrophication. In the former case, treatment begins with aeration to remove odors by purging dissolved gases and volatile organic compounds. Aeration also oxidizes any Fe^{2+} to Fe^{3+}, which forms $Fe(OH)_3$ and precipitates out. Additional Fe^{3+} or Al^{3+} ions are then added deliberately, usually as the sulfate salt, along with lime to adjust the pH. A voluminous $Fe(OH)_3$ or $Al(OH)_3$ precipitate is produced, which traps solid particles that may be suspended in the water supply. When the precipitate is collected and removed, the water is greatly clarified. If dissolved organic compounds require removal, the water can be passed through a filter of activated charcoal, although this is an expensive and uncommon step. Finally, disinfectant is added—chlorine, chlorine dioxide, or ozone—to kill microorganisms and provide safe drinking water.

Sewage treatment similarly relies to a large extent on settling and screening to remove solids; this physical separation step is called primary treatment. Most municipalities also carry out secondary treatment, which harnesses bacteria to metabolize organic compounds, converting them to CO_2. In this way, the BOD is substantially decreased (see Figure 14.6). If the sewage is not metabolized in this way, then the wastewater BOD

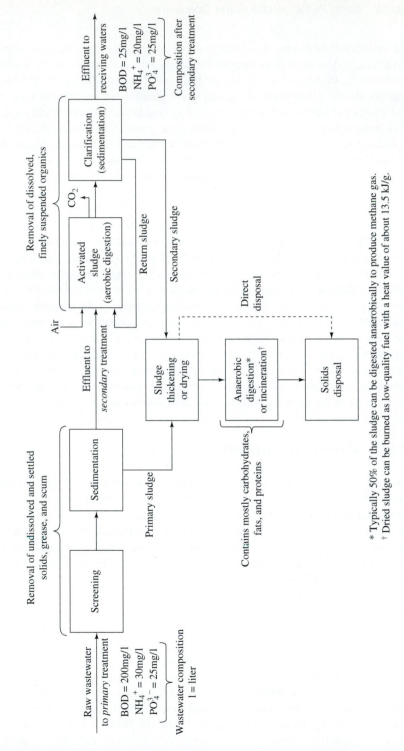

Removal of undissolved and settled
solids, grease, and scum

Removal of dissolved,
finely suspended organics

Effluent to
receiving waters

BOD = 25mg/l
NH_4^+ = 20mg/l
PO_4^{3-} = 25mg/l

Composition after
secondary treatment

Clarification
(sedimentation)

CO_2

Activated
sludge
(aerobic digestion)

Air

Return sludge

Secondary sludge

Effluent to
secondary treatment

Screening

Sedimentation

Primary sludge

Raw wastewater
to primary treatment

BOD = 200mg/l
NH_4^+ = 30mg/l
PO_4^{3-} = 25mg/l

Wastewater composition
l = liter

Sludge
thickening
or drying

Direct
disposal

Anaerobic
digestion*
or incineration†

Solids
disposal

Contains mostly carbohydrates,
fats, and proteins

* Typically 50% of the sludge can be digested anaerobically to produce methane gas.
† Dried sludge can be burned as low-quality fuel with a heat value of about 13.5 kJ/g.

Figure 14.6 Primary and secondary treatment of municipal wastewater.

can overwhelm the oxidizing capacity of receiving waters, leading to anoxic conditions. In secondary treatment, the wastewater is sprayed over a bed of sand or gravel that is covered by aerobic microorganisms, or else agitated with the microbes in a reactor. At the end of this process, the BOD is lowered by as much as 90 percent.

Although microbes convert most of the organic matter to CO_2, they also incorporate some of it into new cells as the culture grows. These cells must be harvested from time to time, and are added to the sludge from the primary settling tank. Sludge disposal is a major issue in municipal sewage treatment. Since it is composed mainly of organic matter, sludge is an excellent fertilizer in principle. In practice, its application to cropland is restricted by the frequent presence of toxic metals that are flushed into the wastewater from domestic and industrial sources, or from urban runoff when storm drains are connected to sewer lines. Alternatively, the sludge can be incinerated and provide energy for heating or electricity production. Another alternative is to convert part of the sludge to methane, a high-quality fuel, by digesting it with anaerobic bacteria. However, poor economics and local opposition often militates against these options. Consequently, a good deal of sludge ends up in landfills. But as landfills become full, pressure for cropland application increases, so the issue of metals hazards and metals removal is again examined more closely. If the metals problem can be solved, sludge can become a valuable resource as fertilizer, instead of an environmental disposal problem.

Although secondary treatment is effective in reducing BOD, it does little to reduce the concentrations of inorganic ions, in particular, NH_4^+, NO_3^-, and PO_4^{3-}. These soluble ions are released with the wastewater, where they can eutrophy the receiving waters. Their removal requires tertiary treatment, in which additional chemical steps are added (see Figure 14.7). Thus, phosphate is removed by precipitation with lime, producing the insoluble mineral hydroxyapetite, $Ca_5(PO_4)_3(OH)$. NH_4^+ can be converted to volatile NH_3 by adding lime to increase the pH, after which the pH is lowered again by CO_2 injection to reprecipitate the lime. Finally, the remaining organic compounds can be filtered out with activated charcoal, and disinfectant can be added to produce quite pure water. These treatment steps add significantly to costs; ammonia stripping in particular is energy-intensive. An alternative process, requiring less energy, is to use nitrifying bacteria to convert NH_4^+ to NO_3^-, and then denitrifying bacteria to convert NO_3^- to N_2. However, these bacteria require careful control of growth conditions. Yet another alternative is to spread the secondary-treated water over a wetland that can efficiently filter out phosphates and nitrates. (Recall in discussion of Figure 13.8 that wetlands on the Danish coast served as a nutrient trap to protect the sea from eutrophication until they were drained.)

14.4   HEALTH HAZARDS

a. Pathogens and disinfection.   The spread of pathogenic microorganisms through the water supply is the most serious pollution hazard to human health. Waterborne pathogens are ubiquitous throughout the world. Even waters that are untouched by humans can be contaminated by animal wastes. Hikers in the wilderness who drink untreated water from seemingly pristine streams often become infected by *Giardia* microbes. The most

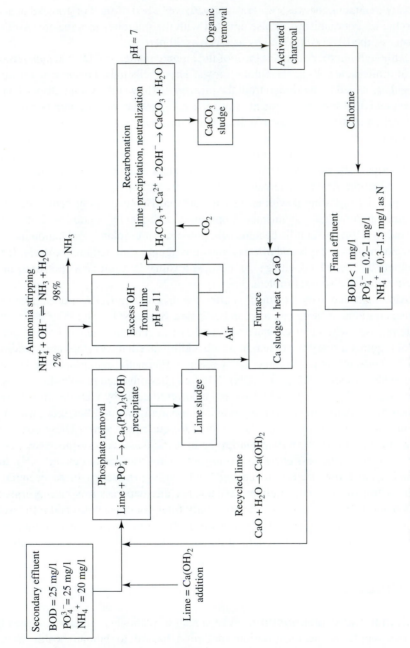

Figure 14.7 Tertiary treatment of municipal wastewater.

serious problems, however, are created by the contamination of drinking water by human wastes. The failure to treat sewage and separate it from drinking water takes an enormous toll in disease around the world. Unsanitary water is one of the most pervasive human problems, especially in developing countries.

Disinfection can take many different forms, one of the oldest of which relies on the removal of pathogens by the earth itself. Water filtered through soil and rock and into deep aquifers is generally free of microbes. Consequently, well water is generally pathogen-free. However, if the water table is shallow, or if the land is heavily laden with human or animal wastes, then the filtering capacity of the soil may be overwhelmed and wells contaminated.

The way to prevent microbial contamination is to keep the water supply as free from wastewater discharges as possible and to treat it with disinfectant. The disinfectants currently in use are ozone, chlorine dioxide, and chlorine. Of these, the most common is chlorine. When chlorine is added to water it disproportionates:

$$Cl_2 + H_2O = HOCl + H^+ + Cl^- \tag{14.1}$$

The HOCl (hypochlorous acid) is an oxidant; the Cl is in the $+1$ oxidation state and is readily reduced to Cl^-. HOCl is the active ingredient of most bleaches, decolorizing fabrics by oxidizing colored molecules. (The colors generally reflect the presence of several double bonds, which are susceptible to oxidation.) Because it is a neutral molecule, HOCl passes through the cell walls of microorganisms and kills them by oxidizing vital molecules. Likewise, ozone and chlorine dioxide are powerful and diffusible oxidants, and kill microbes in the same way.

Chlorine is an effective and relatively inexpensive disinfectant with a proven record of success. Nevertheless, its use has become controversial because it can introduce organochlorine molecules into drinking water. HOCl is not only an oxidant, but also a chlorinating agent. In particular, hydroxybenzenes are readily attacked by HOCl, and are converted to a variety of chlorinated compounds. Although hydroxybenzenes may be present from industrial wastes, they are also found naturally in surface waters as constituents of humic acids. Humic acids are complex molecules that contain benzene rings with a variety of substituents, including hydroxide (see Figure 12.6a,b, p. 296). Although polymeric, some humic acids dissolve in surface or groundwaters and enter the water supply. One of the more abundant reaction products of HOCl with humic acids is chloroform, $CHCl_3$. Most chlorinated water supplies, therefore, have trace levels of chloroform. Chloroform is a suspected liver carcinogen in humans, and there is some epidemiological evidence for a modest increase in the risk of bladder and rectal cancer from drinking chlorinated water.

The chloroform (and most other organic molecules) can be removed from the water supply by filtering it through activated charcoal, but this is an expensive measure. Instead, chlorine is being replaced in some communities by ozone or chlorine dioxide. (Even though the latter contains Cl, it is not an effective chlorinating agent.) These alternative disinfectants are coming into wide use, especially in Europe. They are more expensive than Cl_2, however, because they are too reactive to be stored or shipped, and must be generated on site. O_3 is generated by an electric discharge in air, while ClO_2 is made by oxidizing chlorite ion, ClO_2^- (often with Cl_2, which is reduced in the process to Cl^-). An additional disadvantage of both agents is that they are fast-acting and rapidly decompose. In contrast,

HOCl is less reactive; it acts more slowly and persists for some time in the water supply. Persistence is an advantage in many water systems where old and leaky pipes permit significant infiltration of water from the surroundings. HOCl offers a measure of protection against pathogens that might enter the supply through such leaks. For this reason, even when O_3 or ClO_2 are used as the primary disinfectant, a small quantity of Cl_2 is often added to the water before it enters the water distribution system.

b. Organic and inorganic contaminants.

Surface and groundwaters can be contaminated by migration of chemicals from poorly maintained landfills, industrial waste sites, accidental spills, and leaks in storage tanks, especially those buried in the ground. Regulations for the cleanup of leaky underground oil and gasoline storage tanks have become major preoccupations of homeowners and gas-station owners in the United States. Drinking wells can be contaminated by trace amounts of petroleum fractions, chlorinated solvents, or PCBs (polychlorinated biphenyls—previously used in transformers and pumps; see discussion, Part IV, pp. 429–432). These sparingly soluble organic compounds can escape confinement and migrate through the soil; they often accumulate in underground pools, from which they slowly enter the water table over a long period of time. As discussed elsewhere (see p. 260), leaks from gasoline storage tanks have recently contaminated many wells with the water-soluble additive MTBE, leading to its elimination from the gasoline supply.

Waste sites and spills can also leach metal ions into the water. Their transport is governed by binding equilibria, which are in turn influenced by soil pH and E(w), as discussed in the preceding sections. When these ions enter streams and lakes, some metals such as cadmium and mercury, as well as most lipophillic organic compounds, bioaccumulate in the aquatic food chain and can make fish unsafe to eat (see Part IV, pp. 386–388, 437). Sometimes wells are dug in aquifers that contain high levels of toxic minerals from natural sources, as the tragic arsenic poisonings in Bangladesh (see p. 444) dramatically illustrate.

Remediation of contaminated soils and sediments is extremely difficult. Contaminated water can be extracted by specially dug wells and treated by aeration to volatilize lighter organic compounds, or by charcoal filtration to remove heavier organic compounds and metals. However, because of the volumes involved, this "pump-and-treat" approach is expensive; moreover, its effectiveness is often limited because of slow rates of contaminant transfer from soil reservoirs to the circulating water. Severely contaminated soils and sediments can be dug up or dredged and deposited elsewhere, presumably out of harm's way, but finding an appropriate dump site can be difficult. Moreover, the act of digging or dredging can stir up and transfer significant amounts of contaminated material to the surrounding air and water.

Agricultural areas can have water problems associated with the widespread application of fertilizers, herbicides, and pesticides. Although the trend in herbicide and pesticide use has for some time shifted from long-lived organic compounds to ones that break down fairly rapidly in the environment, some herbicides and pesticides can still accumulate in the groundwater, occasionally threatening farm wells.

Fertilizers can increase the level of nitrate ions in the groundwater. The main nitrate health hazard is "blue baby syndrome," a condition of respiratory failure in babies having excessive nitrate in their diet. Some of the nitrate is reduced by anaerobic bacteria in the stomach to nitrite ion, NO_2^-. The nitrite oxidizes the Fe^{2+} ion in hemoglobin to Fe^{3+}, which is unable to bind O_2. The Fe^{3+}-containing hemoglobin is called *methemoglobin,* and the condition is known as *methemoglobinemia.* In adults the methemoglobin is rapidly reduced back to the Fe^{2+} form, but in babies this process is slow. Nitrate-induced methemoglobinemia is now a rare condition in industrialized countries but remains a concern in developing countries.

Another worry about nitrate is the possibility that the nitrite produced in the stomach can react with amines in the diet to produce N-nitrosamines, R_2NNO (see p. 414). Nitrosamines have been shown to be carcinogenic in animal tests. However, the nitrate levels in drinking water are much lower than in cured meat products or cheeses, to which nitrates are added to inhibit the bacterium that causes botulism. Government agencies have instituted programs to reduce the levels of nitrates and nitrites in food products, and some manufacturers now add vitamins C and E in order to block nitrosamine formation.

SUMMARY

Water is available in abundance on planet Earth, and pure water is supplied continuously by the hydrological cycle, which is powered by the sun. The bounty of fresh water is distributed unevenly around the globe; even where it is abundant, the resource is often mismanaged through profligate use or through contamination from wastes. Access to unpolluted, pathogen-free water is a critical need for much of the world's population.

In soils, the major interactions of water are determined largely by acid-base equilibria established by buffering regimes that vary with soil type. Acidified water passes through the soil, where it is neutralized by limestone or clay minerals. This natural balance is upset when excessive, anthropogenically derived acidification results in acid inputs that exceed the soil's buffering capacity. In this case, acids percolating through the soil are no longer neutralized, and receiving waters become acidified. Because buffering capacities diminish slowly, on a time-scale of decades, long time lags may occur between the onset of pollution and its ultimate effects.

In aquatic ecosystems, water chemistry is determined by the redox potential. Reduced organic carbon is the major biochemical fuel of the biosphere, and its burial, along with the evolution of photosynthesis, has given us our O_2-containing atmosphere. The operation of the biological pump in the oceans is an important determinant of the atmospheric CO_2 level. Although O_2 is the most powerful oxidizing agent, it is only sparingly soluble in water, and is easily depleted when levels of organic carbon are excessive. In the absence of oxygen, microorganisms oxidize reduced carbon, utilizing other environmental oxidants. The choice of oxidant is determined in a sequence of reactions regulated by the redox potential. Water pollution problems arise because some of these reactions cause the release of harmful by-products.

The problem of water pollution is thus one largely of capacity depletion: acid buffer capacity in soil-water interactions, and oxidation capacity in receiving waters. Reduction of water pollution thus requires strategies that focus not only on reductions in emissions of specific pollutants having a direct impact on water quality, but also on maintenance and replenishment of the vital capacities. In some cases this may be achieved by redirecting misplaced resources. For example, provided they are free of toxic chemicals, sewage wastes—instead of being discharged to water bodies—can be applied as fertilizer to the land, which has a much higher oxidizing capacity.

PROBLEM SET

1. The global hydrological cycle is driven by solar evaporation of water on land and sea. Calculate the solar energy required to drive the global hydrological cycle using the data in Figure 10.1, and assuming that heat of evaporation of water (both salt and fresh) at 15°C (the average global temperature) is 44.3 kJ per mole. Compare your answer with the value given in Figure 1.1. Compare it to the global anthropogenic primary energy consumption in 2000 (see Table 1.1).

2. (a) Using data from Table 10.1, calculate the global supply of water per capita in 2025, when the world population is projected to be 8.5 billion. In 2000, the population was 6.07 billion.
   (b) With respect to the demand for water (see Table 10.2), irrigation accounts for 69 percent of global water withdrawals. The global efficiency of irrigation is around 37 percent. Calculate what the demand would have been if global efficiency of irrigation was 70 percent.

3. Consider a corn yield of 7,400 kg per hectare (equivalent to 120 bushels per acre). If 25 kg (one bushel) of corn consumes about 20 m^3 of water during the growing season, what is the ratio of the weight of corn to the weight of water consumed? Where does most of the water end up? Assuming a rainfall of 30 cm/yr, calculate the minimum quantity of irrigation water required per hectare to grow the corn (1 hectare = 10^4 m^2).

4. Due to increasing diversion of the water supply for other purposes, and to frequent droughts, many U.S. cities experience chronic shortages of water for domestic uses. Assume you are the mayor of a city plagued by water shortages, and you decide to focus your efforts on decreasing the *demand* for water rather than increasing the *supply*. Devise a strategy for this "demand management" approach. First, estimate the gallons per capita per day of domestic water use, given that household water use in the United States in 2000 was about 3.90 × 10^{13} liters/yr for a population of around 280 million (1 gallon = 3.785 liters). Second, you learn from a public survey that total domestic water use comprises toilet flushing (38 percent), bathing (31 percent), laundry and dishes (20 percent), drinking and cooking (6 percent), brushing teeth and other miscellaneous uses (5 percent). Flush-toilets in your town use about 20 liters per flush, but water-conserving varieties recommended by the Plumbing Manufacturers Institute average about 13 liters per flush; the average household shower uses about 25 liters of water per minute, but water-saving shower heads are available that spray about 10 liters of water per minute; water-efficient dishwashers and washing machines are available that reduce water use by 25 percent. Devise a long-term plan for reducing water consumption, and calculate a reasonable goal (in terms of gallons of water per capita per day) to be achieved in the next five to ten years.

5. H$_2$S boils at −61°C, H$_2$Se at −42°C, and H$_2$Te at −2°C. Based on this trend, at what temperature would one expect water to boil? Why does water boil at a much higher temperature?

6. If the concentration of atmospheric CO_2 were to double from its current value of 370 ppm, what would be the calculated pH of rainwater (assuming that CO_2 were the only acidic input)? With respect to rising CO_2 levels, do we have to be concerned about enhanced acidity of rain in addition to potential climate warming?

7. **(a)** The total amount of base in a water sample can be determined by titration with standard acid, and is usually reported as the *alkalinity,* in equivalents of acid per liter (eq/L). If the alkalinity of a carbonate-containing sample is 2.0×10^{-3} eq/L, and the pH is 7.0, what are the concentrations of OH^-, CO_3^{2-}, HCO_3^-, and H_2CO_3? (See pp. 279–280 for required equilibrium constants.)

   (b) How do these concentrations change if the pH of the sample increases to 10.0 as a result of photosynthesis by algae (for simplicity, assume no input of CO_2 from the atmosphere). What weight of biomass, (CH_2O), is produced.

8. How much Fe^{2+} could be present in water containing 1.0×10^{-2} M HCO_3^-, without causing precipitation of $FeCO_3$ ($K_{sp} = 10^{-10.7}$)?

9. Describe the three major buffer ranges for neutralizing acidic inputs to soils. For each of the ranges, include in your description: (1) the pH range over which the buffer operates; (2) the major chemical component(s) that participates in the buffering reactions; and (3) the chemical reaction by which H^+ is neutralized.

10. Why is the process of acid deposition beneficial from the point of view of air quality? How does this process transfer a short-term air-pollution problem into long-term problems of soil and water pollution?

11. Why is carbonic acid effective in weathering parent soil rock, but not an important factor in lake acidification?

12. **(a)** With reference to Figure 12.12, assume that the average annual deposition of sulfur in the watershed of Big Moose Lake between 1880 and 1920 was 0.8 g S/m^2, and from 1921 to 1950 it was 2.5 g S/m^2. Calculate the cumulative acid equivalents (eq) per m^2 that were deposited in the watershed over the entire period from 1880 to 1950.

   (b) Assume that the watershed soils buffered acidity via exchange reactions with base cations on clay mineral surfaces. Assume further that base saturation (β) declined from 50 percent in 1880 to 5 percent in 1950. Calculate the total cation exchange capacity (CEC_{tot}), and the buffering capacity ($CEC_{tot} \times \beta$) in 1880 (in units of eq/m^2). Also assume that the silicate buffer rate (br_{Si}), which replenishes the base cations in soil from weathering of silicate, was negligible over this period.

   (c) Do the same calculation as in 12(b), but assume br_{Si} was equal to 0.02 eq m^{-2} yr^{-1} over the time scale from 1880 to 1950. What would you expect the pH of the soil water to be when the base saturation approaches 5 percent?

13. **(a)** Show how the numbers given in Table 12.2 were calculated for natural sources. Make the following assumptions: the pH of unpolluted rainfall is 5.7, and annual global precipitation is 496,000 km^3/yr; production of NO_2 from lightning is 20×10^{12} g N/yr; production of SO_2 from volcanoes is 20×10^{12} g S/yr; and biogenic production of dimethylsulfide (DMS) and H_2S is 65×10^{12} g S/yr.

   (b) Calculate moles of hydrogen ions per year generated from coal burning and smelting, assuming this activity generates about 93×10^{12} g S/yr. Assume combustion processes from all activities generate about 20×10^{12} g N/yr.

   (c) Calculate the acid-neutralizing capacity of the atmosphere due to generation of ammonia (NH_3). Assume that natural and anthropogenic sources each generate about 60×10^{12} g N/yr. Calculate the total *net* generation of H^+/yr from both natural and anthropogenic sources.

(d) Calculate the average moles $H^+ \, m^{-2} \, yr^{-1}$ from natural sources. (The area of the globe is $510 \times 10^{12} \, m^2$.) Calculate the average moles $H^+ \, m^{-2} \, yr^{-1}$ in industrialized regions of the globe. (Assume that the area of industrialized regions is $11.75 \times 10^{12} \, m^2$.)

14. (a) What class of molecules is responsible for most of the reducing power in aqueous environments?

(b) What parameter is a measure of reducing power?

15. Five hundred kg of n-propanol ($CH_3CH_2CH_2OH$) are accidentally discharged into a body of water containing 10^8 liters of H_2O. By how much is the BOD (in milligrams per liter) of this water increased? Assume the following reaction:

$$C_3H_8O + \tfrac{9}{2}O_2 = 3CO_2 + 4H_2O$$

16. A lake with a cross-sectional area of $1 \, km^2$ and a depth of 50 meters has a euphotic zone that extends 15 meters below the surface. What is the maximum weight of the biomass (in grams of carbon) that can be decomposed by aerobic bacteria in the water column of the lake below the euphotic zone during the summer when there is no circulation with the upper layer? The bacterial decomposition reaction is:

$$(CH_2O)_n + nO_2 = nCO_2 + nH_2O$$

The solubility of oxygen in pure water saturated with air at 20°C is 8.9 mg/L; $1 \, m^3 = 1{,}000$ liters.

17. Name the six most important oxidants in the aquatic environment, and how the redox potential regulates their reactivity.

18. (a) If a lake contains high concentrations of dissolved Mn^{2+} and Fe^{2+}, what would be the concentration of dissolved NO_3^- and why?

(b) What environmental effect may accompany reduction of MnO_2 and $Fe(OH)_3$?

19. Write the equations for the oxidation of pyrite (FeS_2) and give two examples of anthropogenic activities that can initiate the oxidation. What is the environmental effect manifested by pyrite oxidation?

20. What is the redox potential of an acid mine water sample having $(Fe^{3+}) = 8.0 \times 10^{-3} \, M$ and $(Fe^{2+}) = 4.0 \times 10^{-4} \, M$, given that E^0 for the Fe^{3+}/Fe^{2+} couple $= 0.77V$?

21. Calculate the equilibrium partial pressure of oxygen (P_{O_2}) in a water sample containing equal concentrations of nitrite (NO_2^-) and ammonia (NH_4^+). For the half-reaction of nitrite to ammonia

$$NO_2^- + 8H^+ + 6e^- = NH_4^+ + 2H_2O$$

$E^0 = 0.892$ volts, and $Eh = 0.340$ volts when $(NO_2^-) = (NH_4^+)$ at pH $= 7$. For the half-reaction involving O_2 reduction

$$4H^+ + O_2(g) + 4e^- = 2H_2O$$

$E^0 = 1.24$ volts.

22. (a) A sample of soil containing both MnO_2 and $Fe(OH)_3$ is in contact with water. From the reduction potentials in Table 13.3, calculate the equilibrium concentration of Fe^{2+} at pH 7, if the Mn^{2+} concentration is $10^{-5} \, M$.

(b) How would the answer to (a) change if the pH were lowered to 5?

23. (a) From the standard free energy of formation of NO_2 (see Table 2.5), calculate the free energy change for the production of NO_2 from N_2 and O_2, the electrochemical potential for the reaction, and the half-reaction standard potential for the reduction of NO_2 to N_2 (recalling that the standard potential for O_2 reduction is 1.24 V).

(b) What is the $E^0(w)$ value for the NO_2 half-reaction?

24. Assume that algae need carbon, nitrogen, and phosphorus in the atomic ratios 106:16:1. What is the limiting nutrient in a lake that contains the following concentrations: total C = 20 mg/L, total N = 0.80 mg/L, and total P = 0.16 mg/L? If it is known that half the phosphorus in the lake originates from the use of phosphate detergents, will banning phosphate builders slow down eutrophication?

25. In anaerobic marine environments, what toxic gas can be generated and by which reaction (name reactants and products)?

26. Explain the "phosphate trap" in the estuary of Chesapeake Bay. Why was a local ban on phosphorus in detergents not particularly helpful in mitigating eutrophication in the estuary?

27. **(a)** Explain why anaerobic freshwater wetlands with high concentrations of organic carbon can serve as natural buffers against sulfates and nitrogen oxides (give reactions).

    (b) When oxidants other than organic carbon are absent from such wetlands, which redox reaction is likely to predominate, and which products will be emitted?

28. The sediments of an estuarine creek in New Jersey contain large amounts of mercury bound as sulfide (with $K = 10^{-52}$) under the prevailing environmental conditions (pH = 6.8; Eh = -230 mV). Environmental scientists have been asked to assess the potential impacts of the polluted sediments. They conclude that the mercury poses no danger in its current state. However, they caution against any action that would expose it to air and increase its redox potential. Explain why the scientists come to this conclusion.

29. Characterize nonpoint-source pollution and give two examples, one in urban areas and one in agricultural areas. Why is it more difficult to control this type of pollution than point-source pollution?

SUGGESTED READINGS

Chapter 10: Water Resources:

United Nations Development Programme, United Nations Environment Programme, World Bank, World Resources Institute (2000). Coastal ecosystems. In *World Resources 2000–2001: People and Ecosystems: The Fraying Web of Life,* pp. 69–86; 163–180 (Washington, DC: The World Resources Institute).

United Nations Development Programme, United Nations Environment Programme, World Bank, World Resources Institute (2000). Freshwater Systems. In *World Resources 2000–2001: People and Ecosystems: The Fraying Web of Life,* pp. 103–118; 193–211 (Washington, DC: The World Resources Institute).

S. Postel (1999). *Pillar of Sand: Can the Irrigation Miracle Last?* (New York: W. W. Norton & Company/Worldwatch Institute).

W. M. Alley, T. E. Reilly, and O. L. Franke (1999). *Sustainability of Ground-Water Resources.* U.S. Geological Survey Circular 1186 (Washington DC: United States Government Printing Office).

W. B. Solley, R. R. Pierce, and H. A. Perlman (1998). *Estimated Use of Water in the United States in 1995.* U.S. Geological Survey Circular 1200 (Washington DC: United States Government Printing Office).

Office of Water (2000). *Liquid Assets 2000: America's Water Resources at a Turning Point.* Report EPA-840-B-00-001 (Washington, DC: U.S. Environmental Protection Agency).

Chapter 11: From Couds to Runoff: Water as Solvent:

D. Langmuir (1997). *Aqueous Environmental Geochemistry* (Upper Saddle River, New Jersey: Prentice Hall).

W. Stumm and J. J. Morgan (1996) *Aquatic Chemistry* (3rd edition) (New York: Wiley).

F. M. M. Morel and J. G. Hering (1993). *Principles and Applications of Aquatic Chemistry* (New York: Wiley).

Chapter 12: Water and the Lithosphere:

W. M. Stigliani (1996). Buffering capacity: Its relevance in soil and water pollution. *New Journal of Chemistry* 20:205–210.

U.S. Environmental Protection Agency (1999). *Progress Report on the EPA Acid Rain Program*, Report EPA-430-R-99-011.

C. T. Driscoll, G. B. Lawrence, A. J. Bulger, T. J. Butler, C. S. Cronan, C. E. Eagar, K. F. Lambert, G. E. Likens, J. L. Stoddard, and K. C. Weathers (2001). Acidic deposition in the United Sates: Sources and inputs, ecosystem effects, and management strategies. *BioScience* 51(3):180–198.

W. M. Stigliani (1988). Changes in valued capacities of soils and sediments as indicators of nonlinear and time-delayed environmental effects. *Environmental Monitoring and Assessment* 10:245–307.

Chapter 13: Oxygen and Life:

V. Smil (2000). Phosphorus in the environment: Natural flows and human interferences. *Annual Reviews of Energy & Environment* 25:53–88.

D. L. Belval and L. A. Sprague (1999). *Monitoring Nutrients in the Major Rivers Draining to Chesapeake Bay.* U.S. Department of Interior, U.S. Geological Survey Water Resources Investigation Report 99-4238. (http://va.water.usgs.gov/online_pubs/WRIR/99-4238/99-4238.html)

Committee on the Causes and Management of Coastal Eutrophication, National Research Council (2000). *Clean Coastal Waters: Understanding and Reducing the Effects of Nutrient Pollution* (Washington, DC: National Academy Press).

R. Howarth, D. Anderson, J. Cloern, C. Elfring, C. Hopkinson, B. Lapointe, T. Malone, N. Marcus, K. McGlathery, A. Sharpley, and D. Walker (1997). *Nutrient Pollution of Coastal Rivers, Bays, and Seas.* Issues in Ecology, Issue No. 7 (Washington, DC: The Ecological Society of America). (http://esa.sdsc.edu/issues.htm)

S. Carpenter, N. F. Caraco, D. L. Correll, R. W. Howarth, A. N. Sharpley, and V. H. Smith (1997). *Nonpoint Pollution of Surface Waters with Phosphorus and Nitrogen.* Issues in Ecology, Issue No. 3 (Washington, DC: The Ecological Society of America). (http://esa.sdsc.edu/issues.htm)

W. F. Ritter and A. Shirmohammadi (eds.) (2001). *Agricultural Nonpoint Source Pollution: Watershed Management and Hydrology* (Boca Raton: Lewis Publishers).

P. H. Lehner, G. P. Aponte Clarke, D. M. Cameron, and A. G. Frank (1999). *Stormwater Strategies: Community Responses to Runoff Pollution* (New York: Natural Resources Defense Council).

Mississippi River/Gulf of Mexico Watershed Nutrient Task Force (2001). *Action Plan for Reducing, Mitigating, and Controlling Hypoxia in the Northern Gulf of Mexico* (Washington, DC: Office of Wetlands, Oceans, and Watersheds, U.S. Environmental Protection Agency)

(http://www.epa.gov/msbasin)

H. T. Odum (2000). *Heavy Metals in the Environment: Using Wetlands for Their Removal* (New York: Lewis Publishers).

Chapter 14: Water Pollution and Water Treatment:

Office of Water (1999). *Understanding the Safe Drinking Water Act.* Report EPA-810-F-99-008 (Washington, DC: U.S. Environmental Protection Agency). (http://www.epa.gov/safewater/sdwa/understand.pdf)

Office of Water (2000). *The Quality of Our Nation's Waters: A Summary of the National Water Quality Inventory: 1998 Report to Congress.* Report EPA-841-S-00-001 (Washington, DC: U.S. Environmental Protection Agency).

Office of Water (2000). *Atlas of America's Polluted Waters*. Report EPA-840-B-00-002 (Washington, DC: U.S. Environmental Protection Agency).

U.S. Geological Survey (1999). *The Quality of Our Nation's Waters—Nutrients and Pesticides*. U.S. Geological Survey Circular 1225 (Washington, DC: United States Government Printing Office).

P. Sampat (2001). Uncovering groundwater pollution. In *State of the World 2001* (New York: W. W. Norton & Company/Worldwatch Books).

R. Johnson, J. Pankow, D. Bender, C. Price, and J. Zogorski (2000). MTBE: To what extent will past releases contaminate community water supply wells. *Journal of Environmental Science & Technology* 34(9):210A–217A.

Office of Solid Waste, U.S. Environmental Protection Agency (1999). *Biosolids* Generation, Use, and Disposal in the United States*. Report EPA-530-R-99-009 (Washington, DC: U.S. Environmental Protection Agency).

M. B. Brennan (April 9, 2001). Waterworks: Research accelerates on advanced water-treatment technologies as their use in purification grows. *Chemical & Engineering News* 79(15):32–38.

Task Force on Biological and Chemical Systems for Nutrient Removal (1998). *Biological and Chemical Systems for Nutrient Removal* (Alexandria, Virginia: Water Environment Federation).

---

*Also known as sewage sludge.

PART IV

BIOSPHERE

CHAPTER 15

NITROGEN AND FOOD PRODUCTION

The final part of this book concerns the biosphere—the realm of living organisms and their interactions. We have already touched on various aspects of the biological world in connection with energy flows and air and water chemistry, but now we focus on several chemical and biological topics that directly affect human and ecological health: food production, nutrition, pesticides, toxic substances, and carcinogenesis. What links these topics is the need of living organisms to take in and process materials from the surrounding world. All organisms have developed ways of ingesting and utilizing substances essential for growth and development. But these same processes of intake and metabolism make living organisms vulnerable to the effects of non-nutritive substances in their environment. When these substances interfere with normal growth or life, we describe them as toxic. But as we shall see, the distinction between essential and toxic substances is often a matter of degree. Moreover, toxic substances come from many sources, including natural ones. Whether a chemical is natural or synthetic, what counts is its reactivity, persistence in the environment, degree of exposure, and influence on biochemistry.

15.1 NITROGEN CYCLE

We first consider food production and the important role of the nitrogen cycle. The central process in biological energy production is the photosynthetic conversion of CO_2 to reduced carbon compounds, which are then used as fuel by all manner of life (see discussions of the carbon cycle on pp. 19–21 and photosynthesis on p. 85). However, organisms need more than carbon; as discussed in the context of aquatic ecosystems, they also require nitrogen, phosphorus, and other elements in smaller amounts (see pp. 323, 374). These elements are normally available in soil in sufficient quantities to support an adequate level of natural plant growth.

Possibilities for increasing productivity beyond natural limits, however, are often constrained by the supply of nitrogen available to the plant. While 80 percent of the atmosphere consists of molecular nitrogen, N_2 is an extremely stable and unreactive form. In order to participate in biological reactions, nitrogen must be *fixed;* that is, it must be combined with other elements. The cycling of nitrogen through the environment is illustrated in Figure 15.1, and quantified in Table 15.1. Some N_2 is fixed nonbiologically through reaction with O_2 at sufficiently high temperatures in combustion or lightning. The nitrogen oxides formed in the atmosphere are converted to nitric acid and washed out in rain, thus providing the soil with a supply of nitrate. Plants can utilize nitrate in the production of protein and other essential organic nitrogen compounds.

But the amount of nitrogen available through the pathway of nitrogen oxides and nitrate is insufficient to support the abundant plant life we know. The majority of naturally occurring nitrogen fixation is accomplished by certain bacteria and blue-green algae (*cyanobacteria*) that are able to reduce N_2 to NH_3. These organisms possess a specialized biochemical apparatus, the *nitrogenase* enzyme complex, the chemistry of which is illustrated in Figure 15.2. The complex consists of two proteins, the "Fe protein," which

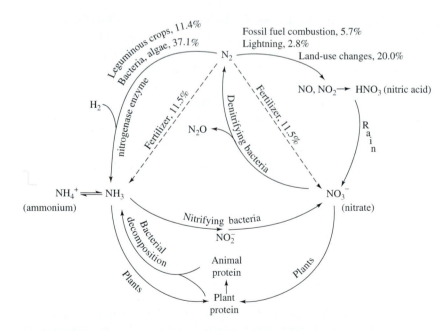

Figure 15.1   The nitrogen cycle, showing percentage contributions to nitrogen fixation from natural and human sources. (Total contribution of synthetic fertilizers, assuming equal shares of ammonia and nitrate, is 23.0 percent.)

TABLE 15.1   GLOBAL SOURCES OF FIXED NITROGEN

	Anthropogenic (teragrams/yr)
Nitrogen fertilizer	83
Legumes & other plants	40
Fossil fuels	20
Land use changes	
• Biomass burning	40
• Wetland draining	10
• Land clearing	20
Total anthropogenic	**213**

	Natural (teragrams/yr)
Lightning	10
Algae, soil bacteria	130
Total natural	**140**

Sources: Vitousek et al. (1997). Human alteration of the global nitrogen cycle: Causes and consequences. *Issues in Ecology* Number 1, Spring 1997 (Washington DC: Ecological Society of America). FAOSTAT (1999). *Agricultural Data, Nitrogenous Fertilizer Consumption.* (Rome, Italy: Food and Agriculture Organization). (http://apps.fao.org)

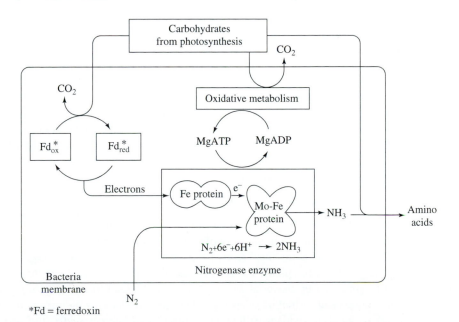

Figure 15.2   The chemistry of the nitrogenase enzyme system in algae and bacteria.

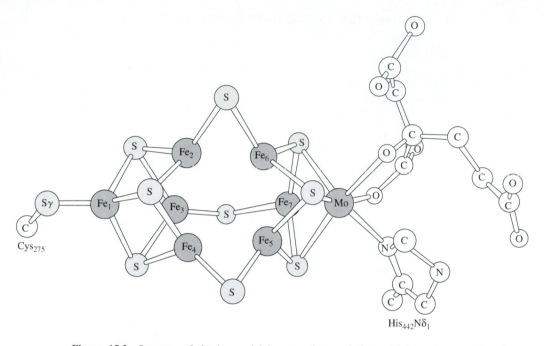

Figure 15.3 Structure of the iron-molybdenum cofactor of the molybdenum-iron protein of nitrogenase. Seven iron and one molybdenum atoms form a cluster with bridging sulfur atoms. The cluster is ligated to the protein by two side chains from the cysteine 275 and histidine 442 residues. In addition, an organic acid, homicitrate, is bound to the molybdenum atom. The cavity in the center is a possible site of binding and activation of N_2.   *Source:* S. J. Lippard and J. M. Berg (1994). *Principles of Bioinorganic Chemistry* (Sausalito, California: University Science Books). Copyright © 1994. University Science Books. Reprinted with permission. All rights reserved.

contains a complex of sulfur and iron atoms, and the "Mo-Fe protein," which contains a complex of sulfur, iron, and molybdenum atoms (see Figure 15.3). The reduction of N_2 is carried out at the active site of the Mo-Fe protein, where six electrons and six protons are added to N_2, producing two molecules of NH_3. The role of the Fe protein is to transfer electrons to the Mo-Fe protein, in coordination with the hydrolysis of MgATP (ATP is *adenosine triphosphate*) to MgADP (ADP is *adenosine diphosphate*), a process that releases energy. Even though the overall reaction

$$N_2 + 3H_2 = 2NH_3 \qquad \Delta G = -94 \text{ kJ/mole} \qquad (15.1)$$

is downhill in energy, the N_2 bond is so strong (941 kJ/mol) that additional energy must be provided by the organism in the form of MgATP to overcome the activation barrier.

Plants can use ammonia directly for their nitrogen source; animals gain their nitrogen by eating plants. When plants and animals die, the reduced nitrogen in their tissues is converted to ammonia by bacterial decomposition and added to the ammonia pool. The ammonia can be used as fuel by other bacteria (*Nitrosomonas*), which convert NH_3 to NO_2^- (nitrite), using O_2 as oxidant. Still other bacteria (*Nitrobacter*) oxidize the nitrite further, to

NO_3^- (nitrate). (The overall process of nitrogen oxidation is called *nitrification.*) Plants can also utilize nitrate as a nitrogen source. Thus, there is a continual cycling between oxidized and reduced forms of fixed nitrogen in soils.

If the fixed nitrogen were not returned to the atmosphere, the atmospheric pool of N_2 eventually would be depleted. But the nitrogen cycle is closed by *denitrifying* bacteria, which use NO_3^- instead of O_2 as the oxidant in metabolic reactions, reducing the nitrate back to N_2 (see Part III, pp. 310–312). Both nitrification and denitrification processes release some N_2O as a side-product. N_2O is a greenhouse gas and the main source of stratospheric NO, a principal actor in the chemistry of O_3 destruction (pp. 164–165, 208–209).

15.2   AGRICULTURE

a. Fertilizer and the green revolution.   The major natural sources of nitrogen for agriculture are the nitrogenase-containing bacteria, some of which grow in symbiosis with a limited variety of crop plants, most notably those of the legume family. These symbiotic bacteria are contained in the root nodules of such legumes as beans, peas, alfalfa, and clover. When the plants die, most of the nitrogen is returned to the soil in fixed form, where it is available for other kinds of plants; a small fraction is returned to the atmosphere via the denitrification reaction. The fertilizing ability of legumes is the reason for crop rotation, an ancient agricultural practice in which legumes are planted alternately with cereals, grains, and other vegetables to maintain the productivity of the nonlegume plants. In the absence of crop rotation, plants that do not fix nitrogen deplete the soil's nitrogen stores quickly, unless these stores are replenished by the addition of fertilizer.

One traditional fertilizer is animal manure, but increasingly in recent decades it has been supplanted by artificial fertilizers produced industrially. The industrial production of nitrogen fertilizer (dashed lines in Figure 15.1) is accomplished via the Haber process, in which reaction (15.1) is carried out over an iron catalyst. Even with a catalyst, the reaction requires high pressures (100 atm) and temperatures (500°C). (In contrast, the nitrogenase enzyme operates at ambient pressure and temperature, but energy input in the form of MgATP, see Figure 15.2, is required.) The resulting ammonia can be injected directly into crop-bearing soils, or, more conveniently, it can be added as ammonium nitrate salt, produced by air oxidation of half the ammonia to HNO_3, which is recombined with the remaining ammonia:

$$NH_3 + 2O_2 = HNO_3 + H_2O \tag{15.2}$$

$$NH_3 + HNO_3 = NH_4NO_3 \tag{15.3}$$

The Haber process requires a source of hydrogen gas. Currently, the most economic process for obtaining hydrogen is from methane reformation:

$$CH_4 + 2H_2O = 4H_2 + CO_2 \tag{15.4}$$

The global production of nitrogen fertilizers has increased dramatically in the past four decades (see Figure 15.4), and is now double the estimated N_2 fixation rate of leguminous crops (83 versus 40 Tg/yr of fixed N). The total of these two anthropogenic sources

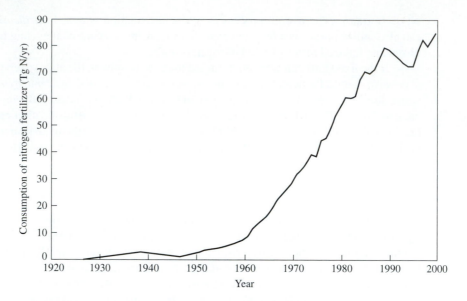

Figure 15.4 Global nitrogen fertilizer consumption from 1920 to 1999 in Tg N/yr (Tg = teragrams = 10^{12} g); cumulative consumption over this 80-year period was 2,179 Tg N, with half of that occurring between 1986 and 1999. *Source:* FAOSTAT (1999). *Agricultural Data, Nitrogenous Fertilizer Consumption* (Rome, Italy: Food and Agriculture Organization). (http://apps.fao.org)

roughly equals the estimated natural microbial N_2 fixation rate of 130 Tg/yr, and the nitrogen oxide contribution from fossil-fuel combustion adds another 20 Tg/yr. In addition, land-use changes such as biomass burning from slash-and-burn agriculture, land clearing, and draining of wetlands result in significant amounts of additional fixed nitrogen. [Land clearing kills trees and other vegetation, which in turn increases the amount of plant protein decomposed to NH_3. Recall that wetlands are a natural sink for nitrates, as they harbor denitrifying bacteria that reduce nitrate to N_2 (see Figure 13.11a,b, p. 329). Removing wetlands, in essence, reduces the biosphere's capacity to denitrify nitrate.] Quantifying the contribution of land-use changes is highly uncertain, but researchers estimate it to be on the order of 70 Tg/yr. Thus, human activities now appear to dominate the global nitrogen cycle, although this was not the case as recently as 1970. The relationship of these fixation rates to the global fluxes and pools of nitrogen is illustrated in Figure 15.5. Human activity has increased the rates of both nitrogen fixation and denitrification. The sum of the increased fixation on land (320 Tg/yr) and increased runoff of fixed nitrogen from land to the oceans (40 Tg/yr), however, does not appear to be entirely balanced by the increase in denitrification on land (160 Tg/yr) and in the oceans (110 Tg/yr). This result suggests that fixed nitrogen is now accumulating in the terrestrial pool (see problem 2, Part IV). The flows in the oceans show that inputs of fixed nitrogen are less than the output via denitrification. The uncertainties are too large to place great stock in this balance, but the result is consistent with the fact that nitrogen is a limiting nutrient, particularly in the deep ocean.

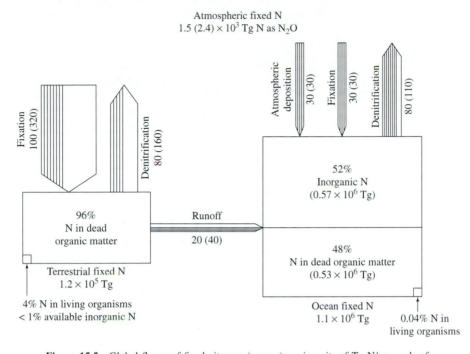

Figure 15.5   Global fluxes of fixed nitrogen (arrows) are in units of Tg N/yr; pools of nitrogen, in atmosphere, land, and oceans, are in units of Tg N. Shaded portions of arrows and the accompanying number (not in parentheses) indicate pre-industrial flows of fixed nitrogen. The arrows in total, and the accompanying numbers (in parentheses) refer to current nitrogen flows. Atmospheric deposition to the oceans includes fixed nitrogen caused by lightning, and net transfers of nitrogen from the land to the oceans via rainfall. *Sources:* Adapted from A. P. Kinsig and R. H. Socolow (1994). Human impacts on the nitrogen cycle. *Physics Today* November, 1994; Vitousek et al. (1997). Human alteration of the global nitrogen cycle: Causes and consequences. *Issues in Ecology* Number 1, Spring 1997 (Washington DC: Ecological Society of America); W. H. Schlesinger (1997). *Biogeochemistry: An Analysis of Global Change* (New York: Academic Press).

Most of the global fixed nitrogen is in the ocean, where it is divided nearly equally between organic and inorganic forms. The inorganic ions, NH_4^+ and NO_3^-, do not accumulate in soils; because they are soluble, they are quickly taken up by plants and bacteria, or are washed out of the soil and transported to the oceans. The terrestrial fixed nitrogen is mainly in organic matter.

The application of fertilizer can dramatically improve agricultural yields (see Figure 15.6a,b). The near quadrupling of yields for corn and other grains between 1950 and 2000 made the U.S. and Canadian Midwest the breadbasket of much of the world. Moreover, crop yields have increased everywhere (see Figure 15.7) as a result of fertilizer availability and improved crop strains.

Improved yields have been particularly important for poor countries, many of which have made substantial strides in expanding agricultural production to meet the nutritional

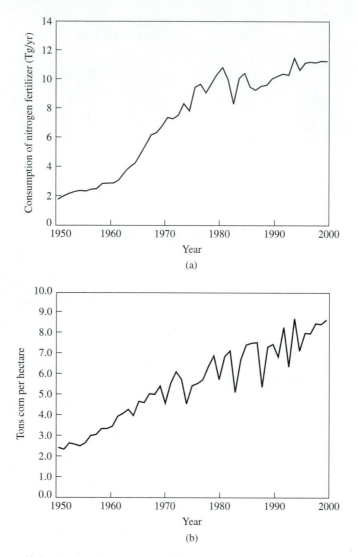

Figure 15.6 (a) Historical trends for the United States in consumption of nitrogen fertilizer (Tg/yr), and (b) yield of corn (tons per hectare). *Sources:* Data for 1950–1960 from archive agricultural database of Worldwatch Institute, Washington, DC; data for 1961–2000 from agricultural database for nitrogenous fertilizer and maize yield, FAOSTAT (2000). (Rome, Italy: Food and Agriculture Organization). (http://apps.fao.org)

needs of their growing populations. Much was accomplished in the 1960s through the "Green Revolution," when plant scientists working in research institutes in Mexico and the Philippines developed new strains of wheat and rice that produce high yields when fertilized. Traditional varieties of grain indigenous to tropical and subtropical countries do not respond well to fertilizer (see Figure 15.8). They grow too tall and fall over in winds and

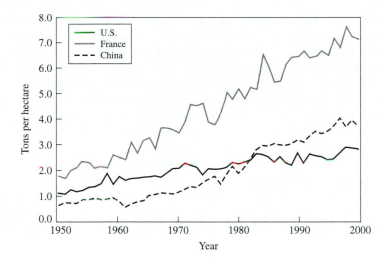

Figure 15.7   Wheat yield per hectare for China, France, and the United States, 1950–2000. *Sources:* Data for 1950–1997 from archive agricultural database of Worldwatch Institute, Washington, DC; data for 1998–2000 from agricultural database for wheat yield, FAOSTAT (2000). (Rome, Italy: Food and Agriculture Organization). (http://apps.fao.org)

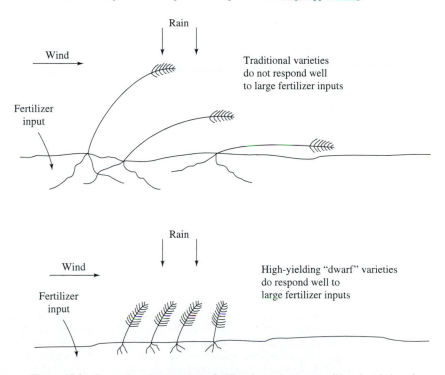

Figure 15.8   Comparison in response to fertilizer inputs between traditional varieties of grain and the dwarf varieties developed during the Green Revolution.

heavy rain. The new strains are dwarf plants with stalks that remain short while the grain matures. An additional benefit of dwarf varieties is that a larger number of plants can be grown per unit-area than traditional, less compact varieties, which have large leaves and extended root networks to absorb the small amounts of nutrients and water often found in tropical soils. However, the higher plant density has increased the required amount of water and made farms more dependent on irrigation (see Figure 10.2, p. 257). In addition, the traditional plants are more resistant to indigenous pests, so the new plants had to be protected with pesticides (see discussion beginning on p. 382). Thus, the Green Revolution had substantial costs as well as benefits.

A major issue is whether the productivity gains of world agriculture can be sustained. Crop yields can be improved only so far by fertilization. When sufficient fertilizer has been added to support plant growth at its maximum rate, then further additions either have no effect or actually inhibit growth. The curves of diminishing returns for corn in various soils are shown in Figure 15.9. Per capita world grain production leveled off around 1980, and

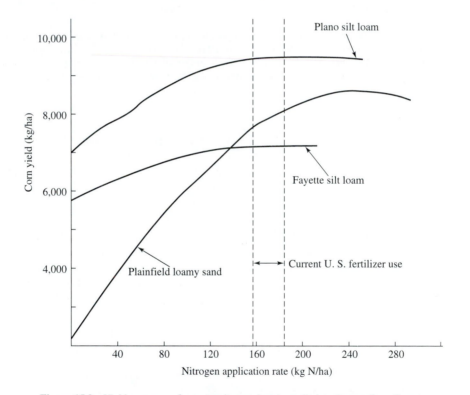

Figure 15.9 Yield response of corn to nitrogen inputs applied to three soils. *Source:* S. L. Oberle and D. R. Keeney (1990). A case for agricultural systems research. *Journal of Environmental Quality* 20: 4–7. Reprinted with permission, copyright © 1990, American Society for Agronomy, Crop Science Society of America, and Soil Society of America. Current U.S. fertilizer use was updated based on fertilizer use for corn in Iowa for the period 1985 to 1995 from data presented in R. A. Ney et al. (1996). *Iowa Greenhouse Gas Action Plan.* University of Iowa Report to the Iowa Department of Natural Resources, December, 1996.

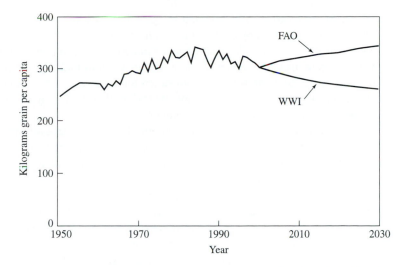

Figure 15.10   Two scenarios for future global per capita production of grain. Data from 1950 to 2000 are from archive agricultural database of Worldwatch Institute, Washington, DC; WWI trajectory from 2000 to 2030 is from L. R. Brown (1994). Facing food insecurity. In *State of the World, 1994,* L. R. Brown et al. (New York: W.W. Norton). FAO trajectory is from FAOSTAT (2000). (Rome, Italy: Food and Agriculture Organization). (http://apps.fao.org)

actually declined slightly during the 1990s. Figure 15.10 shows quite different projections as to whether this decline will continue (WWI—Worldwatch Institute) or recover (FAO—Food and Agricultural Organization of the United Nations). The FAO bases its more optimistic projection on the expectation that crop production will continue to increase in developing countries through higher yields and multiple cropping, and through arable land expansion. Per capita estimates are boosted by the slowing of world population growth, as estimated by the U.N. The decline of the 1990s is attributed to the decline of grain imports to the countries of the former Soviet Union following its collapse, and the reorganization of their agricultural systems; these countries have the capacity to be net grain exporters. The WWI, in contrast, does not expect that yield increases will keep up with population, citing especially the rising pressure on irrigation water, and the decrease in cropland resulting from urbanization and industrialization. Thus agricultural production is sensitive to several offsetting factors, which are difficult to weigh. An important question is the extent to which continuing crop development, including genetically modified crops will improve agricultural productivity, e.g., by improved control of pests (see pp. 401–406).

b. Environmental degradation.   Intensive agriculture carries serious environmental costs. When land is cleared for agriculture, the ecosystem is greatly disturbed. Deforestation is a well-known by-product of agricultural development. In the nineteenth century, much of the eastern United States was deforested and turned into farmland. The eastern mountains proved to be inhospitable for farming, and the thin soil was quickly degraded. Eastern farms closed as agriculture spread to the fertile plains of the Midwest, and the forest gradually recovered. Deforestation now principally affects tropical forests, which are being cleared and converted to agriculture. In the tropics, the problem of soil degradation is aggravated by heavy rains and heat. Under forest cover, the soil is very thin, and once it washes away, only a hard-packed mineral layer remains. Forest recovery takes longer than in temperate climates. In poor countries with rapidly growing populations, the cultivation of marginal lands that are not suited for agriculture produces a vicious cycle in which the newly cultivated land is rapidly degraded, thereby forcing the cultivation of still more marginal lands.

Soil erosion is a major problem for agricultural areas around the world. Wind and water carry off the soil particles that are exposed by farming. Erosion lowers crop productivity because of the loss of nutrients, soil biota, and organic matter, and especially because of the lowered availability of water, which is held less well by eroded soils. In addition, erosion greatly increases the sedimentation of rivers, reservoirs, and harbors. Sedimented stream bottoms prevent several species of fish (including salmon and trout) from reproducing. Severe salmon losses in the Northwest region of the United States have been attributed, in part, to stream sedimentation from agriculture and logging.

Eventually, the soil is reformed by accumulation of dead organic matter and the weathering of minerals, but this takes a long time, and varies widely with locale. Topsoil formation rates have been estimated over a range of 0.5–9 metric tons per hectare (ha) per year, while cropland erosion rates are substantially higher. The FAO reports that 25 billion tons of topsoil are washed away every year from 1.6 billion ha of erodible land worldwide, corresponding to an average erosion rate of 15.6 t/ha/yr. China's Yellow River Basin, one of the most severely affected regions, has an estimated erosion rate of 21 t/ha/yr.

Erosion rates can be reduced substantially by proper land management, such as contouring and terracing of hilly land, and by conservation practices. Iowa reports a reduction in state-wide erosion rates from 23 t/ha/yr in the early 1980s to 15 t/ha/yr in 1992, as a result of conservation tillage and the planting of permanent cover (grasses and trees) through the U.S. Department of Agriculture-supported Conservation Reserve Program. An important development is "no-till" agriculture, in which seeds are injected directly into the root zone (usually after the application of herbicides, see p. 397, to reduce the plant cover), without ploughing the land. Exposure of soil to wind and rain is thereby greatly reduced.

Aside from their effects on the soil, deforestation and crop planting have greatly disrupted ecosystems and reduced biodiversity. In addition, the runoff from fields can pollute aquifers, waterways, and estuaries with pesticides, and overfertilize them with nitrate and phosphate (see pp. 323–328).

15.3   NUTRITION

We now turn our attention to the biochemical pathways through which the food we eat keeps us functioning. This is the chemistry of biological metabolism.

The major nutritional categories are carbohydrates, fats, and proteins (see Figure 15.11). Carbohydrates are sugar molecules linked together in a long chain (see discussion of starch and cellulose, pp. 88–90). Fats are triglycerides of fatty acids, which have long hydrocarbon chains bonded to a glycerol unit. Proteins are composed of strings of amino acids joined by peptide bonds; each amino acid contains a characteristic side chain.

Carbohydrates are sugar polymers

e.g.:

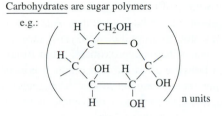

                                             n units

Fats are triglycerides of fatty acids

e.g.:

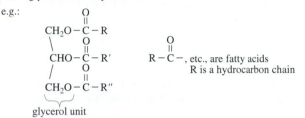

$R-\overset{\displaystyle O}{\overset{\|}{C}}-$, etc., are fatty acids
R is a hydrocarbon chain

glycerol unit

Proteins are polypeptides

e.g.:

Peptide link between two amino acids

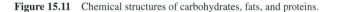

R is an amino acid side chain

One amino Second amino
acid unit   acid unit

Figure 15.11   Chemical structures of carbohydrates, fats, and proteins.

a. Energy and calories.   A simplified flow chart for the main biochemical processes is given in Figure 15.12. Carbohydrates are the immediate product of photosynthesis, and also the immediate source of biological energy in the process of respiration. Most of the energy that we need to keep our various bodily functions going is obtained from the oxidation of carbohydrates. Fats represent a form of biological energy storage. In comparison with carbohydrates, fats contain less oxygen and more carbon and hydrogen,

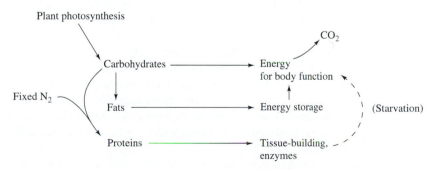

Figure 15.12   Flow chart of biochemical functions of carbohydrates, fats, and proteins.

the oxidation of which is the source of our energy. The energy content of fats is 9 Calories*/g, compared with 4 Calories/g for carbohydrates. Also, fats are immiscible with water, whereas carbohydrates are hydrophilic. Carbohydrates are usually found in association with a quantity of water that is approximately four times their weight. Consequently, the conversion of carbohydrates to fats represents a concentration of energy in lightweight portable form. A person who weighs 70 kg (154 pounds) contains about 16 percent fat. This represents enough energy for bodily requirements for 30 days. If this amount of excess energy were stored as carbohydrate with its associated water, the body weight would have to be 185 kg (406 pounds). When we eat more food than we require for our energy needs, the excess Calories are stored as fat. However, the biological oxidation of fat is slower than that of carbohydrates. When we need energy, we burn off our carbohydrate supply first and then call upon the fat stores.

b. Protein.   Aside from an energy supply, we need to maintain the biochemical machinery of the body itself. This is the realm of protein chemistry. Proteins make up most of the body's structural tissues, and also the myriad enzymes, the biological catalysts that carry out the thousands of reactions that are necessary for the maintenance of life. There are 20 different kinds of amino acids, each with a different chemical side chain (see Figure 15.13). Each kind of protein molecule is made up of a fixed sequence of these amino acids. Most of the amino acids can be synthesized by the body from a variety of starting materials as long as there is an adequate supply of protein nitrogen in the diet. There are, however, eight amino acids (see Figure 15.13) that the body cannot synthesize: valine, leucine, isoleucine, threonine, lysine, methionine, phenylalanine, and tryptophan.

These essential amino acids must be obtained directly from the diet and in amounts that allow us to maintain a proper balance of amino acids overall. For example, the ratio of tryptophan to lysine needs to be $0.6/2.6 = 0.23$, because the average human protein contains 0.6 tryptophans and 2.6 lysines per 100 amino acids (see Table 15.2). If this ratio is exceeded, the amount of protein that can be made by the body is limited by the lysine; the extra tryptophan is simply burned up. The notion of essential amino acid balance is similar to the concept of limiting nutrients, discussed earlier in connection with the growth of algae in water (see p. 323). Whichever essential amino acid is present in the lowest amount, relative to the frequency with which it is used, is the one that limits the total amount of protein produced.

Different food sources contain different amounts of protein; but they also vary in the amino acid composition of their protein. As might be expected, the amino acid composition of animal protein is fairly close to human protein. For this reason, milk and meat provide a rather close approximation to the amino acid balance we need (see Table 15.2). Plant protein, on the other hand, is farther from the human composition. The cereal plants, particularly wheat, are quite deficient in lysine. The lysine frequency in wheat is less than half the human average, which means more than twice as much wheat protein as milk protein is needed to supply human needs. Even though the remaining amino acids are closer to the correct proportions, they cannot be utilized without sufficient lysine. For this reason, wheat

---

*Recall from earlier discussion that a dietary *Calorie* is equal to 1 *kilocalorie* in the metric system.

General formula

Amino acid*	R (side chain)
1. Glycine	—H
2. Alanine	—CH$_3$
3. Serine	—CH$_2$OH
4. Aspartic acid	—CH$_2$COOH
5. Glutamic acid	—CH$_2$CH$_2$COOH
6. Asparagine	—CH$_2$CONH$_2$
7. Glutamine	—CH$_2$CH$_2$CONH$_2$
8. Arginine	—CH$_2$CH$_2$CH$_2$NHC(NH)NH$_2$
9. Cysteine	—CH$_2$SH
10. Tyrosine	—CH$_2$ —⬡— OH

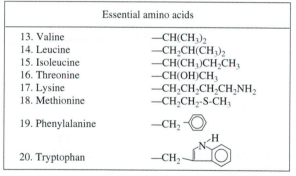

11. Proline

12. Histidine

Essential amino acids	
13. Valine	—CH(CH$_3$)$_2$
14. Leucine	—CH$_2$CH(CH$_3$)$_2$
15. Isoleucine	—CH(CH$_3$)CH$_2$CH$_3$
16. Threonine	—CH(OH)CH$_3$
17. Lysine	—CH$_2$CH$_2$CH$_2$CH$_2$NH$_2$
18. Methionine	—CH$_2$CH$_2$-S-CH$_3$
19. Phenylalanine	—CH$_2$ —⬡
20. Tryptophan	—CH$_2$ — (image)

*1–12 can be synthesized by the human body if sufficient protein nitrogen is present in the diet. 13–20 cannot be synthesized and must be obtained directly from food.

Figure 15.13 The amino acids and their chemical structures.

protein is often referred to as low-quality protein in comparison with the high-quality protein contained in meat, eggs, and milk.

Different plants show different patterns of deviation from the ideal amino acid balance. While wheat is deficient in lysine, beans have lysine in abundance. However, beans are deficient in methionine, which wheat has in relative abundance. In this respect, wheat and beans are complementary. If the two are mixed in equal amounts, the amino acid

TABLE 15.2   ESSENTIAL AMINO ACID CONTENT IN COMMON FOODS

Essential amino acid	Average frequency of occurrence in human protein (out of 100 amino acids)	Out of 100 amino acids, number present in:			
		Cow's milk	Meat	Beans	Wheat
Tryptophan	0.6	0.5	0.5	0.4	0.6
Phenylalanine	3.1	3.8	2.7	2.8	2.8
Lysine	2.6	3.3	3.4	2.8	1.1
Threonine	2.0	2.0	2.0	1.6	1.3
Methionine	3.5	2.8	3.3	2.2	3.2
Leucine	4.0	4.2	3.3	3.1	3.1
Isoleucine	2.5	2.8	2.3	2.2	1.6
Valine	3.2	3.6	2.7	2.5	2.0

Sources: Data from F. E. Deathrage (1975). *Food for Life* (New York: Plenum Press); President's Science Advisory Committee (1967). *The World Food Problem, Report of the Panel on World Food Supply,* Vol. II (Washington, DC: President's Science Advisory Committee).

balance is greatly improved. It takes less total vegetable protein for the mixture than for either one alone to provide human requirements. A diet of half wheat protein and half bean protein provides a protein mixture that is only 10 percent less efficient than milk protein in providing the right balance. It is no accident that beans and rice or wheat products are traditionally eaten together in many parts of the world. In order to be effective complements, the different proteins must be mixed in the same meal, so that they are digested together. Thus, vegetarians must pay attention to balancing different protein sources.

Although meat is an important component in the diets of relatively prosperous nations, most people in the world cannot afford to buy it, and it is possible to do without meat altogether and still remain healthy. Moreover, although meat is a high-quality source of protein, it represents a considerable waste of biological energy because animals typically store a rather small fraction of the food they consume in the form of meat. It takes 10 g of plant protein to produce 1 g of beef protein. Most countries do not have enough plant protein to spare for meat production on a large scale. If there is sufficient grass to support grazing animals, they can make a net contribution to the human food supply. However, in meat-producing countries, it is common to feed animals high-quality plant food such as cereals and soybeans to fatten them more quickly.

Figure 15.14a,b,c compares protein, fat, and Calorie consumption in four regions of the world. Americans consume, on average, 113 grams of protein per day (see Figure 15.14a). This is about twice as high as the recommended daily allowance, and 63 percent of the protein comes from animals. The Chinese diet also slightly exceeds the recommended daily allotment, and derives 34 percent of its protein from meat. The quantity of meat consumed in China has been rising in step with its increasing affluence. Although

Figure 15.14   Geographical comparisons in 1998 of (a) proteins; (b) fats; and (c) dietary Calories (kcal) per day. *Sources:* Food and Agriculture Organization, *Food Balance Sheets* (http://apps.fao.org); *Recommended Dietary Allowances* (10th edition) (Washington, DC: National Academy Press); *Nutrition and Your Health: Dietary Guidelines for Americans,* 5th edition (2000) (Washington, DC: U.S. Department of Agriculture/U.S. Department of Health and Human Services).

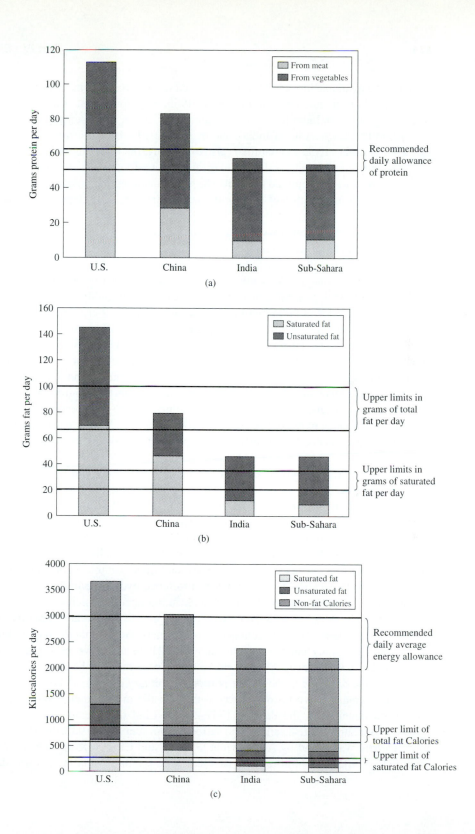

(a)

(b)

(c)

there is little available farmland for growing feed for livestock, China is buying large quantities of feed on the international market. In 1965, per capita meat consumption in China was 8.9 kg, while in India it was 3.6 kg/capita. By 1985 it had risen to 18.9 kg per capita (compared to 4.0 kg/capita in India), and by 1998 it was 46.5 kg/capita (compared to 4.6 kg/capita in India). India's economy is also expanding, but religious and other cultural factors have precluded a major shift toward meat consumption, and in 1998 only 18 percent of its dietary protein was supplied by meat. The Sub-Sahara region of Africa (excluding South Africa) is the poorest region of the world; meat contributes only 20 percent of dietary protein.

A similar pattern emerges for fat consumption (see Figure 15.14b). Americans consume far more fat than the recommended allowances, both in terms of total and saturated fats. The degree of saturated fats consumed affects health. Saturated fats tend to raise blood cholesterol, one of the major risk factors for heart and artery diseases. Meat is the main source of saturated fat, while vegetable oils contain more of the unsaturated variety. Due to the low consumption of meat in India and the Sub-Sahara, their diets are substantially below the upper limits of recommended fat consumption. This is also illustrated in Figure 15.14c, in terms of the recommended amounts of Calories that should be provided by fats. The pattern shows that the U.S. diet exceeds the limits for total Calories consumed, as well as the intake of total fat Calories and saturated fat Calories. In contrast, Indian and Sub-Saharan diets do not exceed any of the three allowances. One could argue from these figures that the American diet suffers from "over-nutrition," and is less healthy than diets in poorer regions of the world. It should also be kept in mind, however, that average per capita diets mask the chronic problems of malnutrition and famine in less-developed regions, especially in the Sub-Sahara.

c. Minerals and vitamins.   The human diet must contain a balanced assortment of building materials for maintaining the biochemical machinery. The needed elements must be present in the correct proportions. They are H, C, N, O, P, S, Na, K, Mg, Ca, Fe, Zn, Cu, Co, Cr, Mo, Se, I, and perhaps other elements in exceedingly small amounts. Moreover, the elements must be in assimilable chemical forms. We are unable to use carbon as CO_2 or nitrogen as NO_3^-, although plants do so. Also, ferric hydroxide is useless as a dietary source of iron because of its insolubility, and many soluble ferric salts are also ineffective, probably because they are converted to ferric hydroxide in the alkaline milieu of the intestine. Microbes and plants produce chelating agents, which extract iron from the ferric hydroxide in their vicinity, but these are not present in animal biochemistry.

Once the elements are assimilated, they are incorporated into the required biological molecules through myriad biosynthetic pathways. There are, however, some needed molecules that the body is incapable of synthesizing. These include the essential amino acids, which must be provided in the proper portions by dietary protein, and the vitamins.

The vitamins have been discovered through the disease states induced when they have been deficient in the diet. In 1747, James Lind discovered that citrus fruit was effective in treating British sailors who suffered from scurvy. In 1932, the active ingredient in citrus fruit was found by Albert Szent-Györgyi and Charles King to be ascorbic acid, vitamin C. Thirteen vitamins are known today. The last one (vitamin B_{12}) was discovered

50 years ago, and it is unlikely that any more will be found. Many people have lived for years on intravenous solutions that contain only the known vitamins and other nutrients.

The vitamins fall into two classes: fat-soluble and water-soluble (see Figure 15.15). The water-soluble vitamins are either enzyme cofactors or are required in cofactor synthesis. For example, niacin provides the pyridine end of NAD (*nicotinamide adenine dinucleotide*), while riboflavin is incorporated into FAD (*flavin adenine dinucleotide*). Both of these cofactors are used to catalyze many biological redox reactions. The roles of the fat-soluble vitamins are more complex. Vitamin A, or retinol, is incorporated into the visual pigment, rhodopsin. Vitamin D is required for the proper deposition of calcium in bones. Vitamin K is involved in the blood-clotting mechanism.

d. Antioxidants.   Vitamins E and A are natural antioxidants, protecting membranes from damage by molecular oxygen. So is β-carotene, the orange pigment of carrots and other vegetables, which is converted to vitamin A in the body (it is two molecules of retinol linked by a double bond, which replaces the OH group and an H atom on the terminal carbon atom). Vitamin C also has antioxidant properties and protects molecules in biological fluids. All of these antioxidants have double bonds, which readily react with free radicals (recall the role of olefins in the chemistry of smog, see p. 231). Figure 15.16 shows the products of successive electron transfers from vitamin C to free radicals.

There is much evidence implicating free-radical damage in the aging process and as a causative factor in a number of diseases, including cancer and heart disease. Radicals are generated in the body as side-products of O_2 reduction; collectively these side-products are sometimes referred to as ROS (reactive oxygen species). The release of partially reduced O_2, superoxide, and peroxide can lead to the production of hydroxyl radicals by transition metal-catalyzed reactions of the kind discussed earlier in connection with transition metal activation of O_2 in raindrops (see p. 192). The extremely reactive hydroxyl radicals attack biological molecules in their vicinity, generating organic radicals, which react further in radical chain reactions. These reactions can lead to mutations and cancer through damage to DNA, and to other disease states through damage to membranes or other critical parts of the biochemical machinery.

Because of the deleterious effects of radicals, organisms have evolved a variety of molecular defenses. These include enzymes that destroy the partially reduced O_2 species before they can generate hydroxyl radicals. These are *superoxide dismutases,* which catalyze superoxide disproportionation:

$$2O_2^- + 2H^+ = O_2 + H_2O_2 \qquad (15.5)$$

and *catalases,* which catalyze peroxide disproportionation:

$$2H_2O_2 = O_2 + 2H_2O \qquad (15.6)$$

Together, these two classes of enzymes ensure that partially reduced forms of oxygen are converted to O_2 and H_2O. This conversion is possible because these partially reduced forms are unstable with respect to O_2 and water; it takes more energy per electron to add one electron to O_2 (superoxide), or two of them (peroxide), than to add four of them (water). There are additional *peroxidase* enzymes, which function to break down peroxides.

Vitamin	Structural formula	Dietary sources	Deficiency symptoms
Fat-soluble vitamins			
Vitamin A	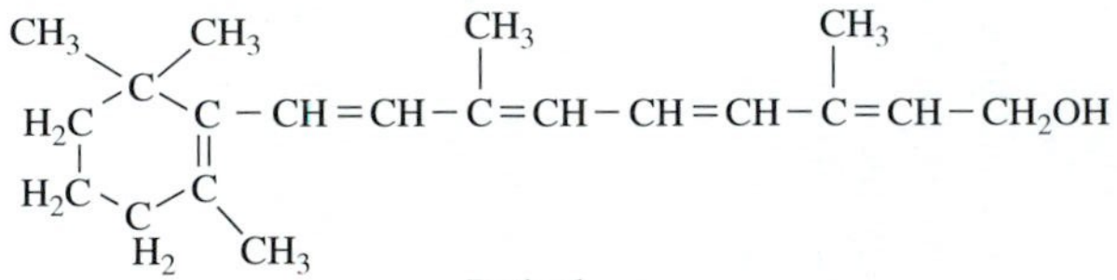 Retinol	Fish-liver oils, liver, eggs, fish, butter, cheese, milk. A precursor, β-carotene, is present in green vegetables, carrots, tomatoes, squash	Night blindness, eye inflammation
Vitamin D	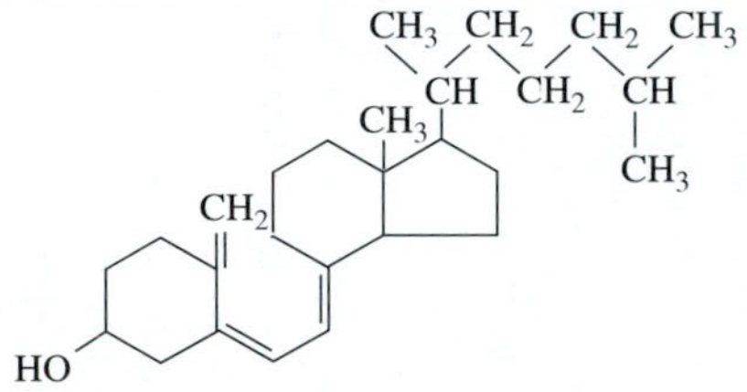 Vitamin D_3	Fish-liver oils, butter, vitamin-fortified milk, sardines, salmon. Body also obtains this compound when ultraviolet light converts 7-dehydrocholesterol in the skin to vitamin D	Rickets, osteomalacia, hypoparathyroidism
Vitamin E	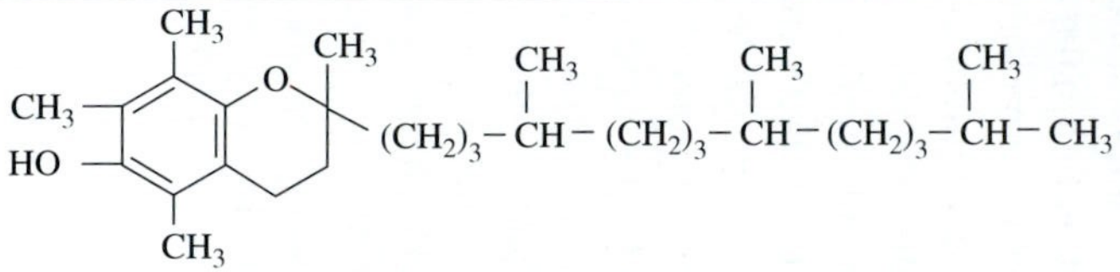 α-Tocopherol	Vegetable oils, margarine, green leafy vegetables, grains, fish, meat, eggs, milk	Anemia in premature babies fed inadequate infant formulas
Vitamin K	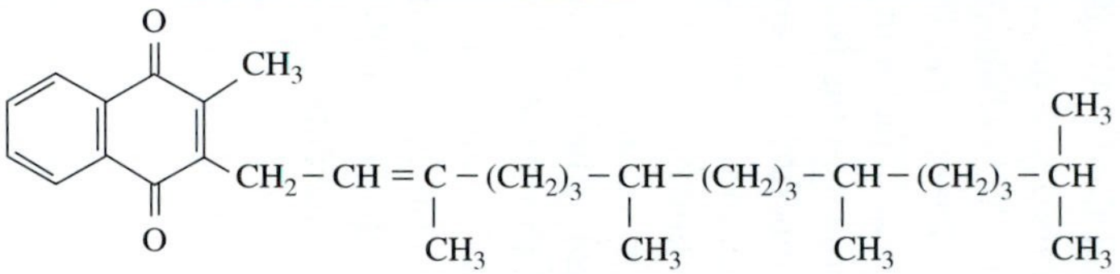 Vitamin K_1	Spinach and other green leafy vegetables, tomatoes, vegetable oils	Increased clotting time of blood, bleeding under skin and in muscles

Figure 15.15 Vitamins: Structures, dietary sources, and deficiency symptoms. *Source:* H. J. Sanders (1979). Nutrition and health. *Chemical and Engineering News* 57 (13): 27–46. Copyright © 1979, American Chemical Society. Reprinted with permission.

Water-soluble vitamins

Vitamin	Structure	Sources	Deficiency
Thiamine (vitamin B_1)	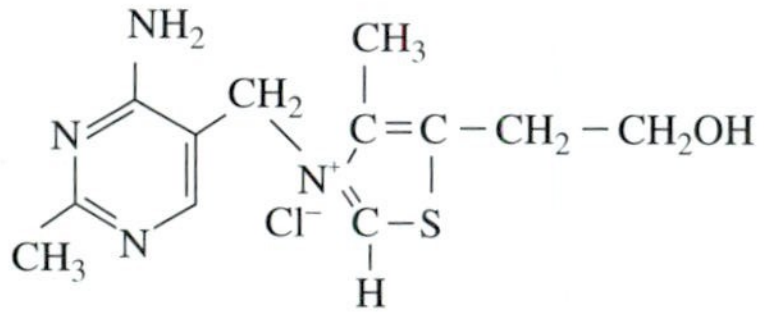 NH_2 CH_3 $C=C-CH_2-CH_2OH$ ... CH_2 N^+ ... Cl^- $C-S$ ... C H ... N N CH_3 Thiamine chloride	Cereal grains, legumes, nuts, milk, beef, pork	Beriberi
Niacin (nicotinic acid)	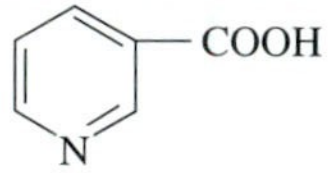 $COOH$ ... N	Red meat, liver, turnip greens, fish, eggs, peanuts	Pellagra
Riboflavin (vitamin B_2)	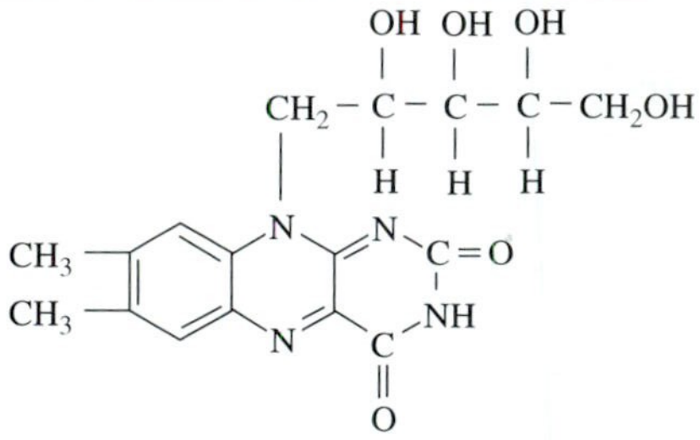 OH OH OH $CH_2-C-C-C-CH_2OH$ H H H N $N-C=O$ NH $C=O$ N CH_3 CH_3	Milk, red meat, liver, green vegetables, whole wheat flour, fish, eggs	Dermatitis, glossitis (tongue inflammation), anemia
Pyridoxine (vitamin B_6)	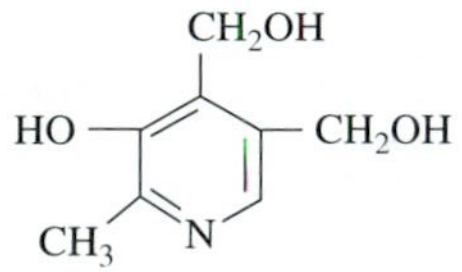 CH_2OH CH_2OH HO N CH_3 Pyridoxol	Eggs, meat, liver, peas, beans, milk	Dermatitis, glossitis, increased susceptibility to infections, irritability, convulsions in infants

Figure 15.15 (*continued*)

Vitamin	Structural formula	Dietary sources	Deficiency symptoms
Pantothenic acid	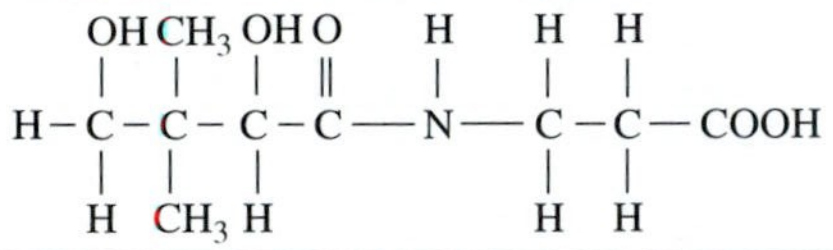 OH CH₃ OH O H H H $H-C-C-C-C-N-C-C-C-COOH$ with CH_3, H substituents	Liver, beef, milk, eggs, molasses, peas, cabbage	Gastrointestinal disturbances, depression, mental confusion
Folic acid (pteroylglutamic acid)	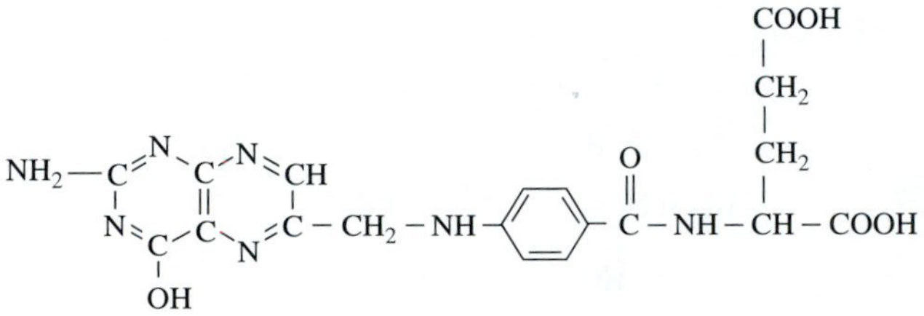 NH_2-C ... ring structure ... $C-CH_2-NH$ — phenyl — $C(=O)-NH-CH-COOH$ with CH_2-CH_2-COOH side chain; OH on ring	Liver, mushrooms, green leafy vegetables, wheat bran	Anemias, gastrointestinal disturbances
Biotin	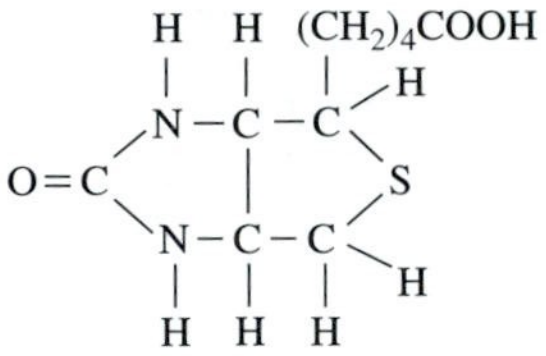 $(CH_2)_4COOH$; bicyclic ring with S and $O=C$ / N-C-N structure	Beef liver, kidney, peanuts, eggs, milk, molasses	Dermatitis

Figure 15.15 *(continued)*

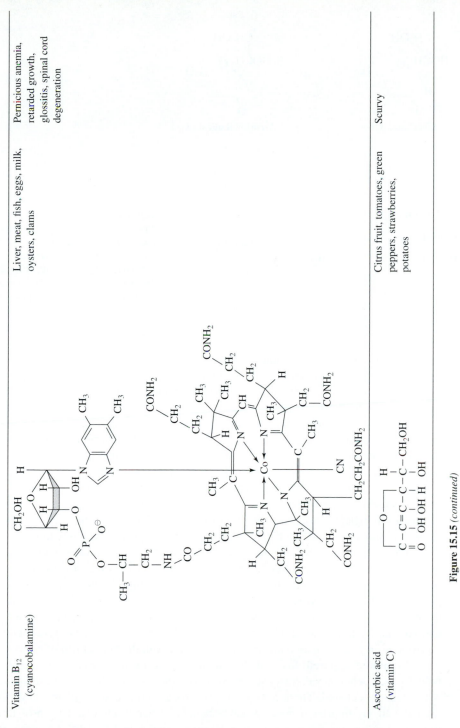

Vitamin B$_{12}$
(cyanocobalamine)

Liver, meat, fish, eggs, milk, oysters, clams

Pernicious anemia, retarded growth, glossitis, spinal cord degeneration

Ascorbic acid
(vitamin C)

Citrus fruit, tomatoes, green peppers, strawberries, potatoes

Scurvy

Figure 15.15 *(continued)*

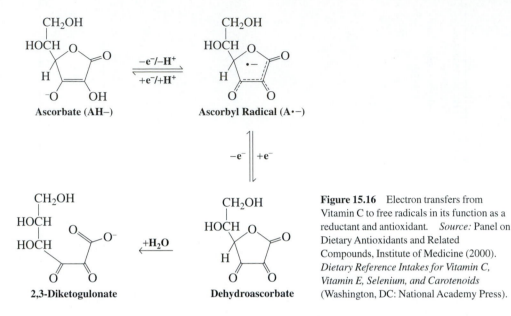

Figure 15.16 Electron transfers from Vitamin C to free radicals in its function as a reductant and antioxidant. *Source:* Panel on Dietary Antioxidants and Related Compounds, Institute of Medicine (2000). *Dietary Reference Intakes for Vitamin C, Vitamin E, Selenium, and Carotenoids* (Washington, DC: National Academy Press).

Another strategy for avoiding radicals is to make sure that redox-active metals, such as iron, copper, and manganese, are tied up in *chelates* (see discussion of chelating agents and detergents, pp. 298–300), so they cannot bind and activate O_2. In biological systems, transition metals do not circulate as free ions, but are bound and chelated to proteins and to smaller organic molecules. This sequestering of metals has been demonstrated to be important for avoiding radicals; in experimental systems, overwhelming the natural chelation system by adding excess metal salts to biological tissues can provoke free-radical damage.

The last line of defense against radicals are the antioxidants, like vitamins E, A, and C, which react sacrificially with radicals before they can reach biological targets. Not all antioxidants in the diet are vitamins, since not all antioxidants are specifically required for essential biological functions. These nonvitamin antioxidants include both naturally occurring molecules and synthetic molecules added to processed foods to preserve them from rancidity caused by oxidative damage. Synthetic antioxidants include butylated hydroxytoluene (BHT) and butylated hydroxyanisole (BHA) (see Figure 15.17). These phenolic compounds react readily with radicals by donating a hydrogen atom; the resulting phenoxyl radical is stabilized by the electron-donating alkyl and methoxy substituents on the benzene rings. Selenium, an essential dietary component in trace amounts, also has an antioxidant function, since it is incorporated into the active sites of the enzymes *glutathione peroxidase* and *thioredoxin reductase,* which function in the breakdown of peroxides.

Minimum required doses of the vitamins have been set by examining the levels below which deficiency diseases set in. However, the optimum level of dietary vitamins is still intensely debated. There have been numerous claims that supplements of vitamins higher than the minimum doses produce beneficial health effects. The most illustrious proponent of vitamin therapy was Linus Pauling, who argued forcefully that the optimum

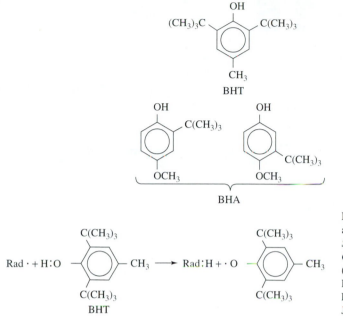

Figure 15.17   Chemical structures of BHT and BHA; reaction of BHT with a radical. *Source:* J. W. Hill and D. K. Kolb (1992). *Chemistry for Changing Times,* 7th Edition (Upper Saddle River, New Jersey: Prentice Hall). Reprinted by permission of Pearson Education, Inc., Upper Saddle River, New Jersey.

level of vitamin C is much higher than the minimum level, and that large doses can prevent colds and even cancer. Clinical studies, however, have not supported this theory. In general, there is a growing body of evidence that antioxidants may decrease the incidence of various diseases, including cancer and heart disease. But even this conclusion has been clouded by results reported in 1994 of a study on 29,000 male smokers in Finland, which showed an 18 percent higher incidence of lung cancer in those who took β-carotene supplements than in those who did not. And a study published in 1995 reported that excessive amounts of vitamin A consumed during the early months of pregnancy increase the risk of birth defects (although β-carotene consumption was not associated with this risk). Apparently, the complexities of metabolism and radical chemistry are such that adding a single component to the diet can have unexpected effects.

These cautionary results have led health authorities to withhold recommendations on antioxidants. At present they are able to conclude only that the diet should contain adequate fresh fruit and vegetables, which supply needed vitamins. Fresh produce also supplies many other natural chemicals that are likely to have beneficial effects; it was recently discovered, for example, that a compound in broccoli (sulforophane) can block tumor growth in experimental rats.

CHAPTER 16

PEST CONTROL

16.1 INSECTICIDES

A serious limitation on the human food supply is that we must share plant foods with insects. It has been estimated that the weight of the world's insect population exceeds that of its human inhabitants by a factor of 12. Only a small fraction of the insect species, about 500 species out of the world total of five million, actually feed on human crops, but these have the potential of doing enormous damage. Indeed, it is estimated that 30 percent of agricultural crops are consumed by insects worldwide.

Moreover, some insect species, including mosquitoes, fleas, and tsetse flies, are carriers of devastating human diseases. The death toll through the ages from such insect-borne diseases as malaria, yellow fever, bubonic plague, and sleeping sickness has been much larger than from warfare. Insect invasions, sometimes on a massive scale, have been a recurrent part of human history.

a. Persistent insecticides: Organochlorines.   Attempts to combat insect pests were relatively ineffective until the development of modern chemical pesticides. The first of these was *para*-dichlorodiphenyltrichloroethane, DDT (see Figure 16.1). Introduced by the Allies during World War II to control typhus and malaria outbreaks, DDT has since saved millions of additional lives through disease-vector control. Its discoverer, the Swiss chemist Paul Muller, won the Nobel Prize for Medicine and Physiology in 1948. After the war, DDT was the first pesticide to come into widespread agricultural use.

In many respects, DDT is an ideal insecticide. It is chemically stable and degrades only slowly under environmental conditions. Having low volatility, it also evaporates slowly, and it is not readily washed away because of its low solubility in water. These three characteristics make it a persistent insecticide. Each application is effective for a long time.

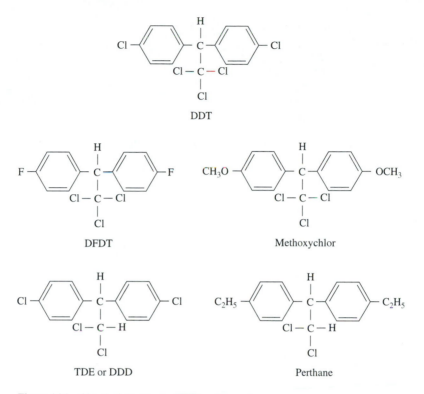

Figure 16.1   Chemical structures of DDT and its analogs.

Because it is hydrophobic, DDT readily penetrates the waxy outer coating of insects and, once inside, quickly paralyzes the insect. DDT acts by binding to the nerve cells of insects in a way that holds open the molecular channels that admit sodium ions, which in turn leads to uncontrolled firing of the nerves. DDT's toxicity to animals, including humans, is low, because animals absorb much less of the chemical in their tissues. It is this combination of persistence and selective toxicity to insects that made DDT such a successful insecticide.

However, it was not long before DDT began to lose its effectiveness because of the buildup of insect resistance; its use had begun to decline by 1960, over a decade before it was banned for most uses in the United States and other industrialized countries. Under heavy DDT application, individual insects that are relatively resistant to it are more likely to survive than those that are susceptible, and new generations have a steadily higher incidence of resistant characteristics. The main resistance factor for DDT is an enzyme called *DDT-ase,* which catalyzes the *dehydrochlorination* (loss of H and Cl atoms) of DDT to form DDE (see Figure 16.2), which has a new double bond at the central carbon atom. Because this atom is now trigonal instead of tetrahedral, DDE has a distinctly different shape than DDT and no longer binds strongly to insect nerve cells. Thus, insects that have evolved the ability to make high levels of DDT-ase can transform DDT into an innocuous molecule.

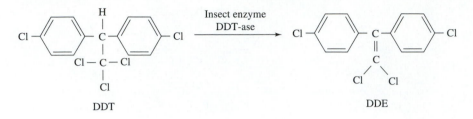

Figure 16.2 Mechanism for insect resistance to DDT.

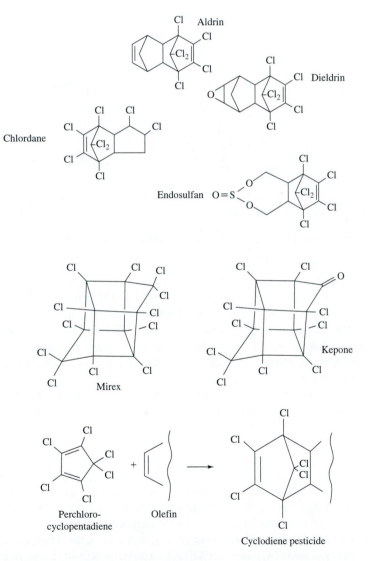

Figure 16.3 Cyclodiene insecticides, formed via the Diels-Alder condensation (bottom) of perchlorocyclopentadiene with oelfinic molecules.

Because of insect resistance, new insecticides were developed to supplement and replace DDT in various applications. The first of these were variations on the DDT theme (see Figure 16.1), molecules that retain the DDT framework but have different chemical substituents that are not as rapidly attacked by resistant insects' DDT-ase. Some substitutions worked while others did not; for example, replacing the chlorine atoms at the ends of the molecule with hydrogen atoms greatly decreased insecticidal activity. Finding effective variations of DDT was a matter of trial and error, because the nature of the binding site on the insect nerve cells was unknown. The more successful variants were brought into widespread use, until resistance to them also built up in the insect population. The nature of resistance means that no insecticide can remain effective for long, and it spurs the drive to develop new insecticides.

It was soon discovered that other organochlorine molecules, quite different from DDT, were also insect neurotoxins. Several of these were products of an addition reaction between perchlorocyclopentadiene and an olefinic molecule (see Figure 16.3). These "cyclodiene" insecticides also came into widespread use, as did toxaphene, a complex mixture resulting from the reaction of the naturally occurring hydrocarbon camphene with chlorine.

FUNDAMENTALS 16.1: MOLECULAR SHAPE AND BIOLOGICAL ACTIVITY

The detoxification of DDT by the DDT-ase insect enzyme is a good example of the critical importance of molecular shape in biology. The dehydrochlorination product, DDE, does not look very different from DDT when drawn as a flat structure (see Figure 16.2). But in three dimensions it is quite different (see Figure 16.4). Even if the chlorobenzene rings are held in approximately the same orientation (they can't be exactly the same, since the angle between them is 109.5° in DDT, but 120° in DDE), the position of the carbon with the two (DDE) or three (DDT) chlorine atoms is quite different. It is in the plane of the bonds to the chlorobenzene rings in DDE, but 54.8° out of the plane in DDT. Thus if the relative positions of these groups of atoms are important for binding to the insect nerve cells in a manner that holds open the sodium channels, then one can see how DDE would be ineffective as an insecticide.

The requirements for binding and activating a biological target such as the insect sodium channels depends on the complementarity of the shapes of the molecule and the target, often called *molecular recognition*. It is not necessarily the case that all the groups of atoms on the molecule participate in molecular recognition. Some of them may simply point away from the binding region of the target, and the precise orientation may then be unimportant. This may be why DDE, although ineffective as a sodium channel blocker, can nevertheless interfere with calcium deposition in eggshells (see next section) just as DDT does. The hormone receptor that regulates calcium deposition differs in shape from the sodium channel blocking site, and has a different profile for molecular recognition.

This kind of difference is found commonly. A molecule can fit a biological target exactly, producing a physiological response. Or it can fit partially, binding to the target without eliciting a response. In this case, the bound molecule can inhibit other molecules, which do

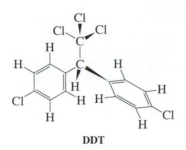

DDT

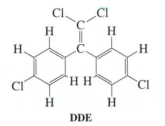

DDE

Figure 16.4   Structures of DDT and DDE. In contrast to DDT, the atoms attached to the C=C unit in DDE are coplanar.

elicit a response, from binding. A blocking molecule is called an *antagonist*, while a molecule that does elicit the physiological response is an *agonist*. Some organic chemicals in the environment have been found in laboratory tests to be either antagonists or agonists, depending on the particular receptor, and on the presence or absence of complementary or competing molecules (*synergistic* effects). This is one reason why evaluating hormonal effects of environmental chemicals (see pp. 418–423) is difficult and contentious.

---

b. Ecosystem effects; bioaccumulation.   DDT's success came at a price. Because insects are part of a very complex web of predator-prey relationships, a broad-spectrum insecticide like DDT is bound to have effects on the entire ecosystem. For example, when DDT was introduced in Borneo, as part of a malaria eradication campaign by the World Health Organization in the 1960s, mosquitoes were suppressed, but so were other insect species, including a wasp that preyed on the caterpillars that lived in the thatched roofs of village houses. Without the wasps, the caterpillar population exploded, and the thatch roofs were consumed. In addition, the dead mosquitoes were eaten by gecko lizards, which became sick and were easy prey for village cats. As a result of eating all the sick lizards, the cats in turn sickened and died, resulting in an explosion of the rat population. The rats ate the local crops and threatened an outbreak of bubonic plague. The Borneo government had to reintroduce cats into the affected region.*

*P. R. Ehrlich and A. H. Ehrlich (1981). *The Causes and Consequences of the Disappearance of Species* (New York: Random House).

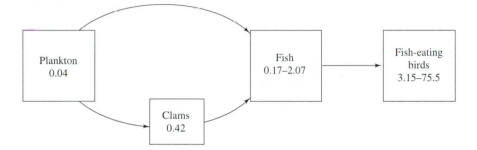

Figure 16.5   Accumulation of DDT in the aquatic food chain (in units of parts per million). *Source:* Adapted from C. A. Edwards (1973). *Persistent Pesticides in the Environment,* 2nd Edition (Cleveland, Ohio: CRC Press). Copyright © 1973 by CRC Press. Reprinted with permission.

The mosquito-lizard-cat connection illustrates the important principle of *bioaccumulation* in *food chains.* Chemicals in the prey become concentrated in the predator if the chemicals are not broken down and excreted rapidly. This is the case for persistent hydrophobic chemicals, which accumulate in the fat tissues of the predator. When the predator is eaten in turn, the chemical concentrates further in the fat of *its* predator. Each link in the food chain provides successive concentration. Figure 16.5 illustrates the accumulation of DDT in a typical aquatic food chain. If the plankton contain 0.04 ppm of DDT, then the clams feeding on the plankton have 10 times as much, while fish can have five times more if they feed on the clams. Finally, the fish-eating birds, at the top of the food chain, can build up quite high levels, up to 75 ppm in their fat tissue.

DDT can have biochemical effects other than neurotoxicity. One of them seems to be disruption of the avian hormonal system that controls calcium deposition during egg formation. As a result, birds having high levels of DDT (or its metabolic product, DDE, which has the same effect) lay eggs with shells that are too thin to endure until hatching. The populations of the peregrine falcon and other birds of prey fell precipitously in the years following the widespread use of DDT.

In 1962, Rachel Carson's book *Silent Spring* brought the ecological dangers of the uncontrolled use of insecticides sharply into public view, and subsequent legal and political actions led to severe restrictions on the use of DDT and other persistent insecticides in many countries during the 1970s. Since then, the threatened bird populations have recovered significantly in these areas. In addition, the accumulation of DDT in human tissues has fallen (see Figure 16.6). There have been worries about the long-term health effects of DDT exposure, although the evidence has been equivocal.

However, many other countries continue to use DDT, particularly in areas where malaria remains endemic. DDT is relatively cheap, and remains reasonably effective in mosquito control. Indeed, stopping DDT use can have dire consequences, as in the case of Sri Lanka, where a DDT-based mosquito control program had reduced the number of reported cases of malaria from 2,800,000 in 1948 to just 17 in 1963. After spraying was stopped in 1964, largely because of the political response to *Silent Spring,* malaria quickly

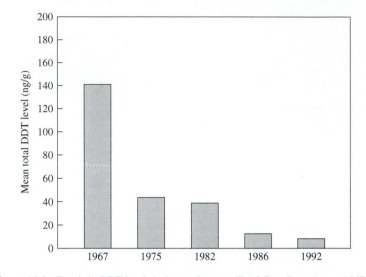

Figure 16.6   Trends in DDT levels in human breast milk of Canadian women, 1967 to 1992.   *Source:* The State of Canada's Environment 1996 (1996). (Ottawa, Canada: Government of Canada).

returned and reached 2,500,000 cases by 1969. A new control program brought some improvement, but malaria remains a serious problem.*

To get around the bioaccumulation problem, it is possible to engineer a molecule that has DDT's effectiveness in some insecticidal applications, but is less persistent. One example is methoxychlor (see Figure 16.1), in which the para-Cl atoms of DDT are replaced by methoxy groups, —OCH_3. The methoxy groups increase water solubility and susceptibility to degradation reactions. Consequently, methoxychlor does not bioaccumulate significantly and continues to be approved for use against flies and mosquitoes. However, the nonpersistent character of methoxychlor makes it more expensive to use since a single application is not as long-acting. As usual, there is no "free lunch" in environmental protection, and difficult choices often have to be made.

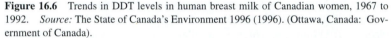

FUNDAMENTALS 16.2: BIOACCUMULATION AND THE PARTITION COEFFICIENT

Organic chemicals accumulate in fat tissues if they are more soluble in fats than in water. Fats are chemically complex, and vary somewhat in composition among different tissues in different organisms. However, it has been found that solubilities in the simple solvent 1-octanol, $CH_3(CH_2)_6CH_2OH$, are close to

---

*K. Mellanby (1992). *The DDT Story* (Farnham, Surrey, UK: British Crop Protection Council).

TABLE 16.1 RELATIVE SOLUBILITIES IN FATS VERSUS WATER FOR SELECTED PESTICIDES

Pesticide	Solubility in water (mg/L)	octanol/water partition coefficient (log K_{ow})
Organochlorine insecticides		
DDT	0.0028	6.0
Aldrin	0.08	5.8
Chlordane	0.20	5.1
Kepone	3.71	4.9
Mirex	0.05	6.4
Organophosphate insecticides		
Parathion	19	3.7
Malathion	144	2.7
Carbamate insecticides		
Carbaryl	73	2.4
Aldicarb	6,017	1.2
Herbicides		
Atrazine	38	2.6
Metolachlor	519	3.0

Source: Taken from average values reported in D. Mackay et al. (2000). *Physical-Chemical Properties and Environmental Fate Handbook.* (Boca Raton, Florida: Chapman & Hall/ CRCnetBASE).

solubilities in fats. The relative solubility in fats versus water is conveniently approximated by the octanol/ water *partition coefficient,* $K_{ow} = (S)_o/ (S)_w$, where $(S)_o$ and $(S)_w$ are the concentrations in octanol and water, respectively, when a compound, S, is equilibrated between the two liquids. Table 16.1 lists K_{ow} values for some pesticides. Values range from $10^{4.9}$ to $10^{6.4}$ for the organochlorine pesticides (see Figures 16.1 and 16.3), which are much more soluble in octanol than in water. This is why they bioaccumulate so readily. The organophosphate insecticides have oxygen and sulfur atoms (see Figure 16.8, p. 391), which are capable of accepting H-bonds from water, as do the nitrogen atoms of carbamates (see Figure 16.8) and the herbicide atrazine (see Figure 16.15, p. 398). These molecules have lower K_{ow} values, $10^{1.2}$ to $10^{3.7}$, and are less bioaccumulative.

The tendency to bioaccumulate can be expressed as the *bioconcentration factor*, which represents the concentration ratio of a chemical in an organism and the water around it. This factor is usually somewhat less than the K_{ow} because it depends on rates of intake and elimination. For example, the K_{ow} of DDE is 10^5, while its bioconcentration factor in fish ranges from 3,000 to 60,000, depending on the species. When K_{ow} is high, the rate of elimination never balances the rate of intake, and bioaccumulation increases with the age of the organism (see Figure 16.7). However, when K_{ow} becomes extremely high (see, for example, mirex in Table 16.1), the chemical tends to adsorb so strongly to particles in the sediment that it is not taken up effectively by organisms.

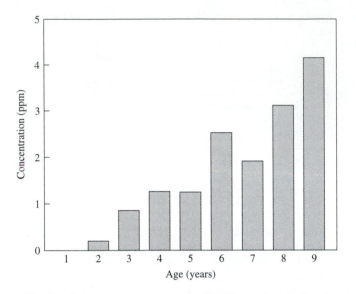

Figure 16.7 Correlation between concentration of DDT in trout in Lake Ontario and age
of trout. *Source: Toxic Chemicals in the Great Lakes and Associated Effects,* Volume 1,
Part 2 (1991). (Ottawa, Canada: Minister of Supply and Services).

c. Nonpersistent insecticides: Organophosphates and carbamates.
Other insecticides have been developed that are nonpersistent because they break down
rapidly into harmless and water-soluble products, once released into the environment. Because they do not last long, they must be highly potent. The two classes of widely used nonpersistent insecticides—organophosphates and carbamates (see Figure 16.8)—are in fact
powerful neurotoxins. Indeed, the organophosphate insecticides are in the same chemical
family as the nerve gas chemical-warfare agents that were developed during and after
World War II.

The organophosphates and carbamates both work by inhibiting the enzyme *acetylcholinesterase,* which hydrolyzes the neurotransmitter *acetylcholine.* Neurotransmitters
are molecules that are released by a nerve cell in order to fire an adjacent nerve cell (see
Figure 16.9) by diffusing across the gap between the cells, called the *synapse,* and binding
to receptors on the second cell. There are many kinds of neurotransmitter molecules, but
the one responsible for firing motor nerve cells in higher life forms is acetylcholine. Once
acetylcholine binds to its receptors, a motor nerve cell will continue to fire until the acetylcholine is broken down by acetylcholinesterase, which is present in the synapse. If acetylcholinesterase is inhibited, then nerve firing continues uncontrollably, leading to paralysis
and death.

Toxicity of organophosphates and carbamates is much lower than the nerve gasses,
but greater than organochlorine insecticides. Some of the most widely used ones, such as
parathion and aldicarb, are highly toxic and have caused death and injury to many agricultural workers. Thus, the environmental advantage of these nonpersistent agents is counter-

(a)
Organophosphates

$$RO - \underset{\underset{OR}{|}}{\overset{\overset{S}{||}}{P}} - X$$

General formula

X represents an SR' or OR' group
R, R' are organic groups
P=S is rapidly oxidized to P=O

examples:

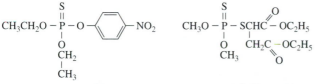

Parathion   Malathion

Chlorpyrifos

(b)
Carbamates

$$RO - \overset{\overset{O}{||}}{C}NHR'$$

General formula

examples:

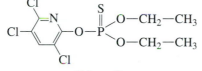

Carbaryl   Aldicarb

Figure 16.8   Chemical structures of organophosphate and carbamate insecticides.

balanced by their health impacts on agricultural workers. In the case of Sri Lanka, mentioned previously, DDT spraying for malaria control was replaced by parathion spraying, which resulted in many deaths among spraying crews, whereas none had been caused by DDT.

Because of health concern, the use of organophosphate and carbamate insecticides is being curtailed. The U.S. EPA has restricted the use of methyl parathion, and has now banned chlorpyrifos (see Figure 16.8), previously the most widely used household insecticide, because of concerns about neorotoxicity in children, arising from laboratory tests with newborn rodents.

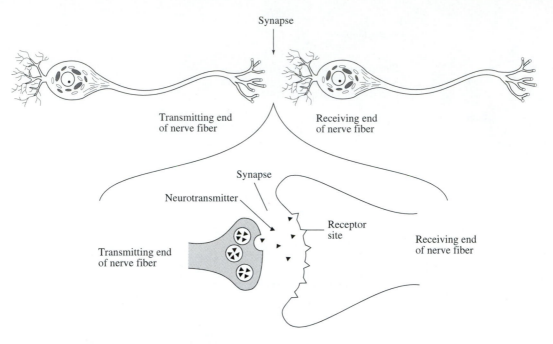

Figure 16.9   The transmission of a nerve impulse by release of neurotransmitter molecules across the synapse. *Source:* P. Buell and J. Gerard (1994). *Chemistry in Environmental Perspective* (Upper Saddle River, New Jersey: Prentice Hall). Reprinted by permission of Pearson Education, Inc., Upper Saddle River, New Jersey.

STRATEGIES 16.1   **Molecular Mechanism of Cholinesterase Inhibitors**

How do the insecticides inhibit cholinesterase? The enzyme works by binding acetylcholine, and then carrying out a displacement reaction on the acetyl group, using the OH group of a serine amino acid residue that is located at the enzyme active site (see Figure 16.10). The *choline* part of the molecule is released in this reaction, leaving the *acetyl* group bound to the enzyme. Next, the enzyme induces a water molecule to carry out a second displacement at the acetyl group, resulting in the release of acetic acid. The enzyme is then ready to carry out a second round of catalysis.

The insecticide molecules trick the enzyme by mimicking acetylcholine. They bind to the active site and induce the serine residue to carry out a displacement reaction, just as acetyl-

choline does. But instead of an acetyl group, the enzyme ends up with a bound *phosphoryl* group in the case of an organophosphate insecticide, or a bound *carbamyl* group in the case of a carbamate insecticide (see Figure 16.10). These groups are much less susceptible to attack by water than is the acetyl group. In both cases, the atom being attacked, carbon in the case of carbamyl and phosphorus in the case of phosphoryl, is attached to additional groups that inhibit the attack by the incoming water molecule. Consequently, the enzyme is blocked by the carbamyl or phosphoryl group from hydrolyzing acetylcholine.

The potency of the inhibitor depends on the rate of the initial displacement reaction, in which the enzyme's active-site serine is cap-

Normal mode of action

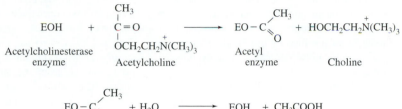

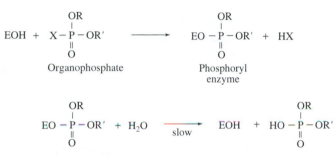

Inhibition by organophosphate insecticide

$$EOH + \underset{\underset{\text{Organophosphate}}{}}{X-\overset{\displaystyle OR}{\underset{\displaystyle O}{\overset{|}{\underset{\|}{P}}}}-OR'} \longrightarrow EO-\overset{\displaystyle OR}{\underset{\displaystyle O}{\overset{|}{\underset{\|}{P}}}}-OR' + HX$$

Phosphoryl
enzyme

$$EO-\overset{\displaystyle OR}{\underset{\displaystyle O}{\overset{|}{\underset{\|}{P}}}}-OR' + H_2O \xrightarrow{\text{slow}} EOH + HO-\overset{\displaystyle OR}{\underset{\displaystyle O}{\overset{|}{\underset{\|}{P}}}}-OR'$$

Inhibition by carbamate insecticide

$$EOH + RO-\overset{\displaystyle O}{\overset{\|}{C}}-N\overset{\displaystyle R'}{\underset{\displaystyle H}{}} \longrightarrow EO-\overset{\displaystyle O}{\overset{\|}{C}}-N\overset{\displaystyle R'}{\underset{\displaystyle H}{}} + ROH$$

Carbamyl
enzyme

$$EO-\overset{\displaystyle O}{\overset{\|}{C}}N\overset{\displaystyle R'}{\underset{\displaystyle H}{}} + H_2O \xrightarrow{\text{slow}} EOH + OH-\overset{\displaystyle O}{\overset{\|}{C}}-N\overset{\displaystyle R'}{\underset{\displaystyle H}{}}$$

Figure 16.10   Mechanisms of cholinesterase inhibition by organophosphate and carbamate insecticides.

tured. The better the *leaving group* (X for the organophosphates, OR for the carbamates, see Figure 16.10), the faster the reaction. For example, fluoride is an excellent leaving group because the fluoride ion is quite stable in water (as opposed to methoxide, for example, which is a strong base and is harder to displace). Organophosphate molecules with fluoride substituents are extremely powerful cholinesterase inhibitors, and the nerve gases are molecules of this class (an example is sarin, methylisopropoxyfluorophosphate).

In order to use organophosphates as insecticides without massively poisoning people and other animals, the reactivity of the organophosphate must be toned down. An effective way to do this is to replace the P=O group with the P=S group (producing a *phosphorothioate*; all the organophosphate insecticides in Figure 16.8 are phosphorothioates). The S atom significantly deactivates the P atom toward attack and slows the rate of reaction with cholinesterase. However, once inside the insect, the S atom is rapidly removed by oxidative enzymes, and the molecule is converted back to a potent organophosphate neurotoxin. Animals also have oxidative enzymes, but at levels much lower than in insects. Thus, the phosphorothioate turns the insect's biochemistry on itself and significantly decreases toxicity toward other species.

It is sometimes possible to introduce added safety by clever molecular engineering. A good example is malathion (see Figure 16.8), whose toxicity is several hundred-fold less than

parathion. Its diesterthiomethyl substituent is a good leaving group, making malathion an effective neurotoxin once the P=S group is replaced by P=O. But the ester groups are rapidly hydrolyzed by *carboxylase* enzymes, and the resulting dicarboxylthiomethyl substituent is a poor leaving group because of its negative charge. The carboxylase enzymes are abundant in animals but not in insects. Thus, the biochemical differences among organisms again are exploited to improve the selectivity of the toxin.

---

d. Natural insecticides.   Many plants have evolved their own chemical defenses against insects, and there is interest in using these natural and environmentally benign insecticides. Usually the plants' molecules are too difficult to extract and/or too complicated to manufacture on a commercial scale. But they may inspire chemists to make new kinds of insecticides. Pyrethroids (see Figure 16.11) are an example. Used by humans for centuries, pyrethrin is obtained from *pyrethrum,* a daisy-like flower. It is available commercially but is of limited use because it is unstable to sunlight; pyrethrins are *too* nonpersistent to be effective. Recently, however, pyrethrin-like insecticides have been developed that are chemically modified to improve their stability in the environment.

A useful weapon in the battle against insects is the soil bacterium *Bacillus thuriengiensis* (abbreviated to Bt), which produces protein toxins that are lethal to a number of plant-eating insects. These toxins are difficult to isolate, but the bacterium itself affords protection when sprayed on crops. Many organic farmers use Bt sprays as a "natural" means of controlling insects. There is currently great controversy over the use of the Bt trait in genetically modified plants, as discussed on pp. 402, 405.

e. Integrated pest management.   Insect resistance remains a problem for all insecticides. Not only do the chemicals become progressively less effective, but they also sometimes remain more effective against natural insect enemies of the target pest, thereby actually making the problem worse than it was at the beginning. This is because the predator species are usually slower to reproduce than the prey species, so resistance takes longer to develop among them. A fairly common pattern is that the introduction of a new insecticide causes an immediate decline in the population of the insect to be controlled, followed a few years later by a population explosion of a strain of the same insect against which the insecticide is no longer effective.

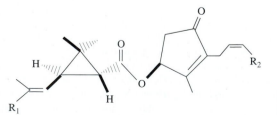

Figure 16.11   General structure of pyrethoids (R_1 and R_2 are alkyl groups).

For this reason, increasing attention has been given to the development of insect-control methods that operate more selectively. For example, research on the biochemistry of insects has led to the discovery of hormones that control their growth and sexual behavior. If applied at the right time, these chemicals can disrupt the insect's life cycle. For example, juvenile hormones (see Figure 16.12a) regulate growth and can disrupt the synchronization of metamorphosis if they are applied externally.

Juvenile hormones: natural insect growth regulators

Example:

$$CH_3-\underset{\underset{CH_3}{|}}{\overset{\overset{R}{|}}{C}}-CH_2-CH_2-CH_2-\underset{\overset{|}{CH_3}}{CH}-CH_2-CH=CH-\underset{\overset{|}{CH_3}}{C}=CH-C\overset{\nearrow O}{\underset{\searrow R'}{}}$$

General formula

Specific formula: Controls stages of growth in:

$R = H$, $R' = OCH_2CH_3$ Potato aphid, cockroach, grain-eating beetles

$R = H$, $R' = OCH_2C \equiv CH$ Green peach aphid, pea aphid, citrus mealybug

$R' = OCH_3$, $R' = OCH(CH_3)_2$ Mosquito, apple maggot, Mediterranean fruit fly

(a)

Pheromones: natural insect sex attractants

Example:

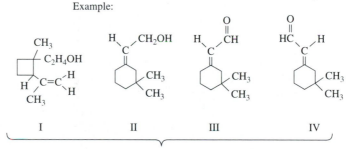

I   II   III   IV

Mixture of molecules emitted by male boll weevil as sex attractant

(b)

Figure 16.12 Use of (a) juvenile hormones and (b) pheromones in insect control. *Sources:* (a) C. Hendrik et al. (1976). Insect juvenile hormone activity of alkyl (2E, 4E)-3,7,11-trimethyl-2,4-dodecadienoates. Variations in the ester function and the carbon chain. *Journal of Agriculture and Food Chemistry* 24:207–218. (b) R. D. Henson et al. (1976). Identification of oxidative decomposition products of the boll weevil pheromone, grandlure, and the determination of the fate of grandlure in soil and water. *Journal of Agriculture and Food Chemistry* 24:228–231. Copyright © 1976 by American Chemical Society. Reprinted with permission.

Many insects rely on *pheromones,* molecules that act as messengers between individuals, guiding them to each other or to their food supply (see Figure 16.12b). Application of sex attractants can confuse the insects and prevent them from finding mates. It is also feasible to use pheromones to bait insect traps that contain high concentrations of toxic chemicals, and are therefore much more effective in killing the insects than broad-scale spraying.

Another technique of insect control is using chemicals or radiation to sterilize large numbers of male insects, which are bred for the purpose. When these sterile insects are released, they mate with the native population but produce no offspring. As a result, the total insect population is substantially reduced. For example, this strategy has been used to control the Mediterranean fruit fly in California. It is also possible to introduce predators artificially to control the population of pests and insects. This solution requires caution, however, because the predator can become so well adapted that it becomes a bigger pest than the target insect.

These newer methods of controlling specific insect species are still under development. They require careful planning and timing, and are much more sophisticated than simply spraying a field with a chemical insecticide. Because they are effective against only one species at a time, they are costlier than applying the broad-spectrum insecticides. Nevertheless, with the increasing environmental costs and decreasing effectiveness of the traditional insecticides, pressure is mounting for their development and use.

These pressures have actually produced a significant decline in insecticide use in the United States since 1975 (see Figure 16.13). Massive spraying has in many cases been

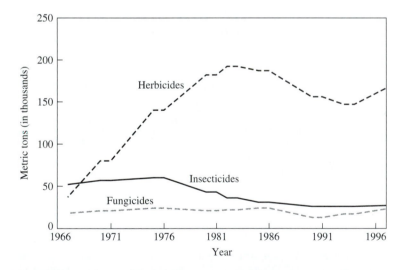

Figure 16.13   Trends in use of herbicide, insecticides, and fungicides in the United States, 1966–1997.   *Sources:* National Research Council (1989). *Alternative Agriculture* (Washington, DC: National Academy Press); Padgett et al. (2000). *Production Practices for Major Crops in U.S. Agriculture, 1990–1997* (Washington, DC: Economic Research Service, USDA, Statistical Bulletin no. 969).

replaced by integrated pest management (IPM) strategies, which utilize a combination of tactics, including crop rotation (which helps to forestall infestation by resistant species), tillage practices, water management, biological controls, and insecticides. Central to IPM is the concept of an optimal treatment threshold; the aim is not to completely eradicate the pest species, but rather to hold its population below a given damage level. IPM has been successfully applied to major crops, including cotton. In the mid-1970s, cotton accounted for over 40 percent of all insecticides used in the United States, but by 1982, the application rate had declined fourfold, from about 6.5 to 1.7 kg of insecticide per hectare.

16.2   HERBICIDES

Weeding out undesirable plants from crops is an essential aspect of food production. In commercial agriculture, manual and mechanical weeding has been widely replaced by herbicides. In addition to saving labor, herbicides are applied in "no-till" agriculture, which minimizes erosion by reducing disturbance of the soil. In this practice, fields are planted without plowing by injecting seeds directly into the ground after weeds have been controlled with herbicides. Broad application of no-till agriculture since the mid-1960s largely explains the dramatic expansion of herbicide use in the United States (see Figure 16.13). The lion's share of this increase (80 percent by the mid-1980s) was attributable to just two crops, corn and soybeans (see Figure 16.14). No-till agriculture requires high levels of herbicides because weeds must be thoroughly eradicated; even a small percentage of weeds remaining after her-

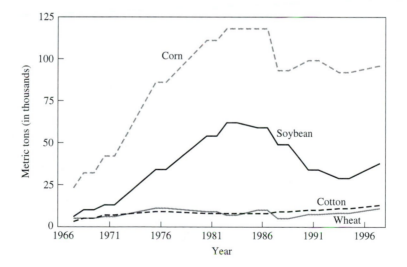

Figure 16.14   Trends in herbicide use for corn, cotton, soybeans, and wheat in the United States, 1966–1997.   *Sources:* National Research Council (1989). *Alternative Agriculture* (Washington, DC: National Academy Press); Padgett et al. (2000). *Production Practices for Major Crops in U.S. Agriculture, 1990–1997* (Washington, DC: Economic Research Service, USDA, Statistical Bulletin no. 969).

bicide application produces enough seeds to restore the weed population and choke off the crop. Nevertheless, herbicide use has declined since the early 1980s, partly because of concerns about health and ecosystem effects similar to those surrounding insecticides.

The largest class of herbicides is the *triazines* (see Figure 16.15), of which the best known is *atrazine*, the main agent used in cornfields. Atrazine does not bioaccumulate significantly, being moderately soluble in water, but it is fairly persistent and often detectable in farm wells. Although it is not very toxic ($LD_{50} = 1,870$ mg/kg, see Chapter 17 and Table 17.1, p. 409), there is some concern about correlations of high well-water concentrations with cancer and birth defects. Atrazine is being replaced in some areas by *metolachlor* (see Figure 16.15), which degrades more rapidly in the field.

Another herbicide, *paraquat* (see Figure 16.15), has been used extensively in marijuana-eradication campaigns. It is quite toxic and is suspected of having produced lung damage in some marijuana smokers. It is also used in developing countries to combat louse infestations, and has produced cases of paraquat poisoning there.

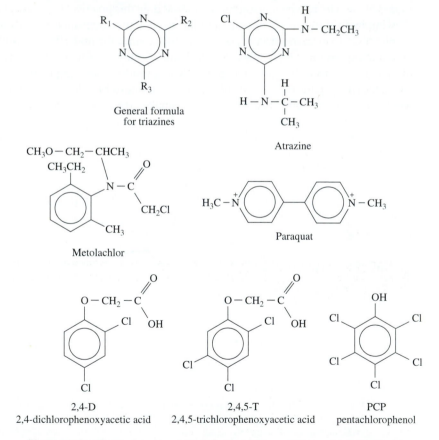

Figure 16.15   Chemical structures of selected herbicides, and the wood preservative pentachlorophenol (PCP).

The chlorophenoxy compounds 2,4-D and 2,4,5-T (see Figure 16.15) are effective herbicides, 2,4-D for weeds in lawns and 2,4,5-T for clearing brush. 2,4-D is still used in large quantities, but 2,4,5-T was banned in the 1970s because of dioxin contamination (see below, p. 423). A mixture of the two herbicides, the infamous Agent Orange, was used extensively as a defoliant during the Vietnam War. Another member of this class of chemical, pentachlorophenol (PCP) (see Figure 16.15), is widely used as a wood preservative.

A notable development has been the introduction of low-toxicity herbicides that break down readily in the environment. The most widely used is glyphosate, [N-(phosphonomethyl)-glycine, $^-HO_3PCH_2N^+H_2CH_2COOH$], sold under the trade name "Roundup." A simple amino acid derivative, glyphosate is water-soluble but clings to soil. Because of its ionic structure, it is attracted to ion-exchange sites on soil particles, and is therefore not washed into groundwater. Glyphosate is metabolized by soil microorganisms and has a half-life in soils of about 60 days, on average. It does not bioaccumulate to any significant extent.

When sprayed on plants, glyphosate enters the leaves and inhibits an enzyme that is required for the synthesis of the aromatic amino acids, phenylalanine, tyrosine, and tryptophan. Blockage of this essential biosynthetic pathway kills the plant. The pathway is different in animals and plants, and animals are unaffected by glyphosate, which they rapidly eliminate. However, all plants are affected, and glyphosate cannot be used to kill weeds selectively among crops. It is effective in clearing all vegetation from an area, and it is increasingly used to prepare fields for no-till agriculture. Once glyphosate has been applied, the field can safely be planted because new plants do not absorb glyphosate from the soil; the molecules are held too tenaciously by the soil particles.

Glyphosate is also useful in combating exotic species of plants that invade a habitat and crowd out native plants. For example, glyphosate was used to restore Wingham Brush, a nine-hectare remnant of Australia's rainforest, in the Manning River Valley of New South Wales. Wingham Brush is an important habitat of the flying fox, a native species of bat. The flying fox population had been threatened by foreign weeds that were smothering the trees where the bats live. These weeds were cleared by the application of glyphosate, allowing the indigenous plants to recover.

STRATEGIES 16.2   **Molecular Mechanism of Glyphosate Inhibition**

How glyphosate kills plants has been worked out in molecular detail. The enzyme it inhibits is 5-enolpyrovylshikimate-3-phosphate (abbreviated EPSP) synthase. EPSP synthase catalyzes the attack of shikimate-3-phosphate (S3P) on phosphoenol pyruvate (PEP) to make EPSP, with the release of a phosphate ion (P_i), (see Figure 16.16). The EPSP product is subsequently converted to the aromatic amino acids required by the plant.

The crystal structure of the EPSP synthase protein reveals that glyphosate binds at a site immediately adjacent to the S3P reactant (see Figure 16.17). Many charged and polar amino acid side chains interact with complementary charged or polar groups on both S3P and glyphosate, holding the two molecules in place. It is proposed that the glyphosate site is also where the PEP reactant binds, allowing attack by S3P. Figure 16.17 shows PEP in its proto-

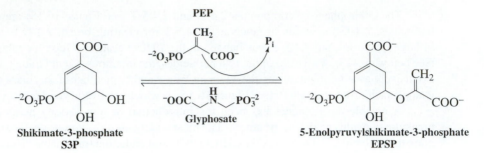

Figure 16.16 Catalytic transfer of the enolpyruvyl moiety from phosphoenol pyruvate (PEP) to shikimate-3-phosphate (S3P) forming the products EPSP and inorganic phosphate. Glyphosate inhibits EPSP synthase in a slowly reversible reaction. *Source:* E. Schönbrunn et al. (2001). Interaction of the herbicide glyphosate with its target enzyme 5-enolpyruvyl-shikimate-3-phosphate synthase in atomic detail. *Proceedings of the National Academy of Science* 98:1376–1380.

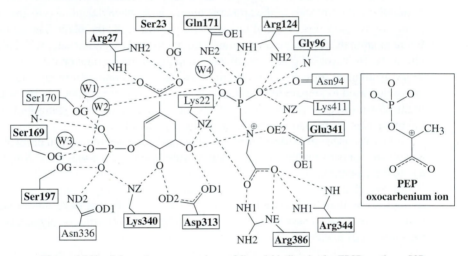

Figure 16.17 Schematic representations of ligand binding in the EPSP synthase S3P glyphosate complex, based on the crystal structure. Ligands are drawn in bold lines. Dashed lines indicate hydrogen bonds and ionic interactions. Strictly conserved residues are highlighted by bold labels. Protein atoms are labeled according to the Protein Data Bank nomenclature. Circled labels W1 to W4 designate solvent molecules. Hydrophobic interactions between S3P and Tyr-200 are omitted. *Source:* E. Schönbrunn et al. (2001). Interaction of the herbicide glyphosate with its target enzyme 5-enolpyruvyl-shikimate-3-phosphate synthase in atomic detail. *Proceedings of the National Academy of Sciences* 98:1376–1380.

nated form; protonation activates it for the reaction. Glyphosate blocks the binding of PEP, and shuts down EPSP production, killing the plant.

However, it has been discovered that certain mutant forms (having one of the amino acids replaced by a different one) of EPSP synthase will not bind glyphosate even though they remain active, indicating that PEP does bind. One of these mutants is Gly96Ala, in which the glycine normally found at position 96 (counting from the amino end of the amino acid chain) is replaced by an alanine. The glycine side chain is only an H atom, while the alanine side chain is a methyl group (see Figure 15.13). Glycine 96 is adjacent to the glyphosate phosphate group (see Figure 16.17), and the replacement of —H by —CH$_3$ apparently crowds the phosphate group enough to prevent glyphosate binding. However, the smaller PEP can still bind and undergo reaction.

The availability of selectively mutated EPSP synthase has made possible the development of glyphosate-tolerant plants (see next section) through genetic engineering. The EPSP gene is replaced by a gene coding for mutant EPSP synthase; the mutant gene is constructed by stringing together the correct sequence of nucleotides (see p. 411). This has allowed the development of crops that can be subsequently treated with glyphosate to control weeds (see next section).

16.3   GENETICALLY MODIFIED ORGANISMS (GMOs)

The continuing revolution in molecular genetics has made it possible to reprogram the genetic instructions of plants, in order to alter their characteristics. Scientists deciphered the genetic code of DNA molecules during the 1950s, and have progressively elucidated the complex biochemical mechanisms whereby genes are expressed. They have learned how to isolate genes and splice them into an organism's complement of DNA (its genome). These scientific discoveries permitted the development of the biotechnology industry.

Gene-splicing was first developed for bacteria, and then for yeast and cultured cells of higher organisms. The technique has allowed drug companies to manufacture medically critical proteins, such as human insulin and growth hormone, by expressing the gene in bacteria, or in other cells, and harvesting the product. Biopharmaceuticals produced from genetically modified cells now represent a large market, and have significantly improved the health of millions of people.

a. GM plants: Actualities and potential.   Plant scientists have learned how to insert foreign genes into the genome of a plant and to produce seeds that carry new traits. The first commercial genetically modified (GM) crops were planted in 1995 and were widely adopted in the next few years. By 1999, 40 million hectares of GM crops had been planted worldwide, including half of the United States' soybean crop, and more than a third of its corn. By 2000, an additional 4 million hectares had been planted, mostly in Argentina and the U.S.; plantings in Canada declined from 4 to 3 million hectares, reflecting resistance to GM foods in European markets (see below). The U.S., Argentina, Canada, and China accounted for 68, 23, 7 and 1 percent of the GM acreage, respectively, with the remaining 1 percent being scattered through several countries.

To date, most commercial GM products involve pest resistance, since pests are such an integral part of the agricultural economy. For example, soybeans and cotton have been engineered to resist the effects of glyphosate herbicide ("Roundup-ready" seeds), so that the glyphosate can be applied to control weeds after the crop is planted (see Strategies 16.2, above). This approach substantially reduces the amount of herbicide required (33 percent for soybeans, for example). Several other combinations of herbicides and plants resistant to them are now available.

Insect resistance has been engineered into another set of crops by splicing in genes for Bt toxins (see p. 394). These plants then produce their own insecticide, thereby reducing or eliminating the need for spraying. Introduction of Bt-cotton is estimated to have reduced the amount of sprayed insecticide by two million pounds in the U.S. The built-in insecticide can also afford a higher degree of protection than spraying. For example, spraying is ineffective against the European corn borer, because the spray does not reach the worms that have burrowed into the plant, but the toxin produced inside a Bt-bearing corn plant is effective. In addition, the harvested GM corn has been found to have lower levels of *mycotoxins* from fungal infestations. This is a health benefit (*aflatoxins* are powerful carcinogens, see p. 414), reducing the level of toxins in the food supply.

An important application of GM technology is protection against plant viruses and pathogenic bacteria, which regularly infest and decimate crops. It has been discovered that resistance to viruses can be induced by incorporating fragments of genes encoding viral or bacterial proteins into the plant's genome, a process similar to vaccination. Apparently the foreign genes trigger a plant defense mechanism that degrades the genetic material of the invading organism. A major success of this strategy was the rescue of the Hawaiian papaya crop from damage by the ringspot virus. About 60 percent of papaya trees are now protected by an introduced viral gene.

In addition to crop protection, genetic engineering can produce improvements in plant quality, e.g., corn or canola with higher oil content, or tomatoes that last longer in storage. These effects are achieved by genetic alteration of the plant's metabolism. For example, longer tomato shelf-life was achieved by turning off the gene for the enzyme *polygalacturonase,* which softens the plant's tissues. (This was accomplished by inserting an "antisense" gene, a reversed copy of the gene coding for the enzyme.)

Moreover, entirely new qualities can be introduced. Researchers have been able to engineer a set of genes into rice that produce β-carotene, the precursor to vitamin A. This "golden rice" (the carotene turns it yellow) could be important to poor countries that depend on rice, because of widespread vitamin A deficiency. UNICEF reports that more than a hundred million children are deficient in vitamin A, and millions lose their eyesight, or suffer fatal infections as a result. (Golden rice is being strongly encouraged by the biotechnology industry, which has supported its development and cleared a number of patent hurdles. However, GM critics consider this effort to be mainly public relations, pointing to the relatively low levels of β-carotene expressed in the rice, the requirement for adequate fat in the diet to absorb the vitamin A, and the possibility of distributing vitamin A in more direct ways to people who need it.)

Other researchers are developing GM bananas that produce oral vaccines for hepatitis B and diarrhea. These diseases are endemic to the tropics, where the vaccine-laced bananas could produce dramatic public health benefits.

b. Resistance to GM foods.   However, the prospects for GM foods are currently dimmed by heated controversy (leavened by some humor, see *New Yorker* cartoon, next page). The commercial production of GM crops has drawn a groundswell of public protest from around the world, but particularly from Europe. As of this writing (July 2001),

"We would like to be genetically modified to taste like Brussels sprouts."

GM foods are widely banned in Europe and Brazil, major food processors are demanding non-GM products from their suppliers, distributors are being required to separate GM from non-GM foods, some farmers are cutting back on GM acreage, and stockmarket values of agricultural biotechnology companies have fallen. The roots of the furor are complex and multifarious. In the following paragraphs we sketch some of the themes in the debate.

1) Moral issues.   There is a deep feeling on the part of many that we should not tamper with our genetic inheritance, particularly when it comes to something as basic as food. Many agree with Prince Charles of England, who said "I happen to believe that this kind of genetic manipulation takes mankind into realms that belong to God, and to God alone." Supporters of biotechnology strongly disagree with this sentiment, pointing out that plant breeders have been engaged in genetic alteration since agriculture began, crossing different strains in order to produce desirable traits. Genetic engineering is a logical extension of this process, and allows for greater precision, by focusing on selected genes, rather than wholesale gene shuffling between organisms. However, some biologists find this argument too facile, and point to inadequate knowledge of the consequences of gene splicing techniques (see below).

Another moral issue has to do with informed consent, since the fact of gene alteration has not, in general, been made known to the consumers of GM foods. Indeed, since GM grains and other commodities have not been handled separately, a great many people have unknowingly consumed GM foods in some degree. Consequently there is insistent demand for labeling of GM-derived products. Food suppliers have argued that labeling is costly (especially because it requires separating the commodities streams), and pointless, since there is no evidence that GM products are any different from equivalent non-GM products. Nevertheless there is wide sentiment that ignoring the issue of informed consent was a mistake. Most European countries now require labeling if more than 1 percent of a product is of GM origin.

Central to this issue is the problem of separating GM and non-GM product streams, and of their inadvertent mixing. In the field, GM plants can invade plantings of non-GM crops through the spread of pollen, and there are many possibilities for mixing of harvested and processed products in the distribution chain. This issue was highlighted by a furor in the autumn of 2000, ignited by the discovery by anti-GM activists using sensitive tests, that some taco shells in U.S. supermarkets contained DNA from StarLink corn, a GM variety engineered to produce an insecticidal protein, Cry9C. Because Cry9C has some character-istics (medium molecular weight, resistance to acid treatment or enzymatic digestion by *proteases,* and the ability to induce an immunological response when tested on rats) in common with human allergens, StarLink had been approved only for animal feed, not for human consumption, although there is no evidence for (or against) actual allergenicity in humans. The discovery of StarLink in taco shells, and subsequently in other consumer products (albeit at very low levels), was evidence that the unapproved GM corn had some-how entered the food products supply chain, and led its producer, Aventis Corp., to suspend distribution and compensate farmers and processors for losses. The cost to the company may end up to be hundreds of millions of dollars, and confidence in agricultural biotech-nology has suffered.

It should be noted that not all GM applications have been controversial. For exam-ple, nearly all hard cheeses are now produced with *chymosin,* a bovine enzyme that is syn-thesized industrially by genetically engineered microorganisms. The GM chymosin has re-placed the use of bovine rennet, a development welcomed by the Vegetarian Society.

2) Political issues. Some of the opposition to GM foods stems from resistance to "globalization" and the power of large corporations, particularly as it plays out in large-scale agriculture. There is a sense that genetic engineering further strengthens the hand of corporate agriculture, whose profits do not necessarily support human needs. Industry's defenders respond that there is no mechanism other than the international market that can effectively disseminate the fruits of biotechnology. However, some sympathetic observers are rueful that the first GM products involved only crop protection, with no tangible bene-fits to consumers.

A particularly damaging incident was the early revelation that the leading developer of GM crops, the Monsanto Corp., planned to adopt genetic technology that would prevent plants from producing viable seeds. The sterile seeds would protect the integrity of the product, which otherwise might alter its traits once in the field. It would also protect the company's investment, since growers would require new seeds every year. However, it was forcefully pointed out that many farmers, especially in poor countries, rely on their harvest for next year's seeds. News of the "terminator gene" created a furor, and the company soon disavowed the technology.

Opposition to GM foods has been intense in Europe, which has long feuded with the United States over agricultural imports. Europeans are especially protective of their food supply, and have tended to view GM crops as one more imposition by U.S. corporations. In addition, they have been suspicious of the adequacy of their own governments vis à vis food protection, especially in the wake of the "mad cow" scare, in which British authorities gave mistaken reassurances that neurodegenerative prion disease could not be transmitted through eating the meat of infected cattle.

3) Health and environment issues.   There are a number of concerns about unintended effects of GM foods on human and ecosystem health.

Allergenicity.   Since the engineered plant produces new protein, there is a potential for allergic response in sensitive individuals; this was the foundation of the StarLink controversy. In one case, a seed company dropped a plan to enrich soybeans with protein from the Brazil nut, which has a higher content of methionine (see discussion of essential amino acids, p. 370), because some people are allergic to Brazil nuts, and tests showed this allergic response to carry over to the engineered soybeans. Thus allergen testing must be a part of GM plant development.

Herbicide resistance.   Herbicide resistance genes can spread from GM plants to other plants via cross-pollination. Indeed, an Alberta farmer unwittingly produced super-resistant canola plants after sowing nearby fields with three different GM canola varieties engineered for resistance to three different herbicides. The super-resistant variety showed up during the next sowing season as stray canola plants that failed to respond to any of the three herbicides. Application of 2,4-D (see pp. 398–399) was required to kill it off. Cross-pollination can be minimized by planting different varieties a sufficient distance apart (175 meters is recommended), but there is concern that herbicide-resistance could be transferred to weedy relatives of the GM crop, thereby producing super-weeds, against which the herbicide would no longer be effective.

Insect resistance.   Insects can evolve resistance mechanisms to toxins produced by the plant, just as they do to externally applied toxins (see pp. 383–385). Indeed, resistance might develop faster since the insects could be exposed continually over the lifetime of the plant, whereas exposure to pesticide sprays is episodic. Current users of Bt-sprays are especially anxious that the widespread planting of Bt-engineered crops will destroy the utility of Bt by inducing resistance. However, resistance can be minimized by proper crop management. The U.S. EPA now requires that growers plant an insect "reservoir," a border of non-GM variants, around the GM plants. The idea is that a sufficient population of non-resistant insects will survive in this border to out-reproduce the resistant individuals that may arise in the GM plot.

Toxicity to non-target insects.   There has been persistent concern about ecosystem effects of insecticide genes. For example, there is some evidence that Bt-plants may alter the populations of soil microbes that degrade the vegetation. A wave of consternation greeted a scientific report, in 1999, that larvae of the Monarch butterfly died when fed on milkweed that had been dusted with pollen from Bt-corn. However, evaluation of field conditions indicates that pollen from Bt-plants is unlikely to drift onto milkweed (the food supply for Monarch larvae) in sufficient quantity to pose a threat to the butterflies.

Underlying all the concerns about the safety of GM crops is uncertainty about the full consequences of gene splicing technology, which is still at an early stage of its development. Many genes work in combination to affect the traits of an organism. The expression of genes is an exceedingly complicated set of biochemical events; these events are coordinated in ways that are not fully understood. The insertion of a foreign gene might disrupt this coordination, with unanticipated consequences. Thus a number of biologists think that commercialization of GM plants has outstripped the knowledge base needed to assure avoidance of long-term harm.

The course of the GM foods debate, and of the resulting regulatory actions, cannot be predicted. However, a slowing of the GM revolution is evident. More research and testing are needed, but complete certainty about the complexities of biology and ecosystem dynamics is not an attainable goal. As usual with new technologies, the risks need to be balanced against the benefits foregone if the technology is not pursued.

CHAPTER 17

TOXIC CHEMICALS

Toxic chemical species have been discussed at many points in this book, and we now direct our attention to the many ways that chemicals can be harmful to living things.

17.1 ACUTE AND CHRONIC TOXICITY

It is useful to distinguish between an acute effect, in which there is a rapid and serious response to a high but short-lived dose of toxic chemical, and a chronic effect, in which the dose is relatively low but prolonged, and a time lag occurs between initial exposure and the full manifestation of the effect. Acute poisons interfere with essential physiological processes, leading to a variety of symptoms of distress, and, if the interference is sufficiently severe, death. Chronic toxins have more subtle effects, often setting in motion a chain of biochemical events that leads to disease states, including cancer.

Sorting out these effects, the province of toxicology and epidemiology, is not an easy matter. The body's biochemistry is extremely complex, and it changes all the time in response to diet, activity, stress, and a variety of environmental factors. There are large differences among individuals, based on variations in genetics and life's circumstances. Moreover, there are stringent limits on the use of humans as experimental subjects, so that most of the available data are from experimental animals or studies of adventitious exposure in the workplace or the environment. Consequently, conclusions about toxic effects are seldom hard and fast, and are frequently modified in light of new studies.

Acute toxicity is relatively easy to gauge. At high-enough levels, the effects of toxins on bodily function are obvious and fairly consistent across individuals and species. These

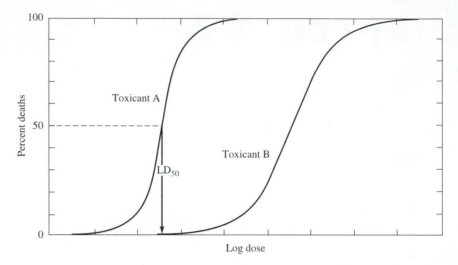

Figure 17.1   Illustration of a dose-response curve in which the response is the death of the organism; the cumulative percentage of deaths of organisms is plotted on the y-axis. *Source:* S. E. Manahan (1991). *Environmental Chemistry*, 5th Edition (Boca Raton, Florida: Lewis Publishers, an imprint of CRC Press). Copyright © 1991 by CRC Press. Reprinted with permission.

levels vary enormously for different chemicals. Almost everything is toxic at some level, and the difference between toxic and nontoxic chemicals is a matter of degree.

The most widely used index of acute toxicity is LD_{50}, the lethal dose for 50 percent of a population. This number is obtained by graphing the number of deaths among a group of experimental animals, usually rats, at various levels of exposure to the chemical, and interpolating the resulting dose-response curve to the dose at which half the animals die (see Figure 17.1). The dose is generally expressed as the weight of the chemical per kilogram of body weight, on the assumption that toxicity scales inversely with the size of the animal. Table 17.1 lists LD_{50} values for several substances, showing nine orders of magnitude variation between the most toxic (botulin toxin, the agent responsible for botulism) and the least toxic (sugar). Among insecticides, we see that DDT is about 30 times less toxic than parathion, but 12 times more toxic than malathion, at least when measured in rats or mice.

Chronic effects are much more difficult to evaluate, especially at the low exposure levels that are likely to be encountered in the environment. In an experimental setting, the lower the dose, the fewer the animals that show any particular effect. To obtain statistically significant results, a study might have to include a prohibitively large number of animals. The only available recourse is to evaluate effects of a series of high doses, and then to extrapolate the dose-response curve to the expected incidence at low doses. But extrapolation may have to extend over several orders of magnitude, and there is no assurance that the actual dose-response function is linear. The biochemical mechanisms that control effects may be different at high and low doses. The controversy over this issue is especially heated in the context of animal testing for cancer (see section 17.2b, pp. 415–418).

TABLE 17.1 LD_{50} VALUES OF SELECTED CHEMICALS

Chemical	LD_{50} (mg/kg)*
Sugar	29,700
Ethyl alcohol	14,000
Vinegar	3,310
Sodium chloride	3,000
Atrazine	1,870
Malathion (insecticide)	1,200
Aspirin	1,000
Caffeine	130
DDT (insecticide)	100
Arsenic	48
Parathion (insecticide)	3.6
Strychnine	2
Nicotine	1
Aflatoxin-B	0.009
Dioxin (TCDD)	0.001
Botulin toxin	0.00001

*For rats or mice

Source: P. Buell and J. Gerard (1994). *Chemistry in Environmental Perspective* (Upper Saddle River, New Jersey: Prentice Hall).

Toxicologists are increasingly turning to biochemical studies, using all the techniques of molecular biology, in order to elucidate the effects of toxicants at the molecular level. The expectation is that a more thorough understanding will provide a better basis for evaluating toxicity risks. Great strides have been made in probing the mechanism of action of various classes of toxicants (the dioxins are a good example, see p. 425), but it is not yet possible to translate this understanding into a quantitative estimate of physiological effects.

The other approach to evaluating health risks is epidemiology, the study of human exposure to chemicals in the workplace or the environment and the effect on the health of a population. Epidemiology, in principle, can provide data that are most directly relevant to risk estimation. The problem is that the variables in epidemiological studies are difficult to control; despite sophisticated statistical analysis, it may be hard to ascertain that a particular effect is not influenced by some other factor, such as smoking or poor diet, rather than by the chemical under study. It is also hard to select a control population without bias, which might skew the estimation of risk. For example, the frequently used method of selecting controls via randomized phone numbers has recently been shown to underrepresent poor people.

Obtaining statistically significant results depends greatly on the sample size, just as in animal studies. Larger numbers are needed when relatively small risks are being evaluated, as is usually the case in environmental exposures. Not surprisingly, results are more reliable when the risk is large than when it is small. For example, the smoking-related risk of lung cancer (see pp. 415–416) is easy to demonstrate statistically, because lung cancer incidence is 10–20 times higher in smokers than in nonsmokers. But the breast cancer risk

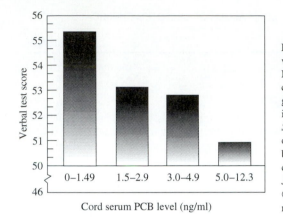

Figure 17.2   Test outcomes (McCarthy verbal test scores) of the 1990 Lake Michigan case study of four-year-old children; the children's scores are graphed versus the PCB concentrations in the umbilical cord serum at birth. *Source:*   J. L. Jacobsen et al. (1990). Effects of *in utero* exposure to polychlorinated biphenyls and related contaminants on cognitive functioning in young children. *Journal of Pediatrics* 116:38–44. Copyright © 1990 by Journal of Pediatrics. All rights reserved. Reprinted with permission.

associated with hormone replacement therapy has been much more difficult to establish, despite laboratory evidence for a connection (see p. 421). Two major studies appeared in 1995, one showing a 1.3- to 1.7-fold increase in the breast cancer risk for women taking estrogen and/or progesterone, and the other showing no added risk.

Both experimental and epidemiological approaches are important for examining a special class of toxic effect that is of increasing concern: prenatal effects on the fetus. The tragedy of birth defects resulting from the introduction of the drug thalidomide in the 1960s sensitized everyone to the possibility of *teratogenic* effects of environmental chemicals in addition to those of drugs. Screening for such effects with experimental animals is now routine. In addition to obvious birth defects, the possibility of developmental deficits resulting from prenatal exposure to toxins is of increasing concern. The occurrence of fetal alcohol syndrome is a flagrant example. But there may be more subtle effects from environmental exposures. For example, a study of families living on the Lake Michigan shore who regularly ate fish caught in the lake found that verbal test scores of four-year-olds decreased noticeably for those with the highest exposure to PCBs (see p. 429) at birth (see Figure 17.2).

17.2 CANCER

Of all the possible effects of chemicals in the environment, none is more feared than cancer, and none has generated more controversy. The public's fear of cancer has driven regulatory agencies to set very low tolerances for many chemicals in various environmental settings, from food to drinking water to toxic waste sites. These standards continue to stir debate; they are claimed to be too lenient by many environmental activists and too strict by manufacturers and others who might have to pay for required cleanups. Because of the uncertainties associated with the available data, as discussed in the preceding section, it is very hard to establish the truth of the matter. In the absence of hard evidence, there is a great deal of room for subjective factors that influence our perceptions of risk.

a. **Mechanisms.**   Cancers occur when cells divide uncontrollably, eventually consuming vital tissues. The normal mechanisms that limit cell growth and division are disrupted. This can happen in many different ways, but the common thread is that mutations occur in the cell's DNA at positions that specify the synthesis of key regulatory proteins. It has been shown that several such mutations are required to *transform* a normal cell into a cancerous one. This requirement explains why there is a long *latency* period, often 20 years or more, between exposure to a cancer-causing substance and the actual occurrence of cancer. Because of the probabilistic nature of mutations, the risk of cancer increases with age. Although children and young adults can and do develop cancers, most cancers are primarily diseases of old age. One of the causes of increasing cancer incidence is simply the increase in life expectancy during the last century.

A mutation occurs when DNA is mistranscribed during cell division. Maintaining the genetic code requires the correct pairing of bases via complementary H-bonding (see Figure 17.3) when a new DNA strand is copied from an old one. If an incorrect base is somehow incorporated into the sequence, then the error will be propagated in succeeding

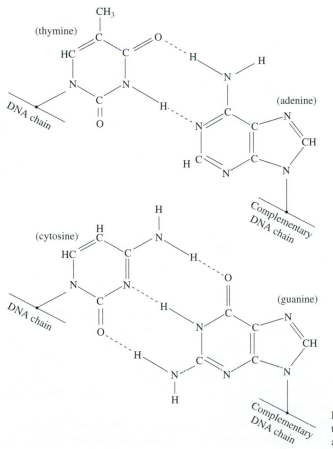

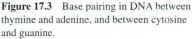

Figure 17.3 Base pairing in DNA between thymine and adenine, and between cytosine and guanine.

generations of the cell. If the incorrect base is part of a gene, then an error is introduced into the protein for which the gene codes, and the protein may misfunction. Mutations occur all the time because the fidelity of DNA transcription cannot be perfect. The normal error rate is very low (about one in 100 million), but it is not zero. Although the mutations are more or less random, there is some probability that they will occur at sites coding for regulatory proteins, and an additional, much smaller probability that enough critical mutations will accumulate to transform a given cell. Since our bodies contain billions of cells, and because we live through many cycles of cell division, it is likely that all of us harbor potentially cancerous cells. But they lead to cancer only rarely because the body has several lines of defense.

The cell itself has a variety of repair enzymes, the job of which is to detect incorrect base pairs and correct them. These enzymes greatly reduce the probability of accumulating enough critical mutations to produce cancer. In addition, the immune system provides powerful protection: cancer cells can be detected and destroyed by virtue of characteristic changes in their surface molecules. Finally, the development of full-blown cancer may require additional biochemical or physiological events. For example, solid tumors require a blood supply in order to grow, and must induce the body to provide a network of blood vessels.

Once in a while, all of these impediments are overcome, and a cancer results. The normally low probability of this happening can be increased by a variety of factors. An important one is genetics. Individuals may inherit a genetic defect that increases the cancer risk. The defect may involve a faulty repair enzyme, so that mutations survive more readily. Or there may be a pre-existing mutation in a gene for one of the regulatory proteins, which increases the odds of accumulating the remaining required mutations. Current genetic research is uncovering a wide range of genes in which mutations increase the risk of developing specific cancers.

Other factors involve exposure to cancer-inducing chemicals (*carcinogens*) or to dietary components that affect this exposure. For example, there is evidence that roughage in the diet protects against colon cancer, probably because the undigested fibers absorb carcinogenic molecules and sweep them out of the colon. Carcinogens can operate in two ways: they can be mutagens, inducing mutations by attacking the DNA bases, or they can be promoters, which increase the cancer probability indirectly. For example, promoters can act by increasing the rate of cell division. The more often cells divide, the greater the probability that cancerous mutations will accumulate. Thus, alcohol is a promoter of liver cancer because its consumption in excessive amounts causes cell proliferation in the liver, which is the organ that handles alcohol metabolism.

There are two requirements for mutagens: 1) they must react with the DNA bases in ways that alter their hydrogen bonding with a complementary base; since the bases are electron-rich, the mutagens tend to be *electrophiles;* and 2) they must gain access to the nucleus where the DNA is located. Many electrophiles are not mutagenic because they react with other molecules and are deactivated before they can reach the nucleus. For this reason, most mutagenic chemicals are not themselves reactive, but are converted into reactive metabolites by the body's own biochemistry.

The body has a variety of ways of ridding itself of foreign chemicals (*xenobiotics*). One of the most important is *hydroxylation* of lipophillic organic compounds. For example,

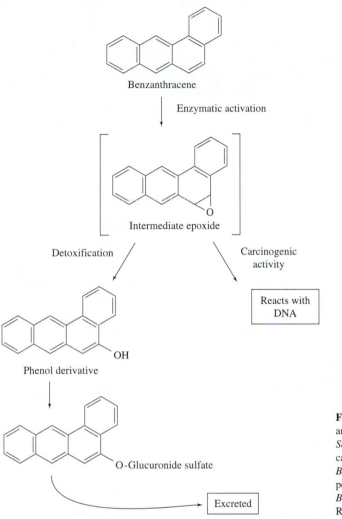

Benzanthracene

Enzymatic activation

Intermediate epoxide

Detoxification

Carcinogenic activity

Reacts with DNA

Phenol derivative

O-Glucuronide sulfate

Excreted

Figure 17.4 Activation of polycyclic aromatic hydrocarbons (PAHs). *Source:* C. Heidelberger (1975). Chemical carcinogenesis. *Annual Review of Biochemistry* 44:79–121. Reprinted with permission, from the *Annual Review of Biochemistry*, Volume 44 © 1975 by Annual Reviews. (http://www.AnnualReviews.org).

when benzanthracene is hydroxylated (see Figure 17.4), not only does a hydroxy group increase water solubility, but it also serves as a point of attachment for other hydrophilic groups such as glucuronide sulfate, which increase the water solubility further and promote excretion by the kidneys. Hydroxylation is accomplished by inserting one of the oxygen atoms of O_2 into a C—H bond, the remaining oxygen atom being reduced to water by supplying two electrons from a biological reductant:

$$O_2 + \text{—C—H} + 2e^- + 2H^+ = \text{—C—O—H} + H_2O \qquad (17.1)$$

This is a tricky reaction because the highly reactive oxygen atom must be generated exactly where it is needed; otherwise it will attack any molecule in its vicinity, adding to the supply of free radicals.

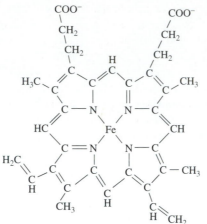

Figure 17.5   The structure of heme.

The reaction is carried out by a class of enzymes, cytochrome P450, that contain a heme group (see Figure 17.5) to bind the O_2 (just as hemoglobin does; see Figure 9.2 p. 218) and an adjacent binding site for the xenobiotic molecule. Despite this juxtaposition of the reactants, the immediate product is sometimes not the hydroxylated molecule but rather an *epoxide* precursor (see Figure 17.4), which is a potent electrophile. Since this precursor is generated inside the cell, it has a chance of diffusing into the nucleus and reacting with the DNA before it rearranges to the hydroxylated product. This is the reason that PAH compounds (see pp. 221–222, 467) like *benzanthracene* are carcinogenic. Another possibility is that the hydroxylated product can itself be a precursor to a reactive agent. For example, hydroxylation of *dimethylnitrosamine* (see Figure 17.6), another carcinogen, releases formaldehyde, leaving an unstable intermediate that is a source of methyl *carbonium* ion (CH_3^+), a powerful electrophile, which can react readily with DNA if generated nearby.

PAHs and nitrosamines are anthropogenic carcinogens, but there are many natural ones as well. *Aflatoxins,* which are complex products of a mold that infests peanuts, corn, and other crops, are powerful carcinogens. Biochemist Bruce Ames, developer of the Ames

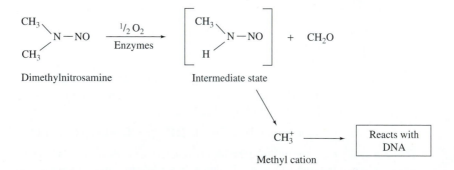

Figure 17.6   Activation of dimethylnitrosamine in the body.

mutagenicity test (see next section), points out that the plants we eat contain natural pesticides, many of which are turning out to be mutagenic when tested. He has estimated that the average American eats 1.5 g per day of natural pesticides, about 10,000 times more than the amount of synthetic pesticide residues. Moreover, although test data on natural pesticides are sparse, about half of those tested in animals are found to cause cancer, a similar percentage to that found for synthetic pesticides. Ames and others have also drawn attention to the high natural level of mutagenesis due to oxidative damage to DNA from the side-products of normal O_2 metabolism (see discussion of antioxidants, p. 375). This research puts the damage caused by synthetic chemicals in the context of the natural background level of DNA damage and repair.

b. Cancer incidence and testing.   Despite extensive epidemiological studies, it is not easy to tease out the contribution of environmental chemicals to cancer incidence, for the reasons mentioned above. For example, even though radon is thought to be a more serious cancer hazard than any other environmental chemical, the studies on radon in houses do not agree as to whether the cancer incidence is elevated when the radon levels are higher than the U.S. EPA's guideline of 4 pCi/l (see pp. 50, 52).

However, some cancer causes are firmly established by epidemiological data. The most striking evidence is the historical data on lung cancer and smoking (see Figure 17.7). A manyfold rise in U.S. lung cancer mortality tracked the increase in cigarette smoking, with a lag of several decades, and this happened in different historical periods for men and

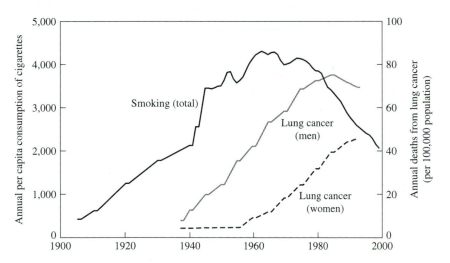

Figure 17.7 Cigarette smoking and lung cancer in the United States; death rates are averages for all ages. *Sources:* Drawn from data given in L. Garfinkel and E. Silverberg (1990). Lung cancer and smoking trends in the United States over the past 25 years. In *Trends in Cancer Mortality in Industrial Countries,* D. L. Davis and D. Hoel, eds. (New York: The New York Academy of Sciences); Centers for Disease Control and Prevention, U.S. Department of Health and Human Services (2000) *Health, United States, 2000* (Hyattsville, MD: National Center for Health Statistics).

women. Smoking accounts for 30 percent of all U.S. cancer deaths (along with 25 percent of fatal heart attacks). Similarly clear is diet's role in cancer, as is strongly suggested by data showing marked changes in the pattern of cancer incidence when people migrate from one part of the world to another (see Figure 17.8). The rates and types of cancers contracted by migrating ethnic groups change when their diets change. It is thought that high levels of salt or smoked fish in the Japanese diet may account for excess stomach cancers, while high fat in the U.S. diet might be responsible for a higher rate of colon cancer. However, the actual contributions of dietary components to cancer incidence (or to protection from cancer) have been hard to pin down.

Data on occupational exposure have firmly implicated several industrial chemicals. For example, vinyl chloride causes liver cancer, benzene causes leukemia, and asbestos causes mesothelioma, a cancer of the lining of the lung. However, exposure of people at large to these chemicals is far lower than in an occupational setting, and the hazard at these lower levels can only be guessed by extrapolation.

Alternatively, carcinogenic risk can be estimated from test data. Since many carcinogens are mutagens, carcinogens can be screened by using the Ames bacterial test. The

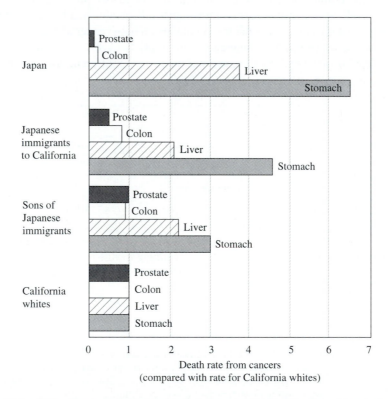

Figure 17.8   Change in incidence of various cancers with migration from Japan to the United States.

suspect carcinogen is administered to mutant bacteria that are unable to grow in the absence of the amino acid histidine in the culture medium. Certain additional mutations will produce a *revertant* organism, capable of growing again in the histidine-deficient medium. The stronger the mutagen, the greater the number of revertant organisms produced. Thus, the number of revertant colonies is a measure of the mutation rate, which can be determined at various concentrations of the test substance. (The test can also be used to monitor complex mixtures for mutagenic activity in order to separate and identify the active ingredient.) Since some chemical compounds are not mutagens until they are metabolically activated, in order to assay carcinogenicity the Ames test requires adding a rat-liver extract, which contains the cytochrome P450 enzymes responsible for oxidative activation of the carcinogen by the hydroxylation mechanisms described in the preceding section. Because bacteria are very different from people, the test cannot distinguish reliably all human carcinogens or evaluate their potency. It is, however, an inexpensive and useful screening method.

The main source of carcinogenicity data has been animal tests, usually involving rats. Cancers are counted over the lifetime of the animal at various doses, and the results are extrapolated to typical exposure levels in order to obtain an estimate of the cancer risk. Because of the need to obtain statistically significant results on a limited number of animals, most of the data are at the *maximum tolerated dose* (MTD), above which acute toxicity symptoms occur.

The use of the MTD has been criticized on the grounds that even in the absence of overt toxic symptoms, there may be significant organ damage and resulting cell proliferation, which increases the cancer probability. There may be additional reasons why conditions at the MTD in animals may have little relevance for human exposure. For example, saccharin carries a carcinogenicity warning when marketed as an artificial sweetener because it was found to cause bladder cancer in male rats at high doses. But subsequent mechanistic research established that these cancers are associated with a protein, α_{2u}-globulin, which is specific to male rats and is not present in humans (or even female rats). The tumors occur when the bladder lining regenerates after erosion by a precipitate of the protein with saccharin in the urine. This mechanism would not occur in humans, even at high doses, and indeed the epidemiological evidence on saccharin is negative.

In addition, there are arguments about how to extrapolate from high to low doses. In the absence of actual data (usually unavailable for the reasons discussed above), the linear model is used, involving a straightforward proportionality of dose and effect. This is thought to be reasonable for mutagens, the effect of which on DNA can be expected to be proportional to the number of molecules. However, promoters of cell proliferation are expected to have a threshold, below which stimulation of cell division would be insufficient to affect cancer incidence. However, since this threshold is usually unknown, it is difficult to incorporate non-linear extrapolations into regulatory limits.

Despite their deficiencies, animal tests can serve as a rough guide for comparing carcinogenic risk from different substances. Ames has proposed an index for this purpose, HERP (human exposure dose/rodent potency). It is calculated by estimating the lifetime exposure for an average person and dividing by the rodent LD_{50} for death by cancer.

TABLE 17.2   COMPARISON OF RISK FROM EXPOSURE TO CARCINOGENS

HERP %*	Risk agent
0.0003	Tap water, 1 L/day (chloroform, 17 μg; average U.S. intake, 1987–1992)
0.0003	Carbaryl (carbamate insecticide, see p. 388), 2.6 μg/day (1990 U.S. average)
0.002	DDT, 14 μg/day (U.S. average before 1972 ban)
0.008	Aflatoxin, 18 ng/day (U.S. average, 1984–1989)
0.03	Orange juice, 140 g/day (d-limonene, 4.3 mg)
0.1	Coffee, 13.3 g/day (caffeic acid, 24 mg)
0.4	Conventional home air, 14 h/day (formaldehyde, 600 μg)
0.5	Wine, 28 g/day (ethanol)
2.1	Beer, 260 g/day (ethanol)
6.8	Butadiene, 66 mg/day (rubber industry workers, 1978–1986)
14	Phenolbarbital, 60 mg/day (one sleeping pill)

*HERP: Human exposure dose/rodent potency. Cancer hazard based on a typical person's average daily exposure to the substances over a lifetime. Calculated as a percentage of the LD_{50} for rats or mice (whichever is more potent), corrected for body weight.

Source: L. S. Gold et al. (2001). *Issues in Environmental Science and Technology* 15:95–128.

Although HERP assumes the applicability of linear extrapolation from high-dose animal test data, it can nevertheless give some idea of the relative magnitude of different risks. The results (see Table 17.2) suggest that exposure to pesticide residues or tap water are much weaker hazards than are such common items in our diet as wine, beer, or coffee.

Because there are currently no viable alternatives to animal tests, they will no doubt continue to be used as a factor in assessing cancer risks. As additional mechanistic insights emerge from biochemical research, they can be factored into the evaluation of the significance of particular tests and may alter experimental protocols.

17.3   HORMONAL EFFECTS

Recently, increasing concern has focused on the biochemical role of environmental chemicals that mimic hormone functions. Hormones are messenger molecules, excreted by various glands, that circulate in the bloodstream and powerfully influence the biochemistry of specific tissues. Hormone activity is initiated by binding to receptor proteins in the target cells.

There are two kinds of hormones, water-soluble and lipid-soluble, with entirely different mechanisms of action. Water-soluble hormones, such as insulin, are peptides and proteins. They bind to receptor proteins embedded in the target cell membrane, analogous to the neurotransmitter receptors (see Figure 16.9). This binding induces the activation of enzymes inside the cell, which catalyze the synthesis of interior messenger molecules; in turn, these *second messengers* bind and activate proteins that control metabolic processes.

The lipid-soluble hormones are *steroids,* derivatives of cholesterol (see Figure 17.9). They diffuse through cell membranes and are picked up at the inside surface by specific receptor proteins that are dissolved in the interior fluid (*cytosol*) of the target cell. Hormone

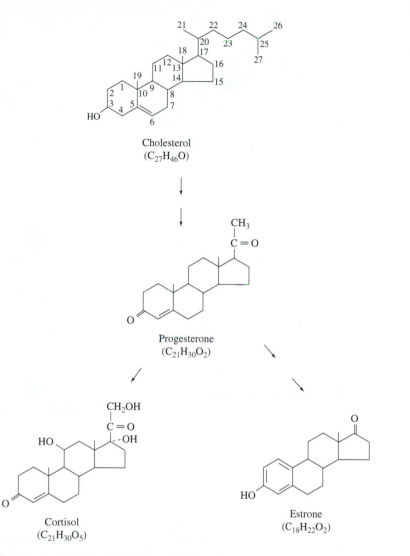

Figure 17.9 Some steroid derivatives of cholesterol.

binding changes the shape of the receptor protein and enables it, after transport to the nucleus, to turn on specific genes (see Figure 17.10). Thus, the steroid hormones act by inducing the synthesis of enzymes and regulatory proteins.

Chemicals from outside the body (*xenobiotics*) can also bind to hormone receptors if they have the proper shape and distribution of electrical charges. This is unlikely to be a problem for peptide hormones because water-soluble xenobiotics are quickly excreted. But lipophillic xenobiotics, which are stored in the fat tissue, might bind to steroid hormone receptors. If the resemblance to the hormone is close enough, the xenobiotic can turn on the

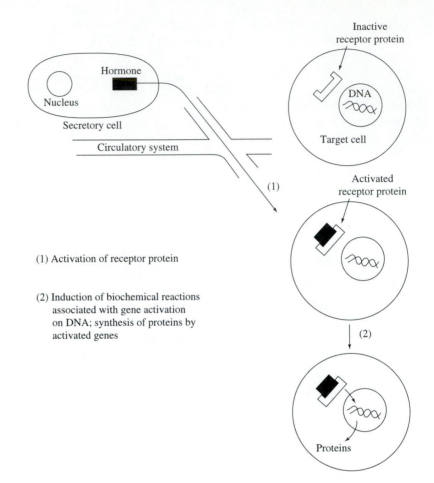

Figure 17.10   Mechanism of steroid hormone function.

same biochemical machinery, but if the resemblance is only partial, then binding may not activate the receptor. In that case, the xenobiotic blocks the hormone and depresses its activity; it is an antihormone (see discussion of agonists and antagonists, p. 386). Either way, there is potential for upsetting the biochemical balance controlled by the hormone. A mechanism of this kind is probably responsible for DDT's disruption of calcium deposition in birds' eggs (see p. 387).

The sex hormones are in the steroid class; estrogens and androgens induce and maintain the female and male sexual systems. They have become a focus of attention because of reports of malformed sex organs in wildlife. In particular, alligators in a Florida lake were found to have impaired reproductive systems (abnormally small penises in the males), low rates of hatching, and high levels of DDE, the breakdown product of DDT, in their tissues. DDE contamination resulted from spills of a DDT-containing pesticide along the lakeshore. Subsequent tests showed that DDE binds to androgen receptors and blocks their activity. A high incidence of intersexes has been found in fish exposed to polluted waters, and the males are found to have *vitellogenin,* a female-specific protein. This evidence of environmental demasculinization has fueled speculation that something similar may be going on in human males, because of reports from a number of clinics that sperm counts among men have been going down for a number of years. But the validity of these data as indicators of male fertility has been questioned.

    Estrogenic xenobiotics have provoked more concern because of the association of estrogen with breast cancer in women. Estrogen binding to receptors in the breast stimulates the proliferation of breast cells; as we have seen in the preceding section, cell proliferation promotes mutagenesis and cancer. A link between breast cancer and estrogen has been established in laboratory animals, and there is a statistical association, albeit equivocal (see discussion on epidemiology above, pp. 409–410) between estrogen therapy and breast cancer incidence. The incidence of breast cancer has been rising, and many environmental chemicals have recently been found in laboratory assays to be estrogenic (see

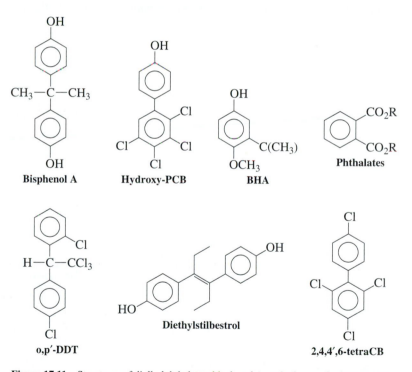

Figure 17.11   Structures of dialkylphthalates, bisphenol A, and other synthetic estrogens. R groups on the phthalate molecule denote a limited number of alkyl groups that render the molecule estrogenically active.   *Source:* Committee on Hormonally Active Agents in the Environment, Board on Environmental Studies and Toxicology, National Research Council (1999). *Hormonally Active Agents in the Environment* (Washington, DC: National Academy Press).

Figure 17.11). These include DDT, the antioxidant BHA, and a variety of organic chemicals that are either used as plasticizers or are products of high-temperature treatment of plastics. Putting these facts together, many people worry that exposure to these chemicals may put women at risk for breast cancer. Skeptics point out, however, that the levels of exposure are low, and that the xenobiotics should be swamped out by the body's own estrogen (although the endogenous estrogen level fluctuates cyclically, while the xenobiotics do not). This is currently an area of active research. A recent epidemiological study from Denmark found a distinct correlation of breast cancer with blood levels of the insecticide dieldrin, but no correlation with levels of DDT, chlordane or kepone.

There has been concern about the widespread use of *dialkylphthalates* and *bisphenol A* (see Figure 17.11) as plasticizers in many products, including containers for food and vinyl medical devices. Laboratory studies show that feeding these compounds to pregnant rats produces abnormalities in sexual development of male offspring. In addition, premature breast development in Puerto Rican girls has been found to correlate with high concentrations of phthalates in their blood.

There are also many naturally occurring estrogenic compounds in the environment (see Figure 17.12), some made by plants (*phytoestrogens*), and others by fungi that infect the plants. Phytoestrogens include lignins and isoflavanoids (e.g., genistein, especially

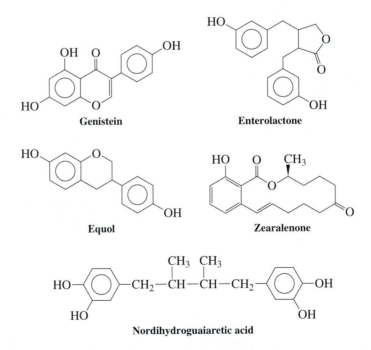

Figure 17.12 Naturally occurring estrogenic compounds. *Source:* Committee on Hormonally Active Agents in the Environment, Board on Environmental Studies and Toxicology, National Research Council (1999). *Hormonally Active Agents in the Environment* (Washington, DC: National Academy Press).

abundant in soy products), or flavanoid-derived compounds (e.g., equol, enterolactone, nordihydroguaiaretic acid), which are common in foods. Fungal metabolites include zearalenone.

17.4   PERSISTENT ORGANIC POLLUTANTS: DIOXINS AND PCBs

There are many organic compounds in the environment, some of them toxic. We have already dealt with oil spills in Part I (see pp. 31–37), air pollutants in Part II (see pp. 216–227), and pesticides and herbicides in preceding sections. There are numerous organic waste products from industrial operations, or from the use and disposal of manufactured products, that may contaminate air, land, and water from effluents, from leakage of waste dumps, or from accidental spills and fires. A worldwide treaty has recently been negotiated to phase out persistent organic pollutants (POPs). The twelve chemicals initially covered by the treaty are: dioxins and furans, polychlorinated biphenyls (PCBs), and nine organochlorine pesticides (aldrin, chlordane, eldrin, dieldrin, heptachlor, hexachlorobenzene, mirex, toxaphene, and DDT—although there is an exemption for developing countries using DDT for malaria control).

Having dealt with pesticides in Chapter 16, we turn our attention to dioxins and furans, and to PCBs.

a. Dioxins and furans.   The term dioxin is shorthand for a family of polychlorinated dibenzodioxins (see Figure 17.13), sometimes abbreviated PCDDs. The polychlorinated dibenzofurans (PCDFs) have a similar structure. These chemicals are not made intentionally, but are formed as contaminants in several large-scale processes, including 1) combustion, 2) paper-pulp bleaching with chlorine, and 3) manufacture of certain chlorophenol chemicals. It was this last process that brought dioxin its initial notoriety as a contaminant of the herbicide 2,4,5-T, a component of Agent Orange. The herbicide was made by reacting chloroacetic acid with 2,4,5-trichlorophenol, which was itself formed by reacting 1,2,4,5-tetrachlorobenzene with sodium hydroxide [reaction sequence (1) near top of Figure 17.13]. During this prior reaction, which was carried out at a high temperature, some of the trichlorophenoxide condensed with itself [reaction (2) of Figure 17.13] to form 2,3,7,8-tetrachlorodibenzodioxin (TCDD); Agent Orange contained about 10 ppm of this material. It was subsequently shown that control of the reaction temperature and of the trichlorophenoxide concentration could keep the TCDD contamination to 0.1 ppm, but it was too late to save the herbicide, which was banned in the United States in 1972.

1) Toxicity.   TCDD turned out to be extraordinarily toxic to laboratory animals, producing birth defects, cancer, skin disorders, liver damage, suppression of the immune system, and death from undefined causes. The LD_{50} in male guinea pigs was only 0.6 μg/kg. In laboratory animals, low doses have been found to be teratogenic and to lead to developmental abnormalities.

Consequently, there was great alarm when TCDD was found at a number of industrial waste sites. In 1983, the U.S. government offered to purchase houses in the town of

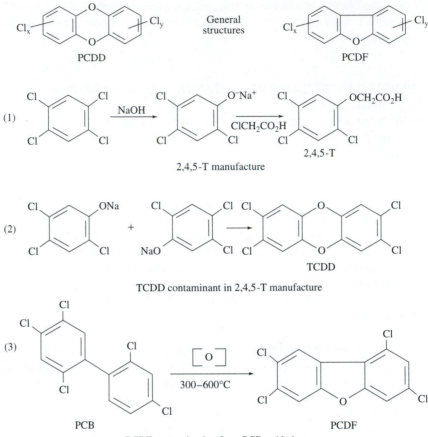

Figure 17.13   Polychlorinated dibenzodioxins (PCDDs), furans (PCDFs), and polychlorinated biphenyls (PCBs); chemical structures and reactions.

Times Beach, Missouri, after its roads were found to have been contaminated by dioxin in waste oil from a 2,4,5-T manufacturer, oil that had been sprayed by a hauler to control dust. The most serious instance of environmental contamination occurred in 1976, when an explosion in a factory in Seveso, Italy, that manufactured 2,4,5-T released a few kg of TCDD into the town and its surroundings.

As evidence on toxicity has accumulated, however, the risk to humans from dioxin has become less clear. The variation in toxicity among species turned out to be large, with LD_{50} values that were orders of magnitude higher in other animals than in guinea pigs (see Table 17.3). In humans, exposure to high levels of PCDDs causes chloracne, a painful skin inflammation, but these levels have only been encountered in accidental industrial exposures. Moreover, although the Seveso contamination produced many wildlife deaths and exposed many people, no serious human health effects were found for many years; nor

TABLE 17.3 ACUTE TOXICITIES OF 2,3,7,8-TETRACHLORODIBENZODIOXIN IN
EXPERIMENTAL ANIMALS

Species	Route	LD_{50} (micrograms per kilogram)
Guinea pig (male)	Oral	0.6
Guinea pig (female)	Oral	2.1
Rabbit (male, female)	Oral	115
Rabbit (male, female)	Dermal	275
Rabbit (male, female)	Intraperitoneal	252–500
Monkey (female)	Oral	<70
Rat (male)	Oral	22
Rat (female)	Oral	45–500
Mouse (male)	Oral	<150
Mouse (male)	Intraperitoneal	120
Dog (male)	Oral	30–300
Dog (female)	Oral	>100
Frog	Oral	1,000
Hamster (male, female)	Oral	1,157
Hamster (male, female)	Intraperitoneal	3,000

Source: Data from F. H. Tschirley (1986). Dioxin. *Scientific American* 254(2):29–35.

have any been tied to the Times Beach contamination. Recently, however, rates for a number of cancers were found to be elevated in the exposed Seveso population, although the small numbers make the statistics uncertain. In addition, men receiving large dioxin exposures from the accident have subsequently fathered fewer sons (38 percent) than daughters, according to a recent study. This finding suggests that dioxin perturbs chemical signals in the reproductive tract of men, consistent with data from animal tests. A study from Finland has found that children exposed to dioxin in their mothers' milk have tooth defects, apparently related to effects on cellular receptors for epidermal growth factor. The U.S. EPA carried out a dioxin assessment, which concluded that dioxin is likely to increase cancer incidence in humans; epidemiological data on industrial workers indicate an association of cancer incidence with increasing exposure levels. The World Health Organization has classified TCDD as a known human carcinogen. However, there is continuing controversy about the magnitude of the risk.

Rapid strides have been made in understanding TCDD's complex biochemistry. The molecule binds strongly to a receptor protein that is present in all animal species. This receptor, called Ah (for aryl hydrocarbon), is activated by a number of planar aromatic molecules (its natural substrate is still unknown); the binding of TCDD is particularly strong, with an equilibrium constant for dissociation of 10^{-11} molar. Like a hormone receptor (see Figure 17.10), the Ah receptor interacts in complex ways with the cell's DNA. One effect is the induction of a cytochrome P450 enzyme (a variant labeled 1A1), which is responsible for hydroxylating a number of xenobiotics, including PAHs (but not TCDD itself, since its chlorine atoms deactivate the ring toward oxidation). There are additional effects on a variety of biochemical pathways, which are currently under study. It remains uncertain, however, whether all of TCDD's toxic effects originate in its binding to the Ah receptor.

2) Paper bleaching and combustion sources.   A variety of PCDDs are formed in small quantities when chlorine is used to bleach paper pulp, probably via chlorination of the phenolic groups in lignin (see lignin structure, p. 24). There has been concern about trace dioxin contamination of paper products, and about bioaccumulation of dioxin in waters receiving paper mill effluents. Dioxin emissions are being reduced by switching from chlorine to chlorine dioxide, which is an oxidant but not a chlorinating agent (see discussion of water disinfection, pp. 343–344).

The main source of dioxin in the environment, however, is combustion. When material containing chlorine is combusted, dioxin is produced in traces; because the volume of material combusted annually is huge, these traces add up to a substantial aggregate environmental load. As might be expected, the dioxin emission rate correlates roughly with the chlorine content of the combustion feed,* although dioxin formation is highly dependent on combustion conditions and the type of pollution controls, if any. It appears that the chlorine need not be organically bound, since wood stoves have been found to produce dioxins, and the chlorine in wood is mostly sodium chloride. The main mechanism of dioxin formation appears to involve reaction of organic fragments in the combustion zone with HCl and O_2. The HCl formation rate might be expected to depend on the form of the chlorine in the combusted material, but this question has not been resolved. Dioxin formation is catalyzed on the surface of fly ash, probably by transition metal ions, and is favored at moderate temperatures, with a maximum at about 400°C. At lower temperatures, the reaction slows down and the products remain adsorbed on the fly ash, whereas at higher temperatures the dioxins are oxidized further.

Combustion produces a wide range of PCDD *congeners* (molecules with the same structure, but with varying numbers and positions of chlorine substituents), as well as PCDFs (see Figure 17.13). In the context of combustion products, "dioxin" means the aggregate of PCDDs and PCDFs, also abbreviated to PCDD/Fs. Both classes of molecules are toxic, but the toxicity varies among the congeners. Toxicity is assumed to be roughly proportional to the strength of binding to the Ah receptor. TCDD is the most toxic of the dioxins; toxicity decreases progressively when chlorine atoms are removed from the 2, 3, 7, and 8 positions, or when they are added to the remaining positions on the rings. These alterations reduce the "fit" of the molecule to the binding site on the Ah receptor. A similar toxicity pattern is observed for the PCDF congeners, but the toxicity is about an order of magnitude lower for the PCDFs than for the PCDDs. In order to gauge the effects of exposure to these chemicals, a scale of international toxicity equivalence factors (I-TEFs) has been established based on toxicity relative to TCDD, which is assigned a value of 1 (see Table 17.4). This factor is 0.1 for 2,3,7,8-PCDF, for example, and 0.001 for the octachloro congeners of either series. With these factors, one can convert the distribution of both classes of molecules (PCDD/Fs) into

---

*V. M. Thomas and T. G. Spiro (1995). An estimation of dioxin emissions in the United States. *Toxicology and Environmental Chemistry* 50:1–37.

TABLE 17.4   INTERNATIONAL TOXICITY EQUIVALENCY
FACTORS FOR PCDDs AND PCDFs

Congener	PCDD series	PCDF series
2,3,7,8	1 (defined)	0.1
1,2,3,7,8	0.5	0.05
2,3,4,7,8		0.5
1,2,3,4,7,8	0.1*	0.1[†]
1,2,3,4,6,7,8	0.01	0.01[‡]
octachloro	0.001	0.001

*same value for 1,2,3,6,7,8- and 1,2,3,7,8,9- congeners

[†]same value for 1,2,3,6,7,8-, 1,2,3,7,8,9-, and 2,3,4,6,7,8- congeners

[‡]same value for 1,2,3,4,7,8,9- congener

Source: N. J. Bunce (1994). *Environmental Chemistry,* Second Edition
(Winnipeg, Canada: Wuerz Publishing, Ltd.).

a single toxicity equivalent quantity (TEQ), expressed in grams of TCDD-equivalents. For
example, 1.0 g each of TCDD and 2,3,7,8-PCDF would have a TEQ value of 1.1 g.

Dioxin inventories have been estimated for a number of countries from data on emission rates for various kinds of combustion and on the total amount of material combusted.
The U.S. distribution of sources, as estimated by the EPA in 1995, is illustrated in Figure 17.14. Municipal- and hospital-waste incinerators have been major sources, at least in
developed countries. But pollution-control devices can cut the incinerator emission rates
dramatically, and these are being widely implemented. A combination of spray dryers and

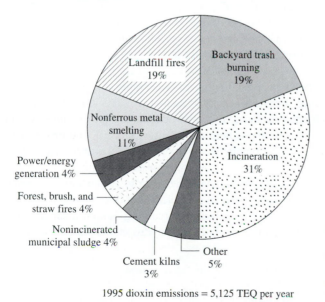

1995 dioxin emissions = 5,125 TEQ per year

Figure 17.14   Dioxin sources in the U.S.,
1995. "Dioxin" is defined as the totality
of seven dioxins and ten furans.
Source: B. Hileman (2001). Reassessing
dioxins. *Chemical and Engineering News*
79(22):25–27.

fabric filters can be effective in removing PCDD/Fs from the incinerator exhaust gases. Ironically, electrostatic precipitators, which were installed on older incinerators to reduce particle emissions, can actually increase dioxin formation, apparently through catalysis by particles lodged on the precipitator surfaces.

Among the numerous other combustion sources, nonferrous metal smelting is significant because organic waste materials are often used for fuel. But the biggest source appears to be open burning of garbage, in backyards and landfills. These fires are difficult to control. Trash burning is banned in most cities, but is widely practiced in rural areas.

3) Natural sources?   The question arises whether there are significant natural sources of dioxins. It has been suggested that forest fires are a major source. The EPA estimated that forest, brush, and straw fires accounted for only 4 percent of dioxin emissions in 1995 (see Figure 17.14), but the amount of biomass burned in such fires is large, and the dioxin emission rate is poorly characterized. Likewise, we know very little about other possible sources in nature. Contrary to popular opinion, organochlorines are not exclusively synthetic, but are widely produced as natural products by a variety of microorganisms.*

Soil organisms produce peroxidase enzymes to break down lignin, and these are capable of incorporating chloride ions into carbon-chlorine bonds. It is not known to what extent PCDD/Fs might result from this natural chemistry, but dioxins have been found in compost piles.

Could natural sources of dioxins outweigh anthropogenic ones? The sedimentary record suggests not. It is possible to establish the trend in the deposition rate over time by analyzing the dioxin profile of cores extracted from lake bottoms. Sediment from Siskiwit Lake has been examined in this way (see Figure 17.15); the lake is located on an island in Lake Superior, far from any pollution source. Dioxins must have reached Siskiwit Lake by long-range transport through the atmosphere. The dioxin deposition rate is found to have increased eightfold between 1940 and 1970, the period of great expansion in the industrial use of chlorine. In contrast, forest fires in the U.S. actually diminished by more than a factor of four in the same period, thanks to more effective fire-control measures. Since 1970, the dioxin deposition rate has declined by about 30 percent (see Figure 17.15), in parallel with the phaseout of 2,4,5-T spraying, and with improved incinerator technology. These trends seem to rule out predominantly natural sources. It is interesting that the dioxin in sediment is mainly the octachloro congeners, probably because the less chlorinated congeners are selectively volatilized from dioxin-bearing particles during long-range transport.[†]

4) Exposure.   The total U.S. PCDD/F emission rate from combustion is estimated to be about 5 kg/yr TEQ. But this total is spread out over an enormous area, and the atmospheric concentrations are very low. Exposure from breathing dioxin-laden air is minimal, even if one lives next to an incinerator. As with other hydrophobic materials, exposure

---

*G. Grible (1994). The natural production of chlorinated compounds. *Environmental Science and Technology* 25(7):310A–318A.

[†]R. A. Hites (1990). Environmental behavior of chlorinated dioxins and furans. *Accounts of Chemical Research* 23:194–201.

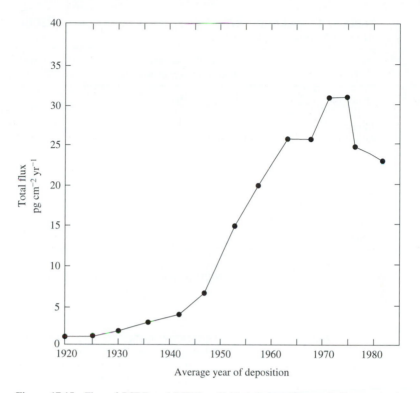

Figure 17.15   Flux of PCDD and PCDF to Siskiwit Lake.   *Source:* J. Czuczwa and R. Hites (1986). Airborne dioxins and dibenzofurans: Sources and fates. *Environmental Science and Technology* 20(2):195–200. Copyright © 1986 by American Chemical Society. Reprinted with permission from ES&T.

to dioxins is determined by bioaccumulation mechanisms (see pp. 386–388). The main exposure route (95 percent) for humans is dietary: meat, dairy products, and fish (see Table 17.5). The dioxins deposit on hay and feed crops consumed by cows, which concentrate the dioxins in their fat tissues. Likewise, fish concentrate dioxins from algae, which absorb dioxins from fallout, and also from local pollution sources such as pulp bleaching plants, or sewage and wastes. As a result, we all have detectable concentrations of dioxins in our fat tissue, although the most prevalent one is the octachloro congener, which is not very toxic (see Table 17.4). The average daily dose of PCDD/Fs is estimated to be roughly 0.1 ng (nanogram = 10^{-9} g) TEQ/day in the United States. This dose is not far from levels at which biochemical effects can be detected in laboratory animals. If the dioxin deposition rate is declining, as the sedimentary record indicates, then the average exposure should also decline.

b. Polychlorinated biphenyls.   As the name implies, polychlorinated biphenyls (PCBs) are made by chlorinating the aromatic compound biphenyl (see Figure 17.13 for the molecular structure of a specific PCB). A complex mixture results, with variable numbers

TABLE 17.5   AVERAGE CONTENT OF 2,3,7,8-TETRACHLOROBENZODIOXIN
IN THE AMERICAN FOOD SUPPLY

Food	TCDD concentration in picograms per gram (pg/g)*	Average TCDD intake (pg/person/day)*
Ocean fish	500	8.6
Meat	35	6.6
Cheese	16	0.31
Milk	1.8	0.20
Coffee	0.1	0.04
Ice cream	5.5	0.04
Cream	7.2	0.01
Sour cream	10	0.01
Cottage cheese	2.1	0.01
Orange juice	0.2	0.01
Total		**15.9**

*1 *picogram* (pg) $= 10^{-12}$ gram.

Source: Data from S. Henry et al. (1992). Exposures and risks of dioxin in the U.S. food supply. *Chemosphere* 25:235–238.

TABLE 17.6   DIOXIN CONGENERS DETECTED IN HUMAN FAT TISSUE

Congener	Average concentration (pg/g)	Standard deviation (pg/g)
2,3,7,8-tetrachloro	11	8
1,2,3,7,8-pentachloro	24	12
1,2,3,6,7,8-hexachloro	172	74
1,2,3,7,8,9-hexachloro	22	9
1,2,3,4,6,7,8-heptachloro	232	181
octachloro	1037	712

Source: Data from G. L. LeBel et al. (1990). Polychlorinated dibenzodioxins and dibenzofurans in human adipose tissue samples from five Ontario municipalities. *Chemosphere* 21(12):1465–1475.

of chlorine atoms substituted at various positions of the rings; a total of 209 congeners are possible. PCBs were manufactured in the United States from 1929 to 1977, with a peak production of about 100,000 tons a year in 1970. They were used mainly as the coolant in power transformers and capacitors because they are excellent insulators, are chemically stable, and have low flammability and vapor pressure. In later years they were also used as heat-transfer fluids in other machinery, and as plasticizers for polyvinylchloride and other polymers; they found additional uses in carbonless copy paper, as de-inking agents for re-cycled newsprint, and as weatherproofing agents. As a result of industrial discharges and the disposal of all these products, PCBs were spread widely in the environment.

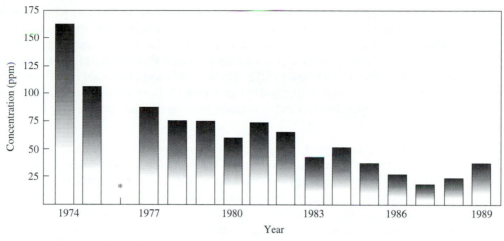

* Indicates no data available for 1976

Figure 17.16 The concentrations of PCBs in herring gull eggs at the Toronto shoreline of Lake Ontario (1974–1989). *Source:* Ministry of Supplies and Services (1990). *The State of Canada's Environment* (Ottawa, Canada: Ministry of Supplies and Services).

Because PCBs are chemically stable, they persist in the environment, and because they are lipophillic, they are subject to bioaccumulation, as are DDT and dioxins. PCB concentrations at the top of the food chain are significant in many localities. For example, herring gull eggs on the shores of Lake Ontario contained more than 160 ppm of PCBs in 1974 (see Figure 17.16). Since then, however, the level has declined by a factor of five, reflecting the termination of PCBs in all open uses, those in which disposal cannot be controlled. Production was drastically curtailed in 1972 and halted in 1977. PCB-containing transformers continue in service, but their disposal is regulated, and spent PCBs are stored or incinerated.

As in the case of dioxin, PCB health effects are hard to pin down. Occupational exposure has mainly produced cases of chloracne. There are, however, two instances of community poisoning by accidental PCB contamination of cooking oil, in Japan (1968) and Taiwan (1979). Thousands of people who consumed the oil suffered a variety of illnesses, including chloracne and skin discoloration, as well as low birth weight and elevated mortality of infants of exposed mothers. It was subsequently discovered that the oil was also contaminated with PCDFs, which are formed when PCBs are subjected to high temperatures [reaction (3) in Figure 17.13]; the PCBs that were mixed into the oil had been used as heat-exchange fluids in the deodorization process for the oil. Most of the toxic effects were attributed to the PCDFs rather than the PCBs.

In laboratory studies, PCBs are less toxic than PCDDs and PCDFs, but they probably operate by the same mechanism, binding to the Ah receptor. The most toxic PCBs are those that have no Cl atoms in the ortho positions of the ring, and can therefore adopt a coplanar configuration of the rings, as in PCDDs and PCDFs. Coplanarity is inhibited in ortho-substituted biphenyls by the steric interaction of the substituent with the ortho H atoms on the other ring. If substituents occupy three or four of the ortho positions, they bump into each other, and the rings are necessarily twisted away from each other. PCBs with this substitution pattern are the least toxic. Even if PCBs are less toxic to humans and other animals than PCDDs and PCDFs, they are much more abundant in the environment. Studies like the one discussed on p. 410 (see Figure 17.2), which indicates a connection between PCB exposure *in utero* and subsequent learning deficits, are cause for concern.

c. Global transport.   Organic pollutants move about in the environment by a variety of mechanisms. They are carried in the fat tissues of migrating animals and birds, and they drift through the air and along waterways with dust particles to which they adsorb. For many organic compounds, the chief mechanism for long-range transport is volatilization. If the compound has a reasonable vapor pressure, its molecules are volatilized when warmed by the sun, and condensed again when the atmosphere cools; in between they are transported by the winds.

Since temperatures are highest at the equator and coldest at the poles, volatilized molecules migrate steadily to higher latitudes. They volatilize and condense repeatedly, each time moving northward (or southward). This process has been likened to a global distillation. The consequence is that many organic pollutants concentrate in the Arctic region, thousands of kilometers from their sources. (Although the Antarctic region is subject to the same physics, there are fewer pollution sources in the southern hemisphere.)

Arctic pollution depends on the pollutant's vapor pressure. If this is high enough, the molecules never deposit, and continue to circulate in the atmosphere until they are destroyed, usually by reaction with hydroxyl radicals. Thus benzene and naphthalene do not accumulate at higher latitudes. But PAHs (polyaromatic hydrocarbons) having three or more rings do accumulate, because their lower vapor pressure induces condensation at low temperatures. If the vapor pressure is very low, as it is for the heavily chlorinated insecticide mirex (see pp. 384, 423), for example, migration becomes insignificant. For intermediate cases, such as most dioxins and PCBs, the vapor pressure is low enough that the molecules accumulate mostly in temperate regions, rather than the Arctic.

Nevertheless, Arctic peoples and animals are at risk of exposure to these molecules because of the high fat content of their diets. The pollutants that do arrive in the northern regions are bioaccumulated in the food chain (see pp. 386–388), and are stored in fat tissues. PCB levels up to 90 ppm have been found in the fat of polar bears, and breast milk is higher in PCBs for women in far northern areas than in temperate areas.

17.5   TOXIC METALS

The biosphere has evolved in close association with all the elements of the periodic table, and indeed, organisms harnessed the chemistry of many metal ions for essential biochemical functions at early stages of evolution. As a result, these elements are required for viability, although in small doses. When the supply of an essential element is insufficient, it limits the viability of the organism, but when it is present in excess, it exerts toxic effects,

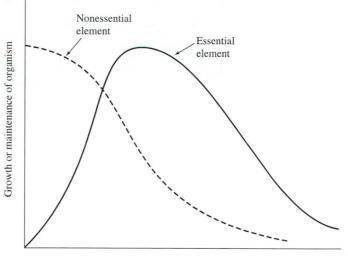

Figure 17.17   Dose-response curves for essential and nonessential elements in metabolic processes.

and viability is again limited. Thus, there is an optimum dose for all essential elements (see Figure 17.17).

   This optimum varies widely for different elements, however. For example, iron and copper are both essential elements, but we harbor about 5 g of the former in our bodies and only 0.08 g of the latter. Toxicity is low for iron but high for copper. Toxicity varies because the chemistry of the element varies. Thus, copper is generally present as Cu^{2+} and forms strong complexes with nitrogenous bases, including the histidine side chains of proteins. In contrast, neither Fe^{2+} nor Fe^{3+}, the common iron oxidation states, bind particularly strongly to nitrogenous bases. Copper is therefore more likely than iron to interfere with critical sites in proteins. At higher levels, nevertheless, iron is harmful, partly because it can catalyze the production of oxygen radicals (recall the discussion of antioxidants, p. 375), and partly because excess iron can stimulate the growth of bacteria and aggravate infections. Chromium is also an essential metal, albeit in traces, but it is a powerful carcinogen as well. Carcinogenicity is associated with the highest oxidation state, Cr(VI), and the main concern is chromate pollution from spills and residues of electroplating baths and from chromate emissions from cooling towers where it is used to inhibit corrosion. When toxic doses are compared for different metals (see Table 17.7), a wide variation is seen.

   The breadth of the peak in the viability curve for metals (see Figure 17.17) depends in part on *homeostatic* mechanisms, which have evolved to accommodate fluctuations in the metal availability. For example, excess iron is deposited in a storage protein, *ferritin,* from which it is released as needed. Many metals have no known biological benefit, and for them the curve of viability decreases steadily with increasing dose (see Figure 17.17). The initial part of the curve may be fairly flat, however, if there are biochemical protection mechanisms that can accommodate low to moderate doses. For example, cadmium

TABLE 17.7 RELATIVE MAMMALIAN TOXICITY OF ELEMENTS IN INJECTED DOSES
AND DIETS

Element	Acute lethal doses (LD_{50}) injected into mammals* (mg/kg bodyweight)	Dose in human diet (mg/day)	
		Toxic	Lethal[†]
Ag	5–60	60	1.3k–6.2k
As	6	5–50	50–340
Au	10	—	—
Ba	13	200	3.7k
Be	4.4	—	—
Cd	1.3	3–330	1.5k–9k
Co	50	500	—
Cr	90	200	3k–8k
Cs	1,200	—	—
Cu	—	—	175–250
Ga	20	—	—
Ge	500	—	—
Hg	1.5	0.4	150–300
Mn	18	—	—
Mo	140	—	—
Nd	125	—	—
Ni	110–220	—	—
Pb	70	1	10k
Pt	23	—	—
(^{239}Pu)	1	—	—
Rh	100	—	—
Sb	25	100	—
Se	1.3	5	—
Sn	35	2,000	—
Te	25	—	2k
Th	18	—	—
Tl	15	600	—
U	1	—	—
V	—	18	—
Zn	—	150–600	6k

*Injected into the peritoneum to avoid absorption through the digestive tract; chemical form of the element will affect its toxicity

[†]k signifies thousands of milligrams/day

Source: H. J. M. Bowen (1979). *The Environmental Chemistry of the Elements* (London: Academic Press).

(see pp. 441–444) is bound by *metallothionen,* a sulfur-rich protein in mammalian kidneys. When bound to the protein, cadmium is prevented from reaching critical target molecules. Toxicity increases rapidly if the metallothionen capacity is exceeded.

Cadmium, along with lead, mercury, and arsenic (all of which are of particular environmental concern), is a "soft" Lewis acid (large polarizability), with particular affinity for soft Lewis bases, such as the sulfhydryl side chain of cysteine amino acids (see p. 371). It is likely that the heavy metals exert their toxic effects by tying up critical cysteine residues in proteins, although the actual physiological consequences vary from one metal to another.

All metals cycle naturally through the environment. They are released from rock by weathering and are transported by a variety of mechanisms, including uptake and processing by plants and microorganisms. For example, sulfate-reducing bacteria convert any mercuric ions they encounter to the highly toxic methylmercury (see discussion, p. 437), and other bacteria long ago developed a defense system involving a pair of enzymes, one that breaks the methylmercury bond (*methylmercury lyase*), and another that reduces the resulting mercuric ion to elemental mercury (*mercury reductase*), which volatilizes out of harm's way. Likewise, plants living on soils derived from ore bodies have evolved protective mechanisms that actively transport toxic metals from the root zone up into special compartments (*vacuoles*) in the leaves, where they are sequestered. These plants are now being pressed into service to extract metals from toxic waste sites, in *phytoremediation* schemes.

The natural biogeochemical cycles of the metals have been greatly perturbed by human intervention. Mining and metallurgy are not new developments; they extend back to the Bronze Age. But the scale of metals extraction has increased enormously since the Industrial Revolution (see Figure 17.18). Evidence for a massive increase in the global

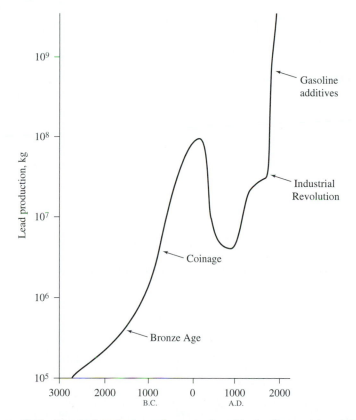

Figure 17.18   Historical production and consumption of lead.   *Source:* Adapted from J. Nriagu (1978). *Biogeochemistry of Lead* (Amsterdam: Elsevier); and P. M. Stokes (1986). *Pathways, Cycling, and Transport of Lead in the Environment* (Ottawa: Royal Society of Canada).

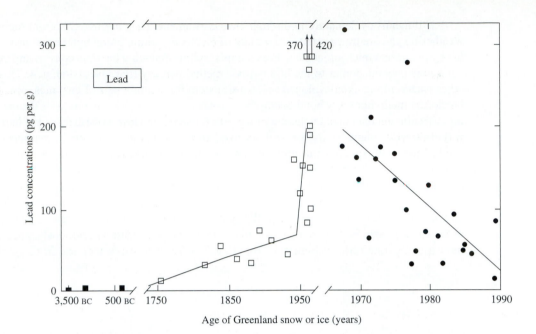

Figure 17.19   Changes in lead concentrations in Greenland ice and snow in units of picograms (10^{-12} g) of lead per gram of ice.   *Source:* C. F. Boutron et al. (1991). Decrease in anthropogenic lead, cadmium, and zinc in Greenland snows since the late 1960s. *Nature* 353:153–156. Copyright © 1991 by *Nature*. Reprinted with permission from *Nature* and authors of journal article.

environmental loading of lead, for example, can be found in the record provided by ice-cores from Greenland (see Figure 17.19); these levels have significantly declined since 1970, thanks to the phasing out of lead additives from gasoline (see p. 447). Large increases in production of several metals between 1930 and 1985 are documented in Table 7.8. Also listed are the large amounts of these metals that are estimated to be dis-

TABLE 17.8   PRIMARY PRODUCTION OF METALS AND GLOBAL EMISSIONS TO SOIL (10^3 t/yr)

Metal	Production in		Global emissions to soil in 1980s
	1930	1985	
Cd	1.3	19	22
Cr	560	9,940	896
Cu	1,611	8,114	954
Hg	3.8	6.8	8.3
Ni	22	778	325
Pb	1,696	3,077	796
Zn	1,394	6,024	1,372

Source: J. O. Nriagu (1988). A silent epidemic of environmental poisoning? *Environmental Pollution* 50:139–161.

persed into the environment and deposited on soils. In some cases (Cd and Hg), the amounts deposited are actually greater than the amount produced by extraction, because there are adventitious sources, such as ore processing for other metals, or the burning of coal, which contains trace concentrations of many metals; in the case of cadmium, traces in phosphate rock, which is mined and incorporated into fertilizer, add up to a significant fraction of the total.

The biogeochemical cycles are completed by sedimentation and burial of the metals in Earth's crust. But this process requires eons, and it is clear that the current massive extraction and dispersal are greatly increasing the amount of metals in circulation. What are the consequences of this buildup for human and ecosystem health? There is no general answer to this question because health effects depend sensitively on the precise exposure routes, not only for the different metals, but for the different forms of a given metal. The physical and chemical state of the metal are all-important for transport mechanisms, and also for *bioavailability*. To exert a toxic effect, metal ions must reach their target molecules, and they may be unable to do so if tied up in an insoluble matrix, or if they are unable to traverse critical biological membranes. In the next sections, we consider these issues for four toxic metals of current concern.

a. Mercury.   The environmental toxicity of mercury is associated almost entirely with eating fish; this source accounts for some 94 percent of human exposure. Sulfate-reducing bacteria in sediments generate methylmercury and release it into the waters above, where it is absorbed by fish from the water passed across their gills or from their food supply. The CH_3Hg^+ ion forms CH_3HgCl in the saline milieu of biological fluids, and this neutral complex passes through biological membranes, distributing itself throughout the tissues of the fish. In the tissues, the chloride is displaced by protein and peptide sulfhydryl groups. Because of mercury's high affinity for sulfur ligands, the methylmercury is eliminated only slowly, and is therefore subject to bioaccumulation when little fish are eaten by bigger fish. The phenomenon is the same as for DDT (see Figure 16.5) and other lipophiles, but the mechanism is different because mercury accumulates in protein-laden tissue (muscle) rather than in fat.

Biomethylation of mercury occurs in all sediments, and fish everywhere have some mercury. But the levels are greatly elevated in bodies of water for which sediments are contaminated by mercury from waste effluents. The worst case of environmental mercury poisoning occurred in the 1950s in the fishing village of Minamata, Japan. A polyvinylchloride plant used Hg^{2+} as a catalyst and discharged mercury-laden residues into the bay, where the fish accumulated methylmercury to levels approaching 100 ppm. Thousands of people were poisoned by the contaminated fish, and hundreds died from it. Those affected suffered numbness in the limbs, blurring and even loss of vision, and loss of hearing and muscle coordination, all symptoms of brain dysfunction resulting from the ability of methylmercury to cross the blood-brain barrier. Likewise, methylmercury can pass from mother to fetus, and a number of Minamata infants suffered mental retardation and motor disturbance before the cause of the poisoning was identified. Based on this incident and others, the recommended limit for mercury in fish for human consumption has been set at 0.5 ppm.

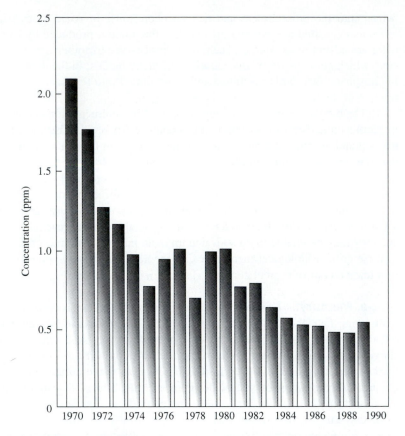

Figure 17.20   Annual variation of mercury concentrations in walleye fish from Lake Saint Clair.   *Source:* Ministry of Supplies and Services (1991). *Toxic Chemicals in the Great Lakes and Associated Effects: Synopsis* (Ottawa, Canada: Ministry of Supplies and Services).

Fortunately, cessation of waste mercury discharge lowers the levels of mercury in local fish, as seen in data for Lake Saint Clair (see Figure 17.20), which is part of the Great Lakes chain. The mercury concentration in walleye fish dropped from 2.0 to 0.5 ppm during the 1970s and 1980s, after the discharge of mercury from chlor-alkali plants was restricted. Even at 0.5 ppm, however, the level hovers at the recommended limit for human consumption. It takes a long time for the biomethylation process to clear the mercury from contaminated sediments.

Much of the world's discharge of mercury into the aqueous environment stems from chlor-alkali plants. These plants manufacture Cl_2 and NaOH, large-volume commodities that are mainstays of the chemical industry. They are produced by electrolysis of aqueous sodium chloride. A carbon anode is used to generate chlorine:

$$2Cl^- = Cl_2 + 2e^- \tag{17.2}$$

while a mercury pool cathode collects metallic sodium as a mercury amalgam:

$$2Na^+ + 2e^- = 2Na(Hg) \tag{17.3}$$

which is then reacted with water in a separate compartment:

$$2Na + 2H_2O = 2NaOH + H_2 \tag{17.4}$$

Reaction (17.4) does not proceed spontaneously because the activity of sodium is depressed in the amalgam, but it can be promoted by applying a small electric current. The purpose of this mercury-mediated two-stage process is to keep the NaOH product free of the NaCl starting material. This can, however, also be accomplished by separating the two electrode compartments using a cation-exchange membrane (see p. 292 for a discussion of ion exchangers), which inhibits the transfer of anions. Although chlor-alkali plants can be retrofitted to greatly reduce mercury discharges, most mercury electrode installations are being phased out and replaced by membrane-based units.

Even though local discharges have caused the most serious mercury contamination, it has been discovered that fish have elevated mercury levels even in lakes that are quite remote from any local source. Thus, mercury is transported over long distances, a consequence of the fact that there are two volatile forms, metallic mercury, Hg^0, and dimethylmercury, $(CH_3)_2Hg$. Both are formed in the same milieu as methylmercury. As mentioned previously, bacteria have a detoxification system that rids their environment of methylmercury by converting it to Hg^0, which is volatilized. And $(CH_3)_2Hg$ is produced in the same biomethylation process as CH_3Hg^+. Both molecules are produced by bacteria, in varying proportions, depending on the pH (see Figure 17.21). The $(CH_3)_2Hg$ is volatilized, while the

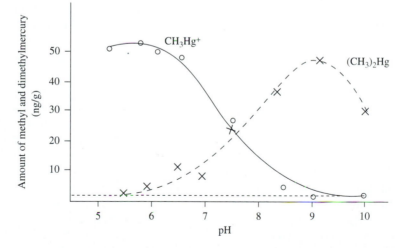

Figure 17.21   Methylation of 100 ppm of Hg^{2+} in sediments over two weeks.   *Source:* I. G. Sherbin (1979). *Mercury in the Canadian Environment,* Report EPS-3-EC-79-6 (Ottawa: Environmental Protection Service Canada).

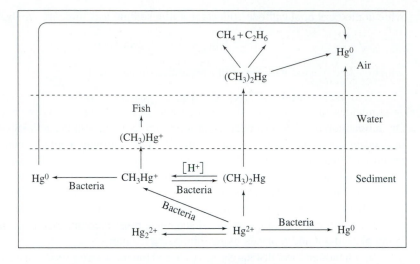

Figure 17.22   The biogeochemical cycle of bacterial methylation and demethylation of mercury in sediments.   *Source:* National Research Council (1978). *An Assessment of Mercury in the Environment* (Washington, DC: National Academy Press).

CH_3Hg^+ is released into the water and is available for bioaccumulation (see Figure 17.22 for a diagram of the mercury cycle). High pH favors $(CH_3)_2Hg$, while low pH favors CH_3Hg^+; the crossover occurs near neutrality. This means that an additional consequence of lake acidification is an increase in the $CH_3Hg^+/(CH_3)_2Hg$ ratio, and therefore an increase in mercury toxification.

Metallic mercury is used in many applications, especially in batteries, switches, lamps, and other electrical equipment; improper use and disposal (batteries in municipal incinerators, for example) add to the global load of mercury vapor. So does the use of mercury to extract gold or silver from ores, a practice used for centuries in Central and South America and applied on a large scale today in the gold fields of Brazil. This process releases massive amounts of mercury into the environment because the extracted gold is recovered by heating the amalgam to drive off the mercury. Large quantities of amalgam washings have contaminated parts of the Amazon River sediment with mercury. This practice is estimated to account for 2 percent of global atmospheric mercury emissions. Only now are simple mercury-condensing hoods being installed to reduce the losses. Moreover, soils in the Amazon Basin contain relatively high concentrations of natural mercury bound to cation exchange sites on humic substances in the organic top soil, see pp. 295–296. Another source of mercury inputs to the river has been slash-and-burn agriculture, which combusts much of the humic material, leaving behind a mobile, mercury-containing residue.

Inorganic mercury is not particularly toxic when ingested because neither the metal nor the ions (Hg_2^{2+} and Hg^{2+} and their complexes) penetrate the intestinal wall effectively. However, Hg^0 is highly toxic when inhaled; in atomic form, it is able to pass through the lung membranes into the bloodstream and across the blood-brain barrier. In the brain it can presumably be oxidized and bound to protein sulfhydryl groups because it produces the same neurological effects as methylmercury. For this reason, individuals should avoid handling elemental mercury, and all spills should be treated (with sulfur, which ties up the mercury atoms) and cleaned up. The Brazilian gold miners suffer serious health problems from elemental mercury released during the amalgam operations, as silver and gold miners have for centuries.

Complexes of phenylmercury, $C_6H_5Hg^+$, have been used as paint preservatives, and as slimicides in the pulp and paper industry, but these uses have now been curtailed. Organomercurials have also been used as fungicides in agriculture and industry, especially in dressings for seed grains. Once in the soil, these compounds break down and the mercury is trapped as insoluble mercuric sulfide. However, hundreds of people died in Iraq from eating bread made from mercury-contaminated flour produced from treated seed grain that had been diverted inadvertently to a flour mill. In the United States, a New Mexico family was poisoned by eating a pig that had been fed mercury-treated seed grain. The family brought action in court, leading the EPA to ban organomercurials for seed treatment. In Sweden and Canada, populations of birds of prey declined after eating smaller birds that had fed on treated seeds. The use of mercury compounds to treat seeds has now been curtailed in Europe and North America.

The natural mercury cycle can also be perturbed indirectly by human activity. Elevated mercury levels have been found in fish living in waters impounded by hydroelectric dams in Quebec and Manitoba. The source of this "pollution" was simply bacterial action on the naturally occurring mercury already present in the newly submerged surface soil.

b. Cadmium.   Cadmium is found in the same column of the periodic table as mercury and zinc, but its chemical properties are much closer to zinc than to mercury. This resemblance to zinc accounts for cadmium's distribution, as well as its particular hazards. Cadmium is always found in association with zinc in Earth's crust, and it is obtained as a side-product of zinc mining and extraction; there are no separate cadmium mines. Moreover, cadmium is always present as a contaminant in zinc products. Indeed, one of the pervasive sources of cadmium in the urban environment is zinc-treated (*galvanized*) steel. The weathering of galvanized steel surfaces produces zinc- and cadmium-laden street dust; though the concentration is low, the total amount of cadmium is substantial. It has been pointed out that proposals to ban cadmium in products (mostly batteries, electroplate, pigments, and plastics stabilizers) in order to reduce environmental exposure might have the opposite effect, because of lowered economic incentive to recover cadmium from zinc-mine residues and from the refined zinc itself.*

Mimicry of zinc is probably why cadmium is actively taken up by many plants, since zinc is an essential nutrient. Most of our cadmium intake is from vegetables and grains in our diet. However, smokers get an extra dose because of the cadmium concentrated in tobacco leaves; heavy smokers have twice as much cadmium in their blood, on average, as nonsmokers. The average cadmium intake per day in the United States is estimated to

*W. M. Stigliani and S. Anderberg (1994). Industrial metabolism at the regional level. In *Industrial Metabolism: Restructuring for Sustainable Development*, R. U. Ayres and U. E. Simonis, eds. (Tokyo: United Nations University Press).

be 10–20 μg, but only a small fraction is absorbed; the absorbed dose is estimated to be 0.4–1.8 μg. Pack-a-day smokers take in an extra 2 μg/day on average, but inhalation greatly increases the fractional absorption. The absorbed dose is thus increased to an estimated 0.9–2.8 μg/day.

There is concern that cadmium buildup in agricultural soils may eventually produce dangerous levels in food. Cadmium inputs to soils are mainly from airborne deposition (wet plus dry) and from commercial phosphate fertilizers, which contain cadmium as a natural constituent of phosphate ore. The cadmium burden could be further increased by the use of fertilizer from sewage sludge, a sludge-disposal measure that is increasingly advocated. Sewage is often contaminated by cadmium and other metals; however, there is some evidence that the cadmium is firmly bound in the sludge and might not be released to growing plants.

The problem of cadmium accumulation has been examined extensively for the heavily industrialized Rhine River Basin, by evaluating the mass flows of cadmium over several decades in a study of the industrial ecology (see pp. 125–128) of the region. It was found that cadmium air emissions declined substantially, thanks to the control of point sources, especially ferrous and nonferrous metal smelters (see Figure 17.23; see Fig-

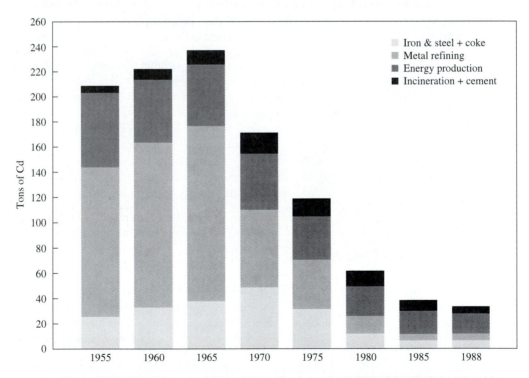

Figure 17.23   Trends in atmospheric emissions of cadmium in the Rhine River Basin by industrial sector (1955–1988). *Source:* S. Anderberg and W. M. Stigliani (1995). Private communication (Laxenburg, Austria: International Institute for Applied Systems Analysis).

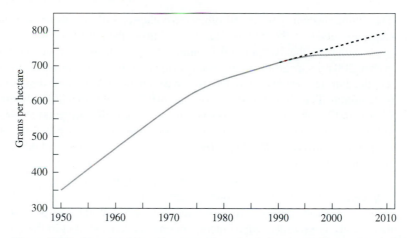

Figure 17.24   Time trend in soil concentration of cadmium in top 20 cm of typical agricultural soil in the Rhine Basin. In projections to the year 2010, solid line assumes cadmium in phosphate fertilizer is eliminated by the year 2000; the dashed line assumes no reduction of cadmium in fertilizer.   *Source:* W. M. Stigliani et al. (1993). Heavy metal pollution in the Rhine Basin. *Environmental Science and Technology* 27(5):786–793. Copyright © 1993 by American Chemical Society. Reprinted with permission from ES&T.

ure 14.5 for trends in aqueous loads of cadmium in the Rhine Basin). In contrast to the reductions in atmospheric and aqueous emissions, concentrations of cadmium in agricultural soils in the basin have increased, and may continue to do so in the future (see Figure 17.24) due to residual inputs from diffuse sources (primarily atmospheric deposition from coal burning and application of phosphate fertilizer). This pattern reflects the fact that the *residence time* of cadmium in soils can be orders of magnitude longer than its lifetime in air or river water, particularly when the pH of the soils is maintained above 6.0, as is usually the case in agricultural soils, owing to additions of lime ($CaCO_3$) (see Figure 12.13, p. 304).

It is estimated that approximately 3,000 tons of cadmium had accumulated in the plow layer of agricultural soils of the basin from 1950 to 1990. Particularly worrisome is a scenario in which the stored cadmium would be released as a result of soil acidification. This could occur if current plans for abandoning large tracts of agricultural land in the basin are implemented. If the abandoned lands were no longer limed, the pH of the soils could drop to as low as 4.0 within several decades. The decrease in pH would result in a rapid release of cadmium out of the topsoil (see Figure 12.13). This occurrence could pose public health problems where heavily polluted soils overlie shallow groundwaters used for drinking.

Soil conditions were certainly a factor in the only known case of wide-spread environmental cadmium poisoning, which occurred in the Jinzu valley of Japan. Irrigation water drawn from a river that was contaminated by a zinc mining and smelting complex led to high levels of cadmium in the rice. Hundreds of people in the area, particularly older women who had borne many children, developed a painful degenerative bone disease called *itai-itai* ("ouch-ouch"), apparently because Cd^{2+} interfered with Ca^{2+} deposition.

Their bones became porous and subject to collapse. Sufferers were estimated to have had a cadmium intake of about 60 μg/day, several times the normal intake.

Although the 70 ppm soil cadmium level in Jinzu was elevated, it has been still higher, 300 ppm, in Shipham, England, a zinc mining locale during the seventeenth to nineteenth centuries. Yet health inventories in Shipham showed only slight effects attributable to cadmium. The Shipham soils have high pH, 7.5, and also a high content of calcium carbonate and hydrous oxides of iron and manganese, which are good absorbers of Cd^{2+} (see entries 3a, 3b, Table 13.2, p. 310). In contrast, the Jinzu soils had low pH, 5.1, and a low content of hydrous oxides. Thus, the cadmium was far more available for plant uptake in Jinzu than in Shipham.

Chronic exposure to cadmium has been linked to heart and lung disease (including lung cancer at high levels) immune system suppression, and liver and kidney disease. As mentioned above, the Cd^{2+}-sequestering protein metallothionen provides protection until its capacity is exceeded. Since metallothionen is concentrated in the kidney, this organ is damaged first by excessive cadmium. The downside of metallothionen protection is that cadmium is stored in the body and accumulates with age, so that damage from long-term exposure becomes irreversible.

c. Arsenic.   Although arsenic is in the same column of the periodic table as phosphorus, and has similar chemistry, it is more easily reduced from the V to the III oxidation level. As(III) (arsenite, AsO_3^{3-}) is more toxic than As(V) (arsenate, AsO_4^{3-}), probably because it binds more readily to sulfhydryl groups on proteins. Indeed, the toxicity of As(V) probably results from its reduction to As(III) in the body. The mechanism of toxicity is uncertain, but recent results suggest that methylation of As(III), a process similar to mercury methylation (see pp. 339–340), produces products that are highly reactive and damage DNA, at least in cultured human cells. In another recent discovery, low levels of arsenic were found to inhibit activation receptors of glucocorticoid hormone, receptors that turn on many genes that suppress cancer and regulate blood sugar; arsenicosis is known to trigger diabetes as well as cancer.

The poisonous properties of arsenic compounds have been known, and exploited, since antiquity, but it is the turn of the twenty-first century that has witnessed inadvertent arsenic poisoning on mass scale. In Bangladesh and the neighboring Indian province of West Bengal, some 70 million people are at risk of poisoning due to high arsenic levels in the groundwater. In West Bengal alone, up to 200,000 people have been diagnosed with arsenicosis. There are many more cases in Bangladesh, where some 4.5 million people may have been drinking arsenic-laced water for many years.

Ironically the problem grew out of a United Nations-sponsored effort, starting in the late 1960s, to provide clean drinking water by sinking tube wells into the shallow aquifer that underlies the region. The effort did in fact improve health greatly by reducing the incidence of waterborne diseases. However, the high arsenic content of the reservoir went unrecognized for years. As many as half of the 4 million tube wells draw water that exceeds the Bangladesh arsenic standard of 50 ppb. (This was the U.S. standard, until 2001 when it was reduced to 10 ppb, a guideline also recommended by the World Health Organization). In the more contaminated areas, arsenic levels routinely exceed 500 ppb.

Arsenic in drinking water is a slow poison. The first symptoms are keratoses, discoloring the skin. Later these develop into cancers, and the liver and kidneys also deteriorate. This process can take up to twenty years. In its early stages arsenicosis is reversible, if arsenic consumption is discontinued, but once cancer begins to develop, there is no effective treatment. Poor nutrition renders the Bangladesh villagers more vulnerable than they otherwise might be.

Why is the aquifer laden with arsenic? The answer is not entirely clear, but the likeliest theory is that it is associated with iron-laden sediments from the rivers that drain into the region (the Ganges, Bramaputra, and Meghna Rivers). Arsenic is found naturally in association with sulfide minerals. (Smelting of gold, lead, copper, and nickel ores can be a source of arsenic pollution. A study in Bulgaria has implicated arsenic from a copper-smelting plant as a cause of a three-fold increase in birth defects in the area.) Trace amounts of arsenic contained in iron sulfide (pyrite) is the most widespread arsenic source, and pyrite weathering leads to the release of arsenate, along with sulfate and ferric hydroxide (see the chemistry of acid mine drainage, p. 305). The sulfate is washed out to sea, but the more highly charged arsenate adsorbs on the ferric hydroxide, and is deposited in river sediments. Typically these sediments contain 2–6 ppb of arsenic, and the alluvial sediments of Bangladesh and West Bengal, with concentrations of 2–20 ppm, are not markedly higher. However, these sediments are also rich in organic matter, and the high BOD leads to reducing conditions, resulting in reduction and solubilization of the iron, with release of the adsorbed arsenate. This is the same scenario as the release of phosphate from sediments under reducing conditions in Chesapeake Bay, discussed on pp. 325–328.

Thus the Bangladesh and West Bengal sediments are not unique in their tendency to release arsenic into the groundwater. This problem is found in many parts of the world, including Chile, Taiwan, and the Southwest region of the U.S. Indeed, an epidemiological study in Taiwan produced a clear link between arsenic levels in well water and skin cancer incidence. However, only in Bangladesh and West Bengal has the arsenic exposure been so widespread and so long-lasting.

The outlook for the victims of arsenicosis is not encouraging. The villagers are desperately poor and have no access to alternative sources of clean water. There is concern that even labelling the tainted wells could backfire if their users revert to polluted surface waters for their needs. New wells could be dug into a deeper, uncontaminated aquifer, and there are several schemes to treat the current well water to remove the arsenic. However, because there are millions of contaminated wells, the cost of any solution is enormous, and there are numerous administrative and political barriers. The World Bank is coordinating a mitigation plan and its funding, but the massive effort could take ten years or more.

d. Lead.   Of all the toxic chemicals in the environment, lead is the most pervasive; it poisons many thousands of people yearly, especially children in urban areas.

1) Exposure.   Unlike cadmium, lead is not taken up actively by plants; nevertheless, it contaminates the food supply because it is abundant in dust and is deposited on food crops, or on food as it is being processed. Food and direct ingestion of dust account for most of the average lead intake, estimated to be about 50 μg/day in the United States. In urban areas or along roadways, the lead levels in dust generally exceed 100 ppm, while in remote rural areas the levels are 10–20 ppm. Ingestion of just 0.1 g of urban dust can provide a 10 μg or higher dose of lead. It does not take much dust, therefore, whether ingested directly or mixed in with food (the two contributions are estimated to be comparable), to account for the average daily intake. Children are particularly at risk because they play in the dust, because they absorb a higher fraction of the lead they take in, and because they are exposed to more lead per unit of body mass.

A significant route of lead exposure in addition to ingestion of food and dust is drinking water. Lead can contaminate water either from lead-based solder used in pipe and fitting connections, or from the pipes themselves, which in older houses and water systems are made of lead. It is generally advisable not to drink the "first draw" water that has been standing overnight in older drinking fountains or the pipes of older buildings. In contact with O_2-bearing water, the metallic lead can be oxidized and solubilized:

$$2Pb + O_2 + 4H^+ = 2Pb^{2+} + 2H_2O \tag{17.5}$$

Since reaction (17.5) consumes two protons per lead ion, the dissolution rate is strongly pH-dependent. The lead hazard is greatest where the water is soft, that is, where there has been little neutralization of the rainwater's acidity (see discussion on p. 298). Hard water, on the other hand, has higher pH; in addition, it has a higher carbonate concentration (because of neutralization by $CaCO_3$). Carbonate precipitates Pb^{2+} as the sparingly soluble $PbCO_3$, which inhibits dissolution of the underlying metal in the pipes and fittings. Some water-supply districts, especially those with soft water and old lead pipes, now add phosphate to the drinking water in order to form a similarly protective coating of lead phosphate. Lead test-kits are commercially available for measuring the lead levels in one's own drinking water.

A related hazard is the practice, now largely discontinued, of using lead solder to seal food and drink in "tin" cans. When the cans are opened and the contents exposed to air, lead can be mobilized into the contents, particularly if they are acidic. Likewise, lead can leach into food and drink stored in pottery if, as is common, its glaze contains lead oxide. It is particularly hazardous to drink fruit juices (because of their acidity) or hot drinks (because the rate of dissolution increases with temperature) from such vessels. Although lead-free glazes are now the rule among pottery manufacturers, lead-glazed pottery remains a significant source of dietary lead.

If lead in dust is the major exposure route, how does the lead get there? House paint is an important source. Renovating old homes is hazardous unless care is taken to contain the dust from layers of old paint. Lead salts are brightly colored and have been widely used as pigments and paint bases. Lead chromate, $PbCrO_4$, provides the yellow coloring for striping on roads and for school buses, while the "red lead" oxide, Pb_3O_4, is the base for the corrosion-resistant paints on bridges and other metal structures. The hydroxycarbonate, $Pb_3(OH)_2(CO_3)_2$, is "white lead," which was widely used as the base of indoor paints, but has now been replaced by titanium dioxide, TiO_2. Nevertheless, older buildings, particularly in the United States, where lead was banned from indoor paint only in 1971 (much of Europe banned it in 1927), still have leaded paint on the walls. Dust and paint chips from the walls are the main source of indoor lead exposure. Leaded paints have also been widely used on building exteriors, and weathering raises the lead levels in the dust outside the building. Although less dangerous than lead inside the house, the lead outside is still a matter of concern because children often play in the dust, and ingest it, or track it inside.

The other major source of lead exposure is leaded gasoline. As discussed in connection with gasoline additives (see pp. 238), adding tetraethyl- or tetramethyl-lead to gasoline improves its octane rating by scavenging radicals and inhibiting pre-ignition. These compounds are themselves toxic; they are readily absorbed through the skin, and in the liver they are converted to trialkyl-lead ions, R_3Pb^+, which, like methylmercury ions, are neurotoxins. However, a much greater threat to public health is the lead that spews out of the tailpipe and into the atmosphere. Most of the lead is emitted in small particles of PbX_2 ($X = Cl$ or Br), formed by reaction with ethylene dichloride or dibromide, added to the gasoline to prevent buildup of lead deposits inside the engine. These particles can travel far on air currents, and are no doubt responsible for the sharp upturn in the lead content of Greenland ice from around 1950 (see Figure 17.19), as automotive traffic expanded greatly around the world in the ensuing decades. However, most of the particles settle out not far from where they are generated, contaminating the dust near roadways and in urban areas with lead. Lead levels in urban dust correlate strongly with traffic congestion. (See the difference in average blood lead concentrations in Figure 17.25 between Stockholm, with a population of over 1 million, and Trelleborg, Sweden, with a population of about 25 thousand.)

Lead additives were phased out of use in the United States starting in the early 1970s, when catalytic converters were introduced for pollution control, because lead particles in the exhaust gases deactivate the catalytic surfaces. By 1990, Brazil and Canada had phased out leaded gasoline, and many other countries, including Argentina, Iran, Israel, Mexico, Taiwan, Thailand, and most European nations, significantly reduced the lead concentration in leaded gasoline (from about 1 g/L to 0.15–0.3 g/L). The use of leaded gasoline is declining further in Europe and Mexico because all new cars are required to have catalytic converters. But leaded gasoline consumption continues in much of the rest of the world, and almost no unleaded gasoline is available in many parts of Africa, Asia, and South America. The global impact of the shift away from leaded gasoline is reflected in the steadily decreasing lead levels in the Greenland ice record since 1970 (see Figure 17.19). The current deposition rate is now approaching pre-automobile levels. This trend could be reversed, however, if leaded gasoline is used in developing countries to meet rapidly escalating demands for automobile transport.

Although the original motivation for removing lead from gasoline was to protect catalytic converters, an additional clear benefit was the reduction in human exposure to lead. Everywhere that lead has been phased out, there has been a steady drop in blood lead levels of the populace (see Figure 17.25). Other sources of lead, particularly leaded solder in food cans, have also been curtailed in the same period. (In the figure, this is especially apparent for Christchurch, New Zealand, which shows a nearly 50 percent drop in blood lead concentrations while gasoline lead concentrations remained unchanged at about 0.84 g/L, reflecting the gradual removal of lead-soldered cans, and a reduction in consumption of canned foods.) Nonetheless, the strongest correlations of blood lead concentrations are

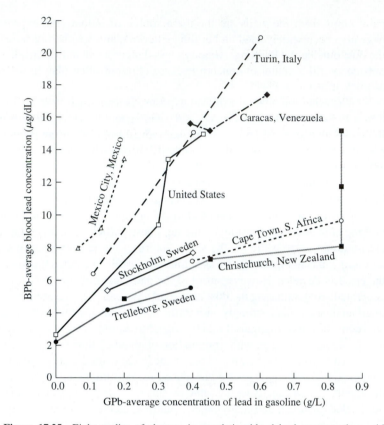

Figure 17.25 Eight studies of changes in population blood lead concentrations with changes in concentration of lead in gasoline. Adapted from V. M. Thomas et al. (1999). Effects of reducing lead in gasoline: An analysis of the international experience. *Environmental Science and Technology* 33:3942–3948. Copyright © 1999 by American Chemical Society. Reprinted with permission from ES&T.

with gasoline lead levels. Whereas average blood lead levels of 15 μg/dL (dL = deciliter; since 1 dL of blood weighs about 100 g, 15 μg/dL is about 1.5 ppm) were common before 1980, they are tending toward 3 μg/dL in regions that have eliminated lead in gasoline.*

2) Bioavailability. Once ingested, the lead may be absorbed or eliminated. The fractional absorption is estimated to be 7–15 percent in adults, on average, and 30–40 percent in children. However, the extent of absorption depends on the chemical and physical state of the lead. Large particles of relatively insoluble lead minerals are likely to pass unaltered through the stomach and intestines, whereas small particles of soluble lead

*V. M. Thomas, R. H. Socolow, J. J. Fanelli, and T. G. Spiro (1999). Effects of reducing lead in gasoline: An analysis of the international experience. *Envirn. Sci. Technol.* 33:3942–3948.

compounds are readily absorbed. No doubt this is why people living in the vicinity of lead-bearing mine tailings tend not to have elevated lead levels: the lead in the tailings occurs mostly as chunks of lead sulfide or oxide. However, people living near lead smelters do have somewhat elevated lead levels; the smelters emit fine particles of more reactive lead oxide phases. The data in Figure 17.25 suggest that the lead from gasoline is readily absorbed since it accounted for such a high percentage of average blood lead levels in earlier years. The emitted lead is in the form of soluble lead-halide particles, which are highly absorbable. Upon contact with moisture-laden soils, the halides may be converted to $Pb(OH)_2$ and then into larger particles of PbO, which are less bioavailable. This process can account for the fact that lead blood levels correlate with the phase out of leaded gasoline without a significant time lag, even though lead levels remain elevated in the dust near roadways.

3) Toxicity.   Lead toxification is as old as human history; the use of lead in artifacts dates as far back as 3800 B.C. The Greeks realized that drinking acidic beverages from lead containers could result in illness. But apparently the Romans did not, for they sometimes deliberately added lead salts to overly acidic wines to sweeten the flavor. The bones of Romans have much higher lead levels than do those of modern humans, and some historians speculate that chronic lead poisoning contributed to the downfall of the Roman Empire.

Once absorbed in the body, lead enters the bloodstream and moves from there to soft tissues. In time it is deposited in bones, because Pb^{2+} and Ca^{2+} have similar ionic radii. The lead content of bones increases with age; when bone matter dissolves, as can happen in illness or old age, the lead is remobilized into the bloodstream and can produce added toxic effects. These effects are expressed primarily in the blood-forming and nerve tissues (see Figure 17.26). Lead inhibits the enzymes involved in the biosynthesis of heme, the iron-porphyrin complex (see Figure 17.5) that binds to hemoglobin and serves as the binding site for O_2 (see Figure 9.2, p. 218). In particular, lead interferes with the *ferrochelatase* enzyme, which inserts iron into the porphyrin; instead, zinc is inserted. The resulting accumulation of zinc-porphyrin can be detected by its characteristic fluorescence emission (iron-porphyrin does not fluoresce), providing a sensitive indirect indicator of lead exposure. High exposures produce anemia due to iron-porphyrin deficiency.

The biochemical mechanism for lead's effects on nerve cells is uncertain, but diminution of nerve conduction velocity can be detected at relatively low blood lead levels (see Figure 17.26), while higher levels are associated with nerve degeneration. Even at quite low levels, possibly as low as 5 μg/dL, lead exposure in several epidemiological studies was found to be associated with impairments in growth, hearing, and mental development of children. In 1984, 17 percent of all U.S. children, and 30 percent of children in inner cities, were estimated to have blood lead levels greater than 15 μg/dL. Since then exposure has diminished, as noted above. In 1991, 8.9 percent of U.S. children aged one to five were estimated to have levels greater than 10 μg/dL, still representing a large population (1.7 million) at risk of experiencing developmental defects. The latest available data from the Centers for Disease Control and Prevention, covering a period from 1991 to 1994, showed that 890,000 children nationwide had blood levels exceeding 10 μg/dL. This exposure is almost

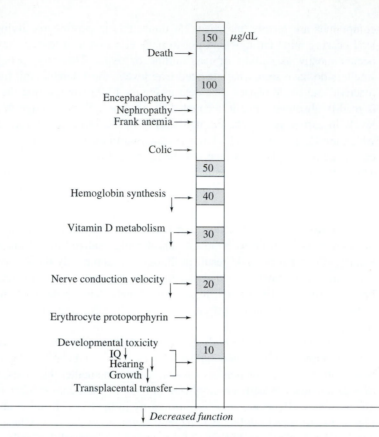

Figure 17.26 Lowest observed effect levels of inorganic lead in children (μg/dL). *Source:* Agency for Toxic Substances Disease Registry (1988). *The Nature and Extent of Lead Poisoning in Children in the United States: A Report to Congress* (Atlanta, Georgia: ATSDR).

entirely the result of ingesting lead-based paint chips in older homes. Nearly 22 percent of poor children living in inner-city homes built before 1946 have elevated blood lead levels. The government goal is to reduce the number of at-risk children by 35 percent by 2003, when the next national survey will be completed. There is also great concern about exposure *in utero* because lead crosses the placenta and can interfere with fetal development.

The precise molecular mechanisms of lead toxicity have not been pinned down, but probably involve lead's ability to bind to nitrogen and sulfur ligands, thereby interfering with the function of critical proteins (like ferrochelatase). This ability is also utilized in *chelation therapy* for lead toxicity. Lead can be cleared from the body by intravenous injection of chelating agents (see Figure 17.27). The chelators compete with protein binding sites, and the resulting Pb^{2+}-chelate complexes are excreted by the kidneys. Since the chelating agents can also bind metal ions other than lead, they are administered as the

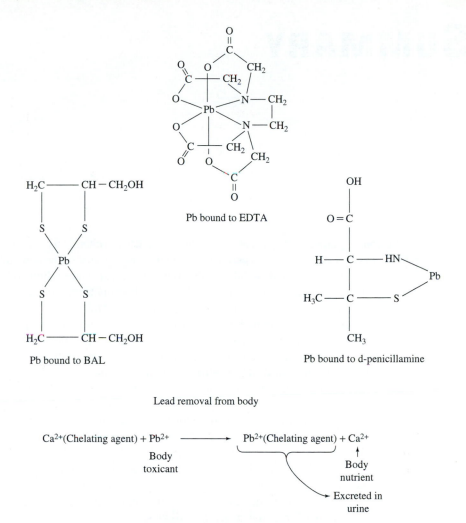

Pb bound to EDTA

Pb bound to BAL

Pb bound to d-penicillamine

Lead removal from body

Ca²⁺(Chelating agent) + Pb²⁺ ⟶ Pb²⁺(Chelating agent) + Ca²⁺

Body
toxicant

Body
nutrient

Excreted in
urine

Figure 17.27   Chelating agents for removing lead from the body.

Ca^{2+} complex, in order to avoid stripping calcium and other more weakly bound metals from the body. The strongly binding Pb^{2+} displaces the Ca^{2+} and is removed selectively.

Lead poisoning is also common among waterfowl, which die from ingestion of lead shot and lead fishing sinkers. In the early 1980s, an estimated 2–3 percent of the autumn population of waterfowl in the United States were dying each year from lead poisoning. Birds that preyed on the waterfowl were likewise exposed to lead. A study of 1,429 bald eagle deaths in the United States between 1963 and 1984 found that 6 percent were attributable to lead poisoning. Lead shot is now banned for waterfowl hunting in the United States and Canada.

SUMMARY

The biosphere plays an integral role in regulating the natural chemical cycles on the planet, but this function is vulnerable to perturbations by human activities. The animal kingdom is dependent on plants for food, and the productivity of plants is largely limited by the availability of nitrogen; "fixed" nitrogen is available naturally only from nitrogenase-containing bacteria (except for the small amount of N_2 oxidized by lightning). The expanding human population is dependent on agricultural productivity gains, which have been made possible by the production of nitrogen-containing fertilizer and the development of new plant varieties that can accommodate intensive fertilization. Commercial fertilizer now accounts for about 23 percent of the world's fixed nitrogen, and human-induced land-use changes contribute about another 20 percent. Thus, human activities have greatly augmented the global nitrogen cycle. The consequences, though not fully understood, include nitrate pollution of groundwater in agricultural areas, facilitation of soil erosion, and a probable contribution to the global increase in emissions of N_2O, a greenhouse gas and major actor in stratospheric ozone depletion.

In addition to total agricultural production, adequate human nutrition depends on a balanced diet comprising a proper complement of essential amino acids and vitamins. Furthermore, antioxidants in the diet may play an important role in increased longevity through their control of free-radical levels in tissues.

Modern agriculture relies not only on fertilizer, but also on synthetic herbicides and insecticides, to boost production and control losses to insects and other pests. In addition, insecticides play a key role in disease prevention through the control of insect vectors. However, their application inevitably breeds resistance among the target insects, leading to a never-ending search for new insecticides. In the meantime, the widespread application of these chemicals has deleterious effects on ecosystems and human health. For these reasons, more sophisticated pest-control strategies are being developed, including biological controls, and chemicals that are targeted to particular pests and that break down quickly into harmless molecules in the environment. The advent of genetically modified plants, harboring their own pesticides or traits for herbicide resistance, has had a major impact on agriculture, although its prospects are somewhat clouded by public reaction against this application of biotechnology.

Organochlorine chemicals are of particular concern because they degrade only slowly in the environment, and, being lipophillic, they bioaccumulate in the food chain by concentrating in fat tissue. Birds of prey are specifically vulnerable because organochlorines interfere with calcium deposition in egg shells. As a result, many organochlorine pesticides, starting with DDT, have been banned in developed countries, although they are still used elsewhere. There are also concerns about human health, which center on cancer and hormonal disruption. One of the most feared chemicals, 2,3,7,8-tetrachlorodibenzodioxin, is not produced deliberately, but was a contaminant of the herbicide 2,4,5-trichlorophenoxyacetic acid. It is also a trace product, along with many congeners, of any combustion process where chlorine and carbon are present in the material being burned. Incinerators are currently the main source of dioxins and the related polychlorinated dibenzofurans, although the use of pollution-control devices can eliminate most releases. Polychlorinated biphenyls, widely used in electrical transformers and other products, are also a concern even though they are no longer produced. These molecules are toxic because they bind to the aryl hydrocarbon receptor present in mammalian cells and activate biochemical events leading to a variety of physiological effects, including the induction of cancer; the consequences for human health at environmental exposure levels are still unclear, however.

Toxicity problems also result from the buildup of metals in the environment. The heavy metals bind strongly to sulfur-containing ligands, and can therefore bind and interfere with critical proteins, although the actual biochemistry varies among different metals. The metals also differ markedly in their environmental chemistries and their exposure routes. Thus, mercury pollution is mainly a problem of exposure through eating fish because of the bioaccumulation of methylmercury that is produced in sediments. Cadmium is a problem of exposure through food crops because of its active uptake by plants. Arsenic in aquifers can contaminate drinking water, and can expose huge numbers of people to long-term poisoning, as has happened in Bangladesh. And lead is mainly a problem of city children ingesting dust laden with lead from weathering paint and the fallout of leaded gasoline. Metals pollution involves a complex interplay between patterns of industrial production, emissions, transport, and biochemistry that requires good chemical detective work to sort out and evaluate.

PROBLEM SET

1. (a) Reductive nitrogen fixation requires the breaking of the triple bond in nitrogen (N≡N) by the reaction:

$$N≡N + 3H_2 \rightarrow 2NH_3 \tag{1}$$

Imagine that the overall reaction occurs by the sequential breaking of single bonds in nitrogen by the reactions:

$$N≡N + H_2 \rightarrow HN=NH \tag{2a}$$

$$HN=NH + H_2 \rightarrow H_2N—NH_2 \tag{2b}$$

$$H_2N—NH_2 + H_2 \rightarrow 2NH_3 \tag{2c}$$

Reaction (1) requires an activation energy of 941 kJ/mol to break the triple bond. The most energy-demanding step in process (2) is reaction (2a), which requires an activation energy of about 527 kJ/mol. Calculate the ratio of the rate constants (k_{2a}/k_1) at 27°C for reactions (1) and (2a) given that k is proportional to $e^{-Ea/RT}$, where Ea is the activation energy, $R = 8.33$ joules/K/mol, and T is measured in degrees Kelvin.

(b) From the value of (k_{2a}/k_1), is it surprising that microorganisms can fix nitrogen at ambient temperatures, whereas the Haber process requires temperatures between 400 and 600°C? Calculate the temperature at which k_1 equals the value of k_{2a} at 27°C.

2. Given the flows of nitrogen shown in Figure 15.5, during pre-industrial times was there any net accumulation in the stocks of nitrogen on land or in the oceans? Is there any accumulation under current nitrogen flows? If so, calculate the net stock changes. Describe the environmental effects of increased nitrogen stocks in the terrestrial biosphere from the point of view of water quality of lakes, rivers, groundwaters, and coastal marine areas (see Part III, pp. 319–328).

3. Figure 15.9 shows the relation between corn yield and nitrogen fertilizer application. With respect to corn grown on "Plainfield loamy sand," estimate the yields for four cases: 1) no fertilizer applied; 2) 100 kg of N applied; 3) 170 kg (current usage) applied; and 4) 200 kg applied. Given that corn is about 1.3 percent nitrogen by weight, calculate the percent of applied nitrogen contained in the harvested corn for the latter three cases. Are we reaching the point of diminishing returns with respect to nitrogen application? Where does the residual nitrogen go?

Problems 4 and 5 refer to the following table:

GRAMS OF CORN AND BEAN PROTEIN REQUIRED TO GIVE
DAILY REQUIREMENTS OF ESSENTIAL AMINO ACIDS

	Corn	Beans
Tryptophan	70.0	47.9
Lysine	72.2	33.0
Threonine	37.5	44.3
Leucine	27.4	44.8
Isoleucine	49.0	39.3
Valine	38.8	44.9
Phenylalanine	35.0	39.3
Methionine	33.3	56.7

4. Given that corn consists of 7.8 percent protein, and only 60 percent of it is absorbed by the digestive tract, calculate the number of grams of corn that must be ingested for an adequate protein intake. There are 368 kilocalories in 100 g of corn. How many kilocalories must be consumed daily in order to get adequate protein? Given that the average American consumes 2,000–3,000 kcal daily, can an all-corn diet supply the proper balance of protein and energy?

5. Given that beans are 24 percent protein, and 78 percent of the protein is absorbed by the digestive tract, calculate the number of grams of corn and beans that must be consumed to supply adequate protein for a person whose diet is half corn and half beans. If there are 338 kcal in 100 g of beans, calculate the number of calories consumed in this diet (see problem 4 for additional data for corn).

6. In 1998, grain consumption in the United States, with a meat-oriented diet, was roughly 0.91 metric tons per capita. This included grain eaten indirectly in the form of livestock products; about 70 percent of the grain produced was fed to livestock. In India, where the diet is basically vegetarian, grain consumption was about 0.19 tons per capita in 1998. In China, a country where diet is in transition from basically vegetarian to meat-oriented, roughly 0.32 metric tons per capita were consumed; 20 percent went to feed animals. China's growing affluence has resulted in a dramatic rise in meat consumption, from about 10 million tons in 1980 to around 47 million in 1998. Assuming that by the year 2025, China, with a projected population of 1.5 billion, has the same per capita grain requirements as did the United States in 1998, calculate the grain demand of China in 2025. How does this compare with a total global grain production in 2000 of 1.84 billion tons? The global population in 2025 is expected to be 8.5 billion. If global grain demand were 0.80 tons per capita, how much grain would be required? How much grain would be required in 2025 if everyone survived on a vegetarian diet?

7. What role do antioxidants play in protecting biological systems? Name two kinds of natural antioxidants. What role do natural chelating agents play in reducing free radical formation?

8. Suppose an insecticide is applied to a crop to eradicate fruit flies. Assume that one out of one million fruit flies possesses an enzyme that breaks down the insecticide into nontoxic metabolic products. Assume further that as the normal fruit flies die off quickly, the population of the resistant flies increases geometrically (that is, 1, 2, 4, 8, . . .). If a new generation occurs every 23.5 days, in how many days will the fruit fly population be restored?

9. Give an example of how a broad-spectrum, persistent insecticide like DDT can disrupt entire ecosystems.

10. **(a)** Consider a *para*-substituted phenyldiethylphosphate insecticide of the following structure:

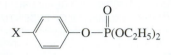

Why does the toxicity of the insecticide increase as the electron-withdrawing capacity of X increases?

(b) The electronegativity value of P is 2.1, that of S is 2.5, and for O it is 3.5. Using this information, explain why phosphothionate insecticides (containing phosphorus-sulfur double bonds rather than phosphorus-oxygen double bonds) are unreactive with cholinesterase, but are activated in the insect's body by oxidizing enzymes. What advantage is there to using sulfur-containing insecticides?

11. Describe alternative ways to control insect populations other than the use of synthetic insecticides.

12. Caffeine is thought to be safe under normal use. However, in large amounts it is poisonous to humans. Assuming that humans and rats are equally sensitive to the toxic effects of caffeine, how many cups of coffee drunk consecutively would be lethal to 50 percent of a group of 150-pound humans? Assume one cup of strong coffee contains about 140 mg of caffeine. Use the data presented in Table 17.1 for the LD_{50} value of caffeine for rats. Is it likely that a person could die from drinking too much coffee?

13. What approaches have been adopted to study chronic diseases such as cancer? What are the limitations of each of these approaches?

14. Why are lipophillic xenobiotics generally more harmful and pervasive than hydrophilic xenobiotics in the environment?

15. Ample evidence indicates that some cancers are caused by exposure to carcinogenic chemicals. Using benzanthracene and dimethylnitrosamine as examples (Figures 17.4 and 17.6), explain how the body actively participates in its own destruction. Correlate your reasoning with the fact that the Ames bacterial test for known carcinogens gives negative results unless a mammalian liver extract is added to the test medium.

16. Explain what a hormone is and describe its mechanism of action. What properties of certain environmental chemicals cause them to disrupt hormonal balance? Give two examples of deleterious health effects attributable to hormonal dysfunction caused by lipophillic xenobiotics.

17. Why do the geometric dimensions of TCDD make it so toxic? Why is toxicity diminished in other congeners where chlorines are removed from the 2, 3, 7, and 8 positions, or when chlorines are added to remaining positions on the rings?

18. Name the uses of PCBs before they were banned in the late 1970s. With reference to Figure 17.2, how would you surmise that PCBs ended up in umbilical cord serum of pregnant women living near Lake Michigan?

19. In the human body, the half-life of methylmercury is 70 days, and for Hg^{2+} it is six days. Why is there such a marked difference in half-lives, and why is methylmercury so much more toxic than inorganic mercury? What is the maximum body accumulation of each type of mercury at a constant ingestion rate of 2 mg Hg/day? (Hint: the maximum concentration in the body can be calculated from the following equation:

$$C_{max}/C_0 = e^{-A}[1/(1 - e^{-A})]$$

where C_{max} is the maximum concentration, C_0 is the initial concentration per time step, $t_{1/2}$ is the half-life, and $\lambda = 0.693/t_{1/2}$.)

20. Relate how mercury accumulation in the food chain led to the Minamata disaster. Why is the conversion to methylmercury the key step in mercury toxicity?

21. Describe how acidification acts to increase the risks to the environment and human health posed by cadmium, lead, and mercury.

22. Describe the main routes of exposure of lead. Why are children most susceptible to lead poisoning? How can lead poisoning be treated?

Suggested Readings

W. H. Schlesinger (1997). *Biogeochemistry: An Analysis of Global Change* (2nd edition). (San Diego: Academic Press).

G. C. Daily, S. Alexander, P. R. Ehrlich, L. Goulder, J. Lubchenco, P. A. Matson, H. A. Mooney, S. Postel, S. H. Schneider, D. Tilman, and G. M. Woodwell (1997). *Ecosystem Services: Benefits Supplied to Human Societies by Natural Ecosystems*. Issues in Ecology, Issue No. 2 (Washington, DC: The Ecological Society of America). (http://esa.sdsc.edu/issues.htm)

S. Naeem, F. S. Chapin III, R. Costanza, P. R. Ehrlich, F. B. Golley, D. U. Hooper, J. H. Lawton, R. V. O'Neill, H. A. Mooney, O. E. Sala, A. J. Symstad, and D. Tilman (1997). *Biodiversity and Ecosystem Function in Maintaining Natural Life Support Processes*. Issues in Ecology, Issue No. 4 (Washington, DC: The Ecological Society of America). (http://esa.sdsc.edu/issues.htm)

D. Tilman and C. Lehman (2001). Human-caused environmental change: Impacts on plant diversity and evolution. *Proceedings of the National Academy of Sciences, USA* 96:5433–5440.

S. R. Palumbi (2001). Humans as the world's greatest evolutionary force. *Science* 293:1786–1790.

Chapter 15: Nitrogen and Food Production

P. M. Vitousek, J. Aber, R. W. Howarth, G. K. Likens, P. A. Matson, D. W. Schindler, W. H. Schlesinger, and G. D. Tilman (1997). *Human Alteration of the Global Nitrogen Cycle: Causes and Consequences*. Issues in Ecology, Issue No. 1 (Washington, DC: The Ecological Society of America). (http://esa.sdsc.edu/issues.htm)

C. R. Frink, P. E. Waggoner, and J. H. Ausubel (1999). Nitrogen fertilizer: Retrospect and prospect. *Proceedings of the National Academy of Sciences, USA* 96:1175–1180.

R. H. Socolow (1999). Nitrogen management and the future of food: Lessons from the management of energy and carbon. *Proceedings of the National Academy of Sciences, USA* 96:6001–6008.

P. A. Matson, R. Naylor, I. Ortiz-Monasterio (1998). Integration of environmental, agronomic, and economic aspects of fertilizer management. *Science* 280:112–115.

Food and Agriculture Organization (2000). *The State of Food and Agriculture: Lessons from the Past 50 Years*. (Rome: Sales and Marketing Group, FAO). (http://www.fao.org/es/ESA/sofa.htm)

P. A. Matson, W. J. Parton, A. G. Power, M. J. Swift (1997). Agricultural intensification and ecosystem properties. *Science* 277:504–509.

G. P. Robertson, E. A. Paul, R. R. Harwood (2000). Greenhouse gases in intensive agriculture: Contributions of individual gases to the radiative forcing of the atmosphere. *Science* 289:1922–1925.

K. G. Cassman (1999). Ecological intensification of cereal production systems: Yield potential, soil quality, and precision agriculture. *Proceedings of the National Academy of Sciences, USA* 96:5952– 5959.

D. Tilman, J. Fargione, B. Wolff, C. D'Antonio, A. Dobson, R. Howarth, D. Schindler, W. H. Schlesinger, D. Simerloff, D. Swackhamer (2001). Forecasting agriculturally driven global environmental change. *Science* 292:281–284.

D. Tilman (1999). Global environmental impacts of agricultural expansion: The need for sustainable and efficient practices. *Proceedings of the National Academy of Sciences, USA* 96:5995–6000.

P. Pinstrup-Andersen, R. Pandya-Lorch, and M. W. Rosegrant (1999). *World Food Prospects: Critical Issues for the Twenty-First Century.* 2020 Vision Food Policy Report (Washington, DC: International Food Policy Research Institute). (http://www.ifpri.org/)

G. Gardner and B. Halweil (2000). Escaping hunger, escaping excess. *World Watch Magazine*: July/August: 24–35.

Dietary Guidelines Advisory Committee (2000). *Dietary Guidelines for Americans 2000.* (Washington DC: U.S. Departments of Agriculture, and Health and Human Services). (http://www.nal.usda.gov/fnic/dga)

Standing Committee on the Scientific Evaluation of Dietary Reference Intakes and Its Panel on Dietary Antioxidants and Related Compounds, Institute of Medicine (1998). *Dietary Reference Intakes: Proposed Definition and Plan for Review of Dietary Antioxidants and Related Compounds* (Washington, DC: National Academy Press).

Standing Committee on the Scientific Evaluation of Dietary Reference Intakes and Its Panel on Dietary Antioxidants and Related Compounds, Institute of Medicine (2000). *Dietary Reference Intakes for Vitamin C, Vitamin E, Selenium, and Carotenoids* (Washington, DC: National Academy Press).

Chapter 16: Pest Control

R. N. Mack, D. Simberloff, W. M. Lonsdale, H. Evans, M. Clout, and F. Bazzaz (1997). *Biotic Invasions: Causes, Epidemiology, Global Consequences and Control.* Issues in Ecology, Issue No. 5 (Washington, DC: The Ecological Society of America). (http://esa.sdsc.edu/issues.htm)

A. S. Moffat (2001). Finding new ways to fight plant disease. *Science* 292:2270–2273.

D. K. Letourneau and B. Goldstein (2001). Pest damage and arthropod community structure in organic vs. conventional tomato production in California. *Journal of Applied Ecology* 38:557–570.

Chapter 17: Toxic Chemicals

U.S. Environmental Protection Agency (2001). 1999 Toxic Release Inventory, Public Data Release (Washington, DC: Office of Environmental Information). (http://www.epa.gov/tri/tri99/pdr/index.htm)

D. G. Crosby (1998). *Environmental Toxicology and Chemistry* (Oxford, UK: Oxford University Press).

Technology Planning and Management Corporation (2001). *Report on Carcinogens, Ninth Edition: Carcinogen Profiles 2000* (Revised January 2001) (Research Triangle Park, North Carolina:

National Institute of Environmental Health Sciences, Department of Health and Human Services, Public Health Service, National Toxicology Program).

Committee on Hormonally Active Agents in the Environment, National Research Council (1999). *Hormonally Active Agents in the Environment.* (Washington, DC: National Academy Press).

Committee on Health Effects of Waste Incineration, National Research Council (1999). *Waste Incineration and Public Health* (Washington, DC: National Academy Press).

T. Harner (1997). Organochlorine contamination of the Canadian Arctic, and speculation on future trends. *International Journal of Environment and Pollution* 8:51–73.

J. Dolbec, D. Mergler, C-J. Sousa Passos, S. Sousa de Marais, and J. Lebel (2000). Methylmercury exposure affects motor performance of a riverine population of the Tapajos River, Brazilian Amazon. *International Archives of Occupational and Environmental Health* 73:195-203.

M. Roulet, M. Lucotte, N. Farella, G. Serique, H. Coelho, C-J. Sousa Passos, E. de Jesus da Silva, P. Scavone de Andrade, D. Mergler, J-R.D. Guimaraes, and M. Amorim (1999). Effects of recent human colonization on the presence of mercury in Amazonian ecosystems. *Water, Air, and Soil Pollution* 112:297–313.

M. Roulet, M. Lucotte, A. Saint-Aubin, S. Tran, I. Rheault, N. Farella, E. de Jesus da Silva, L. Dezencourt, C-J. Sousa Passos, G. Santos Soares, J-R.D. Guimaraes, D. Mergler, and M. Amorim (1998). The geochemistry of mercury in Central Amazonian soils developed on the Alter-do-Chao Formation on the Lower Tapajos River Valley, Para State, Brazil. *Science of the Total Environment* 223:1–24.

J. M. Benoit, C. C. Gilmour, and R. P. Mason (2001). The influence of sulfide on solid-phase mercury bioavailability for methylation by pure cultures of *Desulfobulbus propionicus* (1 pr 3). *Environmental Science & Technology* 35:121–126.

A. H. Welch, S. A. Watkins, D. R. Helsel, and M. F. Focazio (2000). *Arsenic in Ground-Water Resources of the United States.* U.S. Geological Survey, National Analysis of Trace Elements, Fact Sheet FS-063-00 (Washington, DC: U.S. Department of Interior).

R. Wilson (2001). *Chronic Arsenic Poisoning: History, Study and Remediation* (Cambridge, Massachusetts: Harvard University).
(http://phys4.harvard.edu/~wilson/arsenic_project_introduction.html)

V. M. Thomas, R. H. Socolow, J. J. Fanelli, and T. G. Spiro (1999). Effects of reducing lead in gasoline: An analysis of the international experience. *Environmental Science & Technology* 33:3942–3948.

APPENDIX

ORGANIC STRUCTURES

Structures like the one in Figure 2.2 (see p. 22) are how we represent the organic molecules of nature. This appendix offers a brief review of the principles of molecular structure among organic chemicals, to aid the reader in interpreting the structural diagrams found throughout the book.

HYDROCARBONS; ALKANES

Hydrocarbons are organic compounds that contain only carbon and hydrogen. Carbon has four valence electrons (the *valence* electrons are the outer shell of electrons that are available for sharing with other atoms), which it can share with up to four other atoms. These shared electrons are the chemical bonds. They are the glue holding a molecule together. The simplest hydrocarbon is CH_4, *methane,* which is the main constituent of natural gas.

Carbon can also share electrons with other carbon atoms. If we replace one of the H atoms in methane with a C atom, and add three more H atoms, the result is *ethane,* C_2H_6, whose structure can be represented by drawing lines between the atoms sharing electrons:

Replacing an H atom in ethane with another C atom and three more H atoms produces *propane*, C_3H_8:

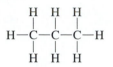

The next members of this series of molecules are *butane*, C_4H_{10}, *pentane*, C_5H_{12}, *hexane*, C_6H_{14}, and so on. Each time a C atom is added, so are two H atoms, and the general formula for this series, called the *alkanes*, is C_nH_{2n+2}.

BRANCHED CHAINS; ISOMERS

However, when we come to butane, there are two ways of arranging the bonds:

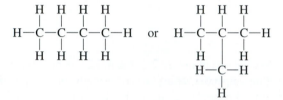

The fourth C atom can be added to the end of the chain of three C atoms, or it can be added to the middle. The two molecules have the same chemical formula, but because the bonding arrangement is different, they have different physical properties (e.g., different melting and boiling points), and different chemical properties (different reactivity). We call them *isomers*. The first isomer is called *n-butane* (*n*- for "normal"; all straight-chain alkanes carry this prefix) and the second *isobutane*.

As the number of C atoms increases, the number of possible isomers goes up rapidly. To distinguish among these isomers, a systematic naming scheme has been developed. The name is based on the longest chain of C atoms in the molecule, and the position of substitution of a shorter chain is numbered from the end of the longer chain. For example, the systematic name for isobutane is 2-methylpropane, because the longest chain has three C atoms, and the second C atom of this chain bears a methyl group. The suffix *-yl* indicates a fragment in which a bond to an H atom is replaced by a bond to the rest of the molecule, thus:

$$H_3C\!-\!H \qquad H_3C\!-\!$$
$$\text{methane} \qquad \text{methyl}$$

(It is convenient to lump the H atoms together with the C atoms to which they are attached, making the C—H bonds implicit.)

Another example of a *branched* chain hydrocarbon is 2,2,4-trimethylpentane:

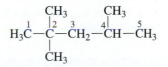

This molecule, also called "iso-octane," is an isomer of *n*-octane. The combustion properties of iso-octane define the "octane scale" of gasoline quality (see p. 237).

The *n*-alkanes are *straight-chain* molecules, while their isomers are *branched-chain* molecules. Organisms manufacture straight-chain hydrocarbons as the main constituents of *fats* and *lipids*. Fats are energy storage molecules, while lipids are the building blocks of biological membranes (see pp. 271, 272, 369–370). Consequently, petroleum contains a substantial proportion of straight-chain hydrocarbons. But it also contains many branched-chain hydrocarbons, because the high pressure and temperature conditions of geological deposits induce rearrangement of the C—C bonds. This process is deliberately continued in the chemical reactors of oil refineries, since hydrocarbons with branched chains have higher octane ratings than those with straight chains (see p. 232).

RINGS

In the process of attaching C atoms to one another, it is possible to form rings, as well as chains. We can have:

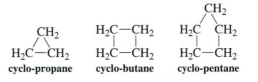

and so on.

Rings can also be joined together to form *polycyclic* molecules. An example is the biological marker molecule, bacteriohopanetetrol, (see Figure 2.2, p. 22) which has four six-membered rings and one five-membered ring, as well as a seven-carbon chain. This complicated molecule is made by bacteria, but animals also make polycyclic molecules, such as cholesterol ($C_{27}H_{46}O$):

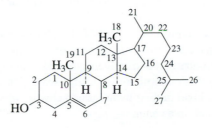

(Numbers indicate the positions of the 27 carbons; not all C—H bonds are shown: carbons 3, 6, 8, 9, 14, 17, 20, and 25 are each bonded to a single hydrogen; carbons 1, 2, 4, 7, 11, 12, 15, 16, 22, 23, and 24 are bonded to two hydrogens; carbons 18, 19, 21, 26, and 27 are bonded to three hydrogens; carbons 5, 10, and 13 have no C—H bonds.)

WORKED PROBLEM A.1: Pentane Isomers

Q. *How many isomers are there of pentane? Draw and name them.*

A. One isomer is obviously the straight chain *n*-pentane:

$$CH_3—CH_2—CH_2—CH_2—CH_3$$

Another isomer is obtained by putting a methyl group in the middle of a chain of four C atoms:

$$H_3C—CH—CH_2—CH_3$$
$$\quad\quad\;|$$
$$\quad\;CH_3$$

This is 2-methylbutane. (Note that it does not matter which of the two middle C atoms is attached to the methyl groups, because you get the same molecule by turning the structure around.) Finally, two methyl groups can both be attached to the middle C of a three-C chain:

$$CH_3$$
$$\;|$$
$$CH_3—C—CH_3$$
$$\;|$$
$$CH_3$$

This is 2,2-dimethylpropane. Another name, equally descriptive, would be tetramethylmethane, recognizing that this molecule can be derived from CH_4 by replacing all four H atoms with methyl groups.

There are no other ways of arranging molecules with the formula C_5H_{12}. Pentane has three isomers.

UNSATURATED HYDROCARBONS

Carbon can share its four valence electrons two or three at a time, forming double or triple bonds (symbolized by double and triple lines, = and ≡). The simplest hydrocarbon with a double bond is *ethene* (commonly called *ethylene*), $CH_2=CH_2$, while the simplest one with a triple bond is *ethyne* (commonly called *acetylene*), CH≡CH. The H atoms in these simplest *alkenes* and *alkynes* can be replaced by C atoms to form more complex molecules. For example, substituting an H atom in ethylene and acetylene with a methyl group produces *propene* (often called *propylene*), $CH_2=CHCH_3$, and *propyne,* CH≡CCH_3, respectively. If there are two or more double bonds or triple bonds, we have a poly-ene, or a poly-yne.

Molecules having C=C or C≡C bonds are said to be *unsaturated*, because the C atoms do not have the full complement of four possible bonding partners. We are all familiar with unsaturated and polyunsaturated cooking oils, because they are less likely than saturated oils to contribute to heart disease. Unsaturated oils contain fat molecules having a double bond in their hydrocarbon chains, while polyunsaturated oils have two or more double bonds in a chain.

MOLECULAR SHAPE

Unsaturated molecules differ in shape from saturated molecules, because the bond angles differ. Pairs of electrons in the valence shell try to keep as far apart as possible. Thus if a C atom has four bonding partners, the bonds extend towards the four corners of a tetrahedron (bond angles of 109.5°). But if there is a double bond, the two electron pairs in the bond are forced into the same region of space, and together they remain as far away as possible from the other two electron pairs in the single bonds. The three bonds, one double and two single, then point toward the corners of an equilateral triangle (bond angles of about 120°). The result is that the two-carbon unit, with its associated single bonds, is planar, Thus, ethylene can be represented as:

Likewise, the electrons in a triple bond are as far away as possible from the pair in the remaining single bonds, giving a bond angle of 180°. A triple bond is colinear with the associated single bonds; acetylene can be represented as

$$H—C≡C—H$$

Another important aspect of geometry is the ease of rotation about a bond. Rotation is easy about single bonds; the parts of the molecule connected by a single bond can adopt various interconvertible orientations. But rotation about a double bond is difficult; it does not occur except at very high temperatures. Consequently, there are two separate orientations for the parts of the molecule connected by a double bond, e.g.:

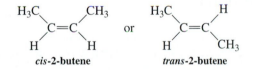

These two orientations are called *cis* and *trans* (methyl groups on the same or opposite side of the double bond). Because the double bond does not rotate, these are *geometric* isomers, which have different physical and chemical properties because their shapes differ.

CARBON FRAMEWORK REPRESENTATIONS

To reduce clutter in organic structural diagrams, one can simply draw the carbon framework as a series of connected lines. For example, *n*-octane and *iso*-octane can be represented:

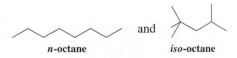

The H and C atoms are implicit in these representations. One understands that the lines connect C atoms, which have as many C—H bonds as are needed to complete their complement of four bonds. The kink between lines represents the fact that saturated C atoms have tetrahedral bond angles.

When the carbon framework contains double or triple bonds, this is indicated by drawing double or triple lines at the appropriate place.

cis-2-butene   2-butyne

(Note that the lines are colinear for 2-butyne, because of the 180° bond angles.) Sometimes the framework representation is used with some of the H atoms left in for clarity, as in the cholesterol structure on p. 463.

WORKED PROBLEM A.2: Name Frameworks

Name the following compounds from the framework representations:

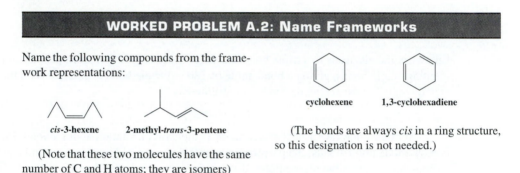

cis-3-hexene   2-methyl-trans-3-pentene

cyclohexene   1,3-cyclohexadiene

(Note that these two molecules have the same number of C and H atoms; they are isomers)

(The bonds are always cis in a ring structure, so this designation is not needed.)

AROMATIC COMPOUNDS

When the six-membered ring has three double bonds, a different kind of molecule, *benzene,* is formed. The double bonds can be drawn in two alternative, but equivalent ways:

benzene

Normally, double bonds are shorter than single bonds, but in benzene all six carbon-carbon bonds are found to be the same length. The electrons end up adopting both bonding arrangements at the same time, a phenomenon termed *resonance* (and indicated by the double-headed arrow above). We often represent this situation by inscribing a circle in the ring.

Because of resonance, the benzene molecule is more stable and less reactive than we would expect for a molecule with three double bonds.

The hydrogen atoms on benzene can be replaced by carbon atoms, giving rise to a large class of molecules called (for historical reasons) *aromatic* molecules. For example, methylbenzene, also called toluene, is a common solvent:

toluene

The lignin molecule in Figure 2.4 on p. 24 is a very complex aromatic molecule.

When two or more of the H atoms on benzene are substituted, the possibility of isomers again arises. For example, there are three isomers of xylene, the molecule obtained by substituting two methyl groups on benzene:

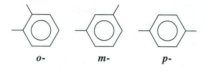

o-   *m-*   *p-*

These are designated *ortho-*, *meta-*, and *para-*, shortened to *o-*, *m-*, and *p-*. (The systematic names, however, are 1,2-, 1,3- and 1,4-dimethylbenzene).

When a benzene molecule is attached to a larger organic molecule, it is called a *phenyl* group. For example, the insecticide DDT is dichlorodiphenyltrichloroethane.

Just as in the case of the saturated rings in polycyclic alkanes, it is possible to fuse benzene rings to form *polycyclic aromatic hydrocarbons [PAH]*, e.g.:

naphthalene   **anthracene**

These PAHs are side-products of incomplete combustion. They are of concern as potent carcinogens (see pp. 221–222, 413). The ultimate PAH is graphite, the structure of which is a continuous sheet of benzene rings (see Figure 2.4).

HETERO-ATOMS; FUNCTIONAL GROUPS

Organic compounds are based on carbon and hydrogen, but often contain atoms of other elements, collectively called hetero-atoms. The elements most commonly encountered are the halogens, oxygen, nitrogen, sulfur, and phosphorus. These hetero-atoms are frequently sites of chemical reactivity, and the chemical groups containing them are called functional groups (see Table A.1). (Double and triple bonds are also functional groups; their electrons are more loosely held than those of single bonds, making them more reactive.)

Halogen atoms form only one bond, and can, in principle, be substituted for any of the H atoms in a hydrocarbon. The general symbol for *organohalides* is RX, where X stands for F, Cl, Br, or I, and R represents the rest of the organic molecule. Organohalides are generally less reactive than the parent hydrocarbon. This is one reason why organochlorine

TABLE A.1   SELECTED ORGANIC FUNCTIONAL GROUPS

Name of class	Functional group	General framework of class*
Alkane	None	R—H
Alkene	—C=C—	R, R′, R″, R‴ on C=C
Alkyne	—C≡C—	R—≡—R′
Halide	—X (X = F, Cl, Br, I)	R—X
Alcohol	—C—O—H	R—OH
Ether	—C—O—C—	R, R′ on O
Aldehyde	—C(=O)—H	R, H on C=O
Ketone	—C(=O)—	R, R′ on C=O
Amine	—C—N—	R—N with R′, R″
Imine	—C=N—	R, R′ on C, =N—R″
Nitrile	—C≡N	R—≡N
Carboxylic acid	—C(=O)—O—H	R—C(=O)OH
Ester	—C(=O)—O—C—	R—C(=O)—O—R′
Amide	—C(=O)—N—	R—C(=O)—N with R′, R″

*For some classes, R, R′, and R″ can be H.

pesticides, such as DDT, are of environmental concern. Because of their low reactivity, they last for a long time, and can build up in the body tissues of animals, including humans (see pp. 386–388).

Oxygen atoms form two covalent bonds, as in H_2O. If an O atom forms a bond with a C atom, as well as with an H atom, we have an *alcohol,* symbolized as ROH. Alcohols are named by adding *-ol* to the end of the name of the organic molecule. The drinkable alcohol is ethyl alcohol, or ethanol, CH_3—CH_2—OH. The complex molecule in Figure 2.2 is called bacteriohopanetetrol because of the four alcohol groups it contains.

If the O atom is bound to two C atoms, we have an *ether,* ROR'. We distinguish R and R', because the two organic fragments that are connected to the O atom need not be the same. An example is:

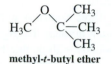

methyl-*t*-butyl ether

abbreviated MTBE. (*t-* stands for *tertiary,* and *t*-butyl is a way of designating a four-carbon group, in which all the bonds from the connecting C atom are to other C atoms.) MTBE is currently used as a gasoline additive intended to cut pollution, but has gained notoriety because of groundwater contamination from gasoline leaks (see pp. 260, 344). Notice that the lignin structure (see Figure 2.4), has many alcohol and ether groups.

The O atom can also form a double bond with a C atom, and the resulting C=O group is called a *carbonyl.* Carbonyl compounds are distinguished by what is attached to the C atom. *Ketones* have two other C atoms, while *aldehydes* have one C and one H atom. Common examples are:

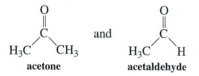

Additional functional groups are *carboxylic acids* and *esters:*

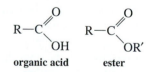

The chemistry of the carbonyl group is altered considerably by having an OH or OR group attached to the C atom. *Polyesters* are an important class of polymer molecules in which hydrocarbons are linked by repeating ester groups into long chains. They are widely used in fabrics and plastic containers.

Finally, two oxygen atoms can share a single bond, while having another single bond each to either H or C. These are *peroxides,* with the general formula R—O—O—R'. If one of the R groups is H, we have a *hydroperoxide,* and the simplest member of the family,

H—O—O—H, is hydrogen peroxide. The O—O bond is weak, and peroxides are a reactive class of molecules. They play an important role in the chemistry of air pollution (see pp. 228, 229).

Since sulfur is in the same group of the periodic table as oxygen, the same valence rules apply, and similar functional groups are found in organic sulfur as in organic oxygen compounds. However, sulfur, being a larger atom, is able to have more than eight electrons in its valence shell when combining with highly electron-seeking (*electronegative*) elements such as oxygen or fluorine. Oxygen atoms can readily add to organic sulfur compounds, producing *sulfones,* R_2SO, *sulfoxides,* R_2SO_2, or *sulfonic acids,* RSO_3H. Although much less abundant than oxygen in biological tissues, sulfur is a significant constituent of coal and petroleum, and is an important contributor to air pollution when these fossil fuels are burned (see pp. 218–220).

Compounds having three bonds to an N atom are *amines*. In addition to *ammonia,* NH_3, they include *primary* amines, RNH_2, *secondary* amines, $R,R'NH$, and *tertiary* amines $R,R',R''N$. Compounds with double-bonded N are *imines,* $R,R'C$=NR'' (the R's can also be H), while compounds with triple-bonded N are *nitriles,* RC≡N.

An important class of compounds is the *amides*, in which an N atom is connected to the C atom of a carbonyl group:

Polyamide polymers (e.g., nylon) are important as fibers and plastics, while protein molecules, which have *amino acids,*

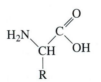

linked head to tail through amide linkages, are nature's own polyamides.

Phosphorus is in the same group of the periodic table as nitrogen, and the phosphorus analogs of amines are *phosphines*. However, phosphorus forms much stronger bonds with oxygen than with carbon or hydrogen. In nature, P atoms are always surrounded by O atoms. The most common form is the phosphate ion, PO_4^{3-}, which can condense into polyphosphate ions, such as *tripolyphosphate* (see pp. 299–300), by forming P—O—P links. Many important biological molecules are phosphate esters, $ROPO_3^{2-}$; the nucleic acids (see p. 411) are polymers held together by *phosphodiester* links, RO—PO_2—OR'.

INDEX

Note: **Boldface** pages locate figures; *italicized* pages locate tables.

Periodic Table of the Elements

Main groups

Transition metals

1 1A									
1 **H** 1.00794	2A								
3 **Li** 6.941	**4** **Be** 9.01218	3 3B	4 4B	5 5B	6 6B	7 7B	8	9 8B	
11 **Na** 22.98977	**12** **Mg** 24.305								
19 **K** 39.0983	**20** **Ca** 40.078	**21** **Sc** 44.9559	**22** **Ti** 47.88	**23** **V** 50.9415	**24** **Cr** 51.996	**25** **Mn** 54.9380	**26** **Fe** 55.847	**27** **Co** 58.933
37 **Rb** 85.4678	**38** **Sr** 87.62	**39** **Y** 88.9059	**40** **Zr** 91.224	**41** **Nb** 92.9064	**42** **Mo** 95.94	**43** **Tc** (98)	**44** **Ru** 101.07	**45** **Rh** 102.90
55 **Cs** 132.9054	**56** **Ba** 137.33	**57** ***La** 138.9055	**72** **Hf** 178.49	**73** **Ta** 180.9479	**74** **W** 183.85	**75** **Re** 186.207	**76** **Os** 190.2	**77** **Ir** 192.2
87 **Fr** (223)	**88** **Ra** 226.0254	**89** **†Ac** 227.0278	**104** **Rf** (261)	**105** **Db** (262)	**106** **Sg** (266)	**107** **Bh** (264)	**108** **Hs** (269)	**109** **Mt** (268)

	58 **Ce** 140.12	59 **Pr** 140.9077	60 **Nd** 144.24	61 **Pm** (145)	62 **Sm** 150.3
***Lanthanide series**					
†Actinide series	90 **Th** 232.0381	91 **Pa** 231.0359	92 **U** 238.0289	93 **Np** 237.048	94 **Pu** (244)

*The discoveries of elements 116 and 118 were announced in 1999, but the disc